Fifth Edition

CONCEPTS
IN
BIOLOGY

Fifth Edition

CONCEPTS
IN
BIOLOGY

Eldon D.   Andrew H.   J. Richard   Frederick C.   Rodney J.
Enger • Gibson • Kormelink • Ross • Smith

Delta College University Center, Michigan

wcb
Wm. C. Brown Publishers
Dubuque, Iowa

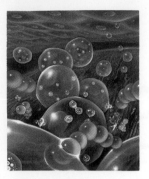

Book Team

Editor *Kevin Kane*
Developmental Editor *Mary J. Porter*
Production Editor *Carla J. Aspelmeier*
Designer *K. Wayne Harms*
Photo Research Editor *Carol M. Smith*
Permissions Editor *Vicki Krug*
Visuals Processor *Reneé Pins*
Marketing Manager *Matt Shaughnessy*

wcb group

Chairman of the Board *Wm. C. Brown*
President and Chief Executive Officer *Mark C. Falb*

wcb

Wm. C. Brown Publishers, College Division

President *G. Franklin Lewis*
Vice President, Editor-in-Chief *George Wm. Bergquist*
Vice President, Director of Production *Beverly Kolz*
National Sales Manager *Bob McLaughlin*
Director of Marketing *Thomas E. Doran*
Marketing Information Systems Manager *Craig S. Marty*
Executive Editor *Edward G. Jaffe*
Production Editorial Manager *Julie A. Kennedy*
Manager of Design *Marilyn A. Phelps*
Photo Research Manager *Faye M. Schilling*

Cover illustration by Scott Barrows/courtesy Upjohn International
Kalamazoo, MI

The cover depicts a "battle" between polymorphonucleocytes
(PMN's—white blood cells that fight infection in the body) and
both streptococci (blue spheres) and staphylococci (magenta
spheres) bacteria. As the bacteria are ingested by phagocytosis, they
are brought inside the PMN where they are destroyed. The orange
particles represent an antibiotic that is present inside the PMN;
these particles destroy the bacteria.

The credits section for this book begins on page 568, and is
considered an extension of the copyright page.

Library of Congress Catalog Card Number: 87–70594

ISBN 0–697–05122–6

Printed in the United States of America by Wm. C. Brown Publishers
2460 Kerper Boulevard, Dubuque, IA 52001

10 9 8 7 6 5

Contents

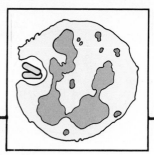

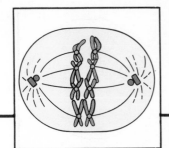

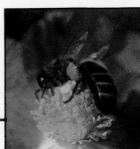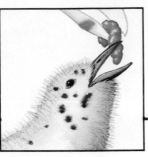

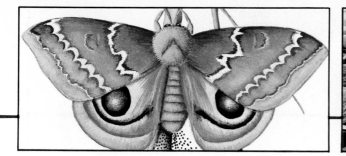

▪ Boxed Readings

Preface

Purpose

The origin of this book is deeply rooted in our concern for the learning of students enrolled in college biology courses. In this fifth edition we have maintained our original goal of writing a book that is both useful and interesting. We hope it is easy to understand and enjoyable to read. Many examples and analogies are used to help put complicated ideas into plain language. We think it is very important that students enjoy learning about some of the concepts in biology, and we hope to stimulate a continuing interest in biology.

Organization

The chapters of *Concepts in Biology* are arranged in a traditional manner. It begins with a discussion of the meaning, purpose, and future of biology as a scientific endeavor. The material then turns to the coverage of biological *concepts* based on an expanding spiral of knowledge. Thus, chemistry is followed by cell biology, nutrition, cell division, reproduction, genetics, ecology, evolution and the diversity and classification of living things. The book progresses from the basic to the complex. Each chapter of the book is built on the foundation of the preceding chapters.

New to This Edition

In response to users of the previous edition, we have modified the material on chemistry, scientific method, behavior, DNA, and genetics significantly. The chapters involved contain recent additional information that provides a more in-depth treatment of this material. In addition, we have reorganized the information on evolution and ecology in chapters 14 through 19; the evolution information now precedes that on ecology. Ecological topics were reorganized to flow from individual organisms, through populations, to ecosystems.

Five new features have been added to this edition. Chapters have been grouped into sections of closely related material. Each set of chapters is introduced with a pertinent overview of the material included in that section. Chapter outlines are presented at the opening of each chapter to assist the student in visualizing the conceptual content of the chapters. Objectives focus the student's attention on specific goals. A "For Your Information" paragraph introduces each chapter. This section contains timely, topical information that relates directly to the chapter. At the end of each chapter, an "Experience This" section has been added to enable the student to apply the knowledge gained with a minimum of equipment and materials. These experiences are intended to supplement the students' learning, but not replace a formal laboratory experience.

In addition, several major changes have been made to increase the usefulness and attractiveness of the text. These changes include the use of full color in areas where it enhances the book's teaching effectiveness. All definitions have been critically reviewed and rewritten when they have been found inadequate. The new glossary enables the students to better incorporate biological vocabulary into their study. Throughout the text a significant number of new boxes, new figures, and new examples have been included to make the text more current and interesting.

Aids to the Reader

The text has a number of features intended to involve the students in an active learning process. These features stimulate student interest as well as help students relate general *concepts* of biology to daily living.

Each chapter contains these elements:

Purpose explains the value of each chapter to the scope and understanding of a complete biology course.

Topical Headings emphasize the essential concepts for understanding biology as a science.

Graphics illustrate a concept or associate a new concept with previously mastered information. Every illustration is used to emphasize a point or to help teach a concept.

Chapter Summary clearly reviews the concepts of each chapter.

Consider This focuses on a question or problem that challenges the student to think logically through a problem and arrive at a conclusion based on the concepts of the chapter.

Questions review the material and help students know if they have mastered the contents of the chapter.

Five parts now divide the chapters into sections of closely related material. Each part begins with an overview of the material contained in that section.

Chapter outlines at the beginning of each chapter present the material to be covered in that chapter.

Learning Objectives are listed at the beginning of each chapter and focus the students' attention on specific goals.

For Your Information introduces each chapter with timely information related to that chapter.

Experience This enables the student to apply knowledge gained from the chapter.

Phonetic Pronunciations

You will notice that new to this edition is a phonetic spelling following each glossary word. This can be helpful if you will learn how to read these symbols.

An unmarked vowel (a, e, i, o, u) at the end of a syllable will have the long sound, as in the word "prey" = PRA.

An unmarked vowel followed by a consonant has the short sound, as in the phonetic spelling of the word "cell" = SEL.

A vowel in the middle of a syllable may have a mark over it to indicate short or long sound. A straight bar ($\bar{a}$) indicates the long sound and a small arc ($\breve{a}$) the short sound. The word "amylase" = am′ĭ-lās shows these two marks plus a slash (′) that tells us to accent the first syllable. Some phonetic spellings may have a double slash (″) as well as a single slash (′). The double slash also indicates an accented syllable but not as heavily accented as with the single slash, as in ob″zer-va′shun.

Writing Style and Readability

The writing is informal and easy to read. The text has been praised by reviewers and adopters for its clear language and effective manner. Particular attention has been given to readability. The *Fry Readability Graph* has been used to verify the appropriateness of the text for Introductory Biology. A number of features enhance the readability of the material:

Boldface focuses student attention on a key term when it is first defined in the text.

Italics emphasize important terms, phrases, names, and titles.

Graphics often in the form of logical flow diagrams, analogy diagrams, and charts.

Chapter Glossaries for immediate reinforcement of the terms necessary for student comprehension of concepts.

Master Glossary located at the end of the text, serves as a single resource for essential terminology used throughout the text.

Useful Ancillaries

Supplementary materials have been designed to provide instructors with a complete educational package. The Instructor's Manual and the Laboratory Manual will help the instructor to plan class activities.

Instructor's Manual provides a rationale for use of text, objectives, explanatory information, answer key for text questions, correlation of text to adapt to differing academic calendars, film sources, testing material, and a supplementary guide to the use of the accompanying laboratory manual.

Laboratory Manual features carefully designed and class-tested exercises with stated objectives.

Transparencies

Fifty-two acetate two- and five-color transparencies are available with the *Concepts in Biology* text. The transparencies are taken from the text and represent the important figures that merit extra visual review and discussion.

QuizPak

wcb QuizPak, a student self-testing program on the microcomputer, is available free of charge to adopters. QuizPak lets students choose a chapter to review, then the computer quizzes them with questions selected for that chapter. QuizPak may be used at a number of work stations simultaneously, runs on the Apple® IIe or IIc, and on the IBM PC, and requires only one disk drive.

TestPak

wcb TestPak, a free, computerized testing service, simplifies testing while offering you flexibility. There are two convenient TestPak options available:

—Use your Apple® IIe or IIc, or IBM PC to pick and choose your test questions, edit them, and add your own questions. We can send you program and test item diskettes for this purpose.

—If you don't have a microcomputer, you can still pick and choose your questions via our call-in/mail-in service. Within two working days of your request, we'll put a test master, a student answer sheet, and an answer key in the mail to you. Call-in hours are 8:30–5:00 CST, Monday through Friday.

Acknowledgments

A large number of people have knowingly or unknowingly helped us write this text. Our families continued to give understanding and support as we worked on this revision. We acknowledge the thousands of students in our classes who have given us feedback over the years concerning the material and its relevancy. They were the best possible source of criticism.

We also wish to thank all who sent us comments and criticisms on the fourth edition of *Concepts in Biology*. Your input was the raw material for this revision.

We gratefully acknowledge the invaluable assistance of many reviewers throughout the development and preparation of the manuscript:

Reviewers of the Fifth Edition
Bonnie Amos, Baylor University
Ronald Basmajian, Merced College
Neal D. Buffaloe, University of Central Arkansas
Lynn Elkin, California State University—Hayward
Richard E. McKeeby, Union College
R. Harvard Riches, Pittsburg State University
E. Russell TePaske, University of Northern Iowa
Robert Wischmann, Lakewood Community College

Reviewers of the Fourth Edition
Harold G. Brotzman, North Adams State College
Terry A. Larson, College of Lake County
James Lipp, Lewistown College Center
Sr. Maureen Webb, Holy Names College
Stanley M. Wiatr, Eastern Montana College

Reviewers of the Third Edition
Frank Bonham, San Diego Mesa College
Richard Boutwell, Missouri Western State College
Clyde L. Britchett, Brigham Young University
Gil Desha, Tarrant County Junior College
Albert J. Grennan, San Diego Mesa College
Robert Kirkwood, University of Central Arkansas
Cathryn McDonald, Jefferson State Junior College
Don Misumi, Los Angeles Trade-Technical College
Robert Romans, Bowling Green State University

Jean Salter, Valencia Community College
Melvin Urschel, Tacoma Community College

Reviewers of the Second Edition
Ferron Andersen, Brigham Young University
Donald A. Denison, Laney College
Steven A. Fink, West Los Angeles College
James Hiser, Normandale Community College
Robert Kirkwood, University of Central Arkansas
Virginia Latta, Jefferson State Junior College
Cathryn McDonald, Jefferson State Junior College
Janice Roberts, Jefferson State Junior College
Robert Romans, Bowling Green State University
Melvin Urschel, Tacoma Community College

Reviewers of the First Edition
Gil Desha, Tarrant County Junior College
Albert J. Grennan, San Diego Mesa College
Rhoda Love, formerly of Lane Community College
Don Misumi, Los Angeles Trade-Technical College
Robert P. Ouellett, Massasoit Community College
Ernest L. Rhamstine, Valencia Community College
John H. Standing, Delaware Valley College
Michael J. Timmons, Moraine Valley Community College

Market Research Respondents

We would like to thank the following adopters of the fourth edition for their help in preparing the current one. Each contributed greatly to our understanding of the relative strengths and weaknesses of the fourth edition by responding to a user's survey of the text.

David W. Allard, Texarkana College
Marjay D. Anderson, Howard University
Rolan E. Anderson, Anoka-Ramsey Community College
Philip J. Arnholt, Concordia College
Ronald Basmajian, Merced College
John Benson, Ohio University
Donald A. Bierle, St. Paul Bible College
Frank L. Bonham, San Diego Mesa College
Virginia Brown, Heritage College
David R. Carisch, Rochester Community College
Jay P. Clymerin, Marywood College
Edith Collins, Indiana Vocational Technical College
Scott L. Collins, University of Oklahoma
Donald W. Cowe, Sumter Area Technical College
Johnnie Driessner, Concordia College
Angie Garrett, Mount St. Mary's College
Elmer E. Gless, Montana College of Mineral Science and Technology
Kenneth R. Hille, Firelands College of Bowling Green State University
Boyd Holdawzy, Ricks College
Ronald Humphrey, Prairie View A&M University
William J. Iams, Sir Wilfred Grenfell College
Daryl H. Johnson, Mississippi County Community College
Elmo A. Law, University of Missouri–Kansas City
Frederick W. Law, Shawnee State University
Rudy Locklear, Robeson Technical College
Ann Lopez, San Jose City College
Perry V. Mack, Bennett College
Richard E. McKeeby, Union College
William D. McNaughton, Oakland Community College
Thomas E. McQuistion, Millikin University

Howard C. Monroe, College of San Mateo
Roger C. Nealeigh, Central Community College
Relda Niederhofer, Firelands College—Bowling Green State University
LeRoy E. Olson, Southwestern College
Bonnie Anne Osif, United Wesleyan College
Robert P. Ouellett, Massasoit Community College
Pat Palanker, Middlesex County College
Herbert D. Papenfuss, Boise State University
Lawrence V. Pion, St. Francis College
Doris S. Powell, University of Baltimore
Linda M. Reeves, Navajo Community College
R. Harvard Riches, Pittsburg State University
Ron L. Rivers, St. John's College
Charles R. Samuelson, Northland Community College
Gary L. Sansom, Yakima Valley Community College
Fred H. Schindler, Central Community College
David W. Schroder, Lincoln College
LeRoy J. Scott, Saint Petersburg Junior College
John D. Sharp, Sr., John Tyler Community College
Keith Smith, Fayetteville Technical Institute
Kenneth J. Smith, Montcalm Community College
Carol Woodard Stewart, Fayetteville Technical Institute
John D. Strehmel, Gateway Technical Institute
John D. Suhr, Concordia College
E. Russell TePaske, University of Northern Iowa
Donald D. Terpening, Ulster County Community College
Jack Thomas, Harbor College
Wayne H. Tinnell, Longwood College
Jane Glasgow Vance, Northeast Alabama State Junior College
Les Whitley, Coastal Carolina College
Ann Williams, Fergus Falls Community College
Ron Young, Bethany Lutheran College
Marua C. Watts, Malcolm X College—Chicago

■ To the Student

The text is designed to make understanding biological principles easier. At the beginning of each chapter is a section entitled *purpose*. This section gives you some hints about how the chapter fits in with the other parts of the book. It directs you to where you are going and lets you know why you are going there. Pay careful attention to the purpose, and you can tell when you have attained your goal and why this goal was set.

Each chapter is subdivided into topics separated by headings. These subdivisions are not chosen by chance, but are logical chunks of material. These subdivisions should make learning more manageable for you.

As is the case with most science classes, you are likely to find biological vocabulary a difficult hurdle to jump. To prevent you from becoming unduly discouraged as you approach this "foreign language," the first time an important term is used in the text it is printed in **boldface.** Each of these *new terms* is defined for you at least three times in the text: first, in the narrative when the term becomes a functional part of biological thought; second, in the chapter glossary at the end of the chapter in which it first appears; and third, in the complete glossary at the end of the book. As you review a chapter, you should mentally define each of the new terms. If you are unsure of the meaning of these terms, check yourself against the definitions in the book. In this edition we have also provided a phonetic pronunciation guide for each glossary term. You will learn to pronounce each term correctly as you learn its meaning.

There are a number of illustrations throughout the text. These illustrations should do more than just attract your attention. Each has been chosen carefully to help you understand a point or help you to tie a point to something you already know. Use these illustrations and their captions to help you learn and understand the ideas presented.

Each chapter ends with a summary. As you finish studying a chapter, read the summary sentence by sentence. Make sure that there is no new information in the summary. If something is new to you, it is because you have not thoroughly studied part of the chapter.

Finally, at the end of the chapter we have presented a thought-provoking situation. It asks you to use your newfound knowledge and previous experience in considering the situation. Most often, there is no one right answer. It stimulates you to think something through and to raise points for discussion. The most valuable aspect of an introductory biology course is not the tidbits of factual information that you gather, but the new ways in which you see yourself and your environment. We hope these thought stimulators will give you practice in using biological information and applying basic biological concepts to real situations.

Following each thought-provoking situation is a series of review questions. These questions can be used in a variety of ways: you might use them to help channel your attention as you study a chapter, or as a review to tell you when you are well prepared for a test over the chapter material. Each of these questions is directly answered in the chapter narrative or in the illustrations. ■

Introduction

It is important to understand the difference between a scientific approach to the study of life and other approaches that are philosophically valid but nonscientific. Even the word ''life'' has different meanings when looked at from a philosophical, legal, or scientific point of view. We need to become acquainted with how scientists perceive the world and what influences their thinking. ■

What Is Biology?

■ Chapter Outline

■ Purpose

This chapter is a general introduction to the nature of science and the significance of biology in your everyday life. It will present a scientist's view of the world and describe what living things are and how they differ from nonliving things. This chapter lays the groundwork to help you understand and answer questions about living things in your environment. While there may be a number of different answers for each question, the answers may not be simple. You will be better able to understand and answer biological questions after you have an understanding of how science works.

For Your Information

As a result of recent, rapid advances in science, most newspapers have added a science page as a weekly feature. These sections are intended to keep the general public aware of the most significant advances in all areas of science. Subjects such as recombinant DNA theory, biological amplification, and punctuated evolution are no longer found exclusively in scientific journals that only the most well-informed, practicing scientist can understand.

Learning Objectives

■ Understand the difference between science and nonscience.

■ Know the steps in the scientific method.

■ Recognize the limits of the scientific method.

■ Differentiate between applied and theoretical science.

■ Know the characteristics that classify matter into the categories of living and nonliving.

■ Understand the value of biology.

■ Be able to give examples of biological problems.

Biological Science in Your Life

Many college students question the need for a biology class if they are not going to major in science. Do you find yourself asking the question, "What am I doing in a biology class?" Consider how your future will be tied to the way in which current biological problems are solved. How will radiation from nuclear power plants affect me or my children? Can sexually transmitted diseases be controlled? When will a cure for cancer be developed? Will water shortages result in mass starvation?

These questions are being raised more frequently today than ever before (figure 1.1). These questions, like many other pertinent questions being asked, deal with biological concepts. There are no easy answers. There are, however, informed ways of looking at problems and arriving at solutions. Because you are a living thing, you are involved in these problems and may be a part of their solution.

As an informed citizen in a democracy, you can have a great deal to say about the solutions to these problems. In a democracy it is assumed that the public is informed enough to make intelligent decisions. This is why an understanding of biological concepts is so important for any person, regardless of his or her vocation. *Concepts in Biology* was written with this philosophy in mind. The concepts covered in this book were selected to help you become more aware of how biology influences nearly every aspect of your life.

Figure 1.1
Biology in Our Everyday Life.

It is very easy to identify biological problems and situations. These news headlines reflect a few of the biologically based issues that face us everyday. While articles such as these seldom propose solutions, they do serve to make the general public aware of situations so that people can begin to explore the possibility of making intelligent decisions about their solutions.

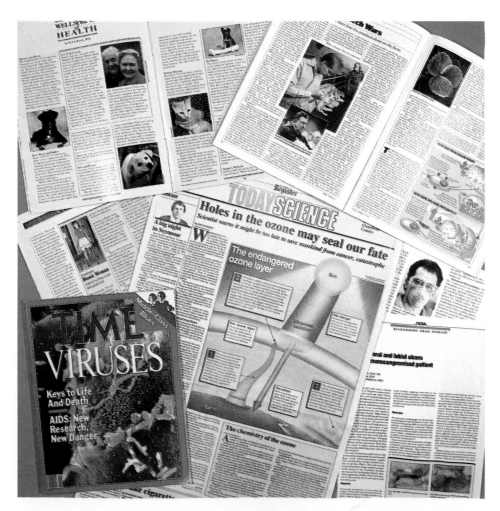

Organization of the Book

Most learning follows a certain pattern. You began to learn to communicate ideas a long time ago. Progressively you learned to make basic sounds, form words, speak in sentences, and communicate abstract ideas. Learning is a stepwise process that builds upon prior knowledge and experiences while new knowledge provides a foundation for future learning. You constantly expand and build your knowledge on the more basic information. The same thing is true of biology. There are certain underlying ideas that you must understand before you can see the whole picture. These ideas need to be approached in an orderly fashion. In biology, the most basic ideas deal with chemical activities and this is your first step. Chemistry is presented in chapters 2 and 3. Just as you had to make sounds before you could make conversation, a knowledge of chemistry is necessary before you can really understand the biology presented later in this book (figure 1.2).

Once you have learned about chemical activity, you will be ready to take the next step in which you are introduced to cells and some of the important things they do. Because organisms are composed of cells, it is logical to study the structure and function of cells before studying how cells interact with each other as a part of an individual organism.

The characteristics of cells and organisms are determined by the chemical code system found within the cells. Any change in this code system could result in differences between individual organisms and possibly lead to the formation of different kinds of organisms. An understanding of individual organisms is necessary since individuals of the same kind make up a population. One of the fascinating characteristics of a population is how its members behave instinctively and socially.

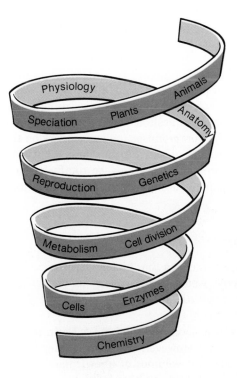

Figure 1.2
Spiral of Knowledge.
This spiral shows how the book is organized. Each topic builds on the material that has previously been covered. As you gain more information from each area, you increase the depth and scope of your understanding of biology.

Figure 1.3
Interactions Between Populations.

Living systems are very complex and interrelated. The grass population influences the population of deer, and they in turn influence the cougar population. A change in any one population will result in changes in the other populations.

Populations also interact with one another in a variety of ways. For example, a population of deer interacts with plant populations by feeding upon them. The deer are, in turn, eaten by cougars (figure 1.3). This book and its organization progress from the basic to the complex: from chemical structure to cell anatomy and physiology, reproduction and heredity, evolution and ecology, and finally to the five kingdoms of living things. The knowledge you gain from this course can prepare you to better understand and appreciate the complexity of biological problems.

The problems raised at the beginning of this chapter are of crucial concern to you. They do not have a single right answer or best solution. Each question involves a certain amount of compromise between what society wants and what your biological self must have. Each of the questions is so complex that few persons can see the "whole picture." They are questions that need input from various points of view and considerable thought, compromise, and active work on the part of many people in order to find acceptable solutions. Furthermore, what we consider to be an acceptable solution to a problem today may be totally unacceptable in the future. Biological knowledge gained through scientific investigation may result in a new and better solution.

Science and Nonscience

You have probably heard the idea that biology has something to do with plants and animals. Most textbooks would define **biology** as the science that deals with life. This appears to be a nice little definition until you begin to think about what the words *science* and *life* mean.

The word science is a noun derived from a Latin term (*scientia*) meaning to have knowledge or to know. However, this is misleading since there are many fields that have accumulated great volumes of knowledge that are not "sciences." The heart of the issues lies with how the knowledge is acquired, not just with the fact that the knowledge exists. **Science** is actually a process or way of arriving at a solution to a problem or understanding an event in nature. The process followed by scientists is known as the scientific method. The **scientific method** is a way of gaining information (facts) about the world around you. What scientists learn by using this approach may help them solve a particular problem or it may just be interesting. The scientific method requires a systematic search for information by observation and experimentation, followed by the formation and testing of a possible solution. Knowledge that is gained by other methods is known as nonscience and is acquired by nonscientific methods.

As information is collected, scientists use certain rules to help place these facts into a framework that helps them better understand the problem. Laws, principles, and rules are developed as scientists begin to see patterns or relationships among a number of isolated facts and the pieces of the puzzle are put into place. Some of the laws are very old, while others are being modified and constructed today. In some cases, the same knowledge may be used in both scientific and nonscientific areas of study. A good example of this is the difference between astronomy, which is a science, and astrology, which is a nonscience. These studies make use of some of the same information. Both astronomers and astrologers observe the sun, moon, and stars, and chart their positions, movements, and interactions. The astronomer makes measurements and observations and organizes this information according to physical laws, such as the law of gravity. In other words, the astronomer observes what is happening and uses various physical laws to organize the millions of bits of information. These laws are continually tested by the addition of new bits of information collected by observation and experimentation. If the new information can fit into the framework constructed, it reinforces the framework. Scientists are continually checking to see if the conclusions that have been drawn are **valid** (meaningful, fit the framework) and **reliable** (giving the same result on successive trials). If the new information does not fit, the formation of a new framework may be necessary.

The astrologer looks at the same sun, moon, and stars and also develops rules about how these celestial bodies relate to one another. However, these rules cannot be tested for accuracy by experimentation. As a matter of fact, an astrologer is usually not very interested in having the rules challenged or tested and usually writes them so that they cannot be tested. An astrologer's basic rule is that the various celestial bodies control the progress of human events. No information can be collected that is valid and reliable and supports this basic rule.

The differences between science and nonscience are often based on the assumptions and methods used to gather and organize information and, most importantly, the testing of these assumptions. The difference between a scientist and a nonscientist is that a scientist continually challenges and tests the principles and laws, whereas a nonscientist may not feel that this is important (figure 1.4).

Once you understand the scientific method, you won't have any trouble identifying astronomy, chemistry, physics, and biology as sciences. But what about economics, sociology, anthropology, history, philosophy, and literature? All of these fields may make use of certain laws that are derived in a logical way, but they are also nonscientific in some ways. Some things are beyond science and cannot be approached using the scientific method. Art, literature, theology, and philosophy are rarely thought of as sciences. They are concerned with beauty, human emotion, and speculative thought rather than with facts and verifiable laws. On the other hand, physics, chemistry, geology, and biology are almost always considered sciences. Music is an area of study in a middle ground where scientific approaches may be used to some extent. "Good" music is certainly unrelated to science, but the study of how

Figure 1.4
Astronomy is a Science.
The astronomer and the astrologer use much of the same information, but they use it in very different ways.

Table 1.1 *Classification of Traditional Fields of Study*

Definitely Science			Definitely Not Science
Geology		Psychology	Astrology
Chemistry	History	Music	Theology
Physics	Economics		Literature
Biology		Sociology	Political science
Astronomy	Anthropology		Philosophy
Geography			Art

the human larynx generates the sound of a song is based on scientific principles. Any serious student of music will study the anatomy of the human voice box and how the vocal cords vibrate to generate sound waves (table 1.1).

Just because scientists say something is true does not necessarily make it true. Everyone makes mistakes and quite often, as new information is gathered, old laws must be changed or discarded. For example, at one time scientists were sure that the sun went around the earth. They observed that the sun rose in the east and traveled across the sky to set in the west. Since scientists could not feel the earth moving, it seemed perfectly logical that the sun traveled around the earth. Once they understood that the earth rotated on its axis, they began to understand that the rising and setting of the sun could be explained in other ways. A completely new concept of the relationship between the sun and the earth developed (figure 1.5).

Although this kind of study seems rather primitive to us today, this change in thinking about the sun and the earth was a very important step in understanding the universe and how the various parts are related to one another. This background information was built upon by many generations of astronomers and space scientists and finally led to space exploration.

Figure 1.5
Knowledge Changes.
Our understanding of the relationship between the earth and the sun has changed considerably as we have gained more information.

Limitations of Science

By definition, science is a way of thinking and seeking information to solve problems. Therefore, the scientific method can be applied only to questions that have a factual basis. Questions concerning morals, value judgments, social issues, and attitudes cannot be answered using the scientific method. What makes a painting great?

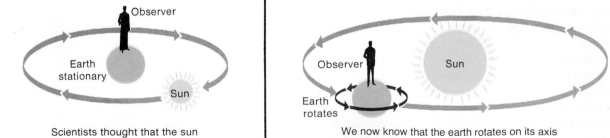

Scientists thought that the sun revolved around the earth.

We now know that the earth rotates on its axis and also revolves around the sun.

What is the best type of music? Which wine is best? What color should I paint my car? These questions are related to values, beliefs, and tastes; therefore, the scientific method cannot be used to answer them. Science does not have, nor does it attempt to provide, all the answers to the problems of the human race. Science is merely one of the tools at our disposal. But what exactly is this process that can lead to giant leaps in understanding our surroundings?

The Scientific Method

People who use the scientific method follow a sequence of thought processes and activities outlined in table 1.2. As with your own thought processes, the minds of scientists bounce from thought to thought and from category to category as they wrestle with the problem at hand. Keep in mind that scientists try not to "re-invent the wheel"; therefore, they do not always start at the beginning. The sequence presented in table 1.2 is idealistic and a simplification of a very complex series of events.

The scientific method usually begins with the *observation* that an unexplained event has occurred repeatedly. The event may be directly observable with the senses or indirectly observable with instruments. If the event only occurs once or is not able to be repeated in an artificial situation, it is impossible to use the scientific method to gain further information about the event and to explain it. The information gained by observing the event is called **empirical evidence.** Empirical evidence is capable of being verified or disproved by further observation.

As scientists gain more empirical evidence about the event, they begin to develop *questions* about it. How does this happen? What caused it to occur? When will it take place again? Can I control the event to my benefit? The formation of the question is not as easy as it might seem since the way the question is asked will determine how you go about answering it. A question that is too broad or too complex may be impossible to answer; therefore, a great deal of effort is put into asking the question in the right way. In some situations, this can be the most time-consuming part of the scientific method since asking the right question is critical to how you look for answers.

Once the question is written, scientists *explore other sources of knowledge* to arrive at an acceptable answer. They turn first to the research of others since there is no sense wasting time and energy answering an already answered question. This usually means a trip to the library or making contact with fellow scientists interested in the same field of study. Even if their particular question has not already been answered, scientific literature and fellow scientists can provide insight into the problem that will help in its solution. It is during this time that a hypothesis is formed.

A **hypothesis** is a possible answer or explanation to a question, must account for all the observed facts, and must be testable. Just as writing a good scientific question is important, hypothesis formation is critical and may be difficult. If the hypothesis does not account for all the observed facts in the situation, doubt will be cast on the work and may eventually invalidate the hypothesis. Doubt may also surround the hypothesis if it is not testable. If a possible answer cannot be proven true, it would only be hearsay and no more valid than a stab-in-the-dark answer. Keep in mind that a hypothesis is an educated guess based on observations and information gained from other knowledgeable sources. In order to determine whether or not the hypothesis is correct, scientists begin the next step in the scientific method—experimentation.

Table 1.2   *The Scientific Method*

Steps in the Scientific Method	Activity	Example
1. Observation	1. Recognize something has happened and that it occurs repeatedly. Empirical evidence gained from experiences or observation.	1. Students in an elementary classroom are stricken with a disease that causes red rash on their faces. This has happened in three major cities every September for the past six years. To treat the disease, physicians prescribed drug X. At the end of five days, all the students in every situation recovered.
2. Question formulation	2. Write many questions about the observations and keep only the ones that will be answerable.	2. Did the medication cure the disease or did the illness go away on its own?
3. Explore alternative resources	3. Go to the library and talk with others to gain knowledge about your question.	3. A search of the medical literature revealed evidence that this particular drug might cure this illness. Consultations with the drug manufacturer, area physicians, and medical microbiologists fail to turn up an answer to the question but they do indicate success in many similar situations.
4. Hypothesis formation	4. Pose a possible answer to your question, realizing that you could be wrong.	4. The drug X is capable of controlling the disease-causing agent.
5. Experimentation	5. Set up an experiment that will allow you to test your hypothesis using a control group and an experimental group. Be sure to collect and analyze the data carefully.	5. To test the cause-and-effect relationship between the medication and the cure, the class is divided into two groups. One group (the experimental) is given the medicine X, while the other group (the control) is given a sugar pill, a placebo. (Sometimes a disease can be cured when any medicine is given, whether it is a beneficial medicine or not. Because the patient *believes* he or she is being treated, the body cures itself. To make sure that this red rash disease is not of that type, the scientist gives some of the class a placebo and the rest of the class the medicine. To make sure that the scientist does not influence the recovery, sometimes the nurse does not know which pill is which.) At the end of five days, all the members of the experimental group are cured and making plans to attend the funerals of everyone else in the control group. By comparing the experimental and control groups, it appears that the medicine cured the disease. Repeat the experiment in other cities with other groups of infected students, collect data, and analyze results. If the results are reliable and valid, proceed to next step.
6. Theory formulation	6. Repeat the experiment in conjunction with others over a long period to challenge your findings. Should the results continue to support your hypothesis, the scientific community will call your hypothesis a theory.	6. Since experimentation with drug X has shown repeatedly that its administration can cure students with red rash disease in five days in comparison to control groups, it is theorized that drug X is able to control this disease.
7. Experimentation	7. Set up an experiment that will allow you to test your theory using a control group and an experimental group. Be sure to collect and analyze the data carefully.	7. Repeat step five, including any new information that has been gained over the years.
8. Law formulation	8. Repeat the experiment in conjunction with others over a long period to challenge your findings. Should the results continue to support your theory, the scientific community will call your theory a scientific law.	8. Since experimentation with drug X has shown repeatedly that its administration can cure students with red rash disease in five days in comparison to control groups, it is a scientific law that drug X is able to control this disease.

An **experiment** is a re-creation of an event or occurrence in a way that enables a scientist to gain valid and reliable empirical evidence. This can be difficult since a particular event may involve a great many separate happenings. To unclutter the situation, scientists have devised what is known as a controlled experiment. A **controlled experiment** allows scientists to compare two situations that are identical in all but one respect. The situation used as the basis of comparison is called the **control group** while the other is called the **experimental group.** The single factor that is allowed to be different in the experimental group but is controlled in the other group is called the **variable.** After the experiment, the new data (facts) gathered are analyzed. If there are no differences between the two groups, the variable evidently did not have a cause-and-effect relationship (i.e., was not responsible for the event). However, if there was a difference, it is likely that the variable was responsible for the difference between the control and experimental groups. The experiment is repeated many times since scientists encourage challenges to their work and admit that no experiment is perfect. During experimentation new information is learned and new questions are formulated that can lead to even more experiments. Some scientists speculate that one good experiment can result in a hundred new questions and experiments.

If the processes of questioning and experimentation continue and evidence continually and consistently supports the original hypothesis, the scientific community will place more support in the hypothesis and it will become an accepted answer. When this happens, the hypothesis becomes a theory. A **theory** is a plausible, scientifically acceptable generalization. An example of a biological theory is the germ theory of disease. This theory states that certain diseases, called infectious diseases, are caused by microorganisms that are capable of being transmitted from one individual to another. As you can see, this is a very broad statement. It is the result of years of questioning, experimentation, and pulling data together. While a hypothesis is an answer to a specific question, a theory encompasses the answers to many complex questions. As a result, there are fewer theories than hypotheses.

However, just because a hypothesis becomes a theory does not mean that testing stops. In fact, many scientists see this as a challenge and exert even greater efforts to disprove the theory, and experimentation continues. Should the theory survive this skeptical approach and continue to be supported by experimental evidence, the theory will become a scientific law. A **scientific law** is a uniform or constant fact of nature. An example of a biological law is the biogenetic law, which states that all living things come from preexisting living things. You can see from this example that laws are even more general than theories and encompass the answers to even more complex questions. Therefore, there are very few scientific laws.

Applied and Theoretical Science

The scientific method has helped us to understand and control many aspects of our natural world. Science is divided into two categories, theoretical and applied, depending on whether experimentation is done to gain knowledge for knowledge's sake, or to better our daily lives. **Theoretical science** is interested in obtaining new information, seeing how it fits or does not fit the "old laws," and writing "new laws" if necessary. Little interest is paid to how the new information may be used in any specific or practical situations. On the other hand, a scientist who actively seeks solutions to practical problems is involved in applied science.

Although they seem different, theoretical and applied science are related. **Applied science** makes practical use of the theories provided by the theoretical scientists by using the new knowledge to solve everyday problems. For example, applied scientists, known as genetic engineers, have altered the chemical code system of small organisms (microorganisms) so that they may produce many new drugs such as antibiotics, hormones, and enzymes. The ease with which these complex chemicals are produced would not have been possible if it had not been for the information gained from the theoretical sciences of microbiology, molecular biology, and genetics (figure 1.6).

Figure 1.6
Knowledge Accumulates.
(a) J. D. Watson and F. W. Crick are theoretical scientists who were interested in the structure of the chemical code system of cells, deoxyribonucleic acid. (b) Little did they know that the information they gained would later be used to alter the genetic makeup of bacteria so that these microorganisms could be used to produce large quantities of such complex chemicals as antibiotics, proteins, and vitamins.

a

b

As another example, Louis Pasteur was interested in the theoretical problem of whether life could be generated from nonliving material. Much of his theoretical work led to practical applications in disease control. His theory that there are microorganisms that cause diseases and decay led to the development of vaccinations against rabies and the development of pasteurization for the preservation of foods (figure 1.7).

What Is Biology?

The science of biology is, broadly speaking, the study of living things. It draws on chemistry and physics for its foundation and applies these basic physical laws to living things. Because there are many kinds of living things, there are many special areas of study in biology. Practical biology—like medicine, crop science, plant breeding, and wildlife management—is balanced by more theoretical biology—such as medical microbiological physiology, photosynthetic biochemistry, plant taxonomy, and animal behavior (ethology). There is also just plain fun biology like insect collecting and bird watching. Specifically, biology is a science that deals with living things and how they interact with all of the things around them.

At the beginning of the chapter, biology was defined as the science that deals with living things. But what does it mean to be alive? You would think that a biology textbook could answer this question very easily. However, this question is more than just a theoretical one, since it has become necessary in recent years to construct some legal definitions of what life is; when it begins and ends. The legal definition of death is important, since it may determine whether or not a person will receive life insurance benefits or if body parts may be used in transplants. In the case of heart transplants, the person donating the heart may be legally "dead" but the heart certainly isn't since it can be removed while the heart still has "life." In other words, there are different kinds of death. There is the death of the whole living unit and the death of each cell within the living unit. A person actually "dies" before every cell has died. Death, then, is the absence of life, but that still doesn't tell us what life is. At this point, we won't try to define life but will describe some of the basic characteristics of living things.

Characteristics of Life

The ability to manipulate energy and matter is unique to living things. Just how this is accomplished can be used to better understand how living things differ from the nonliving. Living things show four characteristics that the nonliving do not display: (1) metabolic processes, (2) generative processes, (3) responsive processes, and (4) control processes.

Metabolic processes are the total of all chemical reactions within an organism. There are three essential aspects of metabolism: (1) *nutrient uptake,* (2) *nutrient processing,* and (3) *waste elimination.* All living things expend energy to take in nutrients (raw materials) and energy from their environment. Many animals take in these materials by eating or swallowing other organisms. Microorganisms and plants absorb raw materials into their cells to maintain their lives. Once inside, nutrients enter a network of chemical reactions. These reactions constantly use energy to process nutrients in order to manufacture component parts, make repairs, and reproduce. However, not all nutrients entering a living thing are valuable to it. There

Figure 1.7
Louis Pasteur and Pasteurized Milk.

Louis Pasteur (1822–1895) performed many experiments while he studied the question of the origin of life, one of which led directly to the food preservation method now known as pasteurization.

Table 1.3 *Characteristics of Life*

Living and nonliving things differ in a number of ways. Some of these differences are shown here.

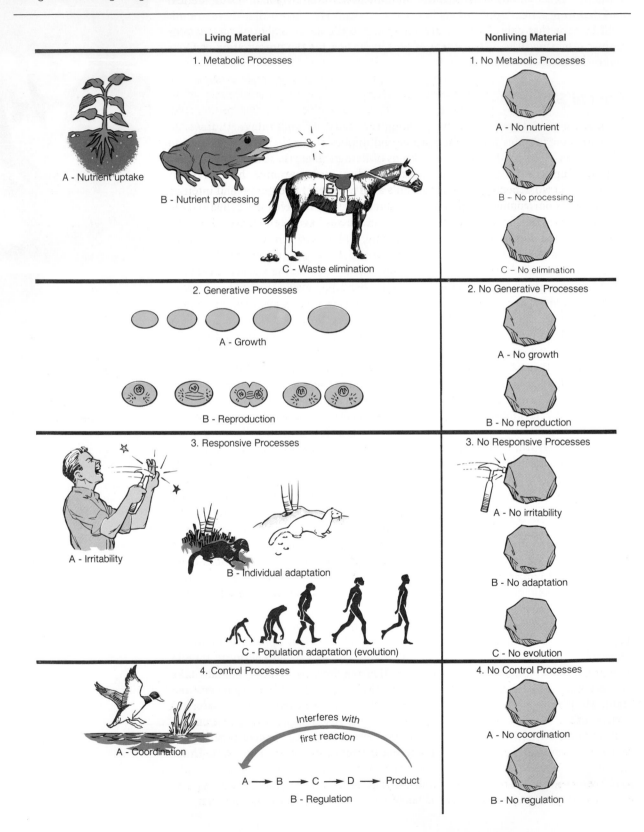

Living Material

1. Metabolic Processes
   A - Nutrient uptake
   B - Nutrient processing
   C - Waste elimination

2. Generative Processes
   A - Growth
   B - Reproduction

3. Responsive Processes
   A - Irritability
   B - Individual adaptation
   C - Population adaptation (evolution)

4. Control Processes
   A - Coordination

   Interferes with first reaction

   A → B → C → D → Product

   B - Regulation

Nonliving Material

1. No Metabolic Processes
   A - No nutrient
   B – No processing
   C – No elimination

2. No Generative Processes
   A - No growth
   B - No reproduction

3. No Responsive Processes
   A - No irritability
   B - No adaptation
   C - No evolution

4. No Control Processes
   A - No coordination
   B - No regulation

may be portions of nutrients that are valuable, but the rest may be useless or even harmful. In that situation, organisms eliminate both matter (solid waste) and energy (heat).

The second group of characteristics of life are known as the generative processes. **Generative processes** are reactions that result in an increase in the size of an individual organism—*growth;* or an increase in the number of individuals in a population of organisms—*reproduction.* During growth, living things add to their structure, repair parts, and store nutrients for later use. However, growth cannot go on indefinitely because as organisms get larger, they become inefficient. The result could be chemical chaos with organisms wasting energy and nutrients. Living things respond to this problem by reproducing. Reproduction is one of the most important functions because it is the only way that living things give rise to offspring. There are a number of ways that organisms can reproduce and guarantee their survival.

Survival also depends on the organism's ability to react to external and internal changes in its environment. The group of characteristics involved are called the **responsive processes.** Three types of responsive processes have been identified: *irritability, individual adaptation,* and *population adaptation* or *evolution.* Irritability is an individual's rapid response to a stimulus, such as a knee jerk reflex. This type of response occurs only in the individual receiving the stimulus and is rapid because the mechanism that allows the response to occur (i.e., muscles, bones, and nerves) is already in place. Individual adaptation is also an individual response but is slower since it requires a genetic action. For example, a weasel changes from its brown summer coat to its white winter coat by turning off the genes that are responsible for brown pigment production. Population adaptation is also known as evolution. Evolution is a slow change in the genetic makeup of a *population* of living organisms. This enables a group of organisms to adapt and better survive threatening changes in their environment over many generations.

The **control processes** of *coordination* and *regulation* constitute the fourth characteristic of life. Control processes are mechanisms that ensure that an organism will carry out all metabolic activities in the proper sequence (coordination) and the proper amount (regulation). All the chemical reactions of an organism are coordinated and linked together in specific pathways. The orchestration of all the reactions ensures that there will be specific stepwise handling of nutrients needed to maintain life. The molecules responsible for coordinating these reactions are known as enzymes. **Enzymes** are molecules produced by organisms that are able to increase and control the rate at which chemical reactions occur. Enzymes also regulate the amount of nutrients processed into other forms. This limits the chance that the organism will die because nutrients will be used too quickly or too much waste will be generated. Table 1.3 summarizes the differences between living and nonliving things.

The Value of Biology

To a great extent, we owe our current high standard of living to biological advances in two areas, food production and disease control. Plant and animal breeders have developed plants and animals that provide better sources of food than the original varieties. One of the best examples of this is the various changes that have occurred in corn. Corn is a grass that produces its seed on a cob. The original corn plant had very small ears that were perhaps only three or four centimeters long. Through selective breeding, varieties of corn with much larger ears and more seeds per cob have been produced. This has increased the yield greatly. In addition, the corn plant has been adapted to produce other kinds of corn, like sweet corn and popcorn.

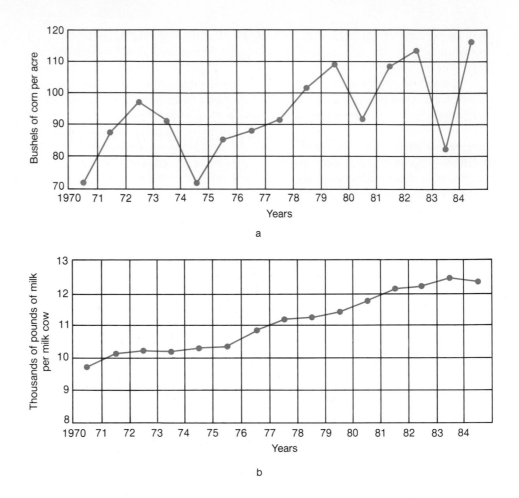

a

b

Figure 1.8 a–b
Food Production in the United States.

Both of these graphs illustrate a steady increase in yield, largely because of changing farming practices and selective breeding programs. (Source: Department of Agriculture, *Agricultural Statistics,* 1985, pp. 30, 316.)

Corn is not an isolated example. Improvements in yield have been brought about in wheat, rice, oats, and other cereal grains. The improvements in the plants along with changed farming practices (also brought about through biological experimentation) have led to a greatly increased production of food (figure 1.8a).

Animal breeders have also had great successes. The pig, chicken, and cow of today are much different animals from those available even one hundred years ago. Chickens lay more eggs, dairy cows give more milk, and beef cattle grow faster (figure. 1.8b). All of these improvements raise our standard of living. One interesting example is the change in the kinds of hogs that are raised. At one time, farmers wanted pigs that were fatty. The fat could be made into lard, soap, and a variety of other useful products. As the demand for the fat products of pigs began to decline, animal breeders began to develop pigs that gave a high yield of meat and relatively little fat. Today, plant and animal breeders can produce plants and animals almost to specifications.

Much of the improvement in food production has resulted from the control of plants and animals that compete with or eat the organisms we use as food. Control of insects and fungi that weaken plants and reduce yields is as important as the invention of new varieties of plants. Since these are "living" pests, biologists have been involved in the study of them also.

There has been fantastic progress in the area of health and disease control. Many diseases such as polio, whooping cough, measles, and mumps can be easily

Figure 1.9
Fire Hazard.
This forest has so much litter on its
floor that a fire could destroy the
entire forest.

controlled by vaccinations or "shots" (box 1.1). Unfortunately, the vaccines have
worked so well that some people no longer worry about getting their shots, and some
of these diseases are reappearing. These diseases have not been eliminated, and people
who are not protected by vaccinations are still susceptible to them.

The understanding of how the human body works has led to treatments that can
control diseases such as diabetes, high blood pressure, and even some kinds of cancer.
Paradoxically, these advances contribute to a major biological problem: the increasing size of the human population.

Biological Problems

Now that you have seen some progress that can be credited to biologists, we will
look at some of the problems that have been created by improperly used biological
principles. For example, the drive to preserve nature has in some cases done more
harm than good. Many of our western forests have been preserved as parks. In order
to preserve the trees, fire was not allowed in these forests. The lack of fire has led
to a dangerous buildup of debris (figure 1.9). Before our involvement, periodic, natural fires cleaned out the debris and helped to preserve the trees from more intense

fires. These fires also eliminated some undesirable species of trees. Our intentions might have been good, but we should have looked at the natural situation before we interfered. A whole generation of people have grown up with the idea that all forest fires are bad.

A second major biological mistake has been the introduction into some countries of exotic (foreign) species of plants and animals. In North America, this has had disastrous consequences in a number of cases. Both the American chestnut and the American elm have been nearly eliminated by diseases that were introduced by accident. Other things have been introduced on purpose, which show shortsightedness or a total lack of understanding about biology. The starling and the English (house) sparrow were both introduced into this country by people who thought that they were doing good. Both of these birds have multiplied greatly and replaced some of the birds that were native to the United States. The gypsy moth is also an introduced species. These animals were brought to the United States by manufacturers in hopes of increasing silk production. When the scheme fell short of their goal and moths were accidentally set free, the moths quickly took advantage of their new environment by feeding on desirable ornamental plants and crops. Even with these examples before us, there are still people who try to sneak exotic plants and animals into the country without thinking about the possible consequences.

The Future of Biology

Where do we go from here? Many problems remain to be solved. Major breakthroughs in the control of the human population are being sought. There is a continued interest in the development of more efficient methods of producing food.

One of the major areas that will receive attention in the next few years will be the relationship between genetic information and diseases such as Alzheimer's disease, stroke, arthritis, and cancer. Many kinds of diseases are caused by abnormal body chemistry. These changes are the result of hereditary characteristics. Curing certain hereditary diseases is a big job. It requires a complete understanding of genetics and the introduction or subtraction of hereditary information from all of the trillions of cells of the organism.

Another area that will receive more attention in the next few years is ecology. The energy crisis is real and pollution is still a problem. The majority of people still need to learn that some environmental changes may be acceptable, whereas other changes will ultimately lead to our destruction. We have two tasks. The first is improving technology and increasing our understanding about how things work in our biological world. The second, and probably the hardest, is educating, pressuring, and reminding people that their actions determine the kind of world the next generation will have.

Summary

The science of biology is the study of living things and how they interact with their surroundings. Science and nonscience can be distinguished by the kinds of laws and rules that are constructed to unify the body of knowledge. Science involves the continuous testing of rules and principles by the collection of new facts. In science these rules are usually arrived at by using the scientific method—observation, questioning, exploring resources, hypothesis formation, experimentation, theory formation, experimentation, and law formation. If the rules are not testable or if no rules are used, it is not science.

Living things show the characteristics of (1) metabolic processes, (2) generative processes, (3) responsive processes, and (4) control processes. Biology has been responsible for major advances in the areas of food production and health. The incorrect use of biological principles has sometimes led to the elimination of useful organisms and the destruction of organisms we wish to preserve. In the future, biologists will study many things. Two areas that are sure to receive attention are the relationship between heredity and disease, and ecology.

Consider This

Many television commercials claim that their products have undergone strict scientific experimentation to prove that they are worth their purchase price. What kinds of tests must have been done to make their statements true, and what more do you need to know to evaluate the truth of their claims? Once you have thought this through, keep track of the number of television advertisements per day that use pseudoscience (false science) to support such claims as "four out of five doctors use brand X. . . ."

Experience This

Take ten minutes to look through your local newspaper and count just how many advertised products have supposedly been "scienced."

Questions

1. What is biology?
2. What is the difference between science and nonscience? Give examples.
3. Why is testing so important in science?
4. What is the difference between theoretical science and applied science?
5. List four characteristics of living things.
6. List three advances that have occurred as a result of biology.
7. List three mistakes that could have been avoided had we known more about living things.
8. The scientific method cannot be used to deny or prove the existence of God. Why?
9. What are controlled experiments? Why are they necessary to support a hypothesis?
10. List the steps in the scientific method.

Chapter Glossary

applied science (ap-plīd si-ens) Science that makes practical use of the theories provided by scientists to solve everyday problems.
biology (bi-ol'o-je) The science that deals with life.
control group (con-trōl' grūp) The situation used as the basis for comparison in a controlled experiment.
controlled experiment (con-trōld' ek-sper'ĭ-ment) An experiment that allows for a comparison of two events that are identical in all but one respect.
control processes (con-trōl pro'ses-es) Mechanisms that ensure that an organism will carry out all metabolic activities in the proper sequence (coordination) and the proper amount.
empirical evidence (em-pir'ĭ-cal ev-i-dens) The information gained by observing an event.

enzymes (en′zīms) Molecules produced by organisms that are able to control the rate at which chemical reactions occur.

experiment (ek-sper′ĭ-ment) A re-creation of an event in a way that enables a scientist to gain valid and reliable empirical evidence.

experimental group (ek-sper-i-men′tal grup) The situation in a controlled experiment that is identical to the control group in all respects but one.

generative processes (jen′uh-ra″tiv pros′es-es) Actions that increase the size of an individual organism (growth), or increase the number of individuals in a population (reproduction).

hypothesis (hi-poth′e-sis) A possible answer to or explanation of a question that accounts for all the observed facts and is testable.

metabolic processes (me-ta-bol′ik pros′es-es) The total of all chemical reactions within an organism; for example, nutrient uptake and processing, and waste elimination.

reliable (re-li′a-bul) Giving the same result on successive trials.

responsive processes (re-spon′siv pros′es-es) Those abilities to react to external and internal changes in the environment; for example, irritability, individual adaptation, and evolution.

science (si′ens) A process or way of arriving at a solution to a problem or understanding an event in nature.

scientific law (si-en-tif′ik law) A uniform or constant feature of nature supported by several theories.

scientific method (si-en-tif′ik meth′ud) A way of gaining information (facts) about the world around you involving observation, hypothesis formation, experimentation, theory formation, and law formation.

theoretical science (the-o-ret′i-kul si-ens) The science interested in obtaining new information for its own sake.

theory (the′-o-re) A plausible, scientifically acceptable generalization supported by several hypotheses and experimental trials.

valid (va′-lid) Meaningful data that fits into the framework of scientific knowledge.

variable (va-r′e-ȧ-bul) The single factor that is allowed to be different.

Cells: Anatomy and Action

T he study of living things has revealed that there is a minimal unit that displays all the characteristics of life, the cell. It is necessary to understand the structure and workings of this most basic unit in order to understand a simple or complex organism. Analysis shows that a cell is composed of many basic chemicals found in nature. Those elements are bound together in special ways and arranged into complex molecules, which are in turn fashioned into more complex structures. Two fundamental cell types have been identified, differing in their chemical composition and structure. They also differ in the ways they perform the chemical reactions required to maintain their existence and reproduce. Biologically important chemical reactions are controlled by the cell and responsible for the breakdown of nutrients and the manufacture of their own essential molecules. Understanding how cells are constructed and how they work enables scientists to control them for the benefit of all. ■

Simple Things of Life

Chapter Outline

Purpose

In order to understand the structure and activities of living organisms, you must know something about the materials from which they are made. In this chapter we will discuss the structure of matter and the energy it contains. As you read this chapter, you should consciously try to build a vocabulary that will help you describe matter.

For Your Information

Baking soda is sodium bicarbonate. It has several uses in the home. You brush your teeth with it to neutralize acids generated by decay-causing bacteria. Because it neutralizes acids, it is used in cooking to make foods less sour. It is also used in baking as a source of carbon dioxide gas to leaven breads and other baked goods.

Learning Objectives

- Understand that all matter is composed of atoms of different elements.

- Explain how molecular motion relates to the laws of thermodynamics and diffusion.

- Know the basic structure of an atom.

- Recognize how isotopes differ from one another.

- Differentiate among ionic, covalent, and hydrogen bonds.

- Describe the relationship among acids, bases, salts, and the pH scale.

- Describe how the three states of matter—liquids, solids, and gases—differ at the molecular level.

- Distinguish among mixtures, solutions, suspensions, and colloids.

- Recognize that atoms may enter into different types of reactions as they become bonded and more stable.

Basic Structures

Everything on earth is part of what we call matter. **Matter** is anything that has weight (mass) and also takes up space (volume). Both of these characteristics depend on the amount of matter you are dealing with; the greater the amount, the greater its mass and volume.

Characteristics that are independent of the amount of matter include density and activity. **Density** is the weight of a certain volume of material; it is frequently expressed as grams per cubic centimeter. For example, a cubic centimeter of lead is very heavy and a cubic centimeter of aluminum is very light. Lead has a higher density than aluminum. The activity of matter depends almost entirely on its composition. All matter is composed of one or more types of substances called elements. **Elements** are the basic building blocks from which all things are made. You already know the names of some of these elements: oxygen, iron, aluminum, silver, carbon, and gold. The sidewalk, water, air, and you are all composed of various types of elements.

Atomic Nucleus

In order to understand the way elements act, we need to understand what they are composed of. The smallest part of an element that still acts like that element is called an **atom.** When we use a **chemical symbol** such as Al for aluminum or C for carbon, it represents one atom of that element. The atom is constructed of three major particles; two of them are in a central region called the **atomic nucleus.** The third type of particle is in the region surrounding the nucleus (figure 2.1). The weight or mass of the atom is concentrated in the nucleus, which is composed of neutrons and protons. One major group of particles located in the nucleus is the **neutrons;** they were named neutrons to reflect their lack of electrical charge. **Protons,** the second type of particle in the nucleus, have a positive electrical charge. **Electrons,** found in the area surrounding the nucleus, have a negative charge.

Figure 2.1
Atomic Structure.
The nucleus of the atom contains the protons and the neutrons, which are massive particles of the atom. The electrons, much less massive, are at distances from the nucleus.

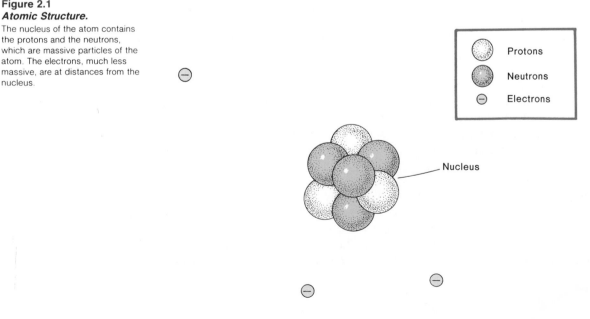

Table 2.1 *Table of Protons and Electrons*

Elements		No. of Protons$^{(+)}$ = Atomic No.	No. of Electrons$^{(-)}$
Carbon	(C)	6	6
Nitrogen	(N)	7	7
Oxygen	(O)	8	8
Sodium	(Na)	11	11
Phosphorus	(P)	15	15
Chlorine	(Cl)	17	17
Potassium	(K)	19	19
Calcium	(Ca)	20	20

Table 2.2 *Comparison of Atomic Particles*

	Protons	Electrons	Neutrons
Location	Nucleus	Outside nucleus	Nucleus
Charge	Positive (+)	Negative (−)	None (neutral)
Number present	Identical to the atomic number	Equal to number of protons	Mass number minus atomic number
Mass	1	1/1,836 Proton mass	1

An atom is always neutral. This means that the atom must have an equal number of positive protons and negative electrons to balance the electrical charges. If you know the number of positive charges in an atom, then you automatically know the number of negative charges (table 2.1).

The atoms of each kind of element have a specific number of protons. For example, oxygen always has eight protons and no other element has that number. Carbon always has six protons. The **atomic number** of an element is the number of protons in an atom of that element; therefore, each element has a unique atomic number. Since oxygen has eight protons, its atomic number is eight. The mass of a proton is 1.67×10^{-24} grams. Since this is an extremely small mass and is awkward to express, chemists have agreed to express it as one **atomic mass unit,** abbreviated as AMU. Neutrons have just slightly more mass but are close enough so that we consider their mass equal to one AMU also (table 2.2).

Although all atoms of the same element have the same number of protons, they do not always have the same number of neutrons. In the case of oxygen, over 99% of the atoms have eight neutrons, but there are others with more or fewer neutrons. Each atom of an element with a particular number of neutrons is called an **isotope.**

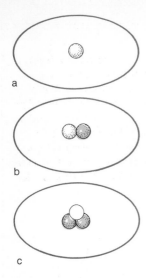

a

b

c

Figure 2.2
Isotopes of Hydrogen.
The most common form of
hydrogen is the isotope that has
1 AMU. It is composed of one
proton and no neutrons. The
isotope deuterium has 2 AMU and
has one proton and one neutron.
Tritium, 3 AMU, has two neutrons
and one proton. Each of these
isotopes has one electron.

The most common isotope of oxygen has eight neutrons, but another isotope of oxygen has nine neutrons. We can determine the number of neutrons by comparing the masses of the isotopes. The **mass number** of an atom is the weight of the protons plus the weight of the neutrons expressed in atomic mass units. The mass number is customarily used to compare different isotopes of the same element. An oxygen isotope with a mass number of sixteen AMUs is composed of eight protons and eight neutrons and is identified as ^{16}O. Oxygen 17, or ^{17}O, has a mass of seventeen AMUs. Eight of these units are due to the eight protons that every oxygen atom has; the rest of the mass is due to nine neutrons ($17-8=9$). Figure 2.2 shows different isotopes of hydrogen.

The **periodic table of the elements** (see box 2.1) lists all the elements in order of increasing atomic number (number of protons). In addition, this table lists the atomic mass number of each element. You can use these two numbers to determine the number of the three major particles in an atom; protons, neutrons, and electrons. Look at the periodic table and find helium in the upper right hand corner (He). The number two is its atomic number; thus, every helium atom will have two protons. Since the protons are positively charged, the nucleus will have two positive charges that must be balanced by two negatively charged electrons. The mass number of helium is given as 4.003. This is the calculated average mass of a group of helium atoms. Most of them have a mass of four, two protons and two neutrons. But some isotopes of helium have three neutrons, so the average mass is slightly greater than a whole number. Generally, you will only need to work with the most common isotope, so the mass number should be rounded to the nearest whole number. If it is a number like 4.003, use 4 as the most common mass. If the mass number is a number like 39.95, use 40 as the nearest whole number. Look at several atoms in the periodic table, determine the number of protons and the number of neutrons in the most common isotopes.

Since isotopes differ in the number of neutrons they contain, one could expect some isotopes to show characteristics that are different from the most common form of the element (table 2.3). For example, there are two isotopes of iodine. The most common isotope of iodine is ^{127}I; it has a mass number of 127. A different isotope of iodine is ^{131}I; its mass number is 131 and it is **radioactive.** This means that it is not stable and that it disintegrates, releasing energy from its nucleus. A radioactive isotope breaks down into a smaller and ultimately more stable atom. As it breaks down, the radioactive isotope of iodine releases energy. The energy can be detected by using photographic film or a Geiger counter. If a physician suspects that a patient has a thyroid gland that is functioning improperly, ^{131}I may be used to help confirm the diagnosis. The thyroid normally collects iodine atoms from the blood and uses them in the manufacture of the body-regulating chemical thyroxin. If the thyroid

Table 2.3   *Comparison of Isotopes*

	Oxygen 16	Oxygen 17	Iodine 127	Iodine 131
Protons	8	8	53	53
Electrons	8	8	53	53
Neutrons	8	9	74	78
Mass number	16	17	127	131

Box 2.1

Traditionally, elements are represented in a shorthand form by letters. For example, the symbol for water, H_2O, shows that a molecule of water consists of two atoms of hydrogen and one atom of oxygen. These chemical symbols can be found on any periodic table of elements. Using the periodic table, we can determine the number and position of the various parts of atoms. Notice that atoms number 3, 11, 19, and so on, are in column one. The atoms in this column act in a similar way since they all have one electron in their outermost layer. In the next column, Be, Mg, Ca, and so on, act alike because these metals all have two electrons in their outermost electron layer. Similarly, atoms number 9, 17, 35, and so on, all have seven electrons in their outer layer. Knowing how fluorine, chlorine, and bromine act, you can probably predict how iodine will act under similar conditions. At the far right in the last column, argon, neon, and so on, all act alike. They all have eight electrons in their outer electron layer. Atoms with eight electrons in their outer electron layer seldom form bonds with other atoms.

THE PERIODIC TABLE of THE ELEMENTS

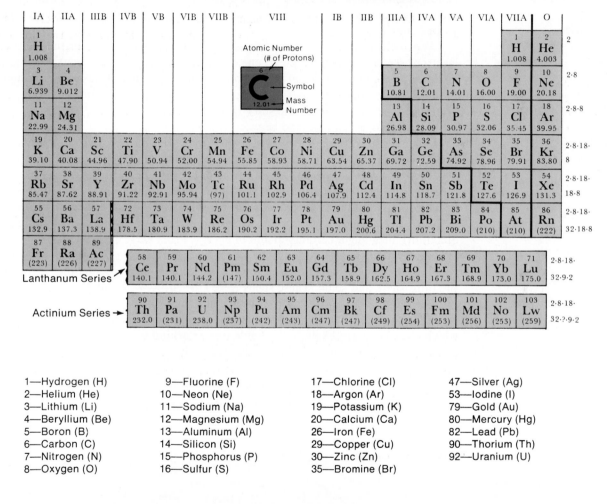

1—Hydrogen (H)
2—Helium (He)
3—Lithium (Li)
4—Beryllium (Be)
5—Boron (B)
6—Carbon (C)
7—Nitrogen (N)
8—Oxygen (O)

9—Fluorine (F)
10—Neon (Ne)
11—Sodium (Na)
12—Magnesium (Mg)
13—Aluminum (Al)
14—Silicon (Si)
15—Phosphorus (P)
16—Sulfur (S)

17—Chlorine (Cl)
18—Argon (Ar)
19—Potassium (K)
20—Calcium (Ca)
26—Iron (Fe)
29—Copper (Cu)
30—Zinc (Zn)
35—Bromine (Br)

47—Silver (Ag)
53—Iodine (I)
79—Gold (Au)
80—Mercury (Hg)
82—Lead (Pb)
90—Thorium (Th)
92—Uranium (U)

gland is working properly to form thyroxin, the radioactive iodine will collect in the gland where its presence can be detected. If no iodine has collected there, the physician knows that the gland is not functioning correctly and can take steps to help the patient.

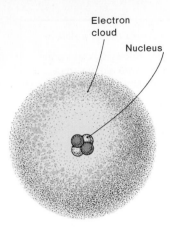

Electron
cloud

Nucleus

Figure 2.3
Electron Cloud.
The electrons are moving so fast around the nucleus they form a cloud around it, rather than an orbit or track. You might think of the electron cloud as hundreds of photographs of an atom. Each photograph shows where an electron was at the time the picture was taken. But when the next picture is taken, the electron is somewhere else. Although we are able to determine where an electron is at a given time, we do not know the path it uses to go from there to where it is the next time we determine its position.

Electron Distribution

Electrons are the negatively charged particles of an atom that balance the charge of the protons in the nucleus. Notice in table 2.2 that the mass of the electron is a tiny fraction of the mass of a proton. This mass is so slight that it usually does not influence the AMU of an element. Although electrons do not have a major effect on the mass of the element, they are important. The number and position of the electrons is responsible for the way the atoms interact with each other.

Electrons are located at some distance from the nucleus (figure 2.3). They are constantly moving at great speeds. Electrons are acted upon by several forces. Since protons and electrons are of opposite charge, electrons are attracted to the central part of the atom. Electrons tend to remain in the neighborhood of the nucleus due to this attraction between the opposite charges. In addition, the electrons are repelled by each other. The speed of the electrons, their attraction to the nucleus, and their repulsion for one another cause the electrons to be located in certain areas around the nucleus.

When chemists first described the atom, they tried to account for the fact that electrons seemed to be traveling at one of several different speeds about the atomic nucleus. They did not travel at intermediate speeds. Because of this, it was thought that electrons had a particular path or orbit that they followed, similar to the path or orbit of the planets about the sun (figure 2.4).

A correlation can be made between the speed of the electron and its distance from the nucleus. Think of swinging a weight on an elastic strip. As you swing the weight around your head, it makes a path a certain distance from you. If you swing it harder to make it go faster, or give it more energy, the path is a bigger circle a greater distance from your head. It was thought that there were only certain paths that electrons could follow. No electrons were thought to go at intermediate speeds between the particular paths. The speeds or paths were called quanta (singular quantum), meaning a certain amount of energy. From early experimental data collected, it was thought that only two electrons could exist in the first quantum or shell, eight electrons could inhabit the second shell, eight (or sometimes eighteen) the third shell, and so on. These shells were labeled, K, L, M, and so forth.

Several decades ago, as more experimental data was gathered and interpreted, we began to think of the K shell not as a particular pathway, but as an area within which electrons were likely to be. Each area or **orbital** is able to hold a maximum of two electrons. Each orbital is designated with a number to indicate the major energy level, and a letter to indicate the shape of the area the electrons occupy. The first orbital is lowest in energy and is designated as *1s*. The *one* indicates the first energy level and the *s* indicates that the electrons tend to be located in a spherical

Figure 2.4
*Planetary Model
of the Atom.*
Several decades ago we thought the electrons revolved around the nucleus of the atom in particular paths or tracks. Each track was labeled with a letter: K, L, M, N, and so on. Each track was thought to contain up to a maximum number of electrons moving at a particular speed. There were no intermediate tracks, so electron tracks were described as quanta of energy.

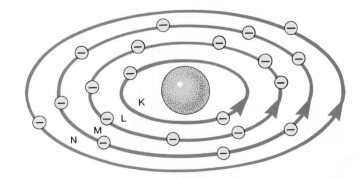

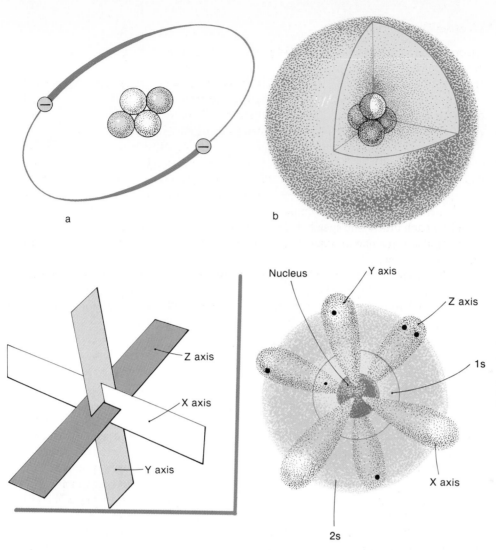

Figure 2.5
Two Models of a Helium Atom.
(a) We thought of helium in the past as a small solar system with two planets (electrons). (b) Now we think of it as a cloud of two electrons moving at great speeds, but staying near the nucleus. The electrons do not necessarily follow a circular orbit or pathway around the nucleus.

a

b

Nucleus

Y axis

Z axis

1s

X axis

2s

Z axis

X axis

Y axis

Figure 2.6
Second Energy Level of Electrons.
The electrons in these clouds are all moving at about the same speed, but since they are farther away from the nucleus, they are moving faster than those in the first energy level. The areas labeled are the spherical orbital and the three propeller-shaped orbitals at right angles to each other.

area. Figure 2.5 is a representation of a helium atom as we thought of it in the past and as we think of it now. Only two electrons are able to be located within each orbital; thus, helium has a *1s* orbital full of electrons. If an atom has more than two electrons, some must be moving faster than others. The area formerly labeled as the *L* shell could hold up to eight electrons. If an atom has ten electrons, two would be in the closest spherical (*1s*) area, and the rest would have to be moving faster. We designate these faster moving electrons as being in the second energy level. We now know that not all of these eight electrons are moving at exactly the same speed. Two of them are moving slightly slower than the rest and are said to be in the *2s* orbital. *Two* indicates the second energy level and *s* the spherically shaped area within which they move. The other six electrons are moving just slightly faster than these. They tend to cluster into three groups as far from each other as possible. We designate these three areas as the *2p* orbitals, *two* for second energy level and *p* for the propeller-shaped area in which the electrons are located. The three propellers are as far apart from each other as possible, yet still have the same center point. See figure 2.6. The orbitals are labeled *2px, 2py,* and *2pz.* The *x, y,* and *z* indicate the orientation of these three propeller-shaped orbitals in space, as far apart as they can be.

Figure 2.7
*Third Energy Level
of Electrons.*
The electrons in these clouds are
all moving at about the same
speed, but since they are farther
away from the nucleus, they are
moving faster than those in the
second energy level. The areas
labeled are the spherical orbital
and the three propeller-shaped
orbitals at right angles to each
other.

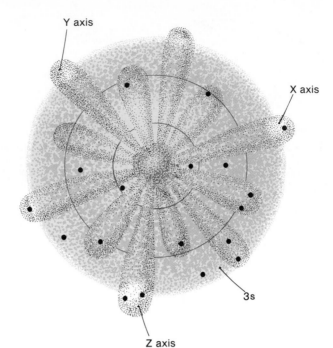

Figure 2.8
Electronic Configuration.
This chart is like a theater seating
chart. The lines represent places
where electrons can be. Each line
can hold a maximum of two
electrons. The electrons are most
likely to be in the lower energy
levels. They will go in an empty
area of the same energy level
before they will occupy an area
that already has a negative
electron in it. The filled in chart is
of the atom potassium, number 19.

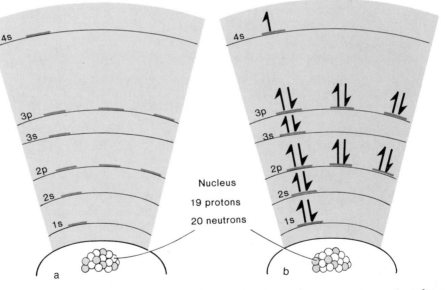

The third energy level (formerly called the *M* shell) we now know has four
different areas that can hold electrons. The areas are *3s, 3px, 3py,* and *3pz* (figure
2.7). You can see how cluttered a graphic representation of an atom with all its
protons, neutrons, and electrons might become as the atom increases in atomic
number. We sometimes think of the atom as a nucleus surrounded by a cloud of
electrons. An easy way to represent the atom is shown in figure 2.8. You must remember the order of the orbitals, but as you diagram an atom you should think of
the forces that are acting on the electrons and position them accordingly.

An atom such as potassium, with nineteen protons and nineteen electrons, would
have two electrons in the first energy level (*1s*). In the second energy level, there
would be two electrons in the *2s* orbital; two electrons in each of the *2p* orbitals;
two electrons in the *3s* orbital; two in each of the *3px, 3py,* and *3pz* orbitals; and
one electron in the *4s* orbital.

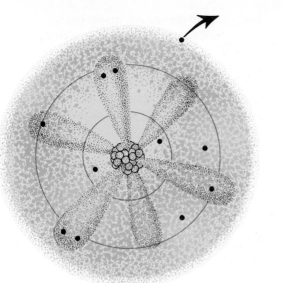

Figure 2.9
A Sodium Ion.
The sodium atom on the left is
losing its electron from the 3s
orbital. When that electron leaves,
the sodium will have an
overabundance of positive charges
so it is called a sodium ion (Na^+).

Sodium ion (Na^+)

11 protons
12 neutrons
10 electrons

Sodium atom (Na)

11 protons
12 neutrons
11 electrons

Ions

Now that you know how to position electrons correctly, it would be convenient if all atoms always followed the rules. Remember that atoms are neutral because they have an equal number of protons and electrons. Certain atoms, however, are able to exist with an unbalanced charge. These unbalanced or charged atoms are called **ions.** The ion of sodium is formed when the sodium atom, with its eleven protons and eleven electrons, allows the outermost electron to escape. We will look at the electron distribution to help explain how and why this happens. The sodium nucleus is composed of eleven positive charges (protons) insulated from each other by twelve neutrons. (The most common isotope of sodium is sodium 23, which has twelve neutrons.) The eleven electrons that balance the charge are most likely positioned as follows: two electrons in the *1s* orbital, two electrons in the *2s* orbital, two electrons in each of the *2px, 2py,* and *2pz* orbitals, and one electron in the *3s* orbital. Focus your attention on the outermost electron. It has more energy than any of the other electrons so it is moving faster. Since it is further from the nucleus than any other electron, the attraction for the positive charges in the nucleus is diminished—this is similar to gravitational attraction; the closer to earth an object is, the greater the gravitational pull (figure 2.9). This electron is the least attracted to the nucleus and is moving the fastest. You can see that the stage is set for this electron to be lost. What remains when the electron leaves the atom is called the ion. In this case, the sodium ion is composed of the eleven positively charged protons, the twelve neutral neutrons, but only ten electrons in orbitals. The fact that there are 11^+ and only 10^- charges means that there is an excess of one positive charge. We still use the chemical symbol Na to represent the ion but we add the $^+$ to indicate that it is no longer a neutral atom, but a charged ion (i.e., Na^+). It is easy to remember that a positive ion is formed because it loses one of its negative electrons.

The sodium ion is relatively stable because its outermost energy level is full. A sodium atom will lose one electron from its third major energy level so that its second major energy level will be full with eight electrons. Similarly, magnesium loses two

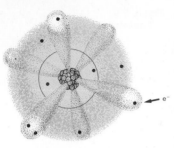

Fluoride ion (F⁻)

9 protons
10 neutrons
9 electrons
1 acquired electron

Figure 2.10
Fluoride Ion.
When the fluorine atom accepts an additional electron, it becomes a negative ion. Negative ions are indicated with a minus sign and also often end in "ide."

electrons from its third major energy level so that the second major energy level will be full with eight electrons. When a magnesium atom (Mg) loses two electrons, it becomes a magnesium ion (Mg^{++}). The periodic table of the elements is arranged so that all atoms in the first column become ions in a similar way. That is, when they form ions, they do so by losing one electron. Each becomes a 1^+ ion. Atoms in the second column of the periodic chart become $^{++}$ ions when they lose two electrons. Those atoms at the extreme right of the periodic table of the elements do not become ions, they tend to be stable as atoms. These atoms are called inert because of their lack of activity. They seldom react since their protons and electrons are equal in number and since they have a full outer shell; therefore, they are not likely to lose electrons.

The column to the left of these gases contains atoms that lack a full outer shell. They all require an additional electron. Fluorine with its nine electrons would have two in the K shell (*1s* orbital) and seven in the L shell (two in *2s,* two in *2px,* two in *2py,* and one in *2pz*). The second major energy level can hold a total of eight electrons. You can see that one additional electron could fit into the *2pz* orbital. Whenever the atom of fluorine can, it will accept an extra electron so that its outermost shell is full. When it does so, it no longer has a balanced charge. When it accepts an extra electron, it has one more negative electron than positive protons; thus, it has become a negative ion (F^-) (figure 2.10).

Similarly, chlorine will form a 1^- ion. Oxygen, in the next column, will accept two electrons and become a negative ion with two extra negative charges (O^{--}). If you know the number and position of the electrons, you are better able to hypothesize whether or not it will become an ion and if it does, whether it will be a positive ion or a negative ion. You can use the periodic table of the elements to help you determine an atom's ability to form ions. This information is useful as we see how ions react to each other.

Chemical Bonds

There are a variety of physical and chemical forces that act on atoms and make them attractive to each other. Each of these results in a particular arrangement of atoms or association of atoms. The forces that combine atoms and hold them together are called **chemical bonds.** There are several types of chemical bonds. They are different from each other in the kinds of attractive forces holding the atoms together. The bonding together of atoms results in the formation of a compound. This **compound** is composed of a specific number of atoms (or ions) joined to each other in a particular way. We generally use the chemical symbols for each of the component atoms when we designate a compound. Sometimes there will be small numbers behind the chemical symbol. This number indicates how many atoms of that particular element are used in the compound. The group of chemical symbols and numbers is termed a **formula;** it will tell you what elements are in a compound and also how many atoms of each element are required. For example, $CaCl_2$ tells us that the compound of calcium chloride is composed of one calcium atom and two chlorine atoms (figure 2.11).

The properties of compounds are very different from the properties of the atoms that make up the compound. Table salt is composed of the elements sodium and chlorine bound together. Both sodium and chlorine are very poisonous materials when they are by themselves. Yet, when they are combined as salt, the compound is an essential, nontoxic, necessary material for living organisms.

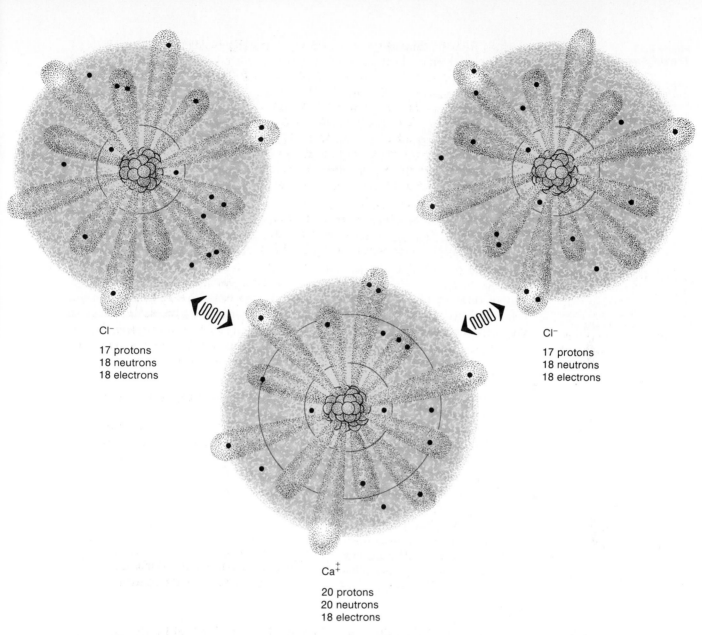

Cl⁻

17 protons
18 neutrons
18 electrons

Cl⁻

17 protons
18 neutrons
18 electrons

Ca⁺

20 protons
20 neutrons
18 electrons

Ionic Bonds

When positive and negative ions are near each other, they are mutually attracted due to their opposite charge. This attraction between ions of opposite charge results in the formation of a stable group of ions. This attraction is termed an **ionic bond.** Compounds that form as a result of attractions between ions are called ionic compounds (see figure 2.11) and are very important in living systems. We can categorize these ionic compounds into three different groups.

Acids, Bases, and Salts

Acids and bases are two classes of biologically important compounds. Their characteristics are determined by the nature of their chemical bonds. When acids are dissolved in water, hydrogen ions (H^+) are set free. The hydrogen ion is positive because it has lost its electron and now has only the positive charge of the proton. An **acid** is any ionic compound that releases a hydrogen ion in a solution. One other

Figure 2.11
Calcium Chloride.

This combination of a calcium ion and two chloride ions makes up the compound calcium chloride. The formula of the compound is $CaCl_2$. Notice that the overabundance of two positive charges on the calcium ion is offset by the two chloride ions, each of which has an overabundance of only one negative charge.

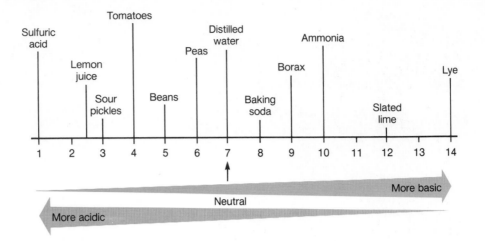

Figure 2.12
The pH Scale.
The concentration of acid (proton donor or electron acceptor) is greatest where the pH number is lowest. As the pH number increases, the concentration of base (proton acceptor or electron donor) increases. At a pH of 7, the concentrations of acid and base are equal.

way of thinking of an acid is that it is a substance able to donate a proton to a solution. This is only part of the definition of an acid. We also think of acids as compounds that act like the hydrogen ion—they attract negatively charged particles. An example of a common acid with which you are probably familiar is the sulfuric acid (H_2SO_4) in your automobile battery.

A **base** is the opposite of an acid in that it is an ionic compound that releases a group known as a **hydroxyl ion** or OH^- group. This group is composed of an oxygen atom and a hydrogen atom bonded together but with an additional electron. The hydroxyl ion is negatively charged. It is a base because it is able to donate electrons to the solution. It can also be thought of as any substance that is able to attract positively charged particles. A very strong base used in oven cleaners is NaOH, sodium hydroxide.

The strength of an acid or base is represented by a number called its **pH** number. The pH scale is a measure of hydrogen ion concentration (figure 2.12). A pH of seven indicates that the solution is neutral and has an equal number of H^+ ions and OH^- ions to balance each other. As the pH number gets smaller, the number of hydrogen ions in the solution increases. The lower the number, the stronger the acid. A number higher than seven indicates that the solution has more OH^- than H^+. As the pH number gets larger, the number of hydroxyl ions increases. The higher the number, the stronger the base.

An additional group of biologically important ionic compounds is called the salts. **Salts** are compounds that do not release H^+, nor do they release OH^-; thus, they are neither acids nor bases. They are generally the result of the reaction between an acid and a base in a solution. For example, when an acid such as HCl is mixed with NaOH in water, the H^+ and the OH^- combine with each other to form water, H_2O. The remaining ions (Na^+ and Cl^-) join to form the salt, NaCl.

$$HCl + NaOH \rightarrow (Na^+ + Cl^- + H^+ + OH^-) \rightarrow NaCl + H_2O$$

The chemical process that occurs when acids and bases react with each other is called **neutralization.** The acid no longer acts as an acid (it has been neutralized) and the base no longer acts like a base (see box 2.2).

Covalent Bonds

In addition to ionic bonds, there is a second strong chemical bond known as a covalent bond. A **covalent bond** is formed by two atoms that share a pair of electrons. A covalent bond should be thought of as belonging to each of the atoms involved.

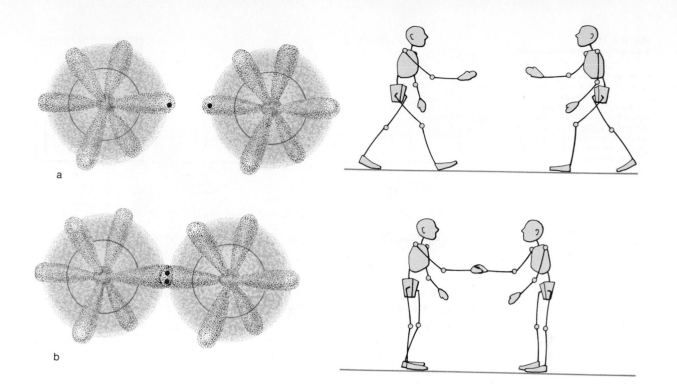

a

b

Visualize the bond as people shaking hands: the people are the atoms, the hands are electrons to be shared, and the handshake is the combining force. Generally, this sharing of a pair of electrons is represented by a single, straight line between the atoms involved (figure 2.13). The reason covalent bonds form relates to the arrangement of electrons within the atoms. If an orbital only contains one electron, it is half full. There are many elements that do not tend to form ions. They will not

Figure 2.13
Covalent Bond.

When two atoms come close enough to each other that their outermost orbitals overlap, an electron from each one can be shared to "fill" that orbital. The shared pair of electrons keeps the two atoms near each other. Frequently this is represented by a straight line between the two atoms. Remember that the line represents two electrons, one from each atom.

Box 2.2
Buffers

In certain situations, it is important that the pH of a solution not be able to change significantly. This is especially true in biological systems where the proper pH can influence the activity of enzymes and other biological chemicals.

In order to make sure that the pH of a system or solution remains constant, materials called buffers are added which absorb or release hydrogen ions. When an acid or base is added to the system, the buffer is there to resist pH change.

One of the most interesting buffer systems in the human body is in the blood. This buffer system consists of weak carbonic acid, H_2CO_3, and bicarbonate, HCO_3^-. These materials are dissolved in the blood plasma. If the blood has an increase of H^+ ions, these ions are able to combine with the bicarbonate ions to form carbonic acid; thus, the excess hydrogen ions are absorbed and the pH does not change. If hydroxyl ions are added to the system, they will join with the hydrogen ions to form water. However, the hydrogen ions will be replaced because some carbonic acid will separate into hydrogen ions and bicarbonate ions. The presence of the carbonic acid and the bicarbonate ions then maintains the stability of the pH in the blood.

Figure 2.14
Atoms in a Water Molecule.
(a) Two atoms of hydrogen and
one atom of oxygen come close
enough to each other to allow their
outermost orbits to overlap. (b) The
partially filled orbitals of each then
contain their full complement of
electrons. (c) The typical way of
indicating the covalent bonds
between the atoms that make up a
molecule of water.

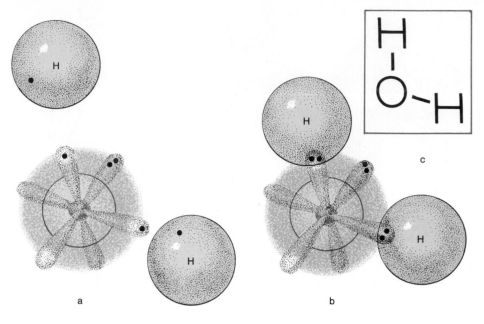

lose electrons, nor will they gain electrons. Instead, these elements get close enough
to other atoms that have unfilled outer orbitals and share electrons with each other.
If the two elements have orbitals that overlap, the electrons can be shared. By sharing
electrons, the unfilled orbitals of each atom will be filled with two electrons. Both
atoms become more stable as a result of the formation of this covalent bond. Figure
2.14 shows diagrams of two hydrogen atoms and one oxygen atom. This figure also
shows the oxygen atom sharing two electrons with each of the two hydrogen atoms
and the typical way the molecule of water is written.

Molecules are defined as the smallest particle of a chemical compound. They
are composed of a specific number of atoms arranged in a particular pattern. For
example, a molecule of water is composed of one oxygen atom bonded covalently
to two atoms of hydrogen. The shared electrons are in the *2P* orbitals of oxygen so
the bonds are almost at right angles to each other. Now that you realize how and
why bonds are formed, it makes sense that only certain numbers of certain atoms
will bond with each other to form molecules. Chemists also use the term molecule
to mean the smallest naturally occurring part of an element or compound. Using
this definition, one atom of iron is a molecule because one atom is the smallest natural piece of the element. Hydrogen, nitrogen, and oxygen tend to form into groups
of two atoms. Molecules of these elements are composed of two atoms of hydrogen,
two atoms of nitrogen, and two atoms of oxygen respectively.

Hydrogen Bonds

Molecules that are composed of several atoms sometimes have an uneven distribution of charge. This may be due to the fact that the electrons involved in the
formation of bonds may be located on one side of the molecule. This makes that
side of the molecule slightly negative, and the other side slightly positive. One side
of the molecule has possession of the electrons more than the other side. When a
molecule is composed of several atoms that have this uneven charge distribution,
the whole molecule may show a positive side and a negative side. We sometimes

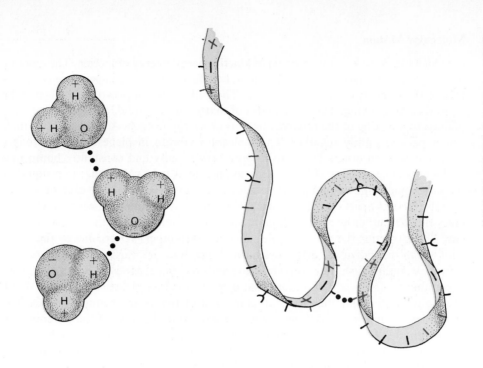

think of these molecules as a tiny magnet with a positive pole and a negative pole. This polarity of the molecule may influence how the molecule reacts with other molecules. When several of these polar molecules are together, they orient themselves so that the partially positive end of one is near the partially negative end of another. This attraction between two molecules is called a **hydrogen bond.** Since hydrogen has the least attractive force for electrons when it is combined with other elements, the hydrogen electrons tend to spend more of their time encircling the other atom's nucleus than its own. The result is the formation of a polar molecule. When the negative pole of this molecule is attracted to the positive pole of another similar polar molecule, the hydrogen will usually be located between the two molecules. Since the hydrogen serves as a bridge between the two molecules, this weak bond has become known as a hydrogen bond.

We usually represent this attraction as three dots between the attracted regions (figure 2.15). This weak bond is not responsible for forming molecules, but is important in determining how groups of molecules are arranged. Water is an example of a polar molecule that forms hydrogen bonds between the molecules. Because of this, individual water molecules are reluctant to separate from each other. They need a large input of energy to become separated. This is reflected in the relatively high boiling point of water. In addition, when a very large molecule, such as a protein or DNA (which is long and threadlike), has parts of its structure slightly positive and other parts slightly negative, these two areas will attract each other and result in coiling or folding of the molecule in particular ways (figure 2.15).

Molecular Energy

Molecules are able to move and to change; therefore, they have a certain amount of energy. While we cannot see the movement of the individual molecules, we can deduce several things about their movement by measuring their activity and noting the results of their movement.

Molecular Motion

All molecules have a certain amount of **kinetic energy,** energy of motion. The amount of energy that a molecule has is related to how fast it moves. **Temperature** is a measure of this velocity or energy of motion. The higher the temperature, the faster the molecules are moving. The three **states of matter,** solid, liquid, and gas, can be explained by thinking of the relative amounts of energy in each. A **solid** contains molecules packed tightly together. The molecules vibrate in place and are strongly attracted to each other. They are moving very rapidly and constantly bump into each other. The amount of kinetic energy in a solid is less than that in a liquid of the same material. A **liquid** has molecules still strongly attracted to each other, but slightly farther apart. As they move, they sometimes slide past each other. This gives the flowing property to a liquid. Still more energetic are the molecules of a **gas.** The attraction the gas molecules have for each other is overcome by the speed with which the individual molecules move. Since they are moving the fastest, their collisions tend to push them further apart and so a gas expands to fill its container. A common example of a substance that displays the three states of matter is H_2O. Ice, liquid water, and steam are all composed of the same chemical—H_2O. The molecules are moving at different speeds in each state because of the difference in kinetic energy. Considering the amount of energy in the molecules of each state of matter helps us explain changes such as freezing and melting. When a liquid becomes a solid, its molecules lose some of their energy; when it becomes a gas, its molecules gain energy.

Mixtures

Whenever several kinds of molecules intermingle with one another but are not chemically bound, we call this a **mixture.** When you add sugar to your cup of coffee, you make a mixture rather than a compound. Because the individual sugar molecules separate from one another in the liquid coffee, we say that they are dissolved or that a solution has been formed. A **solution** is any mixture where the types of molecules are dispersed throughout the system. A **suspension** is similar to a solution, but the dispersed particles are larger than molecular size. A suspension has particles that eventually settle out and are no longer equally dispersed in the system. Dust particles suspended in the air are an example of a suspension. The dust settles out and collects on tabletops and other furniture. A third type of mixture is a **colloid.** This system contains dispersed particles that are larger than molecules but small enough that they do not settle out. Even though colloids are composed of small particles that are mixed together with a liquid such as water, they do not act like solutions or a suspension. In a colloidal system, the molecules form a spongelike network that holds water molecules in place. One unique characteristic of a colloid is that it can become more or less solid depending on the temperature. When the temperature is lowered, the mixture becomes solidified; as the temperature is increased, it becomes more liquid. We speak of these as the gel and sol phases of a colloid. A gelatin dessert is a good example of a colloidal system. If you heat the gelatin, it becomes liquid as it changes to the sol phase. If you cool it again, it goes back to the gel phase and becomes solid. Environmental changes other than temperature can also cause colloids to change their phase. In living cells, this sol-gel transformation can cause the cell to move (figure 2.16).

(a) Solution   (b) Suspension   (c) Colloid

Coffee with sugar

Milk

Chocolate mix which has settled

Jell-O - large particles of gelatin stay dispersed

Figure 2.16
Mixtures.
A mixture is any two materials intermingled with each other. In (a) the mixture is a solution where both materials are equally distributed and remain so; in (b) a suspension is demonstrated where one type of material is larger than molecular size and will eventually settle out; and (c) is a colloid where the large particles remain evenly dispersed.

Diffusion

There is a natural tendency for molecules of different types to completely mix with each other. This is because each of the molecules is moving constantly. Their movement is random and due to the energy found in the individual molecules. As the molecules of one type move about, they tend to disperse from a central location. The other type of molecule also tends to disperse. The result of this random motion is that the molecules are eventually mixed with each other. Remember that the motion of the molecules is completely random. They do not move because of conscious thought, they move because of their kinetic energy. If you follow the path of molecules from a sugar cube placed in a glass of water, you would find that some of the sugar molecules would move away from the cube, while others would move in the opposite direction. However, there would be more sugar moving away from the original cube because there were more there to start with. We generally are not interested in the individual movement, but rather the overall movement. This overall movement is termed **net movement.** It is the movement in one direction minus the movement in the opposite direction. The direction of greatest movement (net movement) is determined by the relative concentration of the molecules. **Diffusion** is the resultant movement; it is defined as the net movement of a kind of molecule from a place where that molecule is in higher concentration to a place where that molecule is more scarce (figure 2.17).

When a kind of molecule is completely dispersed, and movement is equal in all directions, we say that the system has reached a state of **dynamic equilibrium.** There is no longer a net movement, since movement in one direction equals movement in the other. It is dynamic, however, because the system still has energy and the molecules are still moving. The kind of energy we have just dealt with is termed kinetic energy or energy of motion. Energy in the universe remains constant, it can neither be created nor destroyed. This concept is termed the **first law of thermodynamics.** An object that appears to be motionless does not necessarily lack energy. Its individual molecules will still be moving, but the object itself appears to be stationary. An object on top of a mountain may be motionless, but still may contain significant amounts of potential energy. **Potential energy** is defined as the energy an object has

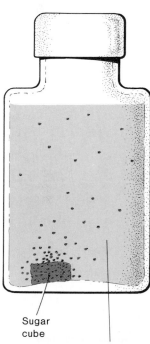

Sugar cube

Water

Figure 2.17
Diffusion.
Random molecular movement causes the molecules to move in all directions; however, more of them move from where they are in greater concentration to where they are in lesser concentration. This net movement is called diffusion.

Figure 2.18
*The Second Law of
Thermodynamics.*
When any form of energy is
converted to another form, some
useful energy is lost in the form of
heat. This light bulb is an example;
electrical energy is converted to
light, but some energy is lost as
heat.

due to its position. If an object were to start rolling down a mountain, the potential energy it contained at the top of the mountain would be converted into kinetic energy. Kinetic and potential energy have many forms, such as light, heat, sound, X rays, radio waves, and electricity. All forms of energy can be interconverted. In biology, we are concerned with many types of energy conversions. Whenever such a conversion takes place, there is always a loss of some "usable" energy. For example, in an ordinary light bulb, electrical energy is converted to usable light energy; however, some heat energy is lost as unusable energy. *Whenever energy is converted from one form to another, some useful energy is lost.* This statement is the **second law of thermodynamics** (figure 2.18).

Reactions

Living systems obey the second law of thermodynamics and are, therefore, constantly losing some of their useful energy in the form of heat. If living things do not receive a constant supply of energy, usually in the form of chemical-bond energy contained in food, they will die. The chemical-bond energy that is so important to living things is not created, but is manipulated by a series of reactions called oxidation-reduction reactions. **Oxidation-reduction reactions** are reactions that deal with the movement of electrons from one atom to another. The **oxidation reaction** is the loss of electrons. We say that an atom has been oxidized when it has fewer electrons than it had before. This atom is also termed a reducing agent because it supplies the electrons to be lost. The **reduction reaction** is the gain of electrons. An atom that has accepted an electron is said to be reduced and is an oxidizing agent since it is an acceptor of electrons (figure 2.19).

An example of an oxidation-reduction reaction may help to make the process clear. We will start with an atom of hydrogen. You know it has one electron in its outermost shell, which it is able to release when it becomes a positive ion. The process of releasing the electron is oxidation. The hydrogen atom has been oxidized because it has fewer electrons than it had before. It is a reducing agent because it provides an electron to some other atom. The electron does not just escape from the hydrogen atom; it is generally pulled off the atom because it is more attracted to something else. The atom that takes the electron, an atom of chlorine for example, is an oxidizing agent. The chlorine, in accepting the electron, becomes reduced. The reactions between acids and bases as they become neutralized to form salt and water is another example of an oxidation-reduction reaction.

Figure 2.19
*Oxidation-Reduction
Reactions.*
The two parts of an oxidation-reduction reaction coexist—that is, you cannot have oxidation without an accompanying reduction. In all oxidation-reduction reactions, the substance that is being oxidized is losing an electron, while the substance that is reduced is gaining an electron.

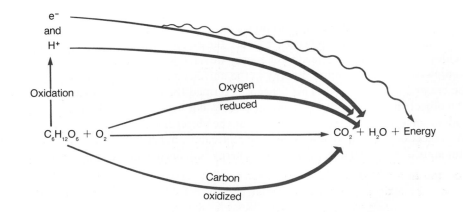

Chapter 2

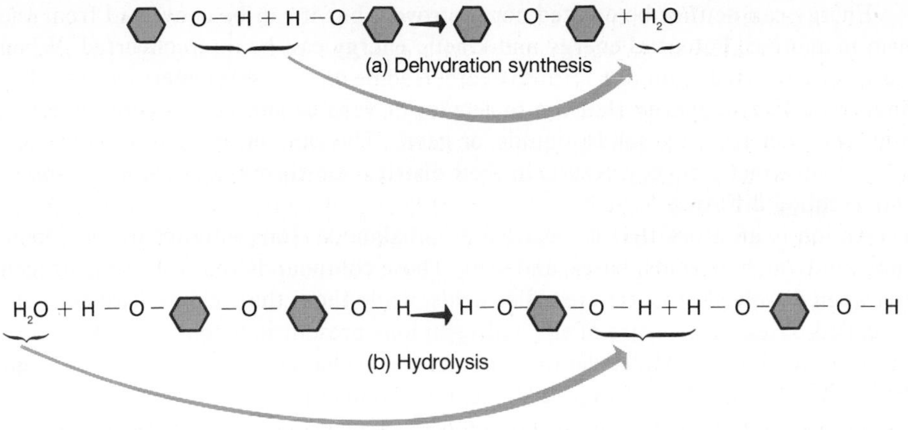

Figure 2.20
*Dehydration Synthesis
Reaction.*
This type of reaction occurs when
two small molecules are bonded
together to form a larger molecule.
A hydrogen from one is removed
and an OH from the other; these
two bond to each other to form
water. The opposite reaction is
called hydrolysis.

The process of shifting electrons from one type of atom or molecule to another can become very elaborate. You will learn in a later chapter that much of the energy you remove from the food you eat becomes available by a series of oxidation-reduction reactions that result in the formation of chemical bonds. The energy contained in these chemical bonds can be used to move, grow, respond, and reproduce.

Another important reaction is called dehydration synthesis. A **dehydration synthesis** reaction results in the joining of two smaller molecules to make one larger molecule and the removal of a molecule of water. The water is formed by removing a hydrogen from one molecule and a hydroxyl group from the other molecule. These then combine to form water (figure 2.20).

The removal of the hydrogen and hydroxyl allows these two smaller molecules to combine and form the larger molecule. The opposite type of reaction, **hydrolysis,** occurs when a large molecule is broken down into two smaller parts by the addition of water.

We frequently use chemical shorthand to express what is going on in a **chemical reaction.** The use of an arrow ($\rightarrow$) indicates a shift in chemical bonds. The arrowhead points to the materials that are produced by the reaction; we call these the **products.** On the other side of the arrow, we generally show the materials that are going to react with each other; we call these the **reactants** (figure 2.20). In the case of a dehydration synthesis reaction, the two smaller molecules are the reactants, the large molecule that is formed and the water released are the products. In a hydrolysis reaction, the water and large molecule are the reactants, the two smaller molecules that are formed are the products.

Summary

All matter is composed of atoms. The atoms contain a nucleus of neutrons and protons. The nucleus is surrounded by moving electrons. There are many kinds of atoms, called elements. These differ from one another by the number of protons and electrons they contain. Each is given an atomic number, based on the number of protons in the nucleus and an atomic mass number, determined by the total number of protons and neutrons. Atoms of an element that have the same atomic number but differ in their atomic mass number are called isotopes. Some isotopes are radioactive, which means that they fall apart and release energy and smaller, more stable particles. Atoms may be combined into larger units called molecules. Two kinds of chemical bonds allow molecules to form—ionic bonds and covalent bonds. A third bond, the hydrogen bond, is a weaker bond that holds molecules together and may also help large molecules maintain a specific shape.

Energy can neither be created nor destroyed, but it can be converted from one form to another. Potential energy and kinetic energy can be interconverted. When energy is converted from one form to another, some of the useful energy is lost. The amount of kinetic energy that the molecules of various substances contain determines whether they are solids, liquids, or gases. The random motion of molecules, due to their kinetic energy, results in their distribution throughout available space. This is called diffusion.

An ion is an atom that is electrically unbalanced. Ions interact to form ionic compounds such as acids, bases, and salts. Those compounds that release hydrogen ions when dissolved in water are called acids, while those that release hydroxyl ions are called bases. A measure of the hydrogen ions present in a solution is known as the pH of the solution. Molecules that interact and exchange parts are said to undergo chemical reactions. The changing of chemical bonds in a reaction may release energy or require the input of additional energy. Three important biological reactions are dehydration synthesis, hydrolysis, and oxidation-reduction.

Consider This

Hydrogen peroxide (H_2O_2) is an antiseptic commonly found in the medicine cabinet. This molecule breaks down into harmless water (H_2O) and reactive oxygen (O_2).

$$2H_2O_2 \rightarrow 2H_2O + O_2$$

The oxygen oxidizes many molecules found in living organisms; thus, disease-causing microorganisms can be destroyed in open wounds. This same reaction takes place in a bottle of hydrogen peroxide over time and results in the loss of antimicrobial properties. Can you describe what happens to the molecules of hydrogen peroxide, water, and oxygen? Can you describe why the bottle will finally contain nothing but water and oxygen? In your description, include diffusion, changes in chemical bonds, and kinetic energy.

Experience This

There are several chemicals available in your home that you can use to demonstrate some of the principles of matter. Dissolve an envelope of unflavored gelatin in a cup of very hot water. Allow it to cool and become solid. Then heat it up again, and cool again. Notice that it changes from solid to liquid and back to solid.

Mix an egg white with a cup of very hot water. Allow it to cool and then warm it up again. Does it also change from solid to liquid and back to solid?

Questions

1. How many protons, electrons, and neutrons are there in an atom of potassium having a mass number of thirty-nine?
2. Diagram an atom showing the positions of electrons, protons, and neutrons.
3. Name three kinds of chemical bonds that hold atoms or molecules together. How do these bonds differ from one another?
4. Diagram two isotopes of oxygen.
5. What does it mean if a solution has a pH number of three, twelve, two, seven, or nine?
6. What relationship does diffusion have to molecular motion?

7. Define the following terms: AMU, atomic number, orbital, 2s, 2pz, and covalent bond.
8. Define the term chemical reaction and give an example.
9. What is the difference between an atom and an element? between a molecule and a compound?
10. How do acids, bases, and salts differ from one another?

Chapter Glossary

acid (ă′sid) Any compound that releases a hydrogen ion (or other ion that acts like a hydrogen ion) in a solution.

atom (ă′tom) The smallest part of an element that still acts like that element.

atomic mass unit (ă-tom′ik mas yu-nit) An expression of the mass of one proton (equal to 1.67×10^{-24} grams).

atomic nucleus (ă-tom′ik nu′kle-us) The central region of the atom.

atomic number (ă-tom′ik num′bur) The number of protons in an atom.

base (bās) Any compound that releases a hydroxyl group in a solution (or other ion that acts like a hydroxyl group).

chemical bonds (kem′ĭ-kal bonds) Forces that combine atoms or ions and hold them together.

chemical reaction (kem′ĭ-kal re-ak′shun) The formation or rearrangement of chemical bonds, usually indicated in an equation by an arrow from the reactants to the products.

chemical symbol (kem′ĭ-kal sim′bol) Represents one atom of an element, such as Al for aluminum or C for carbon.

colloid (kol-loid) Mixture containing dispersed particles that are larger than molecules but small enough that they do not settle out.

compound (kom-pound) A kind of matter that consists of a specific number of atoms (or ions) joined to each other in a particular way and held together by chemical bonds.

covalent bond (ko-va′lent bond) The attractive force formed between two atoms that share a pair of electrons.

dehydration synthesis (de′′hi-dra′shun sin′thĕ-sis) A reaction that results in the joining of smaller molecules to make larger molecules by the removal of water.

density (den′sĭ-te) The weight of a certain volume of a material.

diffusion (dĭ-fiu′zhun) Net movement of a kind of molecule from a place where that molecule is in higher concentration to a place where that molecule is more scarce.

dynamic equilibrium (di-nam′ik e-kwĭ-lib′re-um) The condition where molecular motion continues but the molecules are equally dispersed so movement is equal in all directions.

electrons (e-lek′trons) The negatively charged particles moving at a distance from the nucleus of an atom that balance the positive charges of the protons.

elements (el′ĕ-ments) A kind of matter that consists of only one kind of atom.

first law of thermodynamics (furst law uv thur-mo-di-nam′iks) Energy in the universe remains constant, it can neither be created nor destroyed.

formula (form′yu-lah) The group of chemical symbols that indicate what elements are in a compound and the number of each kind of atom present.

gas (gas) State of matter in which the molecules are more energetic than the molecules of a liquid, resulting in only slight attraction for each other.

hydrogen bond (hi′dro-jen bond) Weak attractive forces between molecules. Important in determining how groups of molecules are arranged.

hydrolysis (hi-drol′ĭ-sis) A process that occurs when a large molecule is broken down into smaller parts by the addition of water.

hydroxyl ion (hi-drok′sil i-on) A group of charged atoms that are released when a base is dissolved in water, (OH^-).

Simple Things of Life   **43**

ionic bond (i-on'ik bond) The attractive force between ions of opposite charge.

ions (i'ons) Electrically unbalanced or charged atoms.

isotopes (i'so-tōps) Atoms of the same element that differ only in the number of neutrons.

kinetic energy (ki-net'ik en'er-je) Energy of motion.

liquid (lik'wid) State of matter in which the molecules are strongly attracted to each other but they are farther apart than in a solid, so they move about each other more freely.

mass number (mas num'ber) The weight of an atomic nucleus expressed in atomic mass units.

matter (mat'er) Anything that has weight (mass) and also takes up space (volume).

mixture (miks'tūr) Several molecules physically near one another but not chemically bound.

molecule (mol'ē-kūl) The smallest particle of a chemical compound, also the smallest naturally occurring part of an element or compound.

net movement (net muv'ment) The movement in one direction minus the movement in the opposite direction.

neutralization (nu'tral-i-za''shun) A chemical reaction involved in mixing an acid with a base; results in formation of a salt and water.

neutrons (nu'trons) Particles in the nucleus of an atom which have no electrical charge, they were named neutrons to reflect this lack of electrical charge.

orbital (or'bi-tal) Area of an atom able to hold a maximum of two electrons.

oxidation reactions (ok''si-da'shun re-ak'shuns) The loss of electrons from the reactant.

oxidation-reduction reactions (ok''si-da'shun re-duk'shun re-ak'shuns) Reactions that deal with the movement of electrons from one atom to another.

periodic table of the elements (pēr-e-od'ik ta-bul uv the el'ē-ments) A list of all of the elements in order of increasing atomic number (number of protons).

pH   Scale used to indicate the strength of an acid or base.

potential energy (po-ten'shul en'er-je) The energy an object has due to its position.

products (prŏ'dukts) New molecules resulting from a reaction.

protons (pro'tons) Particles in the nucleus of an atom that have a positive electrical charge.

radioactive (ra-de-o-ak'tiv) Property of releasing energy or particles from an unstable atom.

reactants (re-ak'tants) Materials that will be changed in a chemical reaction.

reduction reaction (re-duk'shun re-ak'shun) The gain of electrons by a reactant.

salts   Ionic compounds formed from a reaction between an acid and a base.

second law of thermodynamics (sek'ond law uv ther''mo-di-nam'iks) Whenever energy is converted from one form to another some useful energy is lost.

solid   State of matter in which the molecules are packed tightly together; they vibrate in place.

solution (so-lu'shun) Mixture that contains molecules dispersed in the system that will not settle out.

states of matter   Physical condition of matter (solid, liquid, and gas) determined by the relative amounts of energy of the molecules.

suspension (sus-pen'shun) Type of mixture in which the dispersed particles are larger than molecular size so they eventually settle out.

temperature (tem'per-à-chiur) Measure of molecular energy of motion.

3

Organic Chemistry—
The Chemistry of Life

Chapter Outline

Purpose

The chemistry of living things is really the chemistry of the carbon atom and a few other atoms that can combine with carbon. In order to understand some aspects of the structure and function of living things, which will be covered later, you should first learn some basic organic chemistry.

For Your Information

It's not uncommon these days to find food and health stores advertising "natural, organic foods." They may be priced higher than their more common alternatives from supermarkets. These stores also offer products such as protein enriched shampoos, low-cholesterol creams, "natural, organic" skin care products, and many other "natural" dietary additives including "natural" vitamins, lecithin, and cod-liver oil.

Learning Objectives

- Recognize the difference between inorganic and organic molecules.

- Understand the importance or structure in organic molecules.

- Know the major functional groups found in organic molecules.

- Describe the structure and function of the four major groups of organic molecules.

Molecules Containing Carbon

The principles and concepts discussed in chapter 2 apply to all types of matter. Living systems are composed of rather complex molecules that contain carbon atoms in chains or rings. These are called **organic molecules.** In contrast to these complex, carbon-containing molecules, most of the ones we discussed in the last chapter are **inorganic molecules,** because they do not contain carbon atoms in rings or chains.

The original meaning of the terms inorganic and organic is related to the fact that organic materials were thought to be either alive or produced only by living things. Therefore, a very strong link exists between organic chemistry and the chemistry of living things, which is called **biochemistry** or biological chemistry. Modern chemistry has considerably altered the original meaning of the terms organic and inorganic, since it is now possible to manufacture unique, organic molecules that cannot be produced by living things. Many of the materials we use daily are the result of the organic chemist's art. Nylon, aspirin, polyurethane varnish, silicones, Plexiglas, food wrap, Teflon, and insecticides are just a few of the unique molecules that have been invented by organic chemists (figure 3.1).

In other instances, organic chemists have taken their lead from living organisms and have been able to produce organic molecules more efficiently, or in forms that are slightly different from the original, natural molecule. Some examples of these are rubber, penicillin, some vitamins, insulin, and alcohol (figure 3.2).

Figure 3.1
Some Common Synthetic Organic Materials.

These items are examples of useful organic compounds invented by chemists.

Carbon—The Central Atom

All organic molecules, whether they are natural or synthetic, have certain common characteristics. The carbon atom, which is the central atom in all organic molecules, has some unusual properties. Carbon is unique in that it can combine with other carbon atoms to form long chains.

$$-\overset{|}{\underset{|}{C}}-\overset{|}{\underset{|}{C}}-\overset{|}{\underset{|}{C}}-\overset{|}{\underset{|}{C}}-\overset{|}{\underset{|}{C}}-\overset{|}{\underset{|}{C}}-\overset{|}{\underset{|}{C}}-\overset{|}{\underset{|}{C}}-\overset{|}{\underset{|}{C}}-$$

In many cases, the ends of these chains may join together to form ring structures (figure 3.3). Only a few other atoms have this ability. What is really unusual is that these bonding sites are all located at equal distances from one another. If you were

Figure 3.2
**Natural and Synthetic
Organic Compounds.**

Some organic materials, such as
rubber, were originally produced by
plants but are now synthesized in
industry.

to take a rubber ball and stick four nails into it so that they were equally distributed around the ball, you would have a good idea of the geometry involved. These bonding sites are arranged this way because in carbon the four outer orbitals, the *2s, 2px, 2py* and *2pz,* each contain one electron. Since the electrons are all negatively charged, they tend to repel each other and move as far away from each other as possible (figure 3.4). Carbon atoms are usually involved in covalent bonds. Since carbon has four places it can bond, the carbon atom can combine with four other atoms. This

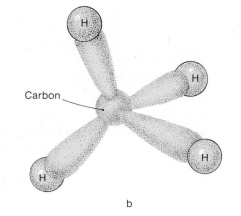

Figure 3.3
Ring Structure.

The ring structure shown here is
formed by joining the two ends of
a chain of carbon atoms.

a

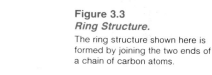

Carbon

b

Figure 3.4
**Bonding Sites of a Carbon
Atom.**

The arrangement of bonding sites
around the carbon is similar to a
ball with four equally spaced nails
in it. Each of the four bondable
electrons inhabits an area as far
away from the other three as
possible. These areas are not
exactly like spherical orbitals, nor
are they exactly like propeller-
shaped orbitals; they are
somewhere in between. Chemists
designate these as hybrid orbitals.

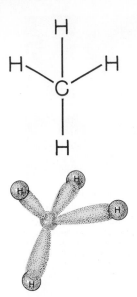

Figure 3.5
Methane Molecule.
A methane molecule is composed of one carbon atom bonded with four hydrogen atoms. These bonds are formed at the bonding sites of the carbon. (Note: for the sake of simplicity, all future diagrams of molecules will be two-dimensional drawings, though in reality they are three-dimensional molecules.) Each line in the diagram represents a covalent bond between the two atoms where a pair of electrons is being shared.

is the case with the methane molecule, which has four hydrogen atoms attached to a single carbon atom. Methane is a colorless and odorless gas usually found in natural gas (figure 3.5).

Some atoms may be bonded to a single atom more than once. This results in a slightly different arrangement of bonds around the carbon atom. An example of this type of bonding occurs when oxygen is attracted to a carbon. Oxygen has two bondable electrons—if it shares one of these with a carbon and then shares the other with the same carbon, it forms a double bond. A **double bond** is two covalent bonds formed between two atoms that share two pairs of electrons. Oxygen is not the only atom that can form double bonds, but double bonds are common between it and carbon. The double bond is denoted by two lines between the two atoms:

$$-\,C = O$$

Two carbon atoms might form double bonds between each other and then bond to other atoms at the remaining bonding sites. Figure 3.6 shows several compounds that contain double bonds.

Although most atoms can be involved in the structure of an organic molecule, only a few are commonly found. Hydrogen (H) and oxygen (O) are almost always present. Nitrogen (N), sulfur (S), and phosphorus (P) are also very important in specific types of organic molecules.

An enormous variety of organic molecules is possible because carbon is able to bond at four different sites, form long chains, and combine with many other kinds of atoms. The types of atoms in the molecule are important in determining the properties of the molecule. The three-dimensional arrangement of the atoms within the molecule is also important. Since most inorganic molecules are small and involve few atoms, there is usually only one way in which a group of atoms can be arranged to form a molecule. There is only one arrangement for a single oxygen atom and two hydrogen atoms in a molecule of water. In a molecule of sulfuric acid, there is only one arrangement for the sulfur atom, the two hydrogen atoms, and the four oxygen atoms.

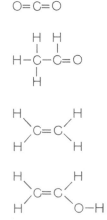

Figure 3.6
Double Bonds.
These diagrams show several molecules that contain double bonds. A double bond is formed when two atoms share two pairs of electrons with each other.

$$H-O-S-O-H$$
(with O above and O below the S)

However,

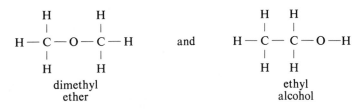

dimethyl ether   and   ethyl alcohol

both contain two carbon atoms, six hydrogen atoms, and one oxygen atom, but they are quite different in their arrangement of atoms and in the chemical properties of the molecules. While the first is an ether, the second is an alcohol. Since the ether and the alcohol have the same number and kinds of atoms, they are said to have

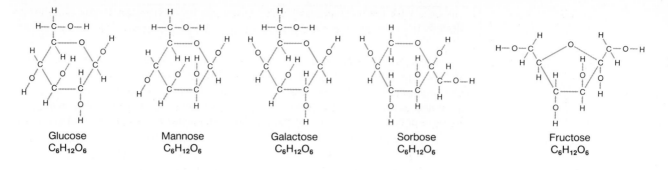

Glucose
$C_6H_{12}O_6$

Mannose
$C_6H_{12}O_6$

Galactose
$C_6H_{12}O_6$

Sorbose
$C_6H_{12}O_6$

Fructose
$C_6H_{12}O_6$

the same **empirical formula,** which in this case is written C_2H_6O. An empirical formula simply indicates the number of each kind of atom within the molecule. When the arrangement of the atoms and their bonding within the molecule is indicated, we call this a **structural formula.** Figure 3.7 shows several structural formulae for the empirical formula $C_6H_{12}O_6$.

Figure 3.7
Six-Carbon Sugars.
Some structural arrangements of six-carbon sugars are represented here. Each has the same empirical formula, but they have different structural formulae. They will also act differently from each other.

Carbon Skeleton and Functional Groups

To help us understand organic molecules a little better we will look at some of their similarities. All organic molecules have a **carbon skeleton,** which is composed of rings or chains of carbons. It is this carbon skeleton in the organic molecule that determines the overall shape of the molecule. The differences between various organic molecules depend on the length and arrangement of the carbon skeleton. In addition, the kinds of atoms that are bonded to this carbon skeleton determine the way the organic compound acts. Specific combinations of atoms called **functional groups** attached to the carbon skeleton determine specific chemical properties. By learning to recognize some of these functional groups, it is possible to identify an organic molecule and to predict something about its activity. Figure 3.8 shows some of the functional groups that are important in biological activity. Remember that a functional group does not exist by itself, but must be a part of an organic molecule (see box 3.1).

Common Organic Molecules

One way to make organic chemistry more manageable is to organize different kinds of compounds into groups on the basis of their similarity of structure or the chemical properties of the molecules. Frequently you will find that organic molecules are composed of subunits that are attached to each other. If you recognize the subunit, then the whole organic molecule is much easier to identify. It is similar to distinguishing between a passenger train and a freight train by recognizing the individual cars in it.

When there are several subunits (monomers) bonded together, the molecule is referred to as a macromolecule or a polymer. The word monomer means a single unit, while the term polymer means composed of many parts. The plastics industry has polymer chemistry as its foundation. The monomers in a polymer are usually

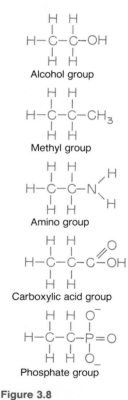

Alcohol group

Methyl group

Amino group

Carboxylic acid group

Phosphate group

Figure 3.8
Functional Groups.
These are some of the groups of atoms that frequently attach to a carbon skeleton. Notice that in each case the carbon skeleton is unchanged, just the group attached to it is changed. The functional group frequently determines how the molecule will act.

Box 3.1
Chemical Shorthand

You have probably noticed that sketching the entire structural formula of a large organic molecule takes a great deal of time. If you know the structure of the major functional groups, you can use several shortcuts to more quickly describe chemical structures.

When multiple carbons with two hydrogens are bonded to each other in a chain, we sometimes write it as follows:

or we might write it as:

$$-CH_2-CH_2-CH_2-CH_2-CH_2-CH_2-CH_2-CH_2-CH_2-CH_2-CH_2-CH_2-$$

or more simply, we may write it as: $(-CH_2-)_{12}$

If the twelve carbons were in a pair of two rings, we probably would not label the carbons, hydrogens, or oxygens unless there was a particular group or point we wished to focus upon. We would probably draw the two six-carbon rings with only hydrogen attached as follows:

Don't let these shortcuts throw you. You will soon find that you will be putting an $-OH$ group onto a carbon skeleton and forgetting to show the bond between the oxygen and hydrogen just like a professional.

---

combined by a **dehydration synthesis reaction.** This reaction results in the synthesis or formation of a macromolecule when water is removed from between the two smaller component parts. For example, when a monomer with an "$-OH$ group" attached to its carbon skeleton approaches another monomer with an available hydrogen, dehydration synthesis can occur. Figure 3.9 shows the removal of water from between two such subunits. Notice that in this case, the structural formulae are used to help identify just what is occurring. However, the chemical equation also indicates the removal of the water. You can easily recognize a dehydration synthesis reaction because the reactant side of the equation shows numerous, small molecules, while the product side lists fewer, larger products and water.

The reverse of a dehydration synthesis reaction is known as hydrolysis. **Hydrolysis** is the process of splitting a larger organic molecule into two or more component parts by the addition of water. Digestion of food molecules in the stomach is an important example of hydrolysis. You can analyze a complex organic molecule and determine its composition or type by identifying its monomeric subunits.

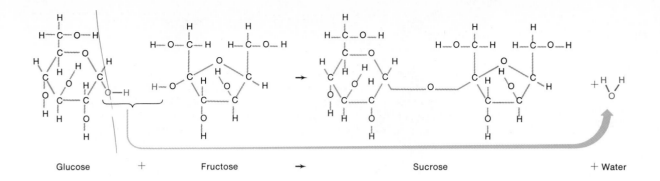

Glucose   +   Fructose   →   Sucrose   + Water

Carbohydrates

One class of organic molecules, **carbohydrates,** is composed of carbon, hydrogen, and oxygen atoms linked together to form monomers called simple sugar or monosaccharide. The empirical formula for a simple sugar is easy to recognize because there are equal numbers of carbons and oxygens and twice as many hydrogens—for example, $C_3H_6O_3$ or $C_5H_{10}O_5$. We usually describe the kinds of simple sugars by the number of carbons in the molecule. The ending **-ose** is a clue that indicates you are dealing with a carbohydrate. A tri**ose** has three carbons, a pent**ose** has five, and a hex**ose** has six. If you remember that the number of carbons equals the number of oxygen atoms and that the number of hydrogens is double that number, these names tell you the empirical formula for the simple sugar. Simple sugars such as glucose, fructose, and galactose provide the chemical energy necessary to keep organisms alive. These simple sugars combine with each other by dehydration synthesis to form **complex carbohydrates** (figure 3.9). When two simple sugars bond to each other, a disaccharide is formed; when three bond together, a trisaccharide is formed (figure 3.10). Generally we call complex carbohydrates larger than this a polysaccharide (many sugar units). In all cases, the complex carbohydrates are formed by the removal of water from between the sugars. Some common examples of polysaccharides are starch and glycogen. Cellulose is an important polysaccharide used in constructing the cell walls of plant cells. Humans cannot digest (hydrolyse) this complex carbohydrate so we are not able to use it as an energy source. Plant cell walls add bulk or fiber to our diet, but no calories.

Simple sugars can be used by the cell as a component of other, more complex molecules, such as the molecule adenosine triphosphate, (ATP). This molecule is important in energy transfer. It has a simple sugar (ribose) as part of its structural makeup. The building blocks of the genetic material (DNA) also have a sugar component (figure 3.18).

Figure 3.9
Dehydration Synthesis Reaction.
In the reaction illustrated here, the two −OH groups form water and the oxygen that remains acts as an attachment site between the two larger sugar molecules. Many structural formulae appear to be complex at first glance, but if you look for the points where subunits are attached and dissect each subunit, they become much simpler to deal with.

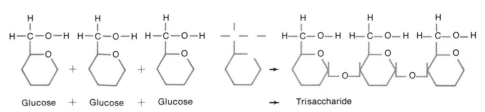

Glucose   +   Glucose   +   Glucose   →   Trisaccharide

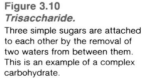

Figure 3.10
Trisaccharide.
Three simple sugars are attached to each other by the removal of two waters from between them. This is an example of a complex carbohydrate.

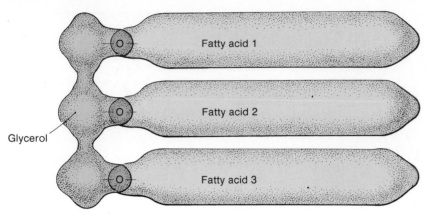

Figure 3.11
Fat Molecule.
The arrangement of the three fatty acids attached to a glycerol molecule is typical of the formation of a fat. The structural formula of the fat appears to be very cluttered until you dissect the fatty acids from the glycerol, then it becomes much more manageable.

Glycerol

Fatty acid 1

Fatty acid 2

Fatty acid 3

Lipids

We generally describe **lipids** as large organic molecules that do not easily dissolve in water. Just like carbohydrates, the lipids are composed of carbon, hydrogen, and oxygen. They do not, however, have the same ratio of carbon, hydrogen, and oxygen in their empirical formula. Generally lipids have very small amounts of oxygen in comparison to the amounts of carbon and hydrogen. Fats, phospholipids, and steroids are all examples of lipids, but they are all quite different from each other in their structure.

Fats are important organic molecules that are used to provide energy. The building blocks of a fat are a glycerol molecule and three fatty acids. The **glycerol** is a carbon skeleton that has three alcohol groups attached to it. Its chemical formula is $C_3H_5(OH)_3$. A **fatty acid** is a long chain carbon skeleton that has a carboxylic acid functional group (figure 3.11). If the carbon skeleton has as much hydrogen bonded to it as possible, we call it **saturated.**

$$H - \overset{\overset{\displaystyle H}{|}}{\underset{\underset{\displaystyle H}{|}}{C}} - \overset{\overset{\displaystyle H}{|}}{\underset{\underset{\displaystyle H}{|}}{C}} - \overset{\overset{\displaystyle H}{|}}{\underset{\underset{\displaystyle H}{|}}{C}} - \overset{\overset{\displaystyle H}{|}}{\underset{\underset{\displaystyle H}{|}}{C}} - \overset{\overset{\displaystyle H}{|}}{\underset{\underset{\displaystyle H}{|}}{C}} - \overset{\overset{\displaystyle H}{|}}{\underset{\underset{\displaystyle H}{|}}{C}} - \overset{\overset{\displaystyle H}{|}}{\underset{\underset{\displaystyle H}{|}}{C}} - \overset{\overset{\displaystyle H}{|}}{\underset{\underset{\displaystyle H}{|}}{C}} - \overset{\overset{\displaystyle H}{|}}{\underset{\underset{\displaystyle H}{|}}{C}} - \overset{\overset{\displaystyle H}{|}}{\underset{\underset{\displaystyle H}{|}}{C}} - \overset{\overset{\displaystyle H}{|}}{\underset{\underset{\displaystyle H}{|}}{C}} - \overset{\overset{\displaystyle H}{|}}{\underset{\underset{\displaystyle H}{|}}{C}} - \text{ACID GROUP}$$

Notice in this structure that at every point, the carbon has as much hydrogen as it can hold. Saturated fats are generally found in animal tissues—they tend to be solids at room temperatures. Some examples of saturated fats are butter, whale blubber, suet, lard, and fats associated with meat such as steak or pork chops.

If the carbons are double bonded to each other at one or more points, the fatty acid is said to be **unsaturated.**

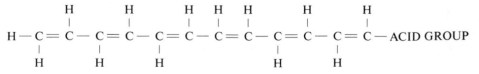

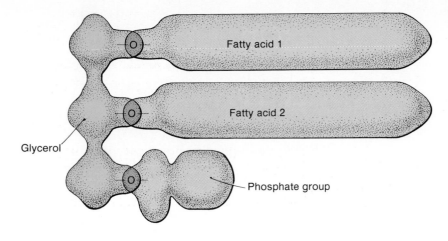

Figure 3.12
A Phospholipid Molecule.
This molecule is similar to a fat but has a phosphate group in its structure. The phosphate group is involved in the dehydration synthesis reaction with the glycerol.

Fatty acid 1

Fatty acid 2

Glycerol

Phosphate group

Notice that in this structure there are several double bonds between the carbons and only about half as much hydrogen as in the saturated fatty acid. Unsaturated fats are frequently plant fats or oils—they are usually liquids at room temperature. Peanut oil, corn oil, and olive oil are considered unsaturated because they have double bonds between the carbons of the carbon skeleton. A polyunsaturated fatty acid is one that has a great number of double bonds in the carbon skeleton. When glycerol and three fatty acids are combined by three dehydration synthesis reactions, a fat is formed. Notice that dehydration synthesis is almost exactly the same as the reaction that caused simple sugars to bond together.

Fats are important molecules for storing energy. There is twice as much energy in a gram of fat as in a gram of sugar. This is important to an organism, since fats can be stored in a relatively small space and still yield a high amount of energy. However, if the cells of an organism need energy, they tap the carbohydrate source before the stored fat. This is evident when a person tries to lose weight. A successful reducing diet is one in which carbohydrate and fat intake is greatly reduced. The body uses up the available carbohydrates and then begins to use the stored fat as a source of energy. Since fat has twice as much energy per gram as carbohydrates, the process of losing weight is often slow.

Fats in animals also provide protection from heat loss. Some animals have a layer of fat under the skin that serves as an insulating layer. The thick layer of blubber in whales, walruses, and seals prevents the loss of internal body heat to the cold, watery environment in which they live. This same layer of fat, together with the fat deposits around some internal organs—such as the kidneys and heart—serves as a cushion that protects these organs from physical damage.

Phospholipids are a class of water-insoluble molecules that resemble fats but contain a phosphate group (PO_4) in their structure (figure 3.12). One of the reasons phospholipids are important is that they are a major component of membranes in cells.

Steroids, a third group of lipid molecules, are characterized by their arrangement of interlocking rings of carbon. They often serve as hormones that aid in regulating body processes. One steroid molecule that you are probably familiar with

Figure 3.13
Some Common Steroids.
These two compounds, cholesterol
and vitamin D, are examples of
molecules normally produced by
the human body.

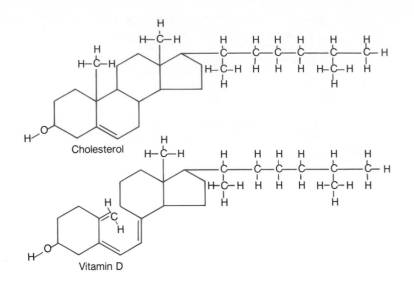

Cholesterol

Vitamin D

is cholesterol. Cholesterol has been implicated in many cases of atherosclerosis—the cholesterol is deposited on the inner walls of arteries and causes narrowing. This places an extra burden on the heart as it pumps blood through smaller and smaller tubes. On the other hand, cholesterol is necessary for the manufacture of vitamin D. Cholesterol molecules in the skin react with ultraviolet light to produce vitamin D, which assists in the proper development of bones and teeth. Figure 3.13 shows a cholesterol molecule and a vitamin D molecule.

A large number of steroid molecules are hormones. Some of them regulate reproductive processes such as egg and sperm production (see chapter 11), while others regulate things such as salt concentration of the blood. Athletes have been using certain hormonelike steroids to increase their muscular bulk. Although the medical community is uncertain that these chemicals increase strength of the athlete, they agree that harmful side effects of their use include liver disfunction, sex characteristic changes, changes in blood chemistry, and possibly death.

Proteins

Proteins are polymers made up of monomers known as amino acids. An **amino acid** is a short carbon skeleton that contains an amino group (a nitrogen and two hydrogens) on one end of the skeleton and a carboxylic acid group at the other end (figure 3.14). In addition, the carbon skeleton may have one of several different side chains on it. There are about twenty amino acids that are important to cells. All are identical except for the side chain (box 3.2).

The amino acids can bond together by dehydration synthesis reactions. When two amino acids form a bond by removal of water, the nitrogen of the amino group of one is bonded to the carbon of the acid group of another. This bond is termed a **peptide bond** (figure 3.15).

Any amino acid can form a peptide bond with any other amino acid. They fit together in a specific way with the amino group of one bonding to the acid group of the next. You can imagine that by using twenty different amino acids as building

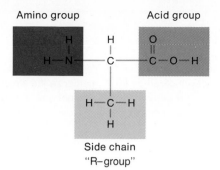

Amino group   Acid group

Side chain
"R–group"

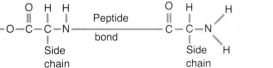

blocks, you can construct millions of different combinations. Each of these combinations is termed a **polypeptide chain.** A specific polypeptide is composed of a specific sequence of amino acids bonded end to end. This is called its primary structure. A listing of the amino acids in their proper order within a particular polypeptide constitutes its primary structure. The specific sequence of amino acids in a polypeptide is controlled by the genetic information of an organism.

The string of amino acids in a polypeptide is likely to twist into a particular shape, a coil or a pleated sheet. These forms are referred to as the secondary structure of polypeptides. For example, some proteins take the form of a helix. This shape is similar to a coiled telephone cord. The helical shape is maintained by hydrogen bonds formed between different amino acid side chains at different locations in the polypeptide. Remember from chapter 2 that hydrogen bonds do not form molecules but result in the orientation of one part of a molecule to another part of a molecule. Similarly, some polypeptides form hydrogen bonds that cause the polypeptide to make several flat folds that resemble a pleated skirt.

It is also possible for a single polypeptide to contain one or more coils and pleated sheets along its length. As a result, these different portions of the molecule can interact to form an even more complex three-dimensional structure. This occurs when the coils and pleated sheets twist and combine with each other. The complex three-dimensional structure formed in this manner is the polypeptide's tertiary (third degree) structure. A good example of tertiary structure can be seen when a coiled phone cord becomes so twisted that it folds around and back on itself in several places.

Frequently several different polypeptides, each with their own tertiary structure, twist around each other and chemically combine. The larger, three-dimensional structure formed by these interacting polypeptides is referred to as the

Box 3.2
Twenty Common Amino Acids

Structure of Amino Acids. All of the individual amino acids drawn here have the amino group to the left, the carboxylic acid group to the right, and the side chain shaded in below. The side chain is usually symbolized as an "R-group." The various R-groups determine the particular amino acid and its activity. You do not need to memorize the structure of these different amino acids.

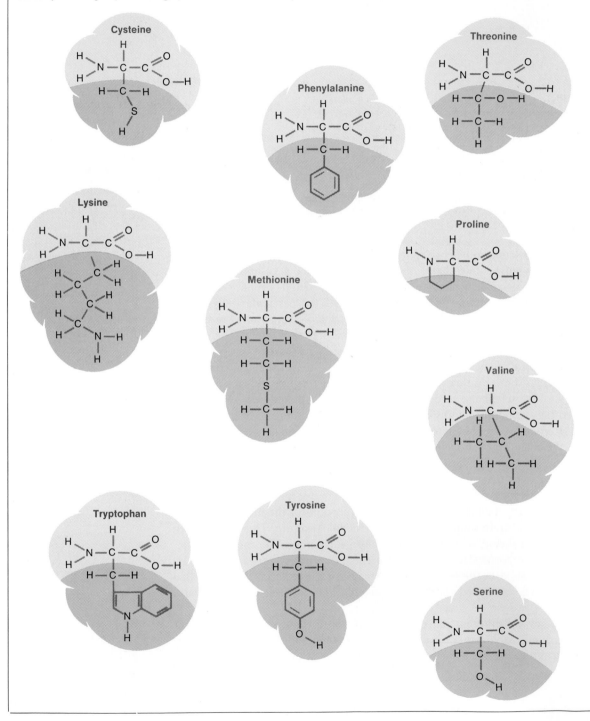

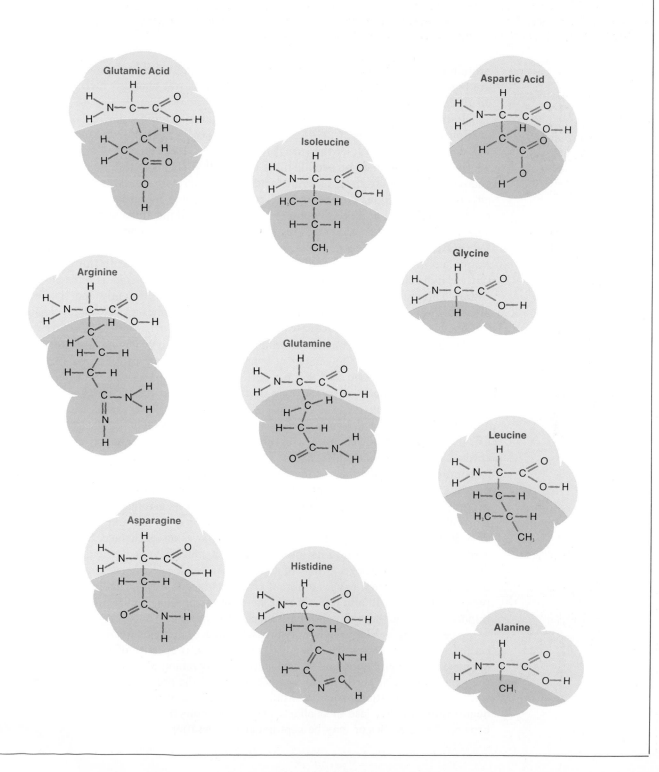

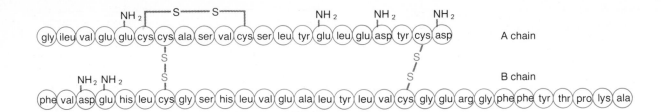

Figure 3.16
Insulin Molecule.
The protein insulin is composed of two polypeptide chains bonded together at specific points by reactions between the side chains of particular amino acids. The side chains of one interact with the side chains of the other and form a particular three-dimensional shape. The bonds that form between the polypeptide chains are called disulfide bonds.

protein's quaternary (fourth degree) structure. The individual polypeptide chains are bonded to each other by the interactions of certain side chains, which can form disulfide bonds (figure 3.16).

Individual polypeptide chains or groups of chains forming a particular configuration are proteins. The structure of a protein is closely related to its function. We will consider two aspects of the structure of proteins; the sequence of amino acids within the protein and the overall three-dimensional shape of the molecule. Any changes in the arrangement of amino acids within a protein can have far-reaching effects on its function. For example, normal hemoglobin found in red blood corpuscles consists of two kinds of polypeptide chains called the alpha and beta chains. The beta chain is 146 amino acids long. If just one of these amino acids is replaced by a different one, the hemoglobin molecule does not function properly. A classic example of this results in a condition known as sickle-cell anemia. In this case the sixth amino acid in the beta chain, which is normally glutamic acid, is replaced by valine. This minor change causes the hemoglobin to fold differently and the red blood corpuscle that contains this altered hemoglobin assumes a sickle shape when the body is deprived of an adequate supply of oxygen.

When a particular sequence of amino acids forms a polypeptide, the stage is set for that particular arrangement to bond with another polypeptide in a certain way. Think of a telephone cord that has curled up and formed a helix. Now imagine that at several irregular intervals along that cord, you have attached magnets. You can see that the magnets at the various points along the cord will attract each other and the curled cord will form a particular three-dimensional shape. You can more closely approximate the complex structure of a protein if you imagine several curled cords, each with magnets attached at several points. Now imagine these magnets as bonding the individual cords together. The globs or ropes of telephone cords approximate the structure of a protein. This shape can be compared to the surface geography of a key. Its shape may appear random, but its arrangement results in a molecule with a particular three-dimensional form. In order for a key to do its job effectively, it has to have particular bumps and grooves on its surface. Similarly, if a particular protein is to do its job effectively, it must have a particular surface geography. This surface geography of a protein can be altered by changing the order of the amino acids that causes different cross linkages to form. Changing environmental conditions will also influence the shape of the protein. Figure 3.17 shows the importance of the three-dimensional shape of the protein.

Energy such as heat or light may break the hydrogen bonds within protein molecules. When this occurs, the chemical and physical properties of the protein are changed and the protein is said to be **denatured.** A common example of this is when the gelatinous, clear portion of an egg is cooked and the protein changes to a white solid. Some medications, such as insulin, are protein and must be protected from denaturation or they lose their effectiveness. For protection they may be stored in brown-colored bottles or may be kept under refrigeration.

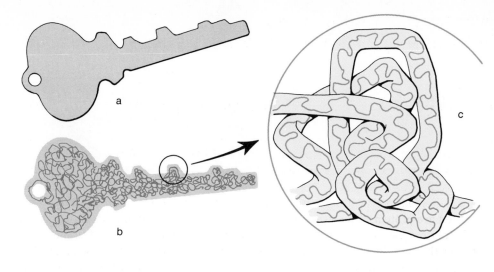

Figure 3.17
Proteins Have a Three-Dimensional Shape.
The specific arrangement of amino acids results in side chains that are available to bond with other side chains. The results are specific three-dimensional proteins that have a specific surface geometry. We frequently compare this three-dimensional shape to the three-dimensional shape of a specific key.

The thousands of kinds of proteins can be placed into two categories. Some proteins are important for maintaining the shape of cells and organisms—they are usually referred to as **structural proteins.** The proteins that make up the cell membrane, muscle cells, tendons, and blood cells are examples of structural proteins. The other kinds of proteins, **regulator proteins,** help determine what activities will occur in the organism. These regulator proteins include enzymes and some hormones. These molecules help control the chemical activities of cells and organisms. Enzymes are important and they will be dealt with in detail in chapter 5. Some examples of enzymes are the digestive enzymes in the stomach and the mouth. Two hormones that are protein in nature and act as regulators are insulin and oxytocin. Insulin is produced by the pancreas and controls the amount of glucose in the blood. If insulin production is too low or if the molecule is improperly constructed, glucose molecules are not removed from the bloodstream at a fast enough rate. The excess sugar is then eliminated in the urine. Other symptoms of excess sugar in the blood include excessive thirst and even loss of consciousness. The disease caused by improperly functional insulin is known as diabetes. Oxytocin, a second protein-type hormone, stimulates the contraction of the uterus during childbirth. It is also an example of an organic molecule that has been produced artificially and is used by physicians to induce labor.

Nucleic Acids

The last group of organic molecules that we will consider are the nucleic acids. **Nucleic acids** are complex molecules that store and transfer information within a cell. They are constructed of fundamental monomers known as **nucleotides.** Each nucleic acid is a polymer composed of nucleotides bonded together. There are eight different nucleotides, but each is constructed of a phosphate group, a sugar, and an organic nitrogenous base.

The two kinds of sugar that can be part of the nucleotide are ribose and deoxyribose. These are five-carbon simple sugars. The phosphate group is attached to the sugar. The nitrogenous base is attached to another part of the sugar. There are five common organic molecules containing nitrogen that are likely to be part of the nucleotide structure. The nitrogen-containing bases are adenine, guanine, cytosine, thymine, and uracil (figure 3.18).

Figure 3.18
Nucleic Acid Components.

The nucleic acids RNA and DNA are composed of these eight building blocks. Note that each building block has a sugar component, a phosphate component, and a nitrogen-containing ring component. These subunits are constructed from the three components by dehydration synthesis reactions.

DNA

RNA

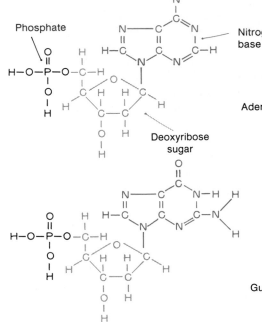

Phosphate

Nitrogenous base

Adenine

Deoxyribose sugar

Ribose sugar

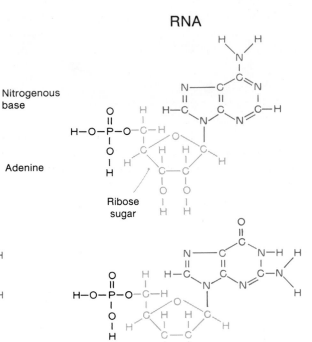

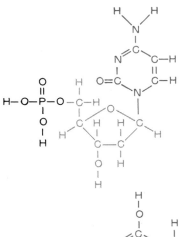

Guanine

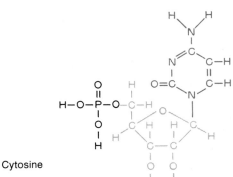

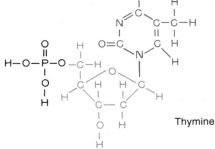

Cytosine

Thymine

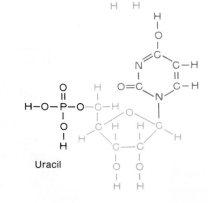

Uracil

The nucleotides can then connect to form long chains by dehydration synthesis reactions. The long chains of nucleotides are of two types—RNA, ribonucleic acid, and DNA, deoxyribonucleic acid. The RNA forms a single polymer, whereas the DNA is generally composed of two matching polymers twisted together and held by hydrogen bonds. These two types of molecules contain the information needed for the formation of particular sequences of amino acids; they determine what kinds of proteins an organism can manufacture. The mechanism of storing and using this information is the topic of chapter 8.

Summary

The chemistry of living things involves a variety of large and complex molecules. This chemistry is based on the carbon atom and the fact that carbon atoms can connect to form long chains or rings. This results in a vast array of molecules. The structure of each molecule is related to its function. Changes in the structure may result in abnormal functions, which we call disease. Some of the most common types of organic molecules found in living things are carbohydrates, lipids, proteins, and nucleic acids. Table 3.1 summarizes the major types of biologically important organic molecules and their roles in living things.

Table 3.1   *A Summary of the Types of Organic Molecules Found in Living Things*

Type of Organic Molecule	Basic Subunit	Function	Examples
Carbohydrate	Simple sugar	Provide energy Provide support	Sugar Cellulose
Lipids 1. Fats	Glycerol and fatty acids	Provide energy Provide insulation Serve as shock absorber	Lard Olive oil Linseed oil Tallow
2. Steroids	A complex ring structure	Some hormones that control the body processes	Testosterone Vitamin D Cholesterol
3. Phospholipids	Glycerol, fatty acids, and phosphorus compounds	Structure of the cell membrane	Cell membrane
Protein	Amino acid	Structure of cells and parts of organisms	Cell membrane Hair Muscle
		Enzymes that regulate chemical reactions	Ptyalin in the mouth
		Some hormones	Insulin
Nucleic acid	Nucleotide	Contains genetic information that controls the cell	DNA RNA

Amino acids and fatty acids are both organic acids. What property must they have in common with inorganic acids such as sulfuric acid? How do they differ? Consider such aspects as structure of molecules, size, bonding, and pH.

Experience This

Look at the labels on five items from your pantry. Make a list of ingredients that you think are organic chemicals. They may be additives to the food, preservatives, coloring materials, or flavor enhancers. They may also be the active ingredients in such things as pesticides or pharmaceuticals. As a result of reading these ingredients, can you identify the class of organic molecules to which each belongs?

Questions

1. Diagram an example of each of the following: amino acid, simple sugar, glycerol, fatty acid.
2. Give an example of each of the following classes of organic molecules: carbohydrate, protein, lipid, and nucleic acid.
3. What is the structural difference between a saturated and unsaturated fat?
4. Describe three different kinds of lipids.
5. What is the difference among primary, secondary, and tertiary structures of proteins?
6. What two characteristics of the carbon atom make it unique?
7. What is the difference between inorganic and organic molecules?
8. What are the functions of the four classes of organic molecules?
9. Describe five functional groups.
10. List three monomers and the polymers that can be constructed from them.

Chapter Glossary

amino acid (ah-mēn'o ǎ'sid) A short carbon skeleton that contains an amino group, a carboxylic acid group, and one of various side groups.

biochemistry (bi-o-kem'iss-tre) The chemistry of living things often called biological chemistry.

carbohydrate (kar-bo-hi'drāt) One class of organic molecules composed of carbon, hydrogen, and oxygen in a ratio of 1:2:1; the basic building block of a carbohydrate is a simple sugar (= monosaccharide).

carbon skeleton (kar'bon skel'uh-ton) Central portion of an organic molecule composed of rings or chains of carbon atoms.

complex carbohydrates (kom'pleks kar-bo-hi'drāts) Macromolecules composed of simple sugars combined by dehydration synthesis to form a polymer.

denature (de-na'chur) Irreversible change of the chemical and physical properties of a protein.

dehydration synthesis reaction (de-hi-dra'shun sin'thuh-sis re-ak'shun) A reaction that results in the formation of a macromolecule when water is removed from between the two smaller component parts.

double bond (dubl bond) A pair of covalent bonds between two atoms formed when they share two pairs of electrons.

empirical formula (em-pēr'i-kal fōr'miu-lah) Chemical shorthand that indicates the number of each kind of atom within a molecule.

fat (fat) A class of water-insoluble macromolecules composed of a glycerol and three fatty acids.

fatty acid (fat-te ǎ'sid) One of the building blocks of a fat, composed of a long chain carbon skeleton with a carboxylic acid functional group.

functional groups (fung'shun-al grūps) Specific combinations of atoms attached to the carbon skeleton that determine specific chemical properties.

glycerol (glis'er-ol) One of the building blocks of a fat, composed of a carbon skeleton that has three alcohol groups (OH) attached to it.

inorganic molecules (in-or-gan'ik mol-uh-kiuls) Molecules that do not contain carbon atoms in rings or chains.

lipids (lī'pids) Large organic molecules that do not easily dissolve in water; classes include fats, phospholipids, and steroids.

nucleic acids (nu'kle-ik ă'sids) Complex molecules that store and transfer information within a cell. They are constructed of fundamental monomers known as nucleotides.

nucleotide (nu''kle-o-tīd') A fundamental subunit of nucleic acid constructed of a phosphate group, a sugar, and an organic nitrogenous base.

organic molecules (or-gan'ik mol'uh-kiuls) Complex molecules whose basic building blocks are carbon atoms in chains or rings.

peptide bond (pep'tīd bond) A covalent bond between amino acids in a protein.

phospholipid (fos''fo-lī'pid) A class of water-insoluble molecules that resemble fats but contain a phosphate group (PO_4) in their structure.

polypeptide chain (pŏ''le-pep'tīd chān) A macromolecule composed of a specific sequence of amino acids.

protein (pro'te-in) Macromolecules made up of amino acid subunits attached to each other by peptide bonds or groups of polypeptides.

regulator proteins (reg'yu-la-tor pro'te-ins) Proteins that influence the activities that occur in an organism—for example, enzymes and some hormones.

saturated (sat'yu-ra-ted) Carbon skeleton of a fatty acid that has as much hydrogen bonded to it as possible.

steroid (stēr'oid) One of the three kinds of lipid molecules characterized by their arrangement of interlocking rings of carbon.

structural formula (struk'chu-ral for'miu-lah) An illustration that shows the arrangement of the atoms and their bonding within a molecule.

structural proteins (struk'chu-ral pro'te-ins) Proteins that are important for holding cells and organisms together, such as the proteins that make up the cell membrane, muscles, tendons, and blood.

unsaturated (un-sat'yu-ra-ted) A carbon skeleton of a fatty acid that contains carbons that are double bonded to each other at one or more points.

4

Cell History, Structure, and Function

▪ Chapter Outline

▪ Purpose

The cell is the simplest structure capable of existing as an individual living unit. Within this unit, certain chemical reactions are required for maintaining life. These reactions do not occur at random, but are associated with specific parts of the many kinds of cells. This chapter deals with certain cellular structures found within most types of cells and discusses their functions.

For Your Information

Here are some tidbits of information about cells:

The smallest cells are the bacteria, some of which are only 0.1 to 0.2 micrometers in diameter. *E. Coli,* a very common bacterium in human intestines, is only 0.5 micrometers long.

The largest cells are the egg cells of birds and mammals. A chicken yolk cell can be 50 millimeters in diameter. The nerve cell from the human spine to the big toe can be 1 meter or more in length.

The animal cells that live the longest are the nerve or muscle cells of long-lived reptiles and mammals. The sea tortoise, which lives two-hundred years, has the same nerve cells at birth as it does at death.

Cells of certain bacteria can divide every fifteen minutes under optimum conditions. This means that if you start with only one bacterial cell and it divides as often as possible, in only twelve hours there could be as many as 281 trillion individual cells.

Learning Objectives

■ Recognize the historical perspective of the development of the cell theory.

■ Describe the molecular structure of a membrane and relate this structure to the process whereby a cell accumulates and releases materials.

■ Describe the processes whereby a cell accumulates some materials and releases others to the environment and the conditions controlling these various processes.

■ Identify the cytoplasmic organelles in most eukaryotic cells.

■ Associate the organelle structure with its major function in eukaryotic cells.

■ Identify the nuclear components of a cell and associate the function with the nuclear structures.

Cell Concept

The concept of a cell is one of the most important ideas in biology because it applies to all living things. It did not develop all at once, but has been added to and modified over many years. It is still being modified today.

Two men in particular made key contributions to the cell concept. Anton van Leeuwenhoek (1632–1723) was one of the first individuals to make and use a microscope (see box 4.1). A **microscope** is an instrument constructed of lenses that enlarge and focus an image of a small object. When van Leeuwenhoek discovered that he could see things moving in pond water using his microscope, his curiosity stimulated him to look at a variety of other things as well. He studied blood, semen, feces, pepper, tartar, and many other things. He was the first to see individual cells and recognize them as living units, but he did not call them cells. The name he gave to these "little animals" that he saw moving around in the pond water was animalcules.

Box 4.1
The Microscope

In order to view very small objects, a magnifying glass is used. A magnifying glass is a lens that bends light in such a way that the object appears larger than it really is. Such a lens might magnify objects ten or even fifty times. Anton van Leeuwenhoek, (1632–1723) a Dutch draper and haberdasher, was one of the first individuals to carefully study magnified cells. He made very detailed sketches of the things he viewed with his simple microscopes and communicated his findings with Robert Hooke and the Royal Society of London. His work stimulated further investigation of magnification techniques and description of cell structure. These first microscopes were developed in the 1600s.

Compound microscopes, developed soon after the simple microscopes, are able to increase magnification by bending light through a series of lenses. One lens, the objective lens, magnifies a specimen that is further magnified by the second lens, known as the ocular lens. With modern technology of producing lenses, use of specific light waves, and immersing an objective lens in oil to collect more of the available light, objects can be magnified one hundred to fifteen hundred times. Microscopes typically available for student use are compound light microscopes.

The major restriction of magnification with a light microscope is the ability of the viewer to distinguish two very close objects as two distinct things. This ability to separate two objects is termed resolution. Think of a picket fence a great distance away from you. You are probably able to distinguish one picket from another because the light easily passes between them. Now imagine the pickets much closer to each other, so close that the fence forms a solid wall. The fact that you can no longer distinguish the individual pickets from each other is because you have reached the limits of the resolving power of your eyes with the available light. If a brighter light is placed behind the picket fence, you might be able to distinguish the individual pickets. Ultimately, however, you will be unable to separate the pickets from each other, regardless of how bright the light used.

In the same way, a microscope has a limited resolving power. If two structures in a cell are very close to each other, you may not be able to distinguish that there are actually two structures close to each other rather than one structure. The limits of resolution of a light microscope are related to the wave lengths of the light being transmitted through the specimen. If you could see ultraviolet light waves with shorter wave lengths, it would be possible to resolve more individual structures.

An electron microscope makes use of this principle because the moving electrons have much shorter wave lengths than visible light. Thus they are able to magnify 200,000 times and still resolve individual structures. The difficulty is, of course, that you are unable to see electrons with your eyes. Therefore, in order to use the electron microscope, the electrons strike a photographic film or television monitor and this "picture" shows the individual structures. Heavy metals scattered on the structures to be viewed with the electron microscope increase the contrast between where there are structures that interfere with the transmission of the electrons and areas where the electrons are transmitted easily. The techniques of preparation of the material to be viewed—slicing the specimen very thinly and focusing the electron beam on the specimen—make electron microscopy an art as well as a science.

The first to use the term "cell" was Robert Hooke (1635–1703) of England, who was also interested in how things looked when magnified. He chose to study thin slices of cork, the tissue from the bark of a cork oak tree. He saw a mass of cubicles fitting neatly together, which reminded him of the barren rooms in a monastery. Hence, he called them cells. As it is currently used, the term **cell** is used to refer to the basic structural unit that makes up all living things. When Hooke looked at cork, the tiny boxes he saw were, in fact, only the cell walls that surrounded the living portions of plant cells. We now know that the cell wall is composed of the complex carbohydrate cellulose and that it is used to provide strength and protection to the living contents of the cell. The cell wall appears to be a rigid, solid layer of material, but in reality it is composed of many interwoven strands of cellulose molecules. Its structure allows certain, very large molecules to pass through it readily, but it acts as a screen to other molecules.

Hooke's use of the term cell was only the beginning. Soon after the term caught on, it was realized that the vitally important portion of the cell was inside the cell wall. This living material was termed **protoplasm,** which means first formed substance. The use of the term protoplasm allowed the living portion of the cell to be distinguished from the nonliving cell wall. As with any new field of study, this early terminology was not complete. Very soon microscopists were able to distinguish two different regions of protoplasm. One type of protoplasm was more viscous and darker than the other; this region, called the **nucleus,** appeared as a central body within a more fluid material surrounding it. **Cytoplasm** is the name given to the more fluid portion of the protoplasm. This set of terms is used in a similar way to the terms egg, yolk, and white. The egg is composed of both the yolk and the white, just as the protoplasm is composed of both the nucleus and the cytoplasm (figure 4.1).

With the development of better light microscopes and ultimately the electron microscope, it was revealed that protoplasm contains many structures called **organelles.** It has been determined that these organelles have particular functions they perform in each cell. The job that an organelle does is related to its structure. Organelles are not simply passive structures for chemical reactions, but are essential to the life of the cell. Each organelle is dynamic in its operation, changing shape and size as it works. They move throughout the cell, and some even self-duplicate. There is a constant turnover of organelles in the cell, because each continuously changes as it carries out its particular function. An organelle may be destroyed during the life of the cell, but lost organelles are rapidly replaced and the total number of a particular organelle remains relatively constant. Most of the organelles are composed of membrane.

In addition to organelles, there are many types of inclusions found in cells. They are collections of materials that do not have a regular structure. Inclusions are concentrations of stored molecules, such as starch, grains, sulfur, and oil droplets. Unlike organelles, which are essential to the survival of a cell, the inclusions are only temporary sites for the storage of nutrients and wastes. Cells put away many of these materials in this form as a food reserve. Their loss from the cell does not result in the death of the cell in which they are manufactured. However, some inclusions may be poisonous to humans. For example, rhubarb leaf cells contain an inclusion composed of the organic acid, oxalic acid. If eaten, these needle-shaped crystals can cause injury by puncturing the kidney and other cell membranes. The presence of this particular inclusion aids in the survival of the rhubarb plant by discouraging animals from eating it.

Figure 4.1
Figure 4.1
*Cells; Basic Structures
of Life.*
The idea that cells are composed
of structures that are common to
all living things is a concept that
has developed over the last four
hundred years from the
''animalcules'' of Van
Leeuwenhoek and the cork ''cells''
of Hooke to the current cell theory.
Cells are composed of living
protoplasm and sometimes
surrounded by a cell wall.
Protoplasm is divided into two
types, nucleus and cytoplasm.

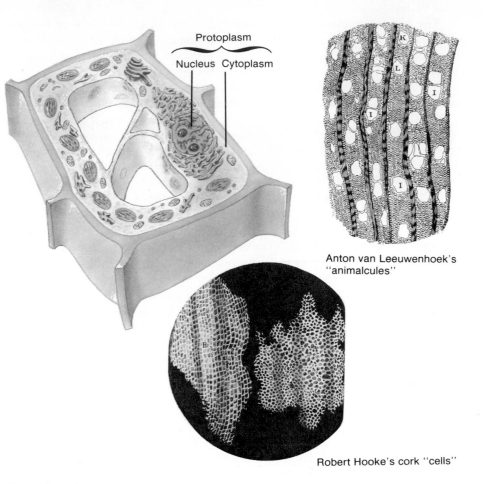

Protoplasm

Nucleus   Cytoplasm

Anton van Leeuwenhoek's
''animalcules''

Robert Hooke's cork ''cells''

Cell Membranes

One feature common to all cells and many of the organelles they contain is a thin
layer of material called membrane. Membrane can be folded and twisted into many
different structures, shapes, and forms. The particular arrangements of membranes
within an organelle is related to the functions it is capable of performing. This is
similar to the way a piece of fabric can be fashioned into a pair of pants, a shirt,
sheets, pillowcases, or a rag doll. All cellular membranes have a fundamental molecular structure that allows them to be fashioned into a variety of different organelles, each with a very specific function.

Cellular membranes are thin sheets of material composed of phospholipids and
proteins. Phospholipid molecules have one end that is soluble in water and another
end that is not. Consequently, when phospholipid molecules are placed in water, they
form a double-layered sheet, with the water-soluble portions of the molecules facing
away from one another. If a phospholipid is shaken in a glass of water, the molecules
will automatically form double-layered membranes. It is important to understand
that the membrane formed is not rigid or stiff. The component phospholipids are in
constant motion as they move with the surrounding water molecules and slide past
one another. The protein component of cellular membranes can be found on either
surface of the membrane or in the membrane among the phospholipid molecules.
Many of the protein molecules are capable of moving from one side to the other.
These proteins can serve as surface catalysts that help with the chemical activities

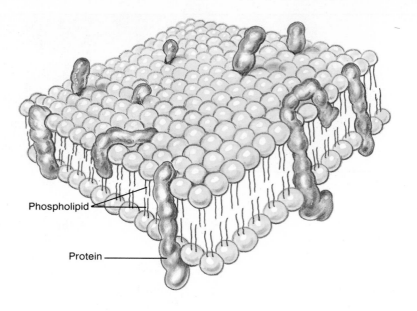

Phospholipid

Protein

Figure 4.2
Membrane Structure.
Membranes in cells are composed
of protein and phospholipids. The
center is the location of two layers
of phospholipid. Each layer is
oriented so that the fatty ends
extend toward each other and the
glycerol portions are on the
outside. The phosphate-containing
chain is coiled near the glycerol
portion. Buried within the
phospholipid layer and/or floating
on it are the globular proteins.
Some of these proteins accumulate
materials from outside the cell,
others act as sites of chemical
activity.

of the protoplasm. Others serve as recognition sites involved in the uptake of materials from the environment. Still others aid in the movement of molecules across the membrane (figure 4.2). In addition to phospholipid and protein, some protein molecules found on the outside surfaces of cellular membrane have carbohydrates or fats attached to them. These combination molecules are important in determining the "sidedness" (inside-outside) of the membrane and help organisms to recognize differences between types of cells. Your body can recognize disease-causing organisms because their surface proteins are different from ours.

Diffusion

Because the cell membrane is composed of phospholipid and protein molecules that are in constant motion, temporary openings are formed that allow small molecules to cross from one side of the membrane to the other. Molecules close to the membrane are in constant motion as well. They are able to move into and out of a cell by passing through these openings in the membrane. Molecular movement is due to the kinetic energy of the individual molecules. Since molecular motion is random, molecules move in all directions unless they are prevented from doing so. However, they tend to move in a direction that will distribute them equally throughout their environment. Molecules tend to move from an area where they are concentrated to an area where they are less concentrated because molecular movement is less inhibited in the area of lesser concentration.

For example, picture a sugar cube that is placed in a glass of water. The sugar molecules are highly concentrated in the cube and are rare in the surrounding water. Sugar molecules will move away from the cube and toward the cube because of their random motion. However, since far more molecules will leave the cube than will return to it, the overall movement of sugar molecules is away from the cube. **Net movement** is the movement of molecules in one direction minus the movement of molecules in the opposite direction. The net movement of a particular kind of molecule from an area of higher concentration to an area of lesser concentration is

Figure 4.3
Concentration Gradient.
Gradual changes in concentrations over distance are called gradients. This bar shows a color gradient with full color at one end to no color at the other end. A gradient is necessary for diffusion to occur, and diffusion results in net movement of molecules from higher concentrations to lower concentrations.

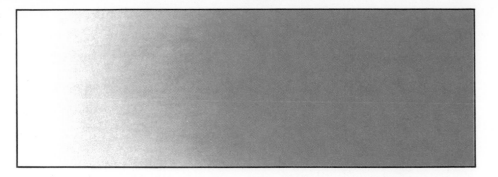

diffusion. If the concentration of a specific kind of molecule is higher on the outside of a cell membrane than on the inside, those molecules will diffuse through the membrane into the cell if the membrane is permeable to the molecules. The rate of diffusion is related to the kinetic energy and size of the molecules. Since diffusion only occurs when molecules are unevenly distributed, relative concentration of the molecules is important in determining how fast diffusion occurs. The difference in concentration of the molecules is known as a **concentration gradient** or **diffusion gradient.** When the molecules are equally distributed, no such gradient exists. For example, if you allow the sugar cube to remain in the glass of water long enough, the sugar molecules will be equally distributed and although molecular motion continues, there is no diffusion. Although the molecules continue to move by molecular motion, there is no net movement from one place to another; therefore, diffusion does not take place (figure 4.3).

Diffusion can take place only as long as there are no barriers to the free movement of molecules. In the case of a cell, the membrane permits some molecules to pass through, while others are not allowed to pass or only allowed to pass more slowly. This permeability is based on size, ionic charge, and solubility of the molecules involved. The membrane does not, however, distinguish direction of movement; therefore, the membrane does not influence the direction of diffusion (figure 4.4).

The process of diffusion is an important method for the exchange of materials between a cell and its environment. Since the movement of the molecules is random, the cell has little control over the process; thus, diffusion is considered a passive process. For example, animals are constantly using oxygen in various chemical reactions. Consequently, the oxygen concentration in cells always remains low. The cells then contain a lower concentration of oxygen in comparison to the oxygen level outside of the cell. This creates a diffusion gradient and the oxygen molecules diffuse from the outside of the cell to the inside of the cell. In large animals, many of the cells are buried deep within the body; if it were not for their circulatory systems, there would be little opportunity for cells to exchange gases directly with their surroundings. The circulatory system is a transportation system within a body composed of blood vessels of various sizes. These vessels carry many different molecules from one place to another. Oxygen may diffuse into blood through the membranes of the lungs, gills, or other moist surfaces of the animal's body. The circulatory system then transports the oxygen-rich blood throughout the body. The oxygen automatically diffuses into cells that are low in oxygen. The opposite is true of the gas, carbon dioxide. Animal cells constantly produce carbon dioxide, and so there is always a high concentration of it within the cells. These molecules diffuse from the cells into the blood where the concentration of carbon dioxide is lower. The blood

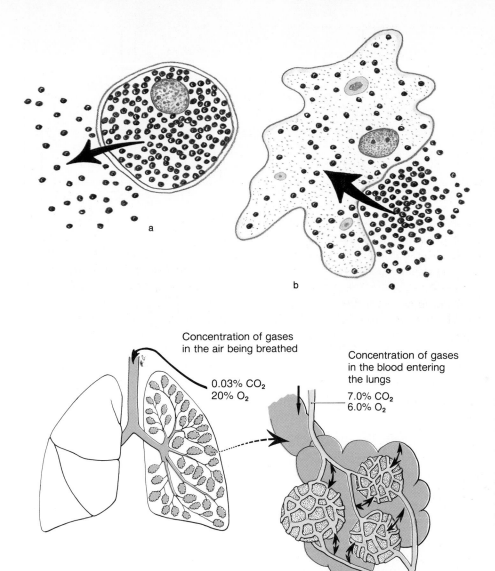

Figure 4.4
Diffusion.
As a result of molecular motion, molecules move from areas where they are concentrated to areas where they are less concentrated. This figure shows (a) molecules leaving a cell by diffusion and (b) molecules entering a cell by diffusion. The direction is controlled by concentration and the energy necessary is supplied by the kinetic energy of the molecules themselves.

a

b

Concentration of gases
in the air being breathed

Concentration of gases
in the blood entering
the lungs

0.03% CO_2
20% O_2

7.0% CO_2
6.0% O_2

Concentration of gases
in the blood leaving
the lungs

5.5% CO_2
14.0% O_2

Figure 4.5
Diffusion in the Lungs.
As blood enters the lungs, it has a higher concentration of carbon dioxide and a lower concentration of oxygen than the air in the lungs. The concentration gradient of oxygen is such that the oxygen diffuses from the lungs into the blood and the concentration gradient of the carbon dioxide is such that it diffuses from the blood into the lungs. These two different diffusions happen simultaneously and the direction of diffusion is controlled by the relative concentrations of the molecules in the blood and in the lung.

is pumped to the moist surface (gills, lungs, etc.) and the carbon dioxide again diffuses into the surrounding environment, which has a lower concentration of this gas. In a similar manner, many other types of molecules constantly enter and leave cells (figure 4.5).

Another characteristic of all membranes is that they are differentially permeable. (The terms selectively permeable and semipermeable are synonyms.) **Differential permeability** means that the membrane will allow certain molecules to pass across it and will prevent others from doing so. Molecules that are able to dissolve in phospholipids, such as vitamins A and D, can pass through the membrane rather easily; however, many molecules cannot pass through at all. In certain cases, the membrane differentiates on the basis of size of the molecules; that is, the membrane allows small molecules, such as water, to pass through and prevents the passage of

Figure 4.6
Osmosis.

Osmosis is one category of diffusion where water molecules are diffusing through a differentially permeable membrane. Pores or holes in the membrane allow water molecules to pass freely, but prevent certain other molecules from passing. The membrane does not necessarily have to be a living membrane. In the case of osmosis in cells, the outer cell membrane is differentially permeable. Diffusion of water follows the rule of movement of water molecules from an area of higher concentration to an area of lower concentration. In the illustration, osmosis results in water moving into the thistle tube and the water level rises.

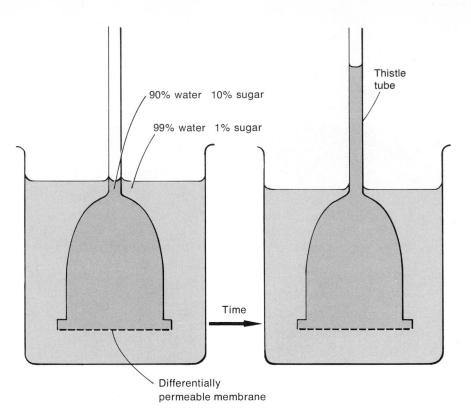

90% water   10% sugar

99% water   1% sugar

Thistle tube

Time

Differentially permeable membrane

larger molecules. The membrane may also regulate the passage of ions. If a particular portion of the membrane has a large number of positive ions on its surface, positively charged ions in the environment will be repelled and prevented from crossing the membrane.

Osmosis

Water is one kind of molecule that easily diffuses through cell membranes. The net movement of water molecules through a differentially permeable membrane is known as **osmosis,** which is a special category of diffusion. Osmosis occurs constantly in living systems. In any osmotic situation, there must be a differentially permeable membrane separating two solutions. The process can be studied using nonliving materials. Figure 4.6 shows a thistle tube that contains a solution of 90% water and 10% sugar. The end of the tube is covered with a differentially permeable membrane. The tube is then submerged in a beaker. In the beaker is a different sugar solution. In this case, 99% water and only 1% sugar. The membrane is differentially permeable in that it allows the water molecules to pass through it easily but blocks the larger sugar molecules. Since there is a higher concentration of water molecules in the beaker (compared to the concentration of water molecules in the thistle tube), the direction of net movement of water molecules is from the beaker into the tube; therefore, the liquid in the tube rises. Be sure you recognize that osmosis is really diffusion in which the diffusing substance is a solvent (usually water) and the regions of different concentrations of the solvent are separated by a differentially permeable membrane.

A proper amount of water is required if a cell is to function efficiently. Too much water in a cell may dilute the cell contents and interfere with the chemical reactions necessary to keep the cell alive. Too little water in the cell may result in a buildup

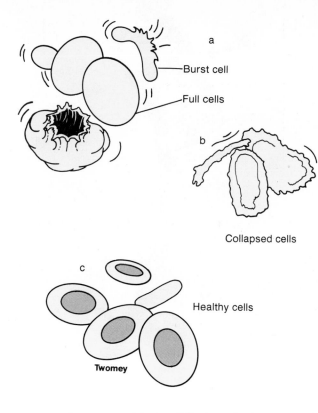

- Burst cell
- Full cells

Collapsed cells

Healthy cells

Twomey

Figure 4.7
Osmotic Influences on Cells.
The three cells sketched here are in three different environments. (a) shows a cell that has accumulated water from the environment because there is a higher concentration outside the cell than in its protoplasm. In (b), water has diffused from the cell to the environment because there was more water in the cell. In (c), water inside the cell and water in the environment are in balance with each other so movement of water into the cell equals movement of water out of the cell.

of poisonous waste products. As with the diffusion of other molecules, osmosis is a passive process since the cell has no control over the diffusion of water molecules. This means that the cell can remain in balance with an environment only if that environment does not cause the cell to lose or gain too much water.

Many organisms have a concentration of water that is equal to their surroundings. This is particularly true of simple organisms that live in the ocean. However, if an organism is going to survive in an environment that has a different concentration of water from its cells, it must expend energy to maintain this difference. Organisms that live in fresh water have a lower concentration of water than their surroundings and tend to gain water by osmosis very rapidly. Organisms whose cells gain water by osmosis must expend energy to eliminate any excess if they are to keep from swelling and bursting (figure 4.7).

Plant cells also experience osmosis. If the water concentration outside of the plant cell is higher than the water concentration inside, more water molecules enter the cell than leave. This creates internal pressure within the cell. But plant cells do not burst since they are surrounded by a rigid cell wall. Lettuce cells that are crisp are ones that have gained water so that there is high internal pressure. Wilted lettuce has lost some of its water to its surroundings so that it has only slight internal cellular water pressure. Osmosis occurs when you put salad dressing on a salad. The dressing has a very low water concentration, so that water from the lettuce diffuses from the cells into the surroundings. Salad that has been "dressed" too long becomes limp and is unappetizing.

So far, we have considered only those situations in which the cell has no control over the movement of molecules. Cells cannot rely solely on diffusion and osmosis, since many of the molecules they require are not able to pass through the membrane or occur in relatively low concentrations in the cell's surroundings.

Figure 4.8
Active Transport.
One possible method whereby
active transport could cause
materials to accumulate in a cell is
illustrated here. Notice that the
concentration gradient is such that
if simple diffusion were operating,
the molecules would leave the cell.
The action of the carrier requires
an input of energy other than the
kinetic energy of the molecules;
therefore, this process is termed
"active" transport.

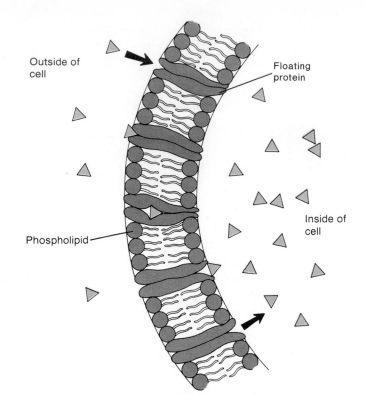

Outside of
cell

Floating
protein

Phospholipid

Inside of
cell

Controlled Methods of Transporting Molecules

Some molecules move across the membrane by combining with specific carrier proteins. When the rate of diffusion of a substance is increased in the presence of a carrier, we call this **facilitated diffusion.** Since this is diffusion, the net direction of movement is in accordance with the concentration gradient. Therefore, this is considered a passive transport method. However, this can only occur in living organisms with the necessary carrier proteins. One example of facilitated diffusion is the movement of glucose molecules across the membranes of certain cells. In order for the glucose molecules to pass into these cells, specific proteins are required to carry them across the membrane. The action of the carrier does not require an input of energy other than the kinetic energy of the molecules.

When molecules are moved across the membrane from an area of *low* concentration to an area of *high* concentration, the cell must be expending energy. The process of using a carrier protein to move molecules up a concentration gradient is called **active transport** (figure 4.8). Active transport is very specific in that only certain molecules or ions are able to be moved in this way, and they must be carried by specific proteins in the membrane. The action of the carrier requires an input of energy other than the kinetic energy of the molecules; therefore, this process is termed "active" transport. For example, some ions, such as sodium and potassium, are actively pumped across cell membranes. Sodium ions are pumped from cells up a concentration gradient. Potassium ions are pumped into cells up a concentration gradient.

In addition to active transport, there are two other methods that are used to actively move materials into cells. **Phagocytosis** is the process that cells use to wrap

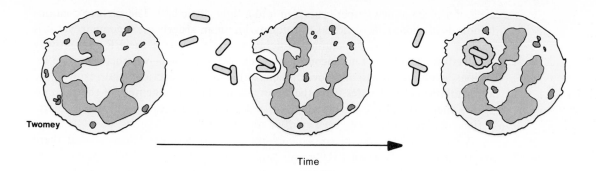

Twomey

Time

membrane around a particle (usually food) and engulf it (figure 4.9). This is the process leucocytes (white blood cells) in your body use to surround invading bacteria, viruses, and other foreign materials. Because of this, these kinds of cells are called phagocytes. When phagocytosis occurs, the material to be engulfed touches the surface of the phagocyte and causes a portion of the outer cell membrane to be indented. The indented cell membrane is pinched off inside the cell to form a sac containing the engulfed material. This sac composed of a single membrane is called a **vacuole.** Once inside the cell, the membrane of the vacuole is broken down, releasing its contents inside the cell. Phagocytosis is used by many types of cells to acquire large amounts of material from their environment. However, if the cell is not surrounding a large quantity of material, but is merely engulfing some molecules dissolved in water, the process is termed **pinocytosis.** In this process, the sacs that are formed are very small in comparison to those formed during phagocytosis. Because of this size difference, they are called **vesicles.** In fact, an electron microscope is needed in order to see them. These two processes, phagocytosis and pinocytosis, differ from active transport in that the cell surrounds large amounts of material with a membrane rather than taking them in through the membrane molecule by molecule. The movement of materials into the cell by either phagocytosis or pinocytosis is referred to as endocytosis. The transport of material from the cell by reverse processes is called exocytosis.

Organelles Composed of Membranes

Now that you have some background concerning the structure and the function of membranes, let us focus on the way cells use the membranes to build specific structural components of their protoplasm. The outer boundary of the cell is termed the **cell membrane** or **plasma membrane.** It is associated with a great variety of metabolic activities including the uptake and release of molecules, sensing stimuli in the environment, the recognition of other cell types, and attachment to other cells and nonliving objects. In addition to the cell membrane, there are many other organelles composed of membranes. Each of these membranous organelles has a unique shape or structure that enables it to be associated with particular functions. One of the most common organelles found in cells is the endoplasmic reticulum.

Endoplasmic Reticulum

The **endoplasmic reticulum, ER,** is a set of folded membranes and tubes throughout the cell. This system of membranes provides a large surface upon which chemical activities take place. Since the ER is an enormous surface area, many chemical reactions can be carried out in an extremely small space. Picture the vast surface

Figure 4.9
Phagocytosis.
This cell is engulfing a large amount of material at one time and surrounding it with a membrane. Pinocytosis, a related process, differs in that smaller amounts of dissolved materials are taken in and placed in a membrane enclosure.

area of a piece of newspaper crumpled into a light, little ball. The surface contains hundreds of thousands of tidbits of information in an orderly arrangement, yet it is packed into a very small volume. Proteins on the surface of the ER are actively involved in controlling and encouraging chemical activities—whether they are reactions involving growth and development of the cell, or reactions that result in the accumulation of molecules from the environment. The arrangement of the proteins allows them to control the sequences of metabolic activities so that chemical reactions can be carried out very rapidly and accurately.

Upon close electron microscopic examination, it becomes apparent that there are two different types of ER—*rough* and *smooth*. The rough ER appears rough because it has ribosomes attached to its surface. Ribosomes are organelles that are nonmembranous organelles associated with the synthesis of proteins from amino acids. Therefore, cells with an extensive amount of rough ER are capable of synthesizing large quantities of proteins—for example, your pancreas cells. Smooth ER lacks attached ribosomes but is the site of many other important cellular chemical activities including fat metabolism and detoxification reactions—for example, your liver cells.

In addition, the spaces between the folded membranes may serve as canals for the movement of molecules within the cell. Some researchers suggest that this system of membranes allows for rapid distribution of molecules within a cell (figure 4.10).

Figure 4.10
Membranous Cytoplasmic Organelles.

Certain structures in the cytoplasm are constructed of membranes. Membranes are composed of protein and phospholipids. The four structures here, ER, Golgi apparatus, vacuoles, and lysosomes, are constructed of simple membranes.

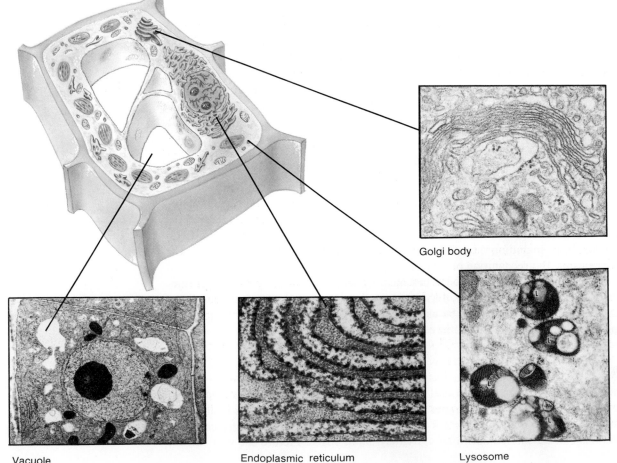

Golgi body

Vacuole

Endoplasmic reticulum

Lysosome

Golgi Apparatus

Another organelle composed of membrane is the **Golgi apparatus.** Even though this organelle is also composed of membrane, the way in which it is structured enables it to perform jobs different from the ER. The typical Golgi is composed of from five to seven flattened, smooth membranous sacs, which resemble a stack of pancakes. The Golgi apparatus is the site of the synthesis and packaging of certain molecules produced in the cell. It is also the place where particular chemicals are concentrated prior to their release from the cell. Some Golgi vesicles are used to transport molecules such as mucous, insulin, and digestive enzymes to the outside of the cell. The molecules are concentrated inside the Golgi and tiny vesicles are budded off the outside surface of the Golgi sacs. These vesicles contain high concentrations of molecules. The vesicles move to the inside of the cell membrane and merge with it. In so doing, the contents are placed outside of the cell membrane.

Another important group of molecules that is necessary to the cell is hydrolytic enzymes. This group of enzymes is capable of destroying proteins and lipids. Since cells contain a great deal of proteins and lipids as membranes and other structures, these enzymes must be controlled in order to prevent the destruction of the cell. The Golgi is the site where these enzymes are converted from their inactive to their active forms and packaged in membranous sacs. These vesicles are budded off from the outside surfaces of the Golgi sacs and given the special name **lysosomes.** The lysosomes are used by cells in four major ways. (1) When a cell is damaged, the membranes of the lysosomes break and the enzymes are released. These enzymes then begin to break down the contents of the damaged cell so that the component parts can be used by surrounding cells. (2) Lysosomes also play a part in the normal development of an organism. For example, as a tadpole slowly changes into a frog, the cells of the tail are destroyed by the action of lysosomes. In humans, while developing, the embryo has paddle-shaped hands. At a prescribed point in the development of the hand, the cells between the bones of the fingers release the enzymes that had been stored in the lysosomes. As these cells begin to disintegrate, the hand with individual fingers begins to take shape. Occasionally this process does not take place and infants are born with "webbed fingers." This developmental defect called syndactylism is surgically corrected soon after birth. (3) In many kinds of cells, the lysosomes are known to combine with food vacuoles. When this occurs, the enzymes of the lysosome break down the food particles into smaller and smaller molecular units. This process is common in one-celled organisms such as *Paramecium*. (4) Lysosomes are also used in the destruction of engulfed, disease-causing microorganisms such as bacteria, viruses, and fungi. As these invaders are taken into the cell by phagocytosis, lysosomes fuse with the phagocytic vacuole. When this occurs, the hydrolytic enzymes of the lysosome move into the vacuole to destroy the microorganisms.

Cells contain many kinds of vacuoles and vesicles. In many cases they are described by their function. Thus, food vacuoles hold food and water vacuoles store water. Specialized water vacuoles called contractile vacuoles are able to forcefully expel excess water that has accumulated in the cytoplasm as a result of osmosis. The contractile vacuole is a necessary organelle in cells that live in fresh water. The water constantly diffuses into the cell and must be actively pumped out. The special containers that hold the contents resulting from pinocytosis are called pinocytic vesicles. In all cases, these simple containers are constructed of a surrounding membrane. In most plants, there is one huge, centrally located vacuole in which water, food, wastes, and minerals are stored.

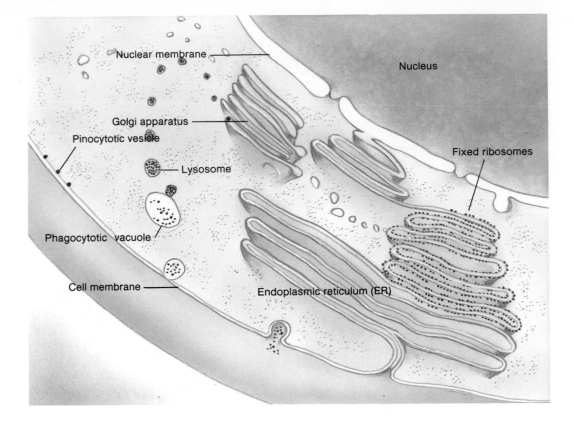

Nuclear membrane

Nucleus

Golgi apparatus

Pinocytotic vesicle

Fixed ribosomes

Lysosome

Phagocytotic vacuole

Cell membrane

Endoplasmic reticulum (ER)

Figure 4.11
Interconversion of Membranous Organelles.

Eukaryotic cells contain a variety of organelles composed of phospholipid and protein. Each has a unique shape and function. Many of these organelles are interconverted from one to another as they perform their essential functions. Cell membranes can become vacuolar membrane or endoplasmic reticulum, which can become vesicular membrane that in turn can become Golgi or nuclear membrane. However, mitochondria and chloroplasts cannot exchange membrane parts.

Nuclear Membrane

The nucleus is a place in the cell, not a solid mass. Just as a room is a place created by the walls, floor, and ceiling, the nucleus is a place in a cell created by the **nuclear membrane.** This membrane separates the nucleoplasm, liquid material in the nucleus, from the cytoplasm. If the membrane was not formed around the genetic material, the organelle we call the nucleus would not exist. Nuclear membrane is formed from many flattened sacs fashioned into a hollow sphere around the genetic material, DNA. The nuclear membrane has large openings in it that allow relatively large molecules to pass.

Energy Converters

All of the membranous organelles described above are able to be interconverted from one form to another (figure 4.11). For example, phagocytosis results in the formation of vacuolar membrane from cell membrane that fuses with lysosomal membrane, which in turn came from Golgi membrane. However, there are two other organelles composed of membranes that are chemically different and incapable of interconversion. The two types of organelles are associated with energy conversion reactions of the cell (figure 4.12).

These organelles are the mitochondrion and the chloroplast. The **mitochondrion** is an organelle resembling a small bag with a larger bag inside that is folded back on itself. These inner folded surfaces are known as the **cristae.** Located on the surface of the cristae are particular proteins and enzymes involved in aerobic cellular respiration. **Aerobic cellular respiration** is the series of reactions involved in the release of usable energy from food molecules by combining them with oxygen molecules. Enzymes are arranged in a sequence on the mitochondrial membrane that speed the breakdown of food molecules. The average human cell contains upwards of ten thousand mitochondria and when properly stained, they can be seen with the compound light microscope. When cells are functioning aerobically, the mitochondria swell with activity, but when this activity diminishes, they shrink and appear as threadlike structures.

A second energy-converting organelle is the **chloroplast.** This membranous sac-like organelle contains the green pigment **chlorophyll.** Some cells contain only one large chloroplast, while others contain hundreds of smaller chloroplasts. It is in this organelle that light energy is converted to chemical-bond energy in the process known

Figure 4.12
Energy-Converting Organelles.
The mitochondria with their inner folds called cristae are the site of aerobic cellular respiration, where food energy is converted to usable cellular energy. The chloroplast, the container of the pigment chlorophyll, is the site of photosynthesis. The chlorophyll, located in the grana, captures light energy that is used to construct organic sugarlike molecules in the stroma. Both of these organelles are constructed of protein and phospholipid membranes.

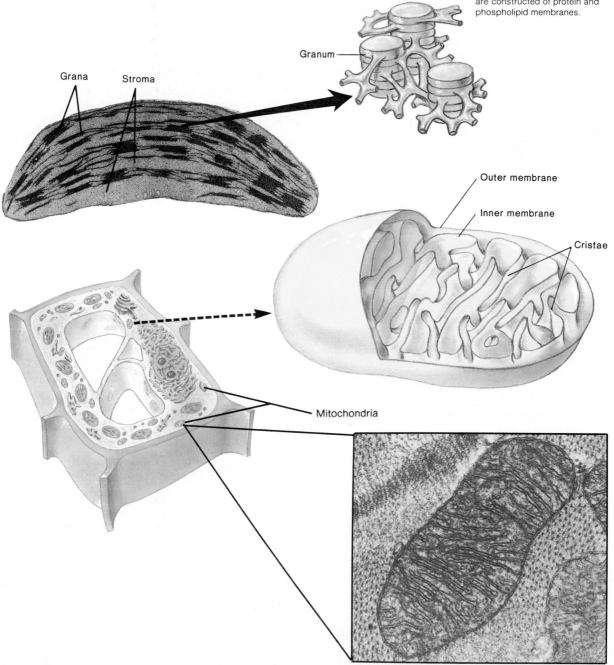

Granum

Grana   Stroma

Outer membrane

Inner membrane

Cristae

Mitochondria

as photosynthesis. Chemical-bond energy is found in food molecules. A study of the ultrastructure—that is, the structures differentiated only by use of the electron microscope—of the chloroplasts shows that the entire organelle is enclosed by a membrane, while other membranes are folded and interwoven throughout. In some areas, concentrations of these membranes are stacked up or folded back on themselves. Chlorophyll molecules are attached to these membranes. These areas of concentrated chlorophyll are called the **grana** of the chloroplast. The space between the grana, which has no chlorophyll, is known as the **stroma** (figure 4.12).

Mitochondria and chloroplasts are different from other kinds of membranous structures in several ways. First, their membranes are chemically different from other membranous organelles; second, they are composed of double layers of membrane—an inner and an outer membrane; third, both of these structures have ribosomes and DNA that are similar to those of bacteria; and fourth, these two structures have a certain degree of independence from the rest of the cell—they have a limited ability to reproduce themselves but must rely on nuclear DNA for assistance.

All of the organelles just described are composed of membranes. Many of these membranes are modified for particular functions. Each membrane is composed of the double phospholipid layer with protein molecules associated with it. These membranous organelles are suspended in a colloid cytoplasm. The cytoplasm is a mixture of a wide variety of molecules including proteins, lipids, and carbohydrates; however, the most abundant material is water. This colloidal suspension converts from the sol to the gel state. Sometimes cytoplasm flows like a thin solution, while at other times it is more firm and jelly-like.

Nonmembranous Organelles

Suspended in the cytoplasm and associated with the membranous organelles are a variety of different kinds of structures that are not composed of phospholipids and proteins arranged in sheets.

Ribosomes

In the cytoplasm are many, very small structures called **ribosomes** that are composed of protein and ribonucleic acid (RNA). Ribosomes function in the manufacture of protein. Each of these ribosomes is composed of two, oddly-shaped subunits, a large one and a small one. The larger of the two subunits is composed of a specific type of RNA associated with several kinds of protein molecules. The smaller is composed of RNA with fewer protein molecules than the large one. These globular organelles are involved in the assembly of proteins from amino acids—they are frequently associated with endoplasmic reticulum as rough ER. Areas of rough ER have been demonstrated to be sites of active production of proteins. Cells actively producing nonprotein materials, such as lipids, are likely to show smooth ER rather than rough ER. Many ribosomes are also found floating freely in the cytoplasm (figure 4.13) wherever proteins are being assembled. Cells that are actively producing protein have great numbers of free and attached ribosomes. The details of how ribosomes function in protein synthesis will be covered in chapter 8.

Microtubules

Another type of nonmembranous organelle, the **microtubules,** consist of small, hollow tubes composed of protein called tubulin. They function throughout the cytoplasm where they provide structural support and enable movement. The microtubules are

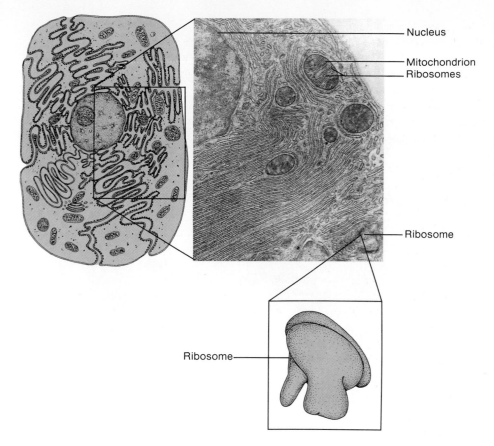

Nucleus

Mitochondrion
Ribosomes

Ribosome

Ribosome

Figure 4.13
Ribosomes.
Each ribosome is constructed of
two subunits of protein and
ribonucleic acid. These globular
organelles are associated with the
construction of protein molecules
from individual amino acids. They
are sometimes located individually
in the cytoplasm where protein is
being assembled or congregated
on the ER. They are so obvious on
the ER using electron micrograph
techniques that when they are
there, we label the ER as rough
ER.

dynamic structures that are capable of being lengthened or shortened by the addition or subtraction of tubulin units (figure 4.14). As a result, there seems to be a constant shifting of microtubular material within a cell. Many of the structures with which microtubules are associated are able to move or grow. Microtubules participate in the movement of chromosomes during nuclear division, in the movement of flagella and cilia, and in the positioning of cellulose molecules during cell wall synthesis. Important structures composed of microtubules are the centrioles.

Centrioles

An arrangement of two sets of microtubules at right angles to each other makes up a structure known as the **centriole.** Each set is composed of nine groups of short microtubules arranged in a cylinder. The other set, likewise composed of nine groups of microtubules, is at a right angle to the first set. The two groups together make up a centriole (figure 4.15). The centriole functions in cell division and will be referred to again in chapter 9. One curious fact about centrioles is that they are present in most animal cells but not in many types of plant cells.

Cilia and Flagella

Many cells have microscopic, hairlike structures projecting from the cell surface, these are **cilia** or **flagella** (figure 4.16). In general, we call them flagella if they are long and few in number, and cilia if they are short and more numerous. They are similar in structure and each functions to move the cell through its environment or move the environment past the cell. They are constructed of a cylinder of nine sets

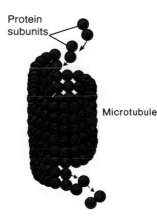

Protein
subunits

Microtubule

Figure 4.14
Microtubules.
Microtubules are hollow tubes
constructed of proteins. The
dynamic nature of the microtubule
is useful in the construction of
certain organelles in a cell such as
centrioles and cilia or flagella.

Figure 4.15
Centriole.
These two sets of short
microtubules are located just
outside the nuclear membrane in
many types of cells. The
micrograph shows an end view of
one of the sets. Magnification is
about 160,000 times.

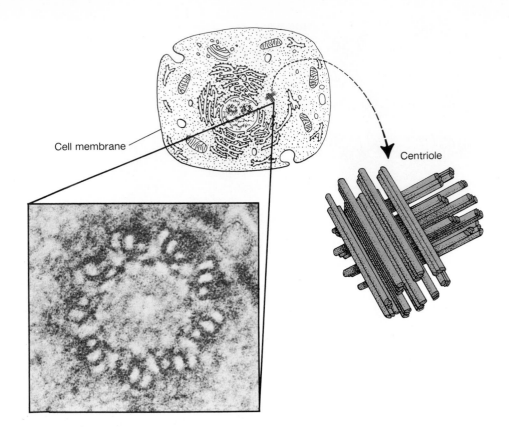

Cell membrane

Centriole

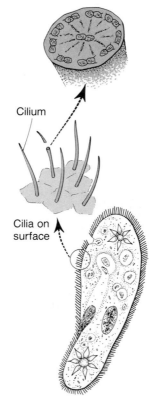

Cilium

Cilia on
surface

Figure 4.16
Cilia and Flagella.
These two structures function like
oars moving the cell through its
environment or moving the
environment past the cell. Cilia and
flagella are constructed of groups
of microtubules. Flagella are longer
than cilia.

of microtubules similar to the centriole, but have an additional two microtubules in the center. These long strands of microtubules project from the cell surface and are covered by cell membrane. When cilia and flagella are sliced crosswise, their cut ends show what is referred to as the 9 + 2 arrangement of microtubules. The cell has the ability to control the action of these microtubular structures, enabling them to be moved in a variety of different ways. Their coordinated actions either propel the cell through the environment or the environment past the cell surface. The protozoan *Paramecium* is covered with thousands of cilia that actively beat a rhythmic motion to move the cell through the water. The cilia on the cells that line your trachea move mucous-containing particles from deep within your lungs.

Inclusions

Inclusions are collections of materials that do not have as well-defined a structure as the organelles we have discussed so far. We lump these structures together as miscellaneous material or **granules**—starch granules and oil droplets, for example. In the past, cell structures such as ribosomes, mitochondria, and chloroplasts were called granules, since their structure and function were not clearly known. As scientists learn more about other unidentified particles in the cells, they will be named and more fully described.

Nuclear Components

As stated at the beginning of this chapter, one of the first structures to be identified in cells was the nucleus. The nucleus was referred to as the cell center. If the nucleus is removed from a cell, it can only live a short time. For example, human red blood corpuscles begin life in bone marrow where they have a nucleus. Before they are

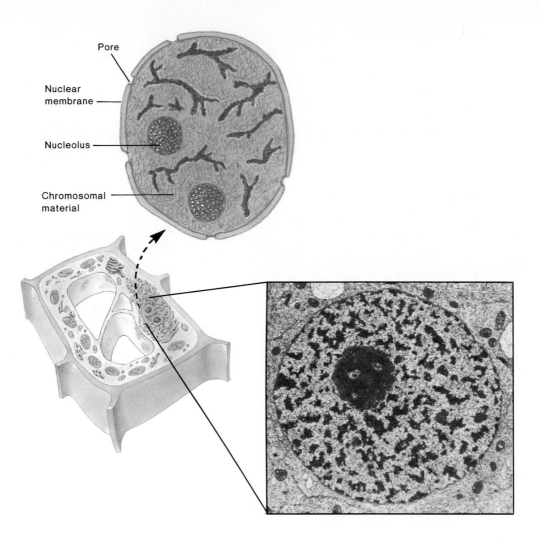

Pore

Nuclear membrane

Nucleolus

Chromosomal material

released into the blood stream to serve as oxygen and carbon dioxide carriers, they lose their nucleus. As a consequence, red blood corpuscles are only able to function for about 120 days before they disintegrate.

When nuclear structures were first identified, it was noted that certain dyes stained some parts more than others. The parts that stained more heavily were called **chromatin,** which means colored material. Chromatin is composed of long molecules of deoxyribonucleic acid (DNA) in association with proteins. These DNA molecules contain the genetic information for the cell. Chromatin is loosely organized DNA in the nucleus. When the chromatin is tightly coiled into shorter, denser structures, we call them **chromosomes** (chromo = color, some = body). Chromatin and chromosomes are really the same molecules but differ in structural arrangement. They contain the blueprints for the construction and maintenance of the cell. In addition to chromosomes, the nucleus may also contain one or more nucleoli. A **nucleolus** is the site of ribosome manufacture. Nucleoli appear to be composed of specific parts of chromosomes that contain the information for the construction of ribosomes. These regions, plus the completed or partially completed ribosomes, are called nucleoli.

The final component of the nucleus is its liquid matrix called the **nucleoplasm.** It is a mixture composed of water and the molecules used in the construction of ribosomes, nucleic acids, and other nuclear material (figure 4.17).

Figure 4.17
Nucleus.

The nucleus, one of the two major regions of protoplasm, has its own complex structure. It is bound by a double membrane that separates it from the cytoplasm. Inside the nucleus are the nucleoli, chromosomes or chromatin material composed of DNA, and the liquid matrix (nucleoplasm). Magnification is about 20,000 times.

Major Cell Types

Not all of the cellular organelles we have just described are located in every cell (figure 4.18). Some cells typically have combinations of organelles that differ from others. For example, some cells have nuclear membrane, mitochondria, chloroplasts, ER, and Golgi, while others have mitochondria, centrioles, Golgi, ER, and nuclear membrane. Other cells are even more simple and lack most of the complex membranous organelles described in this chapter. Because of this fact, biologists have been able to classify cells into two major types, prokaryotic and eukaryotic.

Figure 4.18 a–e
Comparison of Cell Types.
The five types of cells illustrated here indicate the major patterns of construction found in all living things. Note the similarities of all five and the subtle differences among them.

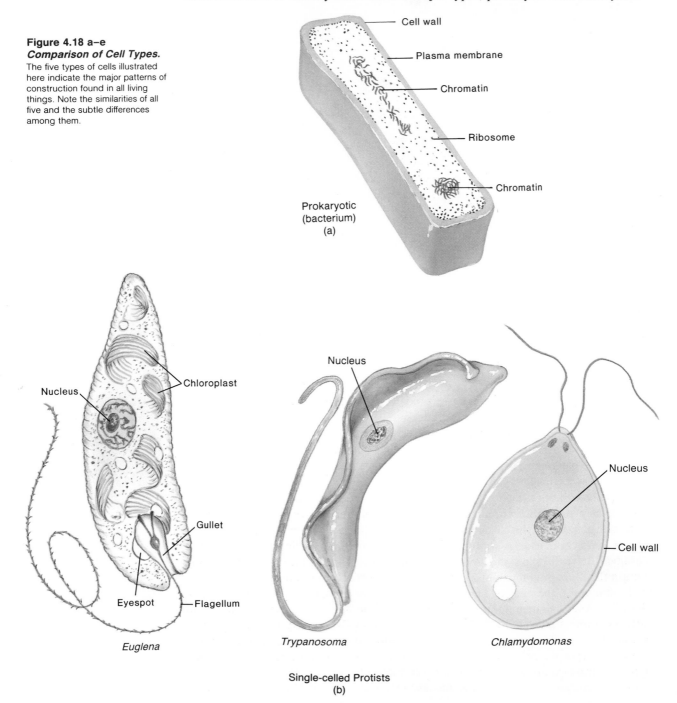

Cell wall

Plasma membrane

Chromatin

Ribosome

Chromatin

Prokaryotic
(bacterium)
(a)

Chloroplast

Nucleus

Gullet

Eyespot

Flagellum

Euglena

Nucleus

Trypanosoma

Nucleus

Cell wall

Chlamydomonas

Single-celled Protists
(b)

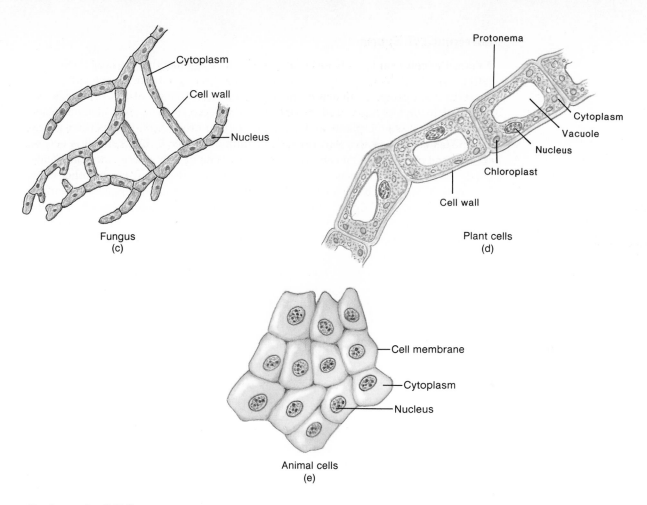

Fungus
(c)

Plant cells
(d)

Animal cells
(e)

Prokaryotic Cell Structure

Prokaryotic cells do not have a typical nucleus bound by a nuclear membrane, nor do they contain mitochondria, chloroplasts, Golgi, or extensive networks of ER. However, prokaryotic cells contain DNA and enzymes and are able to reproduce and engage in metabolism. They perform all of the basic functions of living things with fewer and more simple organelles. Examples of prokaryotic cells include bacteria that cause the diseases tuberculosis, strep throat, gonorrhea, and acne. Other prokaryotic cells are responsible for the breakdown of organic molecules and some are photosynthetic.

One significant difference between prokaryotic and eukaryotic cells is the chemical makeup of their ribosomes. The ribosomes of prokaryotic cells contain different proteins than those found in eukaryotic cells. Prokaryotic ribosomes are also smaller in size. This discovery was important to medicine because many cells that cause common diseases are prokaryotic (bacteria). As soon as differences in the ribosomes were noted, researchers began to look for ways in which to interfere with prokaryotic ribosome's function, but not interfere with the ribosomes of eukaryotic cells. **Antibiotics** such as streptomycin are the result of this research. This drug combines with prokaryotic ribosomes and causes the death of the prokaryote by preventing the production of proteins essential to its survival. Since eukaryotic ribosomes differ from prokaryotic ribosomes, streptomycin does not interfere with the normal function of ribosomes in human cells.

Eukaryotic Cell Structure

Eukaryotic cells contain a true nucleus and most of the membranous organelles described earlier. Eukaryotic organisms can be further divided into several categories based on the specific combination of organelles they contain. The cells of plants, fungi, protozoa and algae, and animals are all eukaryotic. The most obvious characteristic that sets the plants and algae apart from other organisms is their green color. This color indicates that the cells contain chlorophyll. Chlorophyll is necessary for the process of photosynthesis, the conversion of light energy into chemical-bond energy in food molecules. These cells, then, are different from the other cells in that they contain chloroplasts in the cytoplasm. Another distinguishing characteristic of plants and algae is the presence of cellulose in their cell walls. (While some prokaryotic cells have a type of green photosynthetic pigment and carry on photosynthesis, they do so without chloroplasts and use different chemical reactions.)

The group of organisms that have a cell wall but lack chlorophyll in chloroplasts is collectively known as fungi. They were previously thought to be plants that had lost their ability to make their own food, or animals that had developed cell walls. Organisms that belong in this category of eukaryotic cells include yeasts, molds, mushrooms, and fungi that cause human diseases such as athlete's foot, jungle rot, and ringworm. Now we have come to recognize this group as different enough from plants and animals to place them in a separate kingdom.

Eukaryotic organisms that lack cell walls and cannot photosynthesize are placed in separate groups. Organisms that consist of only one cell are called protozoans—examples are *Amoeba* and *Paramecium*. They have all the cellular organelles described in this chapter except the chloroplast; therefore, protozoans must consume food as do the fungi and the multicellular animals.

While the differences in these groups of organisms may seem to set them worlds apart, their similarity in cellular structure is one of the central themes that unifies the field of biology. One can obtain a better understanding of how cells operate by studying specific examples. Since the organelles have the same general structure and function regardless of the kind of cells in which they are found, we can learn more about how mitochondria function in plants by studying how mitochondria function in animals. There is a commonality among all living things with regard to their cellular structure and function. Table 4.1 compares the structure and function of cellular organelles.

Summary

The concept of the cell has developed over a number of years. It passed through a stage where only two regions, the cytoplasm and the nucleus, could be identified. At present, numerous organelles are recognized as essential components of both prokaryotic and eukaryotic cell types. The structure and function of some of these organelles have been compared in table 4.1. This table also indicates whether the organelle is unique to prokaryotic or eukaryotic cells or if it is found in both.

The cell is the common unit of life. We study individual cells and their structure to understand how they function as individual living organisms and as parts of many-celled beings. Knowing how prokaryotic and eukaryotic cell types resemble or differ from each other helps physicians control some organisms dangerous to humans.

Table 4.1   *Comparison of the Structure and Function of the Cellular Organelles*

Organelle	Type of Cell	Structure	Function
Plasma membrane	Prokaryotic Eukaryotic	Typical membrane structure, phospholipid and protein present	Controls passage of some materials to and from the environment of the cell
Granules	Prokaryotic Eukaryotic	Too small to ascribe a specific structure	May have a variety of functions
Chromatin material	Prokaryotic Eukaryotic	Composed of DNA and protein in eukaryotes, but just DNA in prokaryotes	Contains the hereditary information that the cell uses in its day-to-day life and passes on to the next generation of cells
Ribosome	Prokaryotic Eukaryotic	Protein and RNA structure	Site of protein synthesis
Microtubules	Eukaryotic	Hollow tubes composed of protein	Provide structural support and allow for movement
Nuclear membrane	Eukaryotic	Typical membrane structure	Separates the nucleus from the cytoplasm
Nucleolus	Eukaryotic	Group of RNA molecules and genes located in the nucleus	Site of ribosome manufacture and storage
Endoplasmic reticulum	Eukaryotic	Folds of membrane forming sheets and canals	Surface for chemical reactions and intracellular transport system
Golgi apparatus	Eukaryotic	Membranous stacks	Associated with the production of secretions and enzyme activation
Vacuoles	Eukaryotic	Membranous sacs	Containers of materials
Lysosome	Eukaryotic	Membranous container	Isolates very strong enzymes from the rest of the cell
Mitochondria	Eukaryotic	Large membrane folded inside of a smaller membrane	Associated with the release of energy from food; site of cellular respiration
Chloroplast	Eukaryotic (Plants only)	Double membranous container of chlorophyll	Site of photosynthesis or food production in green plants
Centriole	Eukaryotic (Animals only)	Microtubular	Associated with cell division
Contractile vacuole	Eukaryotic (Animals only)	Membranous container	Expels excess water
Cillia and flagella	Prokaryotic Eukaryotic	9+2 tubulin in eukaryotes; different structure in prokaryotes	Movement

Consider This

A primitive type of cell consists of a membrane and a few other cell organelles. This protobiont lives in a sea that contains three major kinds of molecules with the following characteristics:

X	Y	Z
Inorganic	Organic	Organic
High concentration outside of cell	High concentration inside of cell	High concentration inside of cell
Essential to life of cell	Essential to life of cell	Poisonous to the cell
Small and can pass through the membrane	Large and cannot pass through the membrane	Small and can pass through the membrane

With this information and your background in cell structure and function, osmosis, diffusion, and active transport, decide whether or not this protobiont will continue to live in this sea, and explain why or why not.

Experience This

To demonstrate the movement of materials into and out of the cells of living things, you do not necessarily need sophisticated equipment. Usually the equipment will give you quantitative data about how much material has moved.

Using a white potato, cut slices about three-eighths of an inch thick. If you have a balance, you can weigh the slices first—otherwise, try to get a feel of their weight and remember how heavy each slice feels. Take one slice and place it into a very salty solution. Place a second into a sugar water solution and a third in just water.

After ten minutes, remove each slice and again assess its weight, either by heft or using a balance. The differences in weight are due to changes in the water content of the individual cells of the potato. Can you explain why the changes occurred as they did? Protoplasm is mostly water. This fact helps to determine the relative concentrations of water inside and outside of the cells. Can you describe any other differences in these three slices?

Questions

1. Make a list of the membranous organelles of a eukaryotic cell and describe the function of each.
2. Describe how the concept of the cell has changed over the past two hundred years.
3. What three methods allow the exchange of molecules between cells and their surroundings?
4. How do diffusion, facilitated diffusion, osmosis, and active transport differ?
5. What are the differences between the cell wall and the cell membrane?
6. Diagram a cell and show where proteins, nucleic acids, carbohydrates, and lipids are located.
7. Make a list of the nonmembranous organelles of the cell and describe their functions.
8. Define the following terms: cytoplasm, stroma, grana, cristae, chromatin, and chromosome.
9. Why does putting salt on meat preserve it from spoilage by bacteria?
10. In what ways do mitochondria and chloroplasts resemble one another?

active transport (ak-tiv trans-port) A process of using a carrier molecule to move molecules across a cell membrane in a direction opposite of the concentration gradient. The carrier requires an input of energy other than the kinetic energy of the molecules.

aerobic cellular respiration (a''ro'bik sel'yu-lar res''pi-ra'shun) A series of reactions in the mitochondria involved in the release of usable energy from food molecules by combining them with oxygen molecules.

antibiotics (an-te-bi-ot'iks) Drugs that selectively interfere with the function of prokaryotic ribosomes and prevent them from producing the proteins essential for the cell's survival.

cell (sel) The basic structural unit that makes up all living things.

cell membrane (sel mem'brān) The outer boundary membrane of the cell also known as the plasma membrane.

cellular membranes (sel'u-lar mem'brāns) Thin sheets of material composed of phospholipids and proteins. Some of the proteins have attached carbohydrates or fats.

centriole (sen'tre-ōl) Two sets of nine short microtubules, each arranged in a cylinder.

chlorophyll (klo'ro-fil) The green pigment located in the chloroplasts of plant cells associated with trapping light energy.

chloroplast (klo'ro-plast) An energy-converting, membranous, saclike organelle in plant cells containing the green pigment chlorophyll.

chromatin (kro'mah-tin) Areas or structures within the nucleus of a cell composed of long molecules of deoxyribonucleic acid (DNA) in association with proteins.

chromosomes (kro-mo-sōmz) Structures visible in the nucleus that consist of DNA and protein.

cilia (sil'e-ah) Numerous short, hairlike structures projecting from the cell surface that enable locomotion.

concentration gradient (kon''sen-tra'shun gra''de-ent) A gradual change in the number of molecules over distance.

cristae (kris'te) Folded surfaces of the inner membranes of mitochondria.

cytoplasm (si''to-plazm) The more fluid portion of the protoplasm that surrounds the nucleus.

differentially permeable (dif''fer-ent'shul-le per'me-uh-bul) The property of a membrane that will allow certain molecules to pass through it but will interfere with the passage of others.

diffusion (di''fiu'zhun) Net movement of a kind of molecule from an area of higher concentration to an area of lesser concentration.

diffusion gradient (di''fiu'zhun gra''de-ent) A difference in the concentration of diffusing molecules across an area.

endoplasmic reticulum (ER) (en''do-plaz'mik re-tik'yu-lum) Folded membranes and tubes throughout the eukaryotic cell that provide a large surface upon which chemical activities take place.

eukaryotic cells (yu'ka-re-ah''tik sels) One of the two major types of cells; characterized by cells that have a true nucleus as in plants, fungi, protists, and animals.

facilitated diffusion (fah-sil'ī-ta''ted di-fiu'zhun) Diffusion assisted by carrier molecules.

flagella (flah-jel'luh) Long, hairlike structures projecting from the cell surface that enable locomotion.

Golgi apparatus (gol'je ap''pah-rat'us) A stack of flattened, smooth membranous sacs; the site of synthesis and packaging of certain molecules in eukaryotic cells.

grana (gra'nuh) Areas of the chloroplast membrane where chlorophyll molecules are concentrated.

granules (gran'yūls) Materials whose structure is not defined as well as other organelles.

inclusions (in-klu′zhuns) A general term referring to materials inside a cell that are usually not readily identifiable; stored materials.

lysosome (li′so-sōmz) A specialized organelle that holds a mixture of hydrolytic enzymes.

microscope (mi′kro-skōp) An instrument that will produce an enlarged image of a small object.

microtubules (mi′kro-tūb″yūls) Small, hollow tubes of protein that function throughout the cytoplasm to provide structural support and enable movement.

mitochondrion (mi-to-kon′dre-on) A membranous organelle resembling a small bag with a larger bag inside that is folded back on itself; serves as the site of aerobic cellular respiration.

net movement (net muv′ment) Movement in one direction minus the movement in the other.

nuclear membrane (nu′kle-ar mem′brān) Structure surrounding the nucleus that separates the nucleoplasm from the cytoplasm.

nucleoli (singular, nucleolus) (nu-kle′o-li) Nuclear structures composed of completed or partially completed ribosomes and the specific parts of chromosomes that contain the information for their construction.

nucleoplasm (nu′kle-o-plazm) Liquid matrix of the nucleus composed of a mixture of water and the molecules used in the construction of the rest of the nuclear structures.

nucleus (nu′kle-us) Central body that contains the information system for the cell.

organelles (or-gan-els′) Cellular structures that have particular functions they perform in the cell. The function of an organelle is directly related to its structure.

osmosis (os-mo′sis) Net movement of water molecules through a differentially permeable membrane.

phagocytosis (fã″jo-si-to′sis) The process by which the cell wraps around a particle and engulfs it.

pinocytosis (pi″no-si-to′sis) The process by which a cell engulfs some molecules dissolved in water.

plasma membrane (plaz′muh mem′brān) Outer boundary membrane of the cell also known as the cell membrane.

prokaryotic cells (pro′ka-re-ot″ik sels) One of the two major types of cells. They do not have a typical nucleus bound by a nuclear membrane and lack many of the other membranous cellular organelles: example, bacteria.

protoplasm (pro′to-plazm) The living portion of a cell as distinguished from the nonliving cell wall.

ribosomes (ri-bo-sōmz) Small structures composed of two protein and ribonucleic acid subunits involved in the assembly of proteins from amino acids.

stroma (stro-muh) Region within a chloroplast that has no chlorophyll.

vacuole (vak′yu-ōl) Storage container within the cytoplasm of a cell having a surrounding membrane.

vesicles (vĕ′sĭ-kuls) Storage container within the cytoplasm of a cell having a surrounding membrane.

5

Enzymes

◾ Chapter Outline

Catalysts and Enzymes
Mechanisms of Enzyme Action
Environmental Effects on Enzyme Action
Enzymes in Cells

◾ Purpose

Living cells require various chemical reactions to conduct their vital functions. To prevent the malfunction and death of the cell, these reactions must be conducted rapidly and be controlled. The problem is not starting reactions but controlling the rate of the reactions. Regulation of the rates of the many reactions in cells is the task of the enzymes.

For Your Information

As a result of genetic engineering and other advanced research into enzymes, a great variety of products containing enzymes are now available. Meat tenderizer contains enzymes that break down tough protein fibers. Soft contact lenses are cleansed of mucoprotein with enzyme solutions. Pet urine odor is removed from soiled carpets with enzymes. Stained cloths are made bright and white again with enzyme active presoaks and detergents. Hospitals use bacterial enzymes to remove dead skin from burn victims. These and many other enzyme-containing products are only a sample of what is to come.

Learning Objectives

- Differentiate between the collision theory and the transitional molecule theory of enzyme-controlled reactions.

- Explain how enzymes function using both the collision and transitional molecule theories.

- Relate the three-dimensional structure of an enzyme to its ability to catalyze a reaction.

- Describe the influences of environmental factors such as temperature, pH, and concentration on turnover number.

- Define the mechanisms of enzyme competition and inhibition.

Catalysts and Enzymes

All living things require a source of energy and building materials in order to grow and reproduce. These are derived from organic molecules and their chemical bonds. However, since most organic molecules are very stable and do not spontaneously break apart to release their energy or component parts, cells must have a mechanism for encouraging and directing chemical reactions that provide essential energy and building blocks. Most chemical reactions require an input of energy to get them started. This energy is used to make the molecule unstable and more likely to break apart.

Chemists use two different theories to explain how chemical reactions occur. One theory suggests that reactants collide with one another as a result of their natural kinetic energy. However, not all collisions result in changes in the chemical bonds of the reactants. Those collisions that do bring about changes in the chemical bonds are called effective collisions. It is not enough that the two molecules touch or bump into each other; they must both have the correct orientation to each other so that specific chemical bonds come to lie next to one another. You might compare this to fitting two pieces of a jigsaw puzzle together. You must orient the pieces so that the proper side of the one piece will fit into the appropriate part of the second. If they are oriented improperly, they will not fit together. In the collision theory, it is necessary for the two reactants to make enough contact with each other so that eventually some of the collisions will be effective. The energy needed to start a reaction is used to increase the probability of an effective collision. The higher the temperature, the faster the molecules move and the more frequently the reactants will contact each other with the proper orientation (figure 5.1).

Figure 5.1
Collision Theory of Reaction.

According to this theory, the two reactants must randomly collide often enough so that there is an increased probability of a collision that will get the two reacting surfaces together.

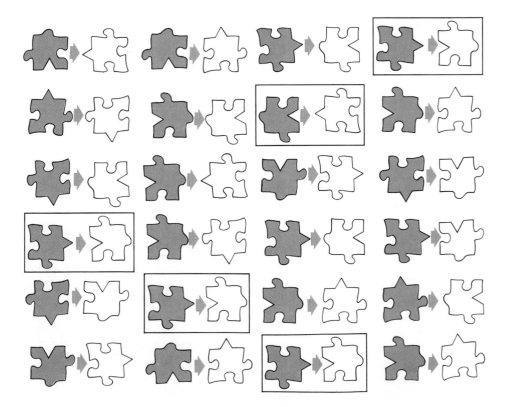

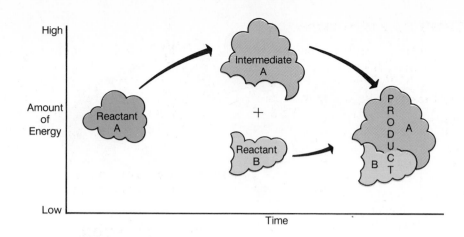

Figure 5.2
Intermediate Product Theory of Reaction.
One way of thinking about how a reaction takes place concerns the formation of an intermediate product. This intermediate product then changes into the final product. We think the intermediate product may require a great input of energy to become formed; therefore, energy must be fed into a reaction.

The second theory that attempts to explain how chemical reactions occur is the transitional molecule theory. This theory proposes that for a reaction to occur, chemical bonds must become altered so that they can be changed more easily. It is believed that the formation of a transitional molecule takes place in order to alter the stability of the chemical bonds that will eventually become involved in the chemical reaction. The transitional molecule is an unstable intermediate that is only temporary and is not the final product of the reaction. In order to form this transitional molecule, it may be necessary to add energy to the reaction (figure 5.2).

The energy required to get a reaction going is called the **activation energy.** This energy may be used to increase the number of effective collisions, or may be required to form the transitional molecule. Some reactions occur naturally at room temperature, since room temperature provides sufficient activation energy. Generally, increasing temperature will increase the rate of the reaction because more individual molecules will have the required activation energy. However, cells cannot tolerate high temperatures. Therefore, they need to have an alternate method of encouraging essential chemical reactions. The activation energy must be reduced.

The rate of a reaction can be increased without increasing the temperature by the addition of chemicals known as catalysts. A **catalyst** is a chemical that speeds up a reaction but is not used up in the reaction. It is able to be recovered unchanged at the end of the reaction. A cell manufactures specific proteins that act as catalysts. A protein molecule that acts as a catalyst to change the rate of a reaction is called an **enzyme.** The genes of the cell contain the information needed to produce specific proteins that can act as enzymes. How the genetic information is used to direct the manufacture of these specific proteins is discussed in chapter 8.

Mechanisms of Enzyme Action

Remember from chapter 3 that protein, composed of a specific sequence of amino acids, will fold and twist to form a specific three-dimensional molecule. The specific shape of the protein is important to enable it to have an effective collision with a reactant. Each enzyme has a very specific three-dimensional shape and is very specific to the kind of reactant with which it can combine. The enzyme must physically fit with the reactant. The molecule to which the enzyme attaches itself is known as the **substrate.** When the enzyme attaches itself to the substrate molecule, a new, temporary molecule is formed known as the **enzyme-substrate complex.** According

Figure 5.3
Alternate Transitional Molecule.
When a reaction occurs in the presence of an enzyme, the amount of energy required to cause the reaction is decreased. One way of explaining how this happens is to theorize a different intermediate product, one that requires less energy to produce. This alternate molecule could only be made in the presence of the enzyme. After this molecule is formed, the reaction progresses to the final product and the enzyme, released from the alternate intermediate, is free to continue catalyzing other reactions.

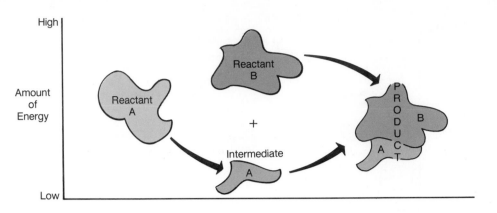

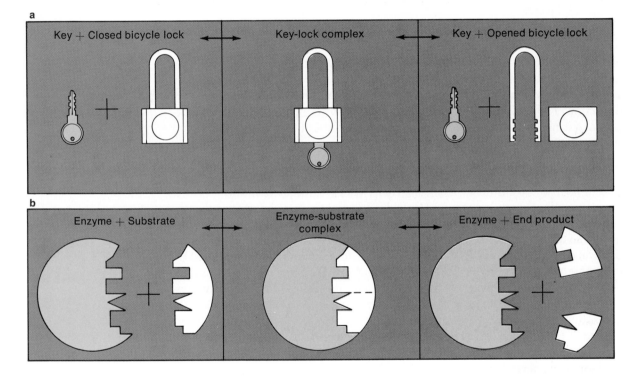

Figure 5.4
Specificity of an Enzyme.
A key will only work with a specific lock because its notches only fit certain arrangements in the tumblers of the lock. The enzyme and the substrate must fit together as a key fits a lock to enable them to operate together. Notice that the key is unchanged after the lock is opened, just as the enzyme is unchanged after the reaction has occurred.

to the effective collision theory, this complex is a temporary molecule that encourages effective collisions. The presence of the enzyme increases the rate of the reaction because more of the random collisions are effective collisions. According to the transitional molecule theory, the enzyme-substrate complex is a transitional molecule (figure 5.3). When the substrate is combined with the enzyme, its bonds are less stable and more likely to be altered and form new bonds. The enzyme is specific because it has a particular surface that can only combine with specific parts of certain substrate molecules. You might think of a key with a specific set of notches that will only work with specific locks. In order for the key to change the tumblers in the lock, it must first be able to physically get close to them. The key must form a key-lock complex. The enzyme must also physically attach itself to the substrate; therefore, it has a specific **binding site** or **attachment site** on the enzyme surface. Figure 5.4 illustrates the specificity of both keys and enzymes. Note that the key in (a) is recovered unchanged after it opens (changes) the lock. In (b) the enzyme is

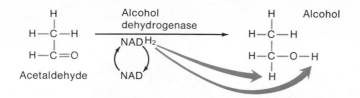

Figure 5.5
Role of Coenzyme.
NADH$_2$ is a coenzyme that works with the enzyme alcohol dehydrogenase. The coenzyme provides the hydrogens necessary for the formation of alcohol. The NAD can then obtain hydrogen from the environment. The presence of the coenzyme makes the enzyme function more efficiently.

specific in that it has a particular surface geometry that matches the geometry of the substrate. (Some people feel that the active site is somewhat flexible and bends or folds to fit the specific structure of the substrate.) Note also that after the enzyme-substrate complex is formed and the reaction (change) takes place, the enzyme is recovered from the reaction unchanged. This means that the enzyme can be used over again. Eventually enzymes wear out and need to be replaced, but generally very small quantities of enzymes are necessary because they are able to be reused.

You would not expect the key that unlocks your car door to also unlock your house. The enzyme amylase, which breaks down starch to glucose, cannot change cellulose to glucose. However, your car key may be used to unlock several similar locks on your car. The same is true of enzymes. A specific enzyme found in the stomach, pepsin, is able to break down many different kinds of similar proteins. The enzyme is specific to the kind of substrate to which it can become attached. It is also specific to the kind of change that will occur to the substrate. The point on the enzyme that causes a specific part of the substrate to change is called the **active site** of the enzyme. This is the point where electrons are shifted to change the bonds that hold the molecule together. The active site may enable a positively charged surface to combine with the negative portion of a reactant. Because the enzyme is specific to both the substrate to which it can attach and the reaction that it can encourage, a unique name can be given to each enzyme. The first part of an enzyme's name is the name of the molecule to which it can become attached. The second part of the name indicates what type of reaction it facilitates. The third part of the name is "ase," which is the ending clue that tells you it is an enzyme.

The enzyme responsible for the dehydration synthesis reactions between several glucose molecules to form glycogen is known as glycogen synthetase. The enzyme responsible for breaking the bond that attaches the amino group to the amino acid arginine is known as arginine aminase. Quite frequently when an enzyme is very common, we begin to shorten its formal name. The salivary enzyme involved in the digestion of starch should be amylose (starch) hydrolase but is generally known as amylase.

Certain enzymes need an additional molecule to enable them to function. This additional molecule is not a protein, but works with the protein to speed up a reaction. This enabling molecule is termed a **coenzyme.** The coenzyme aids the reaction by removing one of the end products or by bringing in part of the substrate. Coenzymes are frequently vitamins. You know that a constant small supply of vitamins is necessary for good health. The reason your cells require vitamins is to serve this coenzyme function. The coenzyme can work with a variety of enzymes; therefore, you need extremely small quantities of vitamins. An example of enzyme-coenzyme cooperation is shown in figure 5.5. Alcoholic fermentation is a series of reactions resulting in the conversion of glucose to ethyl alcohol. The NADH$_2$ acts as a coenzyme, providing two hydrogens for the reaction. The enzyme acetaldehyde

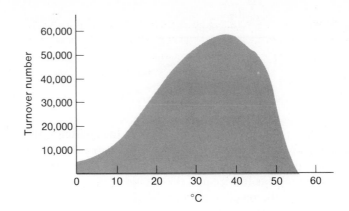

hydrogenase uses the two hydrogens and attaches them to the acetaldehyde. Later the NAD (*n*icotinamide *a*denine *d*inucleotide) will acquire two more hydrogens so that it is essentially unchanged in the reaction. You can see that the presence of the coenzyme is necessary for the enzyme to function properly.

An enzyme will catalyze a reaction going in either direction. It is unable to determine whether the reactants are forming products or if the products are being changed into their original reactants. Therefore, a cell must have some way to control the direction of a reaction. This is determined by the relative amounts of reactants and products. The cell generally uses the product just manufactured for some other job. Therefore, the product doesn't accumulate and is no longer available to combine with the enzyme. Because of this, the reaction goes in one direction only.

Environmental Effects on Enzyme Action

An enzyme forms a complex with one substrate molecule, encourages the reaction to occur, detaches itself, and then forms a complex with another substrate molecule. The number of molecules of substrate that a single molecule of enzyme can react with in a given time under ideal conditions is called the **turnover number.** Sometimes this number seems incredibly large—up to fifty thousand per minute! Without the enzyme, perhaps only fifty or one hundred substrate molecules might be altered in the same time. With this in mind, let's identify the ideal conditions for an enzyme and how these conditions influence the turnover number.

One important environmental condition that affects enzyme-controlled reactions is temperature (figure 5.6). Temperature has two effects on enzymes; it can change the rate of molecular motion and can cause changes in the shape of the enzyme. As the temperature of a system increases, the number of effective collisions between enzymes and substrates increases. This occurs because the increased random movement of the molecules results in greater numbers of collisions, leading to a larger number of effective collisions. You would expect increasing amounts of product to form. This is true up to a point. The temperature at which the rate of enzyme-substrate complex formation is fastest is termed the optimum temperature. Optimum means the best or most productive quantity or condition. In this case, the optimum temperature is the temperature at which the product is formed most rapidly.

As one lowers the temperature below the optimum, molecular motion slows and the rate at which the enzyme-substrate complexes form decreases. Even though the enzyme is still able to operate, it does so very slowly. Therefore, it is possible to preserve foods for long periods by storing them in freezers or refrigerators.

What happens when the temperature is raised above the optimum? In this case, some of the molecules of enzymes are changed in such a way that they can no longer form the enzyme-substrate complex so that the reaction will not be encouraged. If the temperature continues to increase, more and more of the enzyme molecules will become inactive. When heat is applied to the enzyme, it causes permanent changes in the three-dimensional shape of the molecule. The surface geometry of the enzyme will not be recovered even when the temperature is reduced. We can again use the key analogy. When a key is heated above a certain temperature, the metal begins to flow. The shape of the key is changed permanently, and even if the temperature is reduced, the surface geometry of the key is irrevocably lost. When this happens to an enzyme, we say that it has been denatured. A **denatured** enzyme is one whose protein structure has been permanently changed so that it has lost its biological activity. Although egg white is not an enzyme, it is a protein and a good example of what happens when denaturation occurs as a result of heating. As heat is applied to the egg white, it is permanently changed (denatured).

Another environmental condition that influences enzyme action is pH. The three-dimensional structure of the protein leaves certain side chains exposed. These side chains may attract ions from the environment. Under the right conditions, a group of positively charged hydrogen ions may accumulate on certain parts of an enzyme. An environment devoid of these hydrogen ions would not allow this to happen. Thus, variation in the most effective shape of the enzyme could be caused by a change in the number of hydrogen ions present in the solution. Because the environmental pH is so important in determining the shape of protein molecules, there is an optimum pH for each specific enzyme. The enzyme will fit with the substrate only when it is at the proper pH. Many enzymes function best at a pH close to neutral (7.0). However, a number of enzymes perform best at pHs quite different from seven. Pepsin, an enzyme found in the stomach, works well at an acid pH of 1.5 to 2.2, while arginase, an enzyme in the liver, works well at a basic pH of 9.5 to 9.9.

In addition to temperature and pH, the concentration of enzymes, substrates, and products influence the rates of enzymatic reactions. Although the enzyme and the substrate are in contact with one another for only a short period of time, it is possible to have a situation in which all of the enzymes are always occupied by substrate molecules. When this occurs, the rate of product formation cannot be increased unless the number of enzymes is increased. Cells can actually do this by

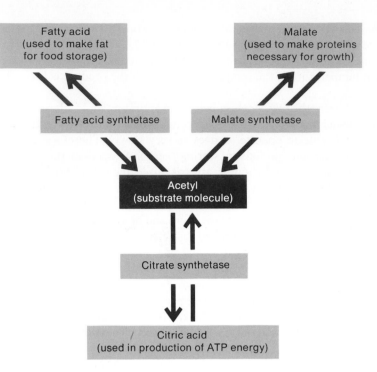

Figure 5.7
Enzymatic Competition.
Acetyl can serve as a substrate for a number of different reactions. Whether it becomes a fatty acid, malate, or citric acid is determined by the enzymes present. Each of the three enzymes may be thought of as being in competition for the acetyl substrate. The cell can partially control which direction the acetyl will take by producing greater numbers of one kind of enzyme and fewer of the other.

synthesizing more enzymes. However, just because there are more enzyme molecules does not mean that any one enzyme molecule will be working any faster. The turnover number stays the same. As the enzyme concentration increases, the amount of product formed increases in a shorter time. A greater number of enzymes are turning over substrates, they are not turning over substrates faster. Similarly, if enzyme numbers are decreased, the amount of product formed declines.

We can also look at this from the point of view of the substrate. As the amount of substrate increases, the amount of product formed increases. However, each enzyme is still only operating at the same turnover number. The increase in product is the result of more substrates available to be changed. If there is a very large amount of substrate, even a small amount of enzyme can change all the substrate to product, it will just take longer. Decreasing the amount of substrate results in reduced product formation because some enzymes will go for long periods without coming in contact with a substrate molecule.

Enzymes in Cells

In any cell, there are thousands of kinds of enzymes. Each is sensitive to changing environmental conditions and each controls specific chemical reactions. These enzymes cooperate with each other to satisfy the cell's needs, such as energy, growth, reproduction, and storage of materials for future use. All of these needs may be met by using the same nutrients. Whenever there are several different enzymes available to combine with a given substrate material, **enzymatic competition** results. The success of any one of these enzyme systems depends on the number of enzymes available and the suitability of the environmental conditions for their optimum operation. As an example, we will look at a case of enzymatic competition that occurs in many organisms (figure 5.7). Acetyl is a substrate that can be used for three different things in a cell. It can be used to generate immediate energy. If a cell requires energy

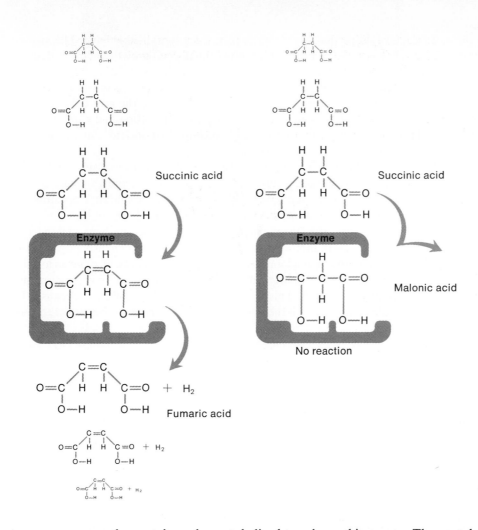

Succinic acid

Succinic acid

Enzyme

Enzyme

Malonic acid

No reaction

Fumaric acid + H₂

Figure 5.8
Enzymatic Inhibition.

The left-hand side of the illustration shows the normal functioning of the enzyme. On the right-hand side, the enzyme is unable to function. This is because an inhibitor, malonic acid, is attached to the enzyme and prevents the enzyme from forming the normal complex with succinic acid. As long as malonic acid is present, the enzyme will be unable to function. If the malonic acid is removed, the enzyme will begin to function normally again. Its attachment to the inhibitor in this case did not do permanent damage to the enzyme.

to move or grow, the acetyl can be metabolized to release this energy. The acetyl can also be converted to an amino acid. If the cell is in need of protein, the acetyl could provide one of the building blocks for the construction of protein. Still another use the cell can make of acetyl is to construct a storage material in the form of fat. How does the cell control which of these three things will happen to the substrate acetyl? It does so by using enzymatic competition. If the cell produces more of the enzyme, citrate synthetase, then these enzymes will have a better chance of combining with the acetyl; therefore, the cell will ultimately release energy from the acetyl. If there are more malate synthetase enzymes, the reaction that is favored is the formation of malate—a step in the direction of constructing a protein. The third alternative is the formation of fatty acid synthetase. When this enzyme outcompetes the other two, the acetyl is used in fat production and storage. You can see that the use the cell makes of the substrate acetyl is directly controlled by the amount and kinds of enzymes it produces. The number and kind of enzymes produced is regulated by genes.

In addition, reactions can be influenced by the presence of other molecules that affect the operation of enzymes. Reactions can be slowed in the presence of particular inhibitor molecues. An **inhibitor** is a molecule that attaches itself to an enzyme, and interferes with its ability to form an enzyme-substrate complex. Certain pesticides are effective because they contain an enzyme inhibitor (figure 5.8). One of the early kinds of pesticides used to spray fruit trees contained arsenic. The arsenic attached itself to insect enzymes and inhibited the normal growth and reproduction of insects. Organophosphates are pesticides that inhibit several enzymes necessary for the operation of the nervous system. When it is incorporated into nerve cells, it disrupts normal nerve transmission and causes the death of the affected organism. Unfortunately, these inhibitors are not selective in their action and can affect humans. In humans, death is usually caused by uncontrolled muscle contractions that result in breathing failure.

Some inhibitors have a shape that closely resembles the normal substrate of the enzyme. The enzyme is unable to distinguish the inhibitor from the normal substrate and so it combines with either or both. As long as the inhibitor is combined with an enzyme, the enzyme is ineffective in its normal role. Some of these enzyme-inhibitor complexes are permanent, such as carbon monoxide-hemoglobin. The carbon monoxide acts as an inhibitor and completely and permanently prevents the normal functioning of the hemoglobin. An inhibitor removes the enzyme as a functioning part of the cell. The reaction that enzyme catalyzes no longer occurs and none of the product is formed. This is termed **competitive inhibition** because the inhibitor molecule competes with the normal substrate for the active site of the enzyme.

We use enzyme inhibition to control disease. The sulfa drugs are used to control a variety of bacteria such as the bacterium *Streptococcus pyogenes,* the cause of strep throat and scarlet fever. The drug resembles one of the bacterium's necessary substrates and so prevents some of the cell's enzymes from producing an essential cell component. As a result, the normal metabolism of the bacterial cell is not maintained and it dies.

Summary

Enzymes are protein catalysts that speed up the rate of chemical reactions—without any significant increase in the temperature—by lowering activation energy. Enzymes have a very specific structure that matches the structure of particular substrate molecules. Actually, the substrate molecule comes in contact with only a specific part of the enzyme molecule, the attachment site. The active site of the enzyme is the place where the substrate molecule is changed. The enzyme-substrate complex reacts to form the end product. The protein nature of enzymes makes them sensitive to environmental conditions, such as temperature and pH, that change the structure of proteins. The number and kinds of enzymes is ultimately controlled by the genetic information of the cell. Other kinds of molecules, such as coenzymes, inhibitors, or competing enzymes, can influence specific enzymes. Changing conditions within the cell shift the enzymatic priorities of the cell by influencing the turnover number.

The following data were obtained by a number of Nobel prize-winning scientists from Lower Slobovia. As a member of the group, interpret the data with respect to:

1. Enzyme activities
2. Movement of substrates into and out of the cell
3. Competition among different enzymes for the same substrate
4. Cell structure

Data

a. A lowering of the atmospheric temperature from 22° C to 18° C causes organisms to form a thick protective coat.
b. Below 18° C, no additional coat material is produced.
c. If the cell is heated to 35° C and then cooled to 18° C, no coat is produced.
d. The coat consists of a complex carbohydrate.
e. The coat will form even if there is a low concentration of simple sugars in the surroundings.
f. If the cell needs energy for growth, no cell coats are produced at any temperature.

There are several cleaning products available that contain enzymes. Select one of these and another household product to test for enzyme activity and specificity in removing stains. Take two clean, white pieces of cloth and stain them with a row of blood (from meat), chocolate, berry juice, grass stain, egg white, cooking oil, and grease. On one stained cloth, put a drop of the enzyme-active detergent in the middle of each stain. On the second cloth, put a drop of ordinary dishwashing detergent. Let the cloths sit for an hour. Wash these two cloths separately and compare results.

1. What is the difference between a catalyst and an enzyme?
2. Describe the sequence of events in an enzyme-controlled reaction.
3. How does changing temperature affect the rate of an enzyme-controlled reaction?
4. Would you expect a fat and sugar molecule to be acted upon by the same enzyme? Why?
5. What factors in the cell can speed up or slow down enzyme reactions?
6. What is the turnover number? Why is it important?
7. What is the relationship between vitamins and coenzymes?
8. What is enzyme competition and why is it important to all cells?
9. What effect might a change in pH have on enzyme activity?
10. Where in a cell would you look for enzymes?

Chapter Glossary

activation energy (ak″tĭ-va′shun en′ur-je) Energy required to start a reaction, it may be used to increase the number of effective collisions or may be required to form the transitional molecule in the progress of the reaction pathway.

active site (ak′tive sĭt) Point on the enzyme where the enzyme causes the substrate to change.

attachment site (uh-tatch′munt sĭt) Specific point on the surface of the enzyme where it can physically attach itself to the substrate.

binding site (bin′ding sĭt) Specific point on the surface of the enzyme where it can physically attach itself to the substrate.

catalyst (cat′uh-list) A chemical that speeds up a reaction but is not used up in the reaction.

coenzyme (ko-en′zĭm) A molecule that works with an enzyme to enable it to function as a catalyst.

competitive inhibition (kum-pet′ĭ-tiv in″hĭ-bĭ′shun) The formation of a temporary enzyme-inhibitor complex that interferes with the normal formation of enzyme-substrate complexes, resulting in a decreased turnover.

denature (de-na′chur) To permanently change the protein structure of an enzyme so that it loses its ability to function.

enzymatic competition (en-zi-mă′tik com-pĕ-ti′shun) Several different enzymes available to combine with a given substrate material.

enzyme (en′zĭm) A specific protein that acts as a catalyst to change the rate of a reaction.

enzyme-substrate complex (en′zĭm sub′strāt kom′pleks) A temporary molecule formed when an enzyme attaches itself to a substrate molecule.

inhibitor (in-hib′ĭ-tōr) A molecule that temporarily attaches itself to an enzyme thereby interfering with the enzyme's ability to form an enzyme-substrate complex.

substrate (sub′strāt) A reactant molecule with which the enzyme combines.

turnover number (turn′o-ver num′ber) Number of molecules of substrate that a single molecule of enzyme can react with in a given time under ideal conditions.

Biochemical Pathways

■ Chapter Outline

■ Purpose

This chapter deals with some of the major chemical reactions that occur in living things. Because these reactions are dependent on one another and occur in specific series, they are commonly referred to as biochemical pathways. An understanding of these biochemical pathways will help you understand how energy is utilized within an organism.

There are hundreds of such pathways, all of which interlink, but we will deal only with those that form the core of all chemical reactions of a living cell. The two major pathways are photosynthesis and cellular respiration.

For Your Information

Several years ago Joni Mitchell, a folksinger, penned the lyrics, "Pave Paradise, Put Up a Parking Lot." In this protest song, she was asking that we consider land development from the standpoint of what we are doing to the earth. Specifically this song was about land development within Hawaii, but has value for development everywhere. Many people are extremely concerned about the loss of tropical rain forests. They feel that these forests provide a significant amount of the oxygen necessary for the survival of all life on earth.

Learning Objectives

■ Associate the major parts of photosynthesis with the ultrastructure of the chloroplast.

■ List the raw materials, products, and describe the processes involved in the two major parts of photosynthesis.

■ List the things a plant can do with the product of photosynthesis.

■ Associate the major parts of aerobic respiration with the ultrastructure of the mitochondrion.

■ List the raw materials, products, and describe the processes involved in the three major parts of aerobic respiration.

■ Describe the processes involved in the alternate pathways known as anaerobic respiration (fermentation).

■ Follow a molecule through the steps of interconversion of fats, proteins, and carbohydrates.

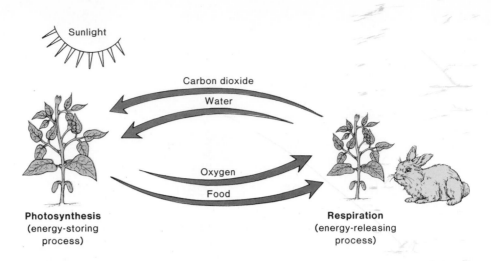

Figure 6.1
The Relationship Between Photosynthesis and Respiration.
The raw materials for photosynthesis—carbon dioxide and water—are the end products of respiration. The raw materials for respiration—food and oxygen—are the end products of photosynthesis. These materials are cycled between those organisms that are involved in the two major pathways, photosynthesis and respiration.

Sunlight

Carbon dioxide

Water

Oxygen

Food

Photosynthesis
(energy-storing process)

Respiration
(energy-releasing process)

Energy and Cells

All living organisms require energy to conduct the many functions necessary to sustain life. The source of this energy for cells is the chemical bonds of food molecules. Cells can be thought of as chemical factories that conduct a variety of chemical reactions. Much of what an organism does can be explained in terms of the chemical activities of its cells. Reactions within cells are frequently linked to each other because the products of one reaction are used as the reactants for another. A series of enzyme-controlled reactions linked together is termed a **biochemical pathway.** Some organisms (green plants and algae) are capable of trapping sunlight energy and converting it to chemical bonds in food molecules. This important energy transformation is the source of all food and energy for all organisms, including animals. Green plants and algae do not need to eat since they are able to manufacture food molecules. These food molecules contain the chemical bonds necessary for their energy-requiring processes. Organisms that are able to make their food molecules from sunlight and inorganic raw materials are **autotrophs** (self-feeders). Organisms that eat food to obtain energy are termed **heterotrophs** (different feeders).

The process of converting sunlight energy to chemical-bond energy, **photosynthesis,** is one of the major biochemical pathways. In this series of biochemical reactions, plants produce food molecules such as carbohydrates for themselves as well as for all the other organisms on earth. **Cellular respiration,** a second major biochemical pathway, is a series of reactions during which cells release the chemical-bond energy from food and convert it into usable forms. All organisms must be able to make this energy conversion regardless of the source of their food molecules. Whether organisms manufacture food or take it in from the environment, they use food molecules as a source of energy. The process of energy conversion is essentially the same in all organisms (figure 6.1).

Within cells, particular biochemical pathways are carried out in specific organelles. These organelles are the location of certain enzymes necessary to control the reactions. In addition, the organelles provide surfaces upon which chemical reactions occur. Chloroplasts are the site of photosynthesis and mitochondria are the site of most of the reactions of cellular respiration. As is the case with most things in life, there are exceptions. Prokaryotic cells lack mitochondria and chloroplasts, yet some are capable of processes very similar to the cellular respiration and photosynthesis processes found in eukaryotic cells.

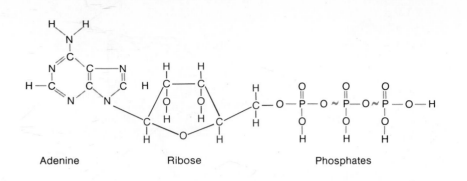

Adenine   Ribose   Phosphates

Figure 6.2
Adenosine Triphosphate (ATP).
A molecule of ATP consists of a molecule of adenine, ribose, and three phosphate groups. The two end phosphate groups are bonded together by high-energy bonds. When the bonds are broken, they release an unusually great amount of energy; therefore, they are known as "high-energy" bonds. These bonds are represented by the curved lines. The ATP molecule is considered to be an energy carrier.

A biochemical pathway may consist of several steps, each involving a molecular change and an energy change. Different chemical bonds have different amounts of chemical energy. If the products of a reaction do not have the same amount of energy as the reactants do, energy is either released or required. Some chemical reactions—like the burning of methane—may have a net release of energy, while others—like the synthesis of sugar by plants—require an input of energy. Cells can encourage reactions that require an input of energy by coupling energy-requiring reactions with others that yield net energy. These reactions are often called **coupled reactions.**

The Currency of the Cell

The second law of thermodynamics states that some usable energy is lost whenever energy is converted from one form to another. This wasted energy is generally released into the environment. If the conversion is to be useful, the amount of energy lost must not be too great. In a cell, the chemical bonds of a molecule of food are broken, releasing small amounts of energy. While some of the energy is lost as heat to the environment, some of the energy is used to form the energy carrier molecule, ATP. ATP is one of the major molecules involved in coupling energy-requiring reactions with energy-demanding reactions. **Adenosine triphosphate** (**ATP**) is formed from adenine, ribose, and phosphates (figure 6.2). These three are chemically bonded to form AMP, adenosine monophosphate (one phosphate). When a second phosphate group is added to the AMP, a molecule of ADP (diphosphate) is formed. The covalent bond that attaches the second phosphate to the AMP molecule is easily broken to release energy for energy-requiring cell processes. Because this covalent bond is such a readily available source of energy, it is called a **high-energy phosphate bond.** Thus, the bond acts as an energy holder. The ADP, with the addition of more energy, is able to bond to a third phosphate group and form ATP. ATP has two high-energy phosphate bonds represented by curved, solid lines. Both ADP and ATP, because they contain high-energy bonds, are very unstable molecules and readily lose their phosphates. When this occurs, the energy found in their high-energy bonds is released to the environment. Within a cell, enzymes direct this release of energy as ATP is broken down into its components. This channeling of energy enables the cell to better utilize this readily available source of energy.

In order to better understand the idea of high-energy bonds, they are frequently compared to money. In fact, ATP is often called the energy currency of the cell. Both money and the energy in ATP can be converted into many things that are necessary for everyday living. A single one-dollar bill can be spent for certain items.

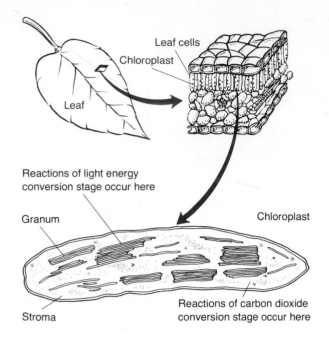

Figure 6.3
Plant Structure and Photosynthesis.

The major structure associated with photosynthesis in plants is the leaf. Certain cells within the leaf contain large numbers of chloroplasts. It is within the chloroplasts that the light energy is converted to chemical-bond energy. It is here also that the carbohydrate is constructed from atmospheric carbon dioxide, water, and the energy from the sun.

A single two-dollar bill can purchase more, but is still only a single bill. The two-dollar bill is the same size and weight as a one-dollar bill but has twice the purchasing power. You might think of the high-energy phosphate bond as a two-dollar bill. A bond that represents more energy than an ordinary covalent bond. The curved lines in the diagram of an ATP molecule represent the high-energy bond. Breaking a high-energy chemical bond yields about twice as much energy as breaking an ordinary covalent bond. The ATP molecule, with two high-energy bonds, acts as a tiny reservoir of energy.

Photosynthesis

The high-energy phosphate bond in ATP can be manufactured from the breakdown of other organic molecules or be formed from captured light energy during the process of photosynthesis. The molecule **chlorophyll** is the green pigment that is directly involved in the light-trapping and energy conversion process. The energy in ATP is used in later steps to combine water and carbon dioxide to form a simple sugar molecule.

During the manufacture of simple sugar, oxygen is released into the atmosphere. The plant can then use the simple sugar to construct other kinds of molecules, provided there are a few additional raw materials, such as minerals and nitrogen-containing molecules. The generalized chemical equation for the process of photosynthesis may be written:

$$\text{sunlight energy} + 6CO_2 + 6H_2O \xrightarrow[\substack{\text{chlorophyll and enzymes in chloroplasts}}]{\text{helped by}} C_6H_{12}O_6 + 6O_2$$

sunlight energy · carbon dioxide · water · chlorophyll and enzymes in chloroplasts · simple sugar · oxygen

For most plants, the entire process of photosynthesis takes place in the leaf where the cells contain large numbers of chloroplasts (figure 6.3). Chloroplasts are oblong, membranous closed bags containing many thin, flat disks called *thylakoids*. These

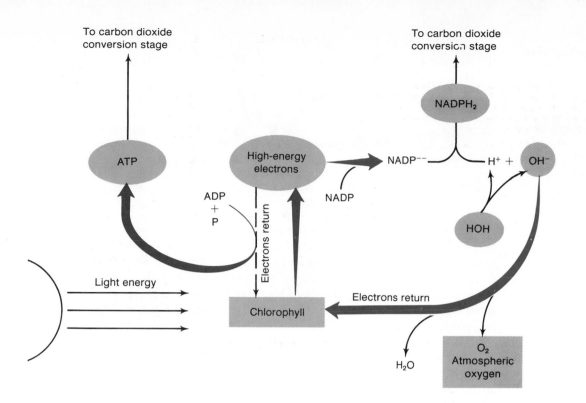

Figure 6.4
Light Energy Conversion Stage of Photosynthesis.

In the light energy conversion stage, sunlight is the source of energy. This stage converts the light energy to kinetic energy of excited electrons and then forms ATP molecules to carry the energy. NADPH$_2$ is produced to carry H$_2$, which is produced from water. O$_2$ is a waste product formed from the water molecules. These processes take place inside the chloroplast and are associated with the structures known as the grana.

basic units contain the chlorophylls, some other pigments, and enzymes. The thylakoids are stacked up in many groups called grana (singular granum). The regions between the grana are called the stroma of the chloroplast. Photosynthesis is divided into two stages. The first of the two stages, which we call the **light energy conversion stage,** takes place in the grana. It is here that light energy is converted into chemical-bond energy. The second stage of photosynthesis is known as the **carbon dioxide conversion stage.** This series of reactions occurs outside the grana in the stroma. In this stage, carbon dioxide becomes incorporated into a simple sugar molecule.

Light Energy Conversion Stage of Photosynthesis

The green pigment chlorophyll, which is present in the chloroplasts of plants, is a complex molecule with many loosely attached electrons. Some of the electrons separate from the chlorophyll molecule when they are struck by certain wavelengths of light. The chlorophyll becomes oxidized in this process. The lost electron may then be picked up by other molecules in the cell in a reduction reaction. This portion of photosynthesis is a series of oxidation-reduction reactions (figure 6.4). When the light energy is transferred to the electrons, the electrons move more rapidly and are called excited electrons. These electrons release their newly acquired energy as they return to their previous position in the chlorophyll molecule. After the electrons leave the chlorophyll molecule, they may follow one of two different pathways during the light energy conversion stage of photosynthesis.

In the first pathway, the excited electrons pass through a series of molecules called electron carriers. These electron carriers are called cytochromes. Each absorbs some of the energy of the excited electrons and then passes the electrons to the next cytochrome. Each successive carrier removes more energy from the excited

Figure 6.5
The Fate of Hydroxyl Ions.
In the light energy conversion
stage of photosynthesis, the
hydroxyl ions (OH⁻) combine with
one another to form water
molecules and oxygen gas (O₂). At
the same time, the electrons are
returned to a chlorophyll molecule.

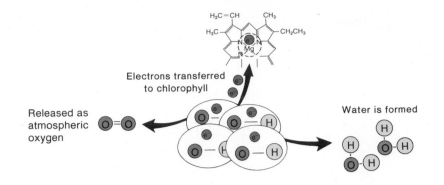

Electrons transferred
to chlorophyll

Released as
atmospheric
oxygen

Water is formed

electrons. These electron carriers are located on the membranes that make up the grana of the chloroplast. The series of electron carriers in the chloroplast takes the energy from the excited electrons and uses it to bind a phosphate to an ADP molecule to form ATP. This energy conversion process begins with sunlight energy exciting the electrons of chlorophyll to a higher energy level. As these excited electrons move from their original position, they are picked up by and moved through a series of cytochrome molecules that extract their energy in a stepwise fashion, transforming it into the chemical bonds of ATP. The electrons, having lost their excess energy, return to their original position in the chlorophyll molecule. Because the electrons that left the chlorophyll molecule eventually return to their original position, this pathway is a complete cycle. This pathway is called cyclic photophosphorylation since the electrons that leave the chlorophyll molecule eventually return to that same molecule (cyclic), the process is stimulated by light (photo), and phosphate is added to ADP to form ATP (phosphorylation).

In the second pathway, the excited electrons are picked up by a different electron carrier known as **NADP** (**n**icotinamide **a**denine **d**inucleotide **p**hosphate), which becomes reduced. Each NADP molecule has the ability to capture two electrons. As a result, a molecule of NADP is converted to a molecule of NADP⁻⁻ that has two negative charges. At the same time, water in the cell is broken into H^+ and OH^- ions. Since the H^+ ions are positively charged, they are attracted to the negatively charged NADP⁻⁻. This results in the formation of $NADPH_2$. The $NADPH_2$ is now carrying two atoms of hydrogen.

The remaining hydroxyl ions (OH⁻) accumulate and combine with each other to form water, oxygen (O_2), and free electrons (figure 6.5). The water becomes part of the cytoplasm, the oxygen is released into the atmosphere, and the free electrons attach to the chlorophyll molecule that had previously lost electrons to the NADP. Since the electrons that left the chlorophyll molecule in this pathway do not return to their original position, but are picked up by NADP, the pathway is noncyclic. The electrons that eventually attach to the chlorophyll molecule come from water. At the completion of the light energy conversion stage, the plant has acquired usable energy in the form of ATP and a source of hydrogen in the form of $NADPH_2$. The energy and hydrogen are necessary for the next stage of photosynthesis, water molecules are used up, and oxygen is released to the environment. The shorthand, chemical equation, for this portion of photosynthesis may be written:

$$2NADP + ADP + P + 4H_2O + \text{sunlight energy} \xrightarrow[\substack{\text{chlorophyll} \\ \text{and enzymes}}]{\text{helped by}} 2\,NADPH_2 + ATP + 2H_2O + O_2$$

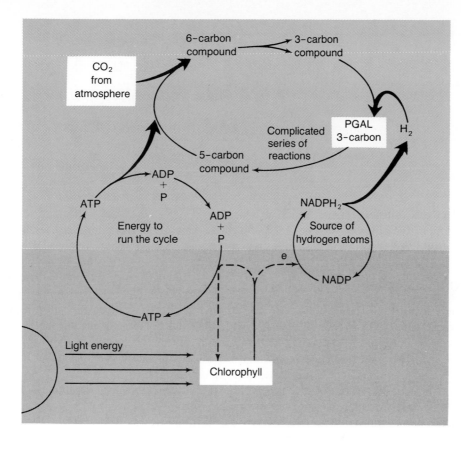

Figure 6.6
The Carbon Dioxide Conversion Stage of Photosynthesis.
During this process, ATP molecules from the light energy stage are used to drive the reactions, which include the incorporation of carbon dioxide molecules and hydrogen atoms into an organic molecule. An important molecule produced is phosphoglyceraldehyde (PGAL). This process takes place in the stroma of the chloroplast.

Carbon Dioxide Conversion Stage of Photosynthesis

The second major series of reactions involved in photosynthesis takes place within the stroma of the chloroplast (figure 6.6). The materials needed for the carbon dioxide conversion stage are ATP, $NADPH_2$, CO_2, and a five-carbon starter molecule called ribulose. The first two ingredients (ATP, $NADPH_2$) are made available from the light energy conversion reactions, the carbon dioxide molecules come from the atmosphere, and the ribulose starter molecule is already present in the stroma of the chloroplast from previous reactions. The major event that occurs in the carbon dioxide conversion stage involves the use of energy from ATP to bond hydrogen from $NADPH_2$ and the carbon dioxide to ribulose in order to ultimately form **PGAL** (**p**hospho**g**lycer**al**dehyde).

The carbon dioxide molecule does not become PGAL directly, but is first attached to the five-carbon starter molecule ribulose to form an unstable six-carbon molecule. This six-carbon molecule immediately breaks down into two three-carbon molecules, which then undergo a series of reactions that involve a transfer of energy from ATP and hydrogen from the $NADPH_2$. The molecules produced from this series of reactions are PGAL. The chemical equation for the CO_2 conversion stage is:

$$CO_2 + ATP + NADPH_2 + \begin{array}{c} \text{5-carbon} \\ \text{starter} \\ \text{(ribulose)} \end{array} \longrightarrow PGAL + NADP + ADP + P$$

Most of the PGAL molecules go through a series of complicated reactions to regenerate ribulose, but some of the PGAL can be considered profit from the process and be used for a variety of other purposes.

Figure 6.7
Uses for PGAL.
The PGAL that is produced as the
end product of photosynthesis is
used for a variety of things. The
plant cell can make sugars,
complex carbohydrates, even the
original five-carbon starter from it. It
can also serve as an ingredient of
fats and amino acids (proteins). In
addition, it provides a major source
of metabolic energy when it is sent
through the aerobic respiratory
pathway.

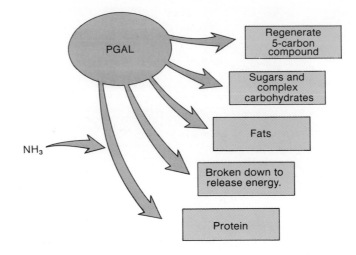

PGAL: An All-Purpose Molecule

There are a number of things the cell can do with PGAL molecules (figure 6.7). Almost 80% of the PGAL formed is used to regenerate ribulose so that photosynthesis can continue. The remaining 20% of the PGAL molecules is used by the plant to make other organic molecules. One of the molecules that can be produced from PGAL is glucose. Glucose molecules can then be combined to form complex carbohydrates such as starch for energy storage or cellulose for the construction of cell wall material. In addition, the carbohydrates might be used as a building block for ATP. The cell may also convert PGAL into a lipid such as oil for storage, phospholipids for cell membranes, or steroids for vitamins. PGAL may become the carbon skeleton for amino acids or may be incorporated into the cell as nucleotides for the synthesis of DNA and RNA. Almost any molecule that a green plant can manufacture begins with this simple PGAL molecule. One additional use of PGAL, which is easy to overlook, is that it can be broken down in cellular respiration. This allows the chemical-bond energy to be released and enables the plant cell to do things that require energy, such as grow and move materials. Figure 6.8 shows the relationship between the two major parts of photosynthesis.

Cellular Respiration

Some organisms are capable of carrying on photosynthesis and storing energy in the chemical bonds of molecules, such as PGAL, glucose, fats, and proteins. Eukaryotic organisms utilize these molecules as a source of energy. To release this energy, plants, as well as animals, fungi, protozoa, and many bacteria, rely on the same basic chemical pathway—**aerobic cellular respiration.** In aerobic cellular respiration, food and oxygen are chemically combined to yield energy and release carbon dioxide and water. In chemical shorthand, this may be written:

$$\underset{\text{glucose}}{C_6H_{12}O_6} + \underset{\text{oxygen}}{6O_2} \longrightarrow \underset{\substack{\text{carbon} \\ \text{dioxide}}}{6CO_2} + \underset{\text{water}}{6H_2O} + \underset{\text{energy}}{36ATP}$$

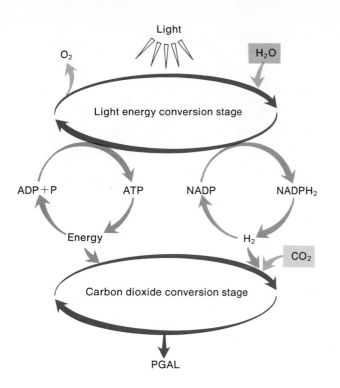

Figure 6.8
Photosynthesis.
The process of photosynthesis is composed of the interrelated stages of light energy conversion and carbon dioxide conversion. The carbon dioxide conversion stage requires the ATP and $NADPH_2$ produced in the light energy conversion stage. The light energy conversion stage in turn requires the ADP and NADP that are released from the carbon dioxide conversion stage. Therefore, each stage is dependent upon the other.

The extraction of usable energy in the form of ATP typically involves three steps: glycolysis, Krebs cycle, and electron transfer system. Each step will be discussed individually, but remember that each is a part of the whole process. Refer to the general equation as you study the individual parts to see how they fit into the whole picture.

Glycolysis

The first stage of the cellular respiration process takes place in the cytoplasm. This first step, known as **glycolysis** (carbohydrate = glyco, splitting = lysis), consists of the enzymatic breakdown of a glucose molecule without the use of molecular oxygen (figure 6.9). Because no oxygen is required, glycolysis is called an **anaerobic** process.

In glycolysis, some energy must be put in to start the process, since glucose is a very stable molecule and will not spontaneously decompose to release energy. For each molecule of glucose entering glycolysis, energy is supplied by two ATP molecules. The phosphates are released from two ATP molecules and become attached to glucose to form phosphorylated sugar, $(P-C_6-P)$. We term this reaction a phosphorylation reaction. It is controlled by an enzyme named phosphorylase. The phosphorylated glucose is then broken down through several enzymatically controlled reactions into two three-carbon compounds, each with one attached phosphate, (C_3-P). These three-carbon compounds are PGAL. (Remember that PGAL is also the end product of photosynthesis. In some situations, a plant will use PGAL manufactured during photosynthesis in cellular respiration.) Each of the two PGAL molecules acquires a second phosphate from a phosphate pool normally found in the cytoplasm. Each molecule now has two phosphates attached $(P-C_3-P)$. A series of reactions follows in which energy is released by breaking chemical bonds, causing each of these three-carbon compounds to lose their phosphates. These high-energy phosphates combine with ADP to form ATP. In addition, four hydrogen atoms

Figure 6.9
Glycolysis.
The glycolytic pathway results in the breakdown of six-carbon sugars under anaerobic conditions Each molecule of sugar releases enough energy to produce a profit of two ATPs. In addition, two molecules of pyruvic acid and two molecules of hydrogen carried as NADH$_2$ are produced.

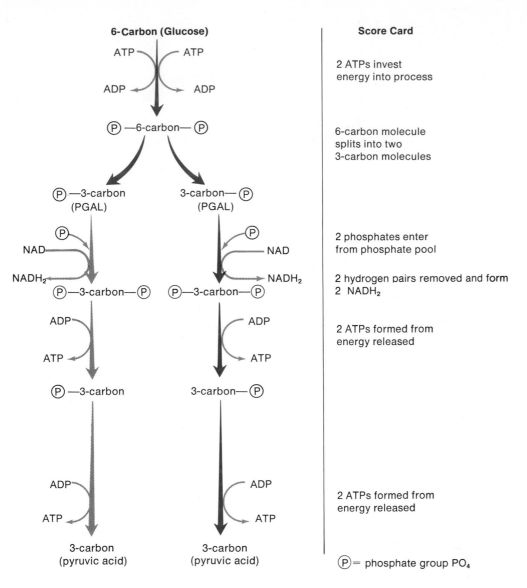

6-Carbon (Glucose)

Score Card

2 ATPs invest energy into process

6-carbon molecule splits into two 3-carbon molecules

2 phosphates enter from phosphate pool

2 hydrogen pairs removed and form 2 NADH$_2$

2 ATPs formed from energy released

2 ATPs formed from energy released

$\textcircled{P}$ = phosphate group PO$_4$

detach from the carbon skeleton (oxidation) and become bonded to two hydrogen-carrier molecules (reduction) known as NAD. **NAD** (**n**icotinamide **a**denine **d**inucleotide) is very similar in structure and function to the NADP, which is the hydrogen carrier used in photosynthesis. The molecules of NADH$_2$ contain a large amount of potential energy that may be released in a usable form in later chemical reactions. The three-carbon molecules that result from glycolysis are called pyruvic acid.

In summary, the process of glycolysis takes place in the cytoplasm of a cell. In this process, glucose undergoes reactions that lead to the formation of four molecules of ATP, two molecules of NADH$_2$, and two three-carbon molecules of **pyruvic acid.** Since two molecules of ATP were used to start the process and a total of four ATPs are generated, each simple sugar molecule that undergoes glycolysis produces a net of two ATP molecules. Glycolysis does not require free oxygen.

If atmospheric oxygen is available, many cells will enter a new pathway by breaking down the pyruvic acid molecules and release even more energy. This series

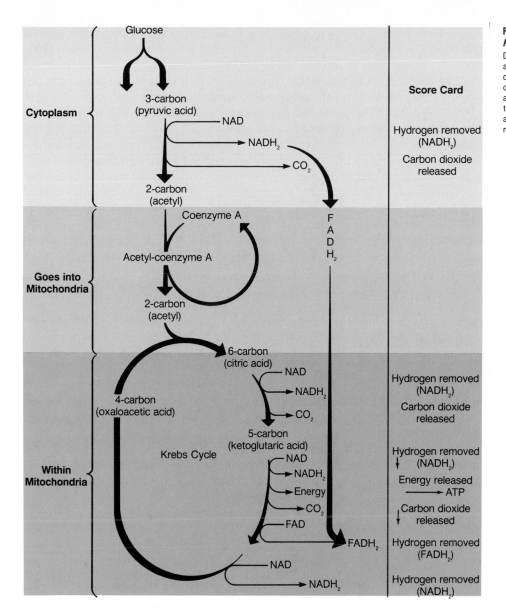

Figure 6.10
Krebs Cycle.
During the Krebs cycle, the pyruvic acid from glycolysis is broken down. The carbon ends up in carbon dioxide, and the hydrogens are carried away to the electron transfer system. Also, small amounts of energy are released to make ATP.

of reactions requires oxygen and is called **aerobic.** The two major pathways that make up the rest of aerobic cellular respiration are the Krebs cycle and the electron transfer system, ETS.

Krebs Cycle

The **Krebs cycle** is a series of oxidation-reduction reactions that complete the breakdown of pyruvic acid produced by glycolysis (figure 6.10). Recall from chapter 4 that the mitochondria are directly involved in the process of energy release. In order for pyruvic acid to be used as an energy source, it must enter the mitochondrion. Once inside, an enzyme converts the three-carbon pyruvic acid molecule to a two-carbon molecule called **acetyl.** When the acetyl is formed, the carbon removed is released as carbon dioxide. In addition to releasing carbon dioxide, each pyruvic acid molecule is oxidized, since it loses two hydrogens that become attached to NAD molecules (reduction) to form $NADH_2$. NAD is serving as a hydrogen carrier.

The carbon dioxide is a waste product that is eventually released by the cell into the atmosphere. The two-carbon acetyl compound temporarily combines with a large molecule called coenzyme A (CoA) to form acetyl-CoA and transfers the acetyl to a four-carbon compound called oxaloacetic acid to become part of a six-carbon molecule. This new six-carbon compound is broken down in a series of reactions to regenerate oxaloacetic acid. In the process of breaking down pyruvic acid, three molecules of carbon dioxide are formed. In addition, five pairs of hydrogens are removed and become attached to hydrogen carriers. Four pairs become attached to NAD and one pair becomes attached to a different hydrogen carrier known as **FAD** (**f**lavin **a**denine **d**inucleotide). As the molecules move through the Krebs cycle, enough energy is released to allow the synthesis of one ATP molecule for each acetyl that enters the cycle. The ATP is formed from ADP and P already present in the mitochondria.

For each pyruvic acid molecule that enters the mitochrondrion and is processed through the Krebs cycle, three carbons are released as three carbon dioxide molecules, five pairs of hydrogens are removed and become attached to hydrogen carriers, and one ATP molecule is generated. When both pyruvic acid molecules have been processed through the Krebs cycle: (1) all of the original carbons from the glucose are released into the atmosphere as six carbon dioxide molecules; (2) all of the hydrogen originally found on the glucose has been transferred to either NAD or FAD to form $NADH_2$ and $FADH_2$; (3) a total of four ATPs have been formed from the addition of phosphates to ADPs. Look back at the general equation for the process and note that one of the products is carbon dioxide.

If you go back through the processes of glycolysis and the Krebs cycle and account for all the hydrogens that have been removed from the original glucose, you find that: (1) the two pairs released during glycolysis are in the cytoplasm attached to $NADH_2$; (2) eight pairs released during the Krebs cycle are inside of the mitochondria attached to $NADH_2$; and (3) two pairs released during the Krebs cycle are in the mitochondria attached to $FADH_2$. Energy release from these attached hydrogens continues on the inner membranes of the mitochondria, the cristae. Remember that these folds of membrane provide a surface for enzymes. It is on these surfaces that the energy from the hydrogens attached to NAD and FAD are released.

Electron Transfer System

The series of reactions in which energy is removed from the hydrogens carried by NAD and FAD is the final stage of aerobic cellular respiration known as the **electron transfer system** (figure 6.11). The reactions that make up the electron transfer system are a series of oxidation-reduction reactions in which the electrons from the hydrogen atoms are passed from one electron carrier molecule to another until they ultimately are accepted by oxygen atoms. The oxygen combines with the hydrogens to form water. The water, a waste product of the process, is released into the cytoplasm. At certain steps along the pathway, energy is released and an ATP molecule can be formed. This system converts the potential energy in the hydrogen-carrying molecules ($NADH_2$ and $FADH_2$) to the energy in ATP.

Let's look at the hydrogens and their carriers in just a bit more detail to account for all of the energy that becomes available to the cell. The two pair of hydrogens

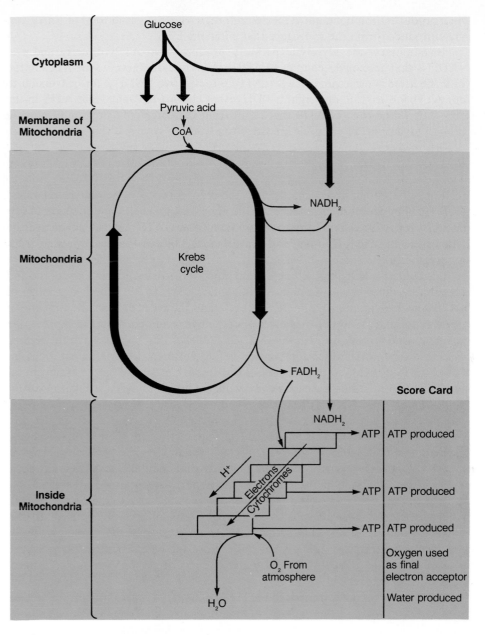

Cytoplasm

Glucose

Pyruvic acid

CoA

Membrane of
Mitochondria

Mitochondria

Krebs
cycle

$NADH_2$

$FADH_2$

Score Card

$NADH_2$

ATP | ATP produced

H^+

Electrons
Cytochromes

Inside
Mitochondria

ATP | ATP produced

ATP | ATP produced

O_2 From
atmosphere

Oxygen used
as final
electron acceptor

H_2O

Water produced

Figure 6.11
Aerobic Respiration.
In this process, glucose, a six-carbon sugar, is completely broken down into carbon dioxide and water and its energy is released in the form of ATP. The total process involves three major portions: glycolysis, the Krebs cycle, and the electron transfer system. This final series of steps known as the ETS requires oxygen. It is the part of the pathway that releases the major quantities of energy.

in the cytoplasm that are carried by $NADH_2$ are converted to $FADH_2$ in order to shuttle them into the mitochondrion. Once they are inside the mitochondria, they follow the same pathway as the other $FADH_2$s. At three points in the series of ox-idation-reductions in the ETS, sufficient energy is released from the $NADH_2$s to produce an ATP molecule. Therefore, twenty-four ATPs are released from these eight pairs of hydrogens carried on $NADH_2$.

The four pairs of hydrogens carried by FAD are lower in energy than those accepted by NAD. When these hydrogens go through the series of oxidation-reduction reactions, they release enough energy to produce ATP at only two points.

They produce a total of eight ATPs; therefore, we have a grand total of thirty-two ATPs produced from the hydrogens that enter the ETS.

In summary, the glycolytic pathway produces two ATPs directly and two $NADH_2$s that are converted to two $FADH_2$s used to manufacture four ATPs in the ETS. The Krebs cycle produces two ATPs directly, two $FADH_2$s for conversion to four ATPs in the ETS, and eight $NADH_2$s that convert to twenty-four ATPs in the ETS. The grand total of ATPs that result from the complete oxidation of glucose through glycolysis, Krebs cycle, and electron transfer system is thirty-six.

Alternative Pathways

Certain cells are not able to go through the entire process of aerobic cellular respiration. They do not have mitochondria or they lack the necessary enzymes to utilize oxygen as a final hydrogen acceptor. Since many cells lack the enzymes of the Krebs cycle and the ETS, they must use alternative biochemical pathways in order to generate ATP. There is a second problem that arises when these enzymes are lacking. As the hydrogen is removed from glucose (oxidized) during glycolysis and is used to reduce NAD, the limited amount of cellular NAD and instability of $NADH_2$ will eventually stop the Krebs cycle and the electron transfer system. An easy way to solve this problem is to transfer the hydrogen to a different, more stable molecule. One alternative pathway that accomplishes this and results in the release of ATP energy in the absence of oxygen is fermentation. Many fermentations include glycolysis followed by some additional steps that depend on the organism involved. Some organisms are capable of producing enzymes that are able to transfer the hydrogen from $NADH_2$ to pyruvic acid. As a result, these cells release ethyl alcohol and carbon dioxide. Other cells generate enzymes that result in the production of other end products, such as acetic acid, lactic acid, acetone, diacetyl, and other organic molecules (figure 6.12).

Although there are many different products that could be formed from pyruvic acid, we will only look at two pathways. **Alcoholic fermentation** is the anaerobic respiration pathway yeast cells follow when oxygen is lacking in their environment. In this pathway, the pyruvic acid is converted to ethanol (two-carbon alcohol) and carbon dioxide. Yeast cells then are only able to generate four ATPs from glycolysis. The cost for glycolysis is still two ATPs; thus, for each glucose molecule a yeast cell oxidizes, it profits by two ATP. The waste products of carbon dioxide and ethanol are useful to humans. In bread making, the carbon dioxide is the important end product since it becomes trapped in the bread dough and makes it rise. The alcohol evaporates during the baking process. In the brewing industry, ethanol is the desirable product produced by yeast cells. Champagne, other sparkling wines, and beer are products that contain both carbon dioxide and alcohol. The alcohol accumulates and the carbon dioxide in the bottle makes it a sparkling (bubbly) beverage. In the manufacture of many wines, the carbon dioxide is allowed to escape so they are not sparkling but "still" wines.

Certain bacteria are unable to use oxygen even though it is available, making aerobic cellular respiration impossible. The pyruvic acid that results from glycolysis is converted to lactic acid by the addition of the hydrogens that had been removed from the original glucose. In this case, the net profit is again only two ATP per glucose molecule. The lactic acid buildup eventually interferes with normal metabolic functions and the bacteria die. We use the lactic acid waste product from these types of anaerobic bacteria when we make yogurt, cultured sour cream, cheeses,

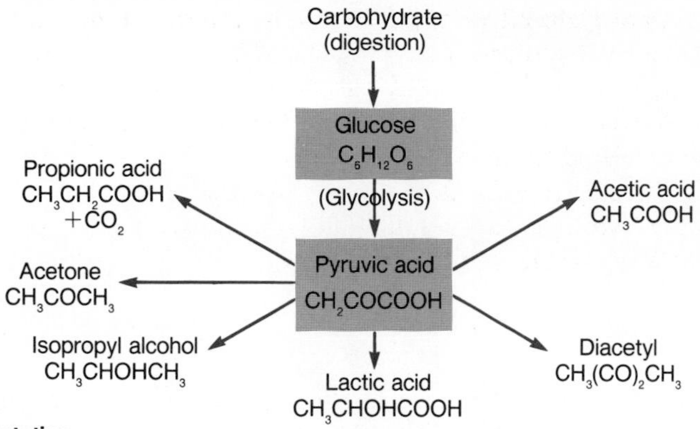

Figure 6.12
A Variety of Fermentations.
This biochemical pathway illustrates the digestion of a complex carbohydrate to glucose followed by the glycolytic pathway forming pyruvic acid. Depending on the genetic makeup of the organism and the enzymes they are able to produce, different end products may be synthesized from the pyruvic acid. The synthesis for these various molecules is the organism's particular way of oxidizing $NADH_2$ to NAD and reducing pyruvic acid to a new end product. Many bacteria use the fermentation process to generate their ATPs and in the process produce a variety of end products that are important in our lives.

Fermentation Product	Possible Source	Importance
Acetic acid	Bacteria: *Leuconostoc* sp. *Acetobacter* sp.	Sours beer Produces vinegar
Diacetyl	Bacteria: *Streptococcus diacetilactis*	Provides fragrance and flavor to buttermilk
Lactic acid	Bacteria: *Lactobacillus bulgaricus*	Aids in changing milk to yogurt
	Human: Muscle cells	Produced when O_2 is limited. Results in pain and muscle inaction.
Isopropyl alcohol	Bacteria: *Clostridium perfringens*	Causes tissue destruction during gas gangrene
Acetone	Bacteria: *Clostridium pasteurinum*	Industrial production for commercial use
Propionic acid $+CO_2$	Bacteria: *Propionibacterium shermani*	Produces the "eyes" and flavor of Swiss cheese

and other fermented dairy products. The lactic acid makes the milk protein coagulate and become puddinglike or solid. It also gives the products their tart flavor, texture, and aroma.

In the human body, different cells have different metabolic capabilities. Red blood cells lack mitochondria and must rely on lactic acid fermentation to provide themselves with energy. Nerve cells can only use glucose aerobically. As long as oxygen is available to muscle cells, they function aerobically. However, when oxygen is unavailable—because of long periods of exercise, or heart or lung problems that prevent oxygen from getting to the cells—the muscle cells make a valiant effort to meet your energy demands and function anaerobically. While your cells are functioning anaerobically, they are building up an oxygen debt. These cells produce lactic acid as their fermentation product. Much of the lactic acid is transported by the bloodstream to the liver where about 20% is metabolized through the Krebs cycle and 80% is resynthesized into glucose. Even so, there is still a buildup of lactic acid in the muscles. It is the lactic acid buildup that makes the muscles tired when exercising (figure 6.13). When the lactic acid concentration becomes great enough,

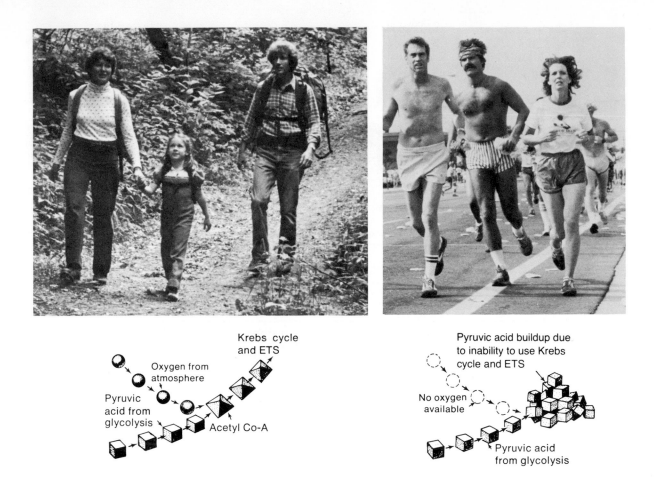

Krebs cycle
and ETS

Oxygen from
atmosphere

Pyruvic
acid from
glycolysis

Acetyl Co-A

Pyruvic acid buildup due
to inability to use Krebs
cycle and ETS

No oxygen
available

Pyruvic acid
from glycolysis

Figure 6.13
Oxygen Starvation.
When oxygen is available to all
cells, the pyruvic acid from
glycolysis is converted into acetyl
Co-A, which is sent to the Krebs
cycle. When oxygen is not
available in sufficient quantities
(due to lack of environmental
oxygen or inability to circulate
oxygen to cells using it), some of
the pyruvic acid from glycolysis is
converted to lactic acid which
builds up in the cells.

lactic acid fatigue results. Its symptoms are cramping of the muscles and pain. Due to the pain, we generally stop the activity before the muscle cells die. As you cool down after a period of exercise, your breathing and heart rate stay high until the oxygen debt is repaid and the level of oxygen in the muscle cells returns to normal. During this period, you are converting some of the lactic acid that has accumulated back into pyruvic acid. The pyruvic acid can now continue through the Krebs cycle and the ETS as you make oxygen available.

Metabolism of Other Molecules

Up to this point we have described the methods and pathways that allow organisms to release the energy tied up in carbohydrates. Frequently cells may lack sufficient carbohydrates, but have other materials from which energy could be removed. Fats and proteins, in addition to carbohydrates, make up the diet of many organisms. These three foods provide the building blocks for the cells and all can provide energy. The pathways that organisms use to extract this chemical-bond energy are summarized here.

Fat Respiration

Fats consist of a molecule of glycerol with three fatty acids attached to it. Before fats can undergo oxidation and release energy they must be broken down into glycerol and fatty acids. The three-carbon glycerol molecule can be converted into PGAL.

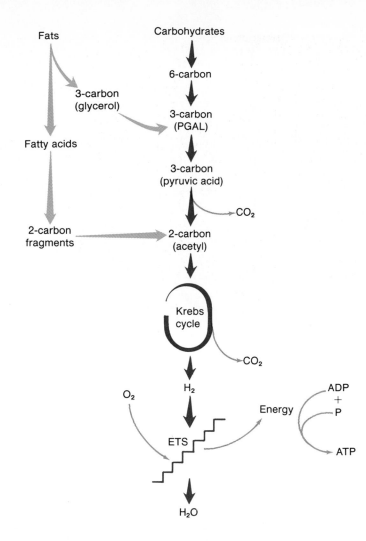

Fats

Carbohydrates

3-carbon (glycerol)

6-carbon

Fatty acids

3-carbon (PGAL)

3-carbon (pyruvic acid)

CO_2

2-carbon fragments

2-carbon (acetyl)

Krebs cycle

CO_2

O_2

H_2

Energy

ADP + P

ETS

ATP

H_2O

Figure 6.14
Fat Metabolism.
Fats must be digested into glycerol and fatty acids before they can undergo respiration. The glycerol is converted into PGAL and then proceeds through the respiratory pathway. Fatty acid molecules are broken into two-carbon fragments that then enter the respiratory pathway as acetyl. The important thing to realize is that fats can be converted directly into units from which energy can be released.

The PGAL can then be inserted into the glycolytic pathway. However, each of the fatty acids must be processed before they can enter the pathway. Each long chain of carbons that makes up the carbon skeleton is hydrolysed into two-carbon fragments. Next, each of the two-carbon fragments is converted into an acetyl molecule. The acetyl molecules are carried into the mitochondria by coenzyme A molecules. If you follow the glycerol and each two-carbon fragment through the cycle, you can see that each molecule of fat has the potential to release several times as much ATP as a molecule of glucose. Notice that each glucose molecule has six pairs of hydrogen, while a typical molecule of fat has up to ten times that number. This is the reason fat makes such a good long-term storage material. It is also why the removal of fat on a weight-reducing diet takes so long! It takes time to use all the energy contained in the hydrogen pairs of fatty acids. On a weight basis, there are twice as many calories in a gram of fat as there are in a gram of carbohydrate. Notice in figure 6.14 that both carbohydrates and fat can enter the Krebs cycle and release energy. Although you require both fat and carbohydrates in your diet, they need not be in precise ratios since your body can make some interconversions. This means that people who eat excessive amounts of carbohydrates will deposit body fat. It also means that people who starve can generate glucose by breaking down fats and using the glycerol to synthesize glucose.

Figure 6.15
Protein Metabolism.
Proteins must be digested into amino acids before they can undergo respiration. The amino acids are converted into various keto acids by removing the amino group. The keto acids enter the respiratory pathway as pyruvic acid or as one of the other acids in the Krebs cycle. The amino group is attached to other waste products and then eliminated.

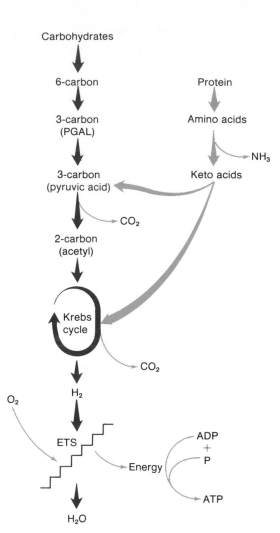

Protein Respiration

Proteins are able to be interconverted just as fats and carbohydrates are. The first step in utilizing protein for energy is to digest the protein into individual amino acids. Each amino acid then needs to have the amino group ($-NH_2$) removed. The remaining carbon skeleton, a keto acid, is changed and enters the respiratory cycle as pyruvic acid or as one of the other types of molecules found in the Krebs cycle. These acids have hydrogens that are part of their structure. As the acids progress through the Krebs cycle and the ETS, the hydrogens are removed and their energy released. The amino group that was removed is converted into ammonia. Some organisms excrete ammonia directly, while others convert ammonia into other nitrogen-containing compounds like urea or uric acid. All of these molecules are toxic and must be eliminated. They are transported in the blood to the kidneys where they are eliminated. In the case of a high-protein diet, you need to increase your fluid intake to allow the kidneys to efficiently remove the urea or uric acid.

When you eat any protein, you are able to digest it into its component amino acids. These amino acids are then available to be used to construct other proteins. If there is no need to construct protein, the amino acids are metabolized to provide energy (figure 6.15) or can be converted to fat for long-term storage. One of the most important ideas you need to recognize from this discussion is that carbohydrates, fats, and proteins can all be used to provide energy. The fate of any type of nutrient in a cell depends on the momentary needs of the cell.

An organism whose daily food energy intake exceeds the daily energy expenditure will convert only the necessary amount of food into energy. The excess food will be interconverted depending on the enzymes present and the needs of the organism at that time. In fact, glycolysis and the Krebs cycle allow molecules of the three major food types (carbohydrates, fats, and proteins) to be interconverted.

As long as a person's diet has a certain minimum of each of the three major types of molecules, the cells' metabolic machinery can interconvert molecules to satisfy its needs. If a person is on a starvation diet, the cells will use stored carbohydrates first. Once the carbohydrates are gone (about two days), cells will begin to metabolize stored fat. When the fat is gone (a few days to weeks), the proteins will be used. A person in this condition is likely to die.

If excess carbohydrates are eaten, they are often converted to other carbohydrates for storage or can be converted into fat. A diet that is excessive in fat results in the storage of fat. Proteins are not able to be stored. If they, or their component amino acids, are not needed immediately, they will be converted into fat, carbohydrates, or energy. This presents a problem for those individuals who do not have ready access to a continuous source of amino acids (i.e., individuals on a low-protein diet). They must convert important cellular components into protein as they are needed. This is the reason why protein and amino acids are considered an important daily food requirement (figure 6.16).

Plant Metabolism

At the beginning of this chapter we considered the conversion of carbon dioxide and water into PGAL through the process of photosynthesis. We described PGAL as a very important molecule because of its ability to be used as a source of energy. Plants and other autotrophs obtain energy from food molecules in the same manner as animals and other heterotrophs. They process the food through the respiratory pathways. This means that plants, like animals, require oxygen for the ETS portion of aerobic cellular respiration. Many people believe that plants only give off oxygen and never require it. This is incorrect! Plants do give off oxygen in the light energy conversion stage of photosynthesis, but in aerobic cellular respiration they use oxygen just like any other organism. During their life span, green plants give off more oxygen to the atmosphere than they take in for use in respiration. The surplus they give off is the source of oxygen for aerobic cellular respiration in both plants and animals. Animals are not only dependent on plants for oxygen, but are ultimately dependent on plants for the organic molecules necessary to construct their bodies and maintain their metabolism (figure 6.17).

Plants, by a series of reactions produce the basic foods for animal life. To produce PGAL, which can be converted into carbohydrates, proteins, and fats, plants require carbon dioxide and water as raw materials. The carbon dioxide and water are available from the atmosphere since they are deposited there as waste products of aerobic cellular respiration. To make the amino acids that are needed for proteins, plants require a source of nitrogen. This is available in the waste materials from animals.

Thus, animals supply raw materials—CO_2, H_2O, and nitrogen—needed by plants, while plants supply raw materials—sugar, oxygen, amino acids, fats, and

Figure 6.16
Interconversion of Fats, Carbohydrates, and Proteins.

Cells do not necessarily utilize all food as energy. One type of food can be changed into another type to be used as raw materials for construction of needed molecules or for storage.

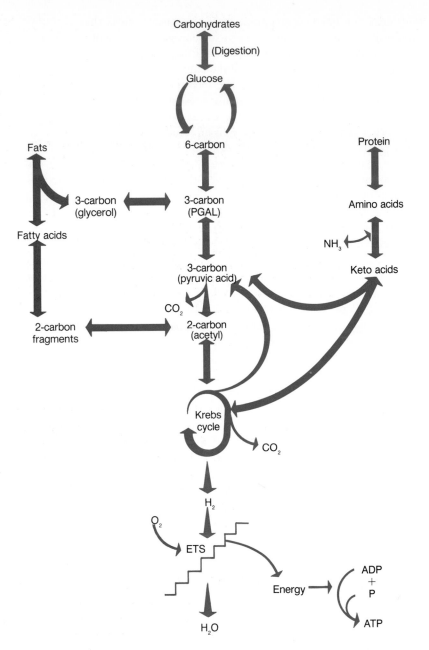

vitamins—needed by animals. This constant cycling is essential to life on earth. As long as the sun shines and plants and animals remain in balance, the food cycles of all living organisms will continue to work properly.

Summary

In the light energy conversion stage of photosynthesis, plants use chlorophyll to trap sunlight to manufacture a source of chemical energy, ATP, and a source of hydrogen, $NADPH_2$. Atmospheric oxygen is released in this stage. The ATP energy

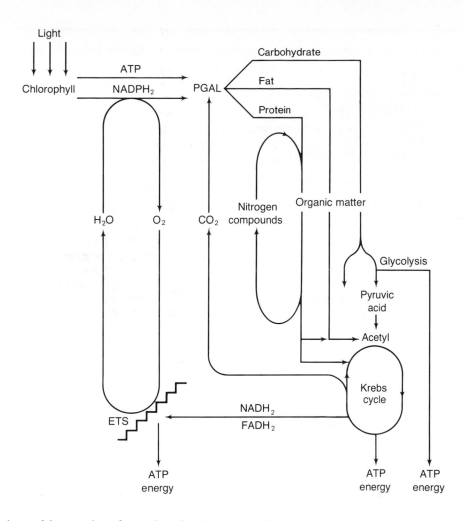

Figure 6.17
Interdependency of Photosynthesis and Respiration.

Plants use the end products of animal respiration—carbon dioxide, water, and nitrogen compounds—to produce various foods. Animals use the end products of plants—food and oxygen—as a source of energy. Therefore, plants are dependent upon animals and animals are dependent upon plants.

is used in a series of reactions in the carbon dioxide conversion stage of photosynthesis to join the hydrogen from the $NADPH_2$ to a molecule of carbon dioxide to form a simple carbohydrate, PGAL.

In subsequent reactions, the plant uses the PGAL as a source of energy and raw materials to make complex carbohydrates, fats, and other organic molecules. With the addition of ammonia, the plant can form proteins.

In the process of respiration, organisms convert foods into energy (ATP) and waste materials (carbon dioxide, water, and nitrogen compounds). Organisms that have oxygen (O_2) available can go through the Krebs cycle and electron transfer system (ETS) and get much more energy per sugar molecule than those that undergo fermentation, since fermenters rely entirely on glycolysis. Glycolysis and the Krebs cycle serve as a molecular interconversion system. Fats, proteins, and carbohydrates are interconverted depending on the needs of the cell.

The waste materials of respiration, in turn, are used by the plant. Therefore, there is a constant recycling of materials between plants and animals. Sunlight supplies the essential initial energy to make the large organic molecules necessary to maintain the forms of life we know.

Consider This

Both plants and animals carry on metabolism. From a metabolic point of view, which is the most complex? Include in your answer the following topics:

1. Cell structure
2. Biochemical pathways
3. Enzymes
4. Organic molecules
5. Autotrophy and heterotrophy

Experience This

For this experience, you will need a large glass jar, with a lid, that you can fit your hand into. Place several cups of pond water with green "scum" into the jar. The green scum is a population of green plantlike organisms that are probably algae. Place the jar in a partially sunny spot. Watch it for the next several days. Note that inside the jar, bubbles develop when sunlight is available. You can collect the bubbles of gas in a test tube filled with water and inverted over the green algae. To determine what kind of gas has been released from the algae remove the test tube from the water and immediately insert a "just blown out" wooden match into it. If the glowing match bursts into flame, the gas collected was oxygen. The release of oxygen by these organisms indicates that photosynthesis was taking place.

Questions

1. What is a biochemical pathway? Give two examples.
2. List four ways in which photosynthesis and aerobic respiration are similar.
3. Photosynthesis is a biochemical pathway that occurs in two stages. What are the two stages, and how are they related to each other?
4. Why does aerobic respiration yield more energy than anaerobic respiration?
5. Even though animals do not photosynthesize, they rely on the sun for their energy. Why is this so?
6. Explain the importance of each of the following:
   NADP in photosynthesis
   PGAL in photosynthesis and in respiration
   Oxygen in aerobic cellular respiration
   Hydrogen acceptors in aerobic cellular respiration
7. In what way does ATP differ from other organic molecules?
8. Pyruvic acid can be converted into a variety of molecules. Name three.
9. Which cellular organelles are involved in the processes of photosynthesis and respiration?
10. Aerobic cellular respiration occurs in three stages. Name these and briefly describe what happens in each stage.

Chapter Glossary

acetyl (ă-sēt′l) The two-carbon remainder of the carbon skeleton of pyruvic acid that is able to enter the mitochondrion.

adenosine triphosphate (ATP) (uh-den′o-sēn tri-fos′fāt) A molecule formed from the building blocks of adenine, ribose, and phosphates. It functions as the primary energy carrier in the cell.

aerobic cellular respiration (a-ro′bik sel′yu-lar res″pi-ra′shun) The biochemical pathway that requires oxygen and converts food such as carbohydrates to carbon dioxide and water. During this conversion it releases the chemical-bond energy as ATP molecules.

alcoholic fermentation (al-ko-hol'ik fur''men-ta'shun) The anaerobic respiration pathway in yeast cells. During this process, pyruvic acid from glycolysis is converted to ethanol and carbon dioxide.

anaerobic respiration (an'uh-ro''bik res''pi-ra'shun) A biochemical pathway that does not require oxygen for the production of ATP.

autotrophs (aw'to-trōfs) Organisms that are able to make their food molecules from sunlight and inorganic raw materials.

biochemical pathway (bi'o-kem''i-kal path'wa) A major series of enzyme-controlled reactions linked together.

carbon dioxide conversion stage (kar'bon di-ok'sid kon-vur'zhun stāj) The second stage of photosynthesis during which inorganic carbon from carbon dioxide becomes incorporated into a sugar molecule.

cellular respiration (sel'yu-lar res''pi-ra'shun) Major biochemical pathway during which cells release the chemical-bond energy from food and convert it into a usable form (ATP).

chlorophyll (klo'ro-fil) A molecule directly involved in the light-trapping and energy conversion process of photosynthesis.

coupled reactions (kup'ld re-ak'shuns) The linkage of a set of energy-requiring reactions with energy-releasing reactions.

electron transfer system (e-lek'tron trans'fur sis'tem) The series of oxidation-reduction reactions in aerobic cellular respiration in which the energy is removed from hydrogens and transferred to ATP.

FAD (**f**lavin **a**denine **d**inucleotide) A hydrogen carrier used in respiration.

glycolysis (gli-kol'i-sis) The anaerobic first stage of cellular respiration that consists of the enzymatic breakdown of a sugar into two molecules of pyruvic acid.

heterotrophs (he'tur-o-trōfs) Organisms that require an external supply of food to provide a source of energy.

high-energy phosphate bond (hi en'ur-je fos-fāt bond) The bond between two phosphates in an ADP or ATP molecule that readily releases its energy for cellular processes.

Krebs cycle (krebs si'kl) The series of reactions in aerobic cellular respiration that results in the production of two carbon dioxides, the release of four pairs of hydrogens, and the formation of an ATP molecule.

light energy conversion stage (lit en'ur-je kon-vur'zhun stāj) The first of the two stages of photosynthesis during which light energy is converted to chemical-bond energy.

NAD (**n**icotinamide **a**denine **d**inucleotide) An electron acceptor and hydrogen carrier used in respiration.

NADP (**n**icotinamide **a**denine **d**inucleotide **p**hosphate) An electron acceptor and hydrogen carrier used in photosynthesis.

PGAL (**p**hospho**g**lycer**al**dehyde) The end product of the carbon dioxide conversion stage of photosynthesis produced when a molecule of carbon dioxide is incorporated into a larger organic molecule.

photosynthesis (fo-to-sin'thuh-sis) A major biochemical pathway in green plants resulting in the manufacture of food molecules.

pyruvic acid (pi-ru'vik ass'id) A three-carbon carbohydrate that is the end product of the process of glycolysis.

Digestion and Nutrition

■ Chapter Outline

■ Purpose

The biochemical pathways described in the previous chapter all occur inside living cells. In humans and many other organisms, food cannot directly enter these metabolic processes. The food eaten must be chemically prepared before it can be used. This chapter deals with digestion, nutrition, and the factors that influence our health.

For Your Information

Some people from Europe, Great Britain, and other parts of the world who visit the United States find us to be a society that is highly self-restricting. For example, we cannot eat what we want because we are afraid of getting fat. They contend that they have no such problems. But what many fail to realize is that what and how much a person is able to eat depends not only upon their genetic makeup, but on their life-style. People from other countries, especially those less advanced than ours, live very different life-styles—a fact that very few Americans appreciate.

Learning Objectives

■ Understand the concept of basal metabolism and how it is determined.

■ Recognize the difference between a complete and incomplete protein.

■ Be able to explain the role of growth factors in metabolism.

■ Know the structures of the human digestive system and the function of each.

■ Be familiar with the four basic food groups, know examples of each, their sources, and the benefits of maintaining good health.

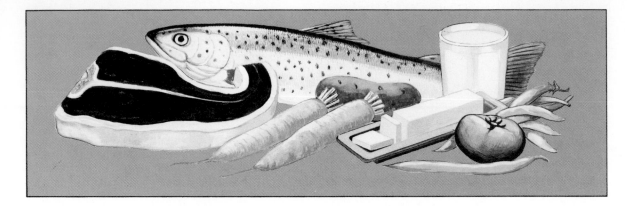

Figure 7.1
Basic Foods.
The kinds of foods shown here contain the five classes of nutrients. Any one of these foods might not contain all of the five necessary nutrients, but together they can supply the body with proteins, carbohydrates, lipids, vitamins, and chemical elements.

Digestive Function

The **digestive tract** is a long tube that passes through the body and has specialized sections that break down large food molecules into smaller molecules. The basic function of this system is to provide the cells of the body with the five essential ingredients that keep them alive and well. These five classes of nutrients are the proteins, carbohydrates, lipids, vitamins, and minerals (figure 7.1). For you to remain in good health, your diet must include foods that contain these five classes of nutrients. Carbohydrates, proteins, and lipids are the sources of energy and raw materials for growth. As noted in the previous chapter on cell metabolism, all these types of molecules may come together in the metabolic "mill" of the cell and be converted from one form to another. But the type and amount of food you take into your body each day affects how these conversions take place.

The digestive system does not absorb all of the protein, carbohydrate, and lipid molecules that flow through it. Of those that are taken in, there is a very specific order for the uptake of nutrients. Usually, carbohydrates are taken into the body first and proteins last. This is also the order in which each of the three types of molecules are respired in the cells. Your cells begin with the respiration of the carbohydrates. About five hours later, your body shifts to burning the fats for energy. The proteins are left for last. If you were to measure the amount of usable energy your body extracts from a gram of each of these types of molecules, it would come out to be 4.0 kilocalories per gram of carbohydrate, 9.0 kilocalories per gram of fat, and 4.0 kilocalories per gram of protein.

Calories and Food

The unit used to measure the amount of energy in food that is available to the body is the **calorie.** A calorie is the amount of heat it takes to raise the temperature of one gram of water one degree Celsius. Do not confuse a calorie with a kilocalorie, a common mistake. One **kilocalorie** is the amount of energy needed to raise the temperature of one *kilo*gram of water one degree Celsius. Remember that the prefix *kilo* means one thousand times the value listed. Therefore, a kilocalorie is one thousand times more heat energy than a calorie. Over the years these terms have been misused since many people changed the term kilocalorie to *Calorie*, written with a capital C. In other words, one thousand calories equal one *Calorie*. As time passed, the "big C" got lost in the shuffle; so when you talk about a person being on a one-thousand calorie diet, you really mean that he or she is on a one million calorie diet

Table 7.1   A Weight-Control Diet—1,500 Kilocalories or 1,500,000 Calories

With any diet, there are general guidelines to follow. Baking, boiling, or broiling meats and fish add the least amount of fat to the diet. Prepare all foods using measured amounts of fats or flour and measure the amount of foods. Remember, high carbohydrate foods (for example, candy, sugars, and beer) and high fat foods (for example, creamed foods) should be avoided.

	Measured	Nonmeasured
Breakfast	Orange juice, ½ cup	Coffee or tea
	Egg, 1	
	Milk, 1 cup	
	Cereal, dry, ¾ cup	
	Cream, sweet, 2 tbsp	
Lunch	Cheese, 1 slice	Salad
	Anchovies, 3 medium	Coffee or tea
	Bread, 1 slice	
	Potato, ½ cup	
	Butter, 1 tsp	
	Apple, 1 medium	
	Pear, 1 small	
	Milk, ½ cup	
Dinner	Roast beef, 3 slices	Broth
	(3 × 2 × ⅛ in.)	Coffee or tea
	Green beans, 1 cup	
	Carrots, ½ cup	
	Bread, 1 slice	
	Butter, 1 tsp	
	Milk, ½ cup	
	Peach, 1 medium	

Table 7.2   Energy Requirements

This list of activities shows the amount of energy expended (measured in kilocalories) if the activity is performed for an hour.

Kinds of Activity	Kilocalories (Per Hour)
Walking up stairs	1,100
Running (a jog)	570
Swimming	500
Vigorous exercise	450
Slow walking	200
Dressing and undressing	118
Sitting at rest	100

(table 7.1). Therefore, you are always really eating one thousand times the calories listed in dieting books! The energy requirements for different types of activity are listed in table 7.2. These are all listed in "real" calories. All the activities include physical exercise since the majority of energy expenditure is through muscular activities.

Figure 7.2
Starving and Stored Foods.

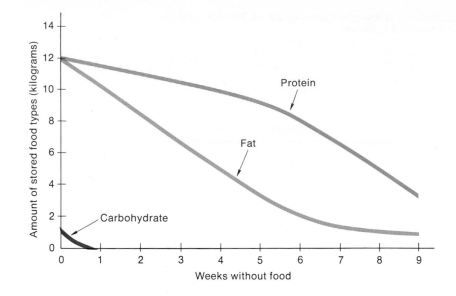

As a person engages in the activities of everyday life, the body constantly uses energy that it acquired from the food eaten. At no time are the food molecules entering the body locked into some part of the cells. They are constantly being changed and exchanged; fat molecules, for example, are completely exchanged about every four weeks. This exchange is easily seen when a person either gains or loses weight. In each case, the molecules of the body are rearranged. The nature of the rearrangement depends on the amount of incoming food and the amount of activity. If a person's activity level increases and eating habits remain the same, there will be a loss of weight. On the other hand, if the activity level drops and eating habits remain the same, the person will gain weight. Of course, there are extremes of both cases. A person who gains a great deal of unnecessary weight is **obese.** Obesity occurs when people take in more food energy than is necessary to meet their activity requirements. Many physically inactive persons become fat simply because it is conventional to eat "three square meals" a day. In some cases it is difficult to avoid eating more food than is really required for good health. For example, most social gatherings have high-calorie snack foods available, and many business people have daily luncheon dates. The solution to obesity is biologically simple; eat the amount of food calories that corresponds to your daily activity level. If you eat less and increase your activity, you will lose weight. Don't fall for fad diets; just cut down on portions and frequency of eating. Rarely does a person have a "glandular problem." But if you honestly try to rid yourself of excess weight and fail, you should check with a physician. Box 7.1 describes three important psychological eating disorders—obesity, bulimia, and anorexia nervosa. These diseases have shown marked increases in recent years.

Very little carbohydrate is stored in your body. If you starve yourself, this small amount will last as a stored form of energy only for about two days. After the stored carbohydrate has been used, your body will begin to use its stored fat deposits as a source of energy; the proteins will be used last. During the early stages of starvation, the amount of fat in the body will steadily decrease, but the amount of protein will drop only slightly (figure 7.2). This can continue only up to a certain point. During about six weeks of this starvation period, the fat acts as a protein protector. You can see the value of this kind of protection when you remember the vital roles that proteins play in cellular metabolism. After six weeks, however, so much fat has been

Table 7.3 *Table for Determining Body Surface Area*

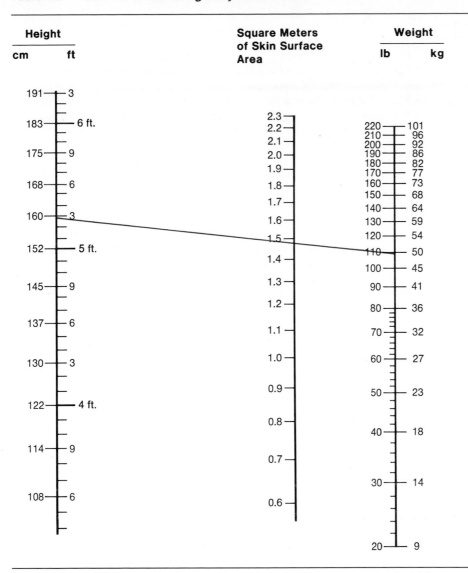

Height		Square Meters of Skin Surface Area	Weight	
cm	ft		lb	kg

Note: From Langley, L. L., et al., *Dynamic Anatomy and Physiology*, 4th ed. © 1974 McGraw-Hill, Inc., New York, NY. Reprinted by permission of McGraw-Hill Book Company.

lost from the body that proteins are no longer protected, and they will begin to be respired by the cells as a source of energy. This results in a loss of proteins from the cells of the body that prevents cells from carrying out their normal functions. When not enough enzymes are available to do the necessary cellular jobs, the cells die.

Basal Metabolism

Everyone requires a certain amount of energy to maintain basic body functions while at rest. Your **basal metabolism** is the measure of this amount of energy and is usually measured in kilocalories. To determine your basal metabolism, calculate your skin surface area (table 7.3). To use table 7.3, locate your height in the left column and

Box 7.1
**Psychological Eating Disorders—
Obesity, Bulimia, and Anorexia
Nervosa**

Eating disorders are grouped into three categories—obesity, bulimia, and anorexia nervosa. All three disorders are founded in psychological problems of one kind or another and are strongly influenced by the culture in which we live. Obesity is probably the disorder familiar to most people and the one that is most publicized. **Obesity** is the condition of being overweight to the extent that a person's health and life span is adversely affected. The majority of people suffering from this condition use food as a psychological crutch. Overeating, which results in an unhealthy, overweight person, is stimulated by outside pressures and the person attempts to cope with problems by overeating. Overeating to solve problems is encouraged by our culture. Americans celebrate almost all occasions with food. Social gatherings of almost every type are considered incomplete without some sort of food and drink. If snacks, usually high-calorie foods, are not made available by the host, many people feel uneasy or even unwelcome. It is also true that Americans and other cultures show love and friendship by sharing a meal. Most of the pictures in the albums of many families have been taken at mealtime. Controlling obesity can be very difficult since it requires a basic change in eating habits, life-style, and value system. For some this may require outside, professional help.

While the majority of cases of obesity are psychologically based, some have been demonstrated to have a biological component. It appears that some obese individuals have a chemical imbalance of their nervous system that prevents them from feeling "full" until they have eaten an excessive amount of food. This imbalance prevents the brain from "turning off" the desire to eat after a

reasonable amount of food has been eaten. Research into the nature and action of this brain chemical indicates that if obese people lacking this chemical receive it in pill form, they can become "full" eating one-quarter less food.

Bulimia is sometimes called the silent killer because it is difficult to detect. Bulimics are usually normal size or overweight. This disorder involves a cycle of eating binges and purges. The cause is thought to be psychological, stemming from depression, low self-esteem, displaced anger, a need to be in control of one's body, or a personality disorder. The cycle usually begins with an episode of overeating followed by elimination of the food by induced vomiting and the excessive use of laxatives and diuretics. Vomiting may be induced physically or by the use of some nonprescription drugs. Case studies have shown that bulimics may take forty to sixty laxatives a day to rid themselves of food. For some, the laxative becomes addictive. Diuretics are also used by bulimics to increase water loss through urination. The binge-purge cycle results in a variety of symptoms that can be deadly. The following is a list of the major symptoms observed in many bulimics:

excessive water loss
diminished blood volume
extreme potassium, calcium, and
 sodium deficiencies
kidney malfunction
increase in heart rate
loss of rhythmic heartbeat
lethargy
diarrhea
severe stomach cramps
damage to teeth and gums
loss of body proteins
migraine headaches
fainting spells
increased susceptibility to infections

Anorexia nervosa is a nutritional deficiency disease characterized by severe, prolonged weight loss. An anorexic person's fear of becoming overweight is so intense that even though weight loss occurs, it does not lessen their fear of obesity and they continue to diet. They even refuse to maintain the optimum body weight for their age, sex, and height. This nutritional deficiency disease is thought to stem from sociocultural factors. Our society's preoccupation with weight loss and the desirability of being thin strongly influences this disorder. Anorexic individuals starve themselves to death. Just turn on your television or radio, or look at newspapers, magazines, or billboards and you can see how our culture encourages people to be thin. Male and female models are thin. Muscle protein is considered to be healthy and fat to be unhealthy. Unless you are thin, so the ads imply, you will never be popular, get a date, or even marry. In fact, you may die early. Are these prophecies self-fulfilling? Our culture's constant emphasis on being thin has influenced many people (primarily women) to lose too much weight and become anorexic. Here are some of the symptoms seen in persons with anorexia nervosa:

decreased heart rate
loss of body proteins
weaker heartbeat
calcium deficiency
osteoporosis
hypothermia (low body
temperature)
hypotension (low blood pressure)
increased skin pigmentation
reduction in size of uterus
inflammatory bowel disease
slowed reflexes
fainting
weakened muscles

Table 7.4 *Kilocalories Per Day Per Square Meter of Skin*

Age	Male	Female	Age	Male	Female
6	1,265	1,217	20–24	984	886
8	1,229	1,154	25–29	967	878
10	1,188	1,099	30–40	948	876
12	1,147	1,042	40–50	912	847
14	1,109	984	50–60	886	826
16	1,073	924	60–70	859	806
18	1,030	895	70–80	828	782

Table 7.5 *Additional Kilocalories Required as Determined by Occupation*
These are general figures and will vary from person to person depending on the specific activities performed in the job.

Occupation	Kilocalories Needed Above Basal Metabolism
Sedentary (student)	500–700
Light work (business persons)	750–1,200
Moderate work (laborer)	1,250–1,500
Heavy work (professional athletes)	1,550–5,000 and up

your weight in the right column. Place a straightedge between these two points. The straightedge will cross the middle column and show your skin surface area in square meters. For example, a person 160 cm tall (5′3″) who weighs 50 kg (110 lb) has 1.5 square meters of skin surface area. The heat production in kilocalories released per square meter of skin varies with a person's age and sex (table 7.4). A 20-year-old, 160 cm, 50 kg female uses 886 kilocalories per day for each square meter of skin. Therefore, her basal metabolism is 1,329 kilocalories per day (1.5 × 886 = 1,329 kilocalories). To calculate your basal metabolism, determine your skin surface area using table 7.3 and multiply this figure by the kilocalorie figure determined from table 7.4. Remember, basal metabolism represents the amount of energy your body requires if you are at rest. Since few of us rest for twenty-four hours a day, we normally require more than the energy needed for basal metabolism. Besides age, sex, weight, and height, basal metabolism depends upon a number of factors such as climate, altitude, physical condition, race, previous diet, time of the year, and occupation. A good general indicator is the type of occupation a person has (table 7.5). If a 20-year-old, 160 cm, 50 kg female were a bank teller, she would need between 750 and 1,200 kilocalories per day above her basal metabolism of 1,329 kilocalories. Therefore, her total daily need would be somewhere between 2,079 and 2,529 kilocalories per day.

Using tables 7.3, 7.4, and 7.5, calculate your daily caloric requirements. If you exceed the caloric requirements for your height, weight, age, sex, and occupation, you will gain weight. For every extra 4,500 kilocalories you consume, you will gain

Table 7.6 *Sources of Essential Amino Acids*
The essential amino acids are required in the diet for protein building and, along with the nonessential amino acids, allow the body to metabolize all nutrients at an optimum rate. Combinations of different plant foods can provide essential amino acids even if the complete protein foods—such as meat, fish, and milk—are not in the diet.

Essential Amino Acid	Food Sources
Threonine	Dairy products, nuts, soybeans, turkey
Lysine	Dairy products, nuts, soybeans, green peas, beef, turkey
Methionine Cysteine	Dairy products, fish, oatmeal, wheat
Arginine (essential to infants only)	Dairy products, beef, peanuts, ham, shredded wheat, poultry
Valine	Dairy products, liverwurst, peanuts, oats
Phenylalanine	Dairy products, peanuts, calves' liver
Tyrosine	Dairy products, calves' liver, peanuts
Leucine	Dairy products, beef, poultry, fish, soybeans, peanuts
Tryptophan	Dairy products, sesame seeds, sunflower seeds, lamb, poultry, peanuts
Isoleucine	Dairy products, fish, peanuts, oats, macaroni, lima beans

454 g (one pound). Therefore, if you eat only 150 extra kilocalories a day for 30 days, you will gain 454 g (1 pound). If you eat 150 fewer kilocalories than you require per day for 30 days, you will convert 454 g (1 pound) of fat to energy and lose that amount of weight.

Nutritional Balance

The amount of food you eat each day should be regulated according to your activities. But even more important is the kind of food you eat. Most carbohydrates can be digested for use in the human body. So can proteins and lipids. However, not all proteins contain the same amino acids. Proteins can be divided into two main groups, the **complete proteins** and the **partial proteins.** Complete proteins contain all the amino acids necessary for good health, while partial proteins lack certain amino acids that the body must have to function efficiently. Table 7.6 lists the ten so-called **essential amino acids.** Without minimal levels of these amino acids in the diet, a person may develop health problems that could ultimately lead to death. In many parts of the world, large populations of people live on diets that are very high in carbohydrates and fats but low in complete protein. This is easy to understand, since carbohydrates and fats are inexpensive to grow and process in comparison to proteins. For example, corn, rice, wheat, and barley are all high-carbohydrate foods (figure 7.3). Corn and its products (meal, flour) contain protein, but it is a partial protein that lacks the amino acid tryptophan. Without this amino acid, many necessary enzymes cannot be made in sufficient amounts to keep the person healthy. One protein deficiency disease is called **kwashiorkor,** and the symptoms are easily seen (figure 7.4). A person with this deficiency has a distended belly, slow growth,

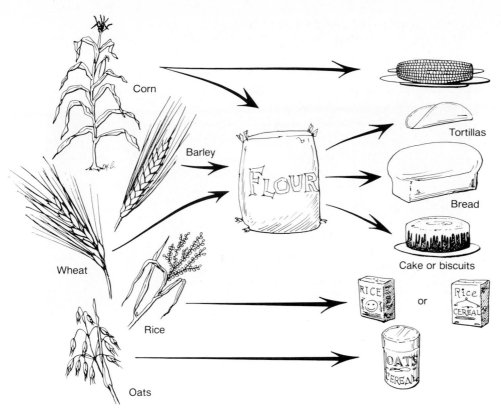

Figure 7.3
Carbohydrates in Your Diet.
Many of the world's food staples
are plant carbohydrates that have
been ground into flours or that are
eaten directly. Combinations of
certain grains can supply the
essential amino acids in a diet.

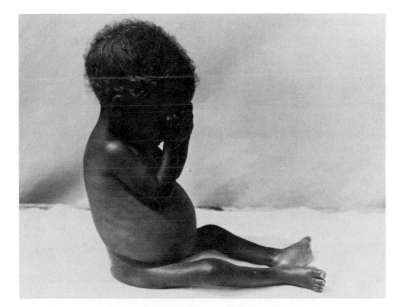

Figure 7.4
Kwashiorkor.

This starving child shows the
symptoms of the protein deficiency
disease kwashiorkor. If treated with
a proper diet containing all amino
acids, the disease can be cured.

slow movement, and is emotionally depressed. If caught in time, brain damage may
be prevented and death averted. This requires a change in diet that includes expensive complete protein. Such a diet might include poultry, fish, beef, shrimp, or
milk. As the world food problem increases, these expensive foods will be in even
shorter supply and will become more and more costly.

Figure 7.5
Hybrid Corn.
Selective breeding of grains can
produce crops that are higher in
certain essential amino acids than
are the common varieties. These
are especially high in the amino
acid lysine.

Biologists have changed many inexpensive crops that were low in complete protein so that they now contain more amino acids. You may have driven past a cornfield next to which was a sign marked "L3369" (figure 7.5). The corn being grown in this cornfield has been produced by biologists because it has an exceptionally high content of the amino acid lysine in comparison to other corn plants. The production of these special corn plants may help supplement the protein portion of the diets of people throughout the world.

Growth Factors

Even if all the proteins, carbohydrates, and lipids are supplied to the cells, a person may not attain good health. Almost all cells require small amounts of certain other materials that aid the functioning of the cell or are a necessary part of the cell's structure. These materials are called **growth factors,** but are more commonly referred to as **vitamins** and **minerals.** All minerals must be acquired through the diet, since the cells cannot manufacture chemical elements. Examples of minerals and their functions are listed in table 7.7. The lack of a mineral results in certain abnormal reactions (table 7.8). Small amounts of vitamins are also needed in the diet. Vitamins are not elements, but organic molecules manufactured by some other organisms. Table 7.9 shows some of the sources of the various vitamins and the roles they play in keeping a person healthy. The lack of a particular vitamin in the diet can result in a **vitamin deficiency disease.** These diseases are listed in table 7.10.

A great deal has been said about the need for vitamin and mineral supplements in diets. Some people claim that supplements are essential, while others claim that a well-balanced diet provides adequate amounts of vitamins and minerals. Supporters of vitamin supplements have even claimed that extremely high doses of certain vitamins can prevent ill health or even create supermen. It is very difficult to evaluate many of these claims, since the functioning of vitamins and minerals and their regulation in the body is not completely understood. In fact, the minimum

Table 7.7 *Sources and Uses of Minerals in the Human Body*

Mineral	Use in Body	Food Source
Calcium	Building of bones and teeth; aids in clotting of blood; regulation of heart, nerve, and muscle activity; enzyme formation; milk production	Asparagus, beans, cauliflower, cheese, cream, egg yolk, milk
Chlorine	Regulation of osmotic pressure; enzyme activities; formation of hydrochloric acid in stomach	Bread, buttermilk, cabbage, cheese, clams, eggs, ham (cured), sauerkraut, table salt
Cobalt	Normal appetite and growth; prevention of a type of anemia; prevention of muscular atrophy	Liver, seafoods, sweetbreads
Copper	Formation of hemoglobin; aids in tissue respiration	Bran, cocoa, liver, mushrooms, oysters, peas, pecans, shrimp
Iodine	Formation of thyroxin; regulation of basal metabolism	Broccoli, fish, iodized table salt, oysters, shrimp
Iron	Formation of hemoglobin; oxygen transport; tissue respiration	Almonds, beans, egg yolk, heart, kidney, liver, meat, soybeans, whole wheat
Magnesium	Muscular activity; enzyme activity; nerve maintenance; bone structure	Beans, bran, brussels sprouts, chocolate, corn, peanuts, peas, prunes, spinach
Phosphorus	Tooth and bone formation; buffer effects in the blood; essential constituent of all cells; muscle contraction	Beans, cheese, cocoa, eggs, liver, milk, oatmeal, peas, whole wheat
Potassium	Normal growth; muscle function; maintenance of osmotic pressure; buffer action, regulation of heart beat	Beans, bran, molasses, olives, oranges, parsnips, potatoes, spinach
Sodium	Regulation of osmotic pressure; buffer action; protection against excessive loss of water	Beef, bread, cheese, oysters, spinach, table salt, wheat germ
Sulfur	Formation of proteins	Beans, bran, cheese, cocoa, eggs, fish, lean meat, nuts, peas
Zinc	Normal growth; tissue respiration	Beans, cress, lentils, liver, peas, spinach

Note: Adapted from Morrison, et al., *Human Physiology*. New York: Holt, Rinehart and Winston, Publishers, 1977, p. 186. Used with permission.

Table 7.8   *Mineral Deficiencies*

Mineral	Symptoms of Deficiency
Calcium	Stunted growth, rickets, convulsions
Chlorine	Reduced appetite, muscle cramps
Cobalt	Not reported for humans
Copper	Rare in humans. Anemia
Iodine	Enlarged thyroid gland (goiter)
Iron	Anemia
Magnesium	Reduced growth, behavioral disturbances, spasms
Phosphorus	Loss of calcium and weakening of bones
Potassium	Weakened muscles, paralysis
Sodium	Reduced appetite, muscle cramps
Sulfur	Associated with deficiency of sulfur-containing amino acids, methionine and cysteine. Kwashiorkor
Zinc	Stunted growth, small sex glands

Note: Data from "The Requirements of Human Nutrition" by Nevin S. Scrimshaw and Vernon R. Young, September 1976, *Scientific American.*

Table 7.9   *Sources and Uses of Vitamins in the Human Body*
These are only a few of the vitamins used in human metabolism. Notice that a number of them have been referred to earlier in the chapters on enzymes and respiration.

Vitamin	Food Source	Use in Body
Thiamine (B$_1$)	Peas, beans, eggs, pork, liver	Coenzyme used in Krebs cycle
Riboflavin (B$_2$)	Milk, whole grain cereals, green vegetables, liver, eggs	Part of coenzyme used in electron transfer system (FAD)
Niacin (nicotinic acid)	Milk, poultry, yeast, cereal	Part of coenzyme used in electron transfer system (NAD)
Pyridoxine (B$_6$)	Most foods	Coenzyme used in synthesis of amino acids
Cyanocobalamin (B$_{12}$)	Meats, dairy products	Used in red-blood-corpuscle formation
Ascorbic acid (C)	Citrus fruits, vegetables	Part of cell cement used to hold cells together
E	Green vegetables, vegetable oils in most foods	Maintains fertility
D[a]	Dairy products, fish oil	Aids in calcium use in bones
A	Dairy products, vegetables	Used in formation of visual pigment; maintains skin (action not known)

[a]Vitamin D came to be known as a vitamin because of a mistaken idea that it is taken in through food rather than formed in the skin on exposure to sunshine. It would be more correct to call it a hormone, but no one does.

Table 7.10 *Vitamin Deficiencies*

Vitamin	Symptoms of Deficiency
B_1	Beriberi—breakdown of nerve cells and muscle, heart failure
B_2	Cracking of skin around the eyes and mouth; skin infections
Niacin	Skin infections, diarrhea, insanity
B_6	Vitamin is so easily obtainable that no deficiency disease has been noted
B_{12}	Pernicious anemia—defective formation of red blood corpuscles
C	Scurvy—small blood vessels break just under the skin and around mouth
E	Sterility and weakness in rats; human disease not fully known
D	Rickets—soft, misshapen bones
A	Night blindness; dry skin; leads to infections of skin

daily requirement of a number of vitamins has not been determined. The information in tables 7.7 and 7.8 should be used in considering the kinds of foods you eat each day.

Basic Food Groups

Planning your diet around the four basic food groups is generally easier than trying to account for the requirements of calories, essential amino acids and fats, and the multitude of vitamins and minerals. These four basic food groups provide a guideline to help you maintain a balanced diet.

You should have some items from each of the following groups of food each day.

Group 1

Meat, poultry, or fish. This group also includes the meat substitutes, such as nuts, beans, peas, and eggs. About two servings from this group is the recommended amount each day. These foods furnish minerals, vitamins, and proteins. Because protein is not readily stored in the human body, it is important that you eat the recommended amount from this group each day. Vegetarians must pay particular attention to acquiring an adequate source of protein since they have eliminated the usual sources from their diet.

Group 2

Fruits and vegetables. This includes all of the raw and cooked fruits and vegetables that you eat. You should eat about four servings each day from this group. They provide minerals, vitamins, and bulk or fiber in your diet. You should try to vary the type of vegetable or fruit you eat so that you get all the different vitamins and minerals available. The fiber in your diet ensures the proper functioning of the digestive tract. There seems to be a correlation between some cancers in the lower intestinal tract and the lack of dietary fiber.

Group 3

Cereals and breads. This group includes cereals and grain products. They provide most of your calorie requirements. You should have four servings from this group each day, which might include bread, breakfast cereals, or the like. These foods also provide some of the B vitamins in your diet.

Group 4

Milk and other dairy products. All of the cheeses, ice cream, yogurt, and milk are in this group. Two servings from this group are recommended each day. These may be two glasses of milk or their equivalent. Dairy products provide minerals such as calcium in your diet, but they also provide vitamins and protein.

Sometimes we ingest materials that may be harmful, such as stimulants, acids, sugar substitutes, and drugs. Stimulants such as the caffeine in coffee, tea, or soft drinks may have adverse effects on the kidneys, the heart, and the nervous system. The acid content of soft drinks and other foods may tax the ability of the stomach to buffer itself and maintain the proper conditions for its enzymes to function best. Sugar substitutes, cyclamates, and saccharine have been linked to some cancer. Finally, medication other than prescription drugs may be harmful. The uncontrolled use of common aspirin may cause stomach upset, excessive acidity, and even complication of ulcers. Alcohol, the most commonly used drug, is a depressant that impairs the normal functioning of the nervous system. The amount of impairment is directly related to the amount of alcohol consumed. Avoid drugs and other harmful materials and plan your diet wisely. Otherwise you are doing yourself a great disservice.

Digestive Anatomy and Physiology

All the processes involved in preparing foods (nutrients) for entry into a cell are called digestion. These events take place outside the cell. Many single-celled organisms, such as bacteria, release enzymes that break down large molecules of food so that they may enter the cell. Bread mold, for example, releases its enzymes into the bread and makes it soft as it breaks down the bread (figure 7.6). Many-celled organisms usually have a large sac or cavity into which the nutrients are placed. Enzymes, which are secreted by some of the cells lining the cavity, are then mixed with the nutrients in this cavity. The digestion of large particles of food into smaller molecules, therefore, occurs outside the cell.

The process of digestion involves hydrolysis reactions in which complex carbohydrates are changed into simple sugars (see chapter 3). Fats are converted to glycerol and fatty acids by hydrolysis also. Proteins are digested or hydrolyzed to smaller chains of amino acids. These smaller food molecules are then taken into the body cells for use in metabolic pathways (figure 7.7). In the case of humans and many other complex animals, the simple digestive sac has been replaced by a very complex and lengthy system of tubes and specialized cells, which make the digestive process much more efficient. This digestive tract has many parts, each of which plays an important role in the breakdown of foods. These parts include the mouth, salivary glands, esophagus, stomach, small and large intestines, liver, gallbladder, pancreas, and anus (figure 7.8).

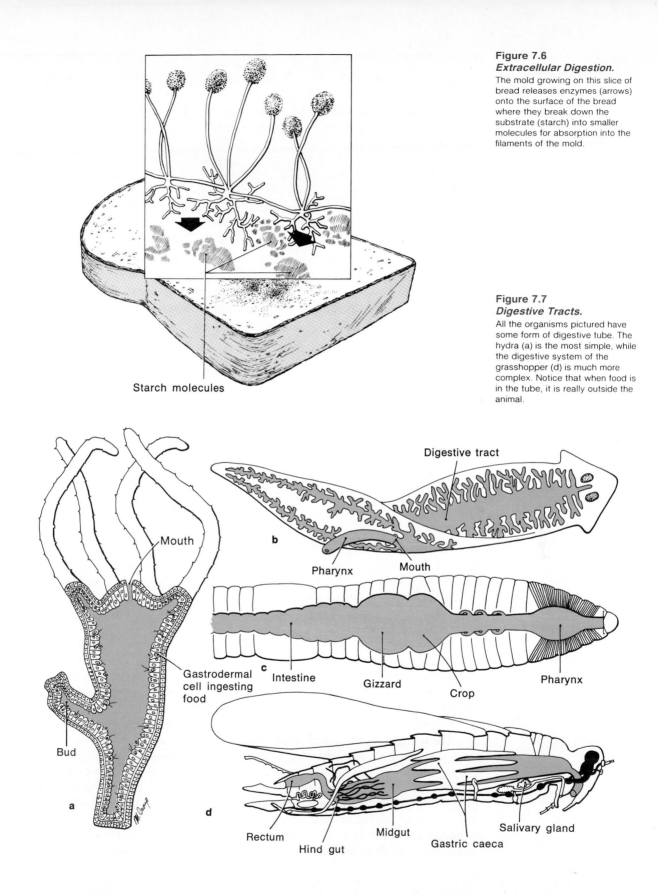

Figure 7.6
Extracellular Digestion.
The mold growing on this slice of bread releases enzymes (arrows) onto the surface of the bread where they break down the substrate (starch) into smaller molecules for absorption into the filaments of the mold.

Starch molecules

Figure 7.7
Digestive Tracts.
All the organisms pictured have some form of digestive tube. The hydra (a) is the most simple, while the digestive system of the grasshopper (d) is much more complex. Notice that when food is in the tube, it is really outside the animal.

Mouth

Digestive tract

b

Pharynx Mouth

c Intestine Gizzard Crop Pharynx

Gastrodermal cell ingesting food

Bud

a

d

Rectum Hind gut Midgut Gastric caeca Salivary gland

Figure 7.8
Human Digestive Tract.
Illustrated here are the digestive
system of a human and the
specialized organs that aid in the
chemical breakdown of foods.

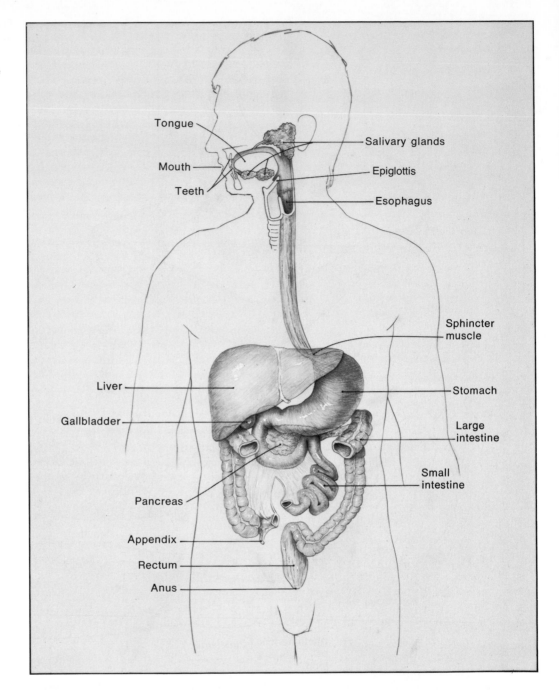

The first part of the human digestive tract is the **mouth.** The mouth, teeth, and tongue break down the incoming foods into smaller pieces that can move easily through the esophagus into the stomach. This breakdown also increases the surface area of the foods so that the digestive enzymes can unite with more of their substrate molecules. During this process, the salivary glands produce the fluid **saliva,** which adds moisture and the enzyme amylase to the food. **Amylase** begins the breakdown of starch molecules to more simple sugars. Saliva also allows many of the flavor molecules of foods to be released, enabling the taste buds on the tongue to let you

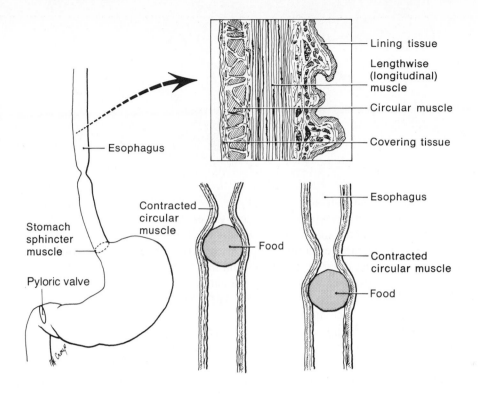

Figure 7.9
***Peristalsis of the
Esophagus.***
A series of rhythmic muscular
contractions called peristalsis
forces food through the
esophagus.

know what you are chewing. Once the food has been chewed, the tongue rolls it into a ball and presses it to the back of the mouth. This pressure acts as a trigger mechanism for swallowing, and the food begins its passage through the next part of the digestive tract, the **esophagus.** Muscles squeeze the ball of food down the esophagus in a wavelike manner (figure 7.9). This sequence of contractions, which moves the food to the stomach, is called **peristalsis.** These waves of contractions occur throughout the digestive tube and move the food all the way through to the anus (the opening at the end of the digestive tract).

Once the food enters the **stomach,** it is usually kept there by a muscle between the stomach and esophagus that remains closed. This muscle, called the **sphincter muscle,** wraps around the esophagus and when contracted or tightened squeezes the tube closed. In some cases, however, the sphincter muscle may be relaxed or forcefully opened by pressures from the stomach. When this happens, a small amount of the contents of the stomach moves back into the esophagus. Since the stomach's contents are very low in pH (acid) and contain many strong enzymes, this backup may cause an irritation in the lower portion of the esophagus. When this occurs, a person feels what is called a "heartburn." In extreme cases of stomach irritation, **vomiting** may take place. The stomach is squeezed by the muscles of the sides of the body, violently forcing the contents of the stomach through the esophagus and mouth.

The Stomach

The stomach receives about 2,000 ml (2 qt) of fluid per day from the cells lining the stomach. This fluid contains a variety of digestive agents including hydrochloric acid and the protein-digesting enzyme **pepsin.** The HCl is responsible for keeping the stomach contents at a low pH of 1.0–3.5. When food reaches the stomach, the

Figure 7.10
Stomach Ulcer.
If the mucous lining in the stomach
is lost, the pepsin and hydrochloric
acid may cause damage to the
inner wall of the stomach. Notice
the ulcer at the end of the pointer.
Should this damage become
extensive enough that it breaks
through the stomach wall, the
contents will enter the body cavity.
This may result in death.

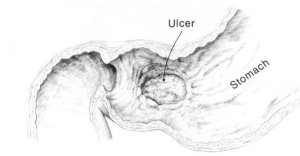

Ulcer

Stomach

stomach muscles contract and relax with regularity to churn the contents. As the food is churned, it mixes with the HCl and the enzyme. The low pH enables milk protein to be coagulated into a more solid material called casein or curd. HCl also produces a very favorable environment for the operation of the enzyme pepsin. Pepsin operates on curd and other protein foods and increases the rate of their breakdown into free amino acids. Because of the large amount of very dense food in the stomach, the pepsin does not reach all the protein. The final breakdown of the proteins is accomplished later in the digestive process.

The digestive enzymes of the stomach operate on any protein. In fact, it is impossible for pepsin to distinguish between the molecules of incoming foods and similar molecules that make up the structure of the stomach. Enzymes work on substrates, and as long as the two (enzyme and substrate) fit together properly, it makes no difference whether the substrate is food or your own cell material! How then does a person keep from digesting himself to pieces? The digestive tract is lined with many types of cells, some of which specialize in producing molecules of a thick, sticky substance called **mucus.** The mucus forms a lining on the inside of the stomach and many other structures and helps to protect the tract from the digestive enzymes and acid.

As digestion occurs, the mucous lining wears away and must constantly be replaced. Some persons fail to produce enough mucus to prevent their own stomach juices from coming in contact with the cells lining the stomach. When this happens, the enzymes and acid react with the cells of the stomach and cause damage to the stomach wall. This is called an **ulcer** (figure 7.10). If the ulcer is the result of too much pepsin being produced by the stomach, it is called a **peptic ulcer.** As you may know, the damage can become quite extensive and may even destroy some of the blood vessels of the stomach wall. When this occurs, the ulcer causes a great deal of pain, begins to bleed, and sometimes causes death.

The Small Intestine

Solid food is kept in the stomach for about three to four hours before it is released to the next part of the digestive tract, the **small intestine.** The small intestine contains an even larger assortment of digestive enzymes and other related molecules than does the stomach (see table 7.11). Some enzymes, such as amylase, continue the processes started earlier, while other enzymes begin the breakdown of other foods into their simpler, more useful parts. This assortment of enzymes and other molecules has three sources: the lining of the intestine, the liver, and the pancreas

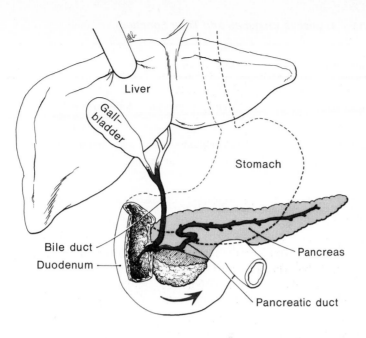

Figure 7.11
Enzymes and Digestion.
Digestive enzymes and other
molecules enter the small intestine
from the liver and pancreas
through tubes or from the lining of
the small intestine.

(figure 7.11). As the food enters the first portion of the small intestine, the **duodenum,** the cells lining the intestine are stimulated to release their enzymes. These enzymes are unable to wash back up into the stomach because of the pyloric valve, a circular muscle that, when constricted, closes off the lower end of the stomach. Both the **gallbladder** and the pancreas release their contents, which flow through tubes to the small intestine. The product of the liver, bile, is produced continuously, while the products of the pancreas are released only as needed. When bile is not needed in the duodenum, it is stored in the gallbladder.

Table 7.12 lists the various digestive enzymes and related molecules, their sources, and the chemical changes they produce in the small intestine. Here, all three main food materials are completely broken down into their simplest forms. Proteins are split into the amino acids. Fat molecules are broken down into fatty acids and glycerol. Carbohydrates are broken down into simple sugars.

The operation of all the intestinal enzymes provides the person with the variety of simple nutrients needed for proper cell metabolism. If any of the enzymes are lacking, the cells of the body do not receive the necessary nutrients, and disruptions in the intestinal tract may occur. A good example of this can be seen in people who are **lactose intolerant.** In some people, the wall of the intestine produces the enzyme **lactase.** Lactase breaks the milk sugar, lactose, into its two simpler components, glucose and galactose. Since lactose is absorbed into the small intestine and enters the blood slower than the glucose or galactose molecules, this enzymatic reaction helps sugar get to cells at a faster rate. Some persons lack this enzyme. When they eat or drink dairy products, the lactose is not digested but accumulates in the intestine. This accumulation would not be a problem, except that the intestine contains an assortment of bacteria capable of fermenting lactose. As the lactose increases, these bacteria ferment the sugar, causing the person to have cramps, gas, growling, and diarrhea. People with lactose intolerance show this reaction to milk thirty to ninety minutes after consuming dairy products. This intolerance should

Table 7.11   *Digestive Enzymes and Their Function*

Production Site	Digestive Function
Mouth (salivary glands)	
Salivary amylase	Starch → Polysaccharide fragments
Stomach	
Gastric lipase	Neutral fat → Fatty acids and glycerol
Pepsin	Protein → Peptides
Pancreas	
Carboxypeptidase	Peptides → Smaller peptides and amino acids
Chymotrypsin	Polypeptides → Peptides
Nuclease	Nucleotides → Nucleotides
Pancreatic amylase	Polysaccharides → Disaccharides
Pancreatic lipase	Fats → Fatty acids and glycerol
Trypsin	Polypeptides → Peptides
Small intestine	
Aminopeptidase	Peptides → Smaller peptides and amino acids
Dipeptidase	Dipeptides → Amino acids
Lactase	Lactose → Glucose and galactose
Maltase	Maltose → Glucose
Nuclease	Nucleic acids → Nucleotides
Sucrase	Sucrose → Glucose and fructose

Table 7.12   *Digestive Enzymes and Their Activity*

Digestible Food Types	Enzymes	Source	Product of Reaction
Proteins	Pepsin	Gastric juice	Smaller proteins
	Trypsin	Pancreatic juice	Smaller proteins
	Carboxypeptidase	Pancreatic juice	Amino acids
	Dipeptidase	Intestinal juice	Amino acids
Carbohydrates	Amylase	Saliva / Pancreatic juice / Intestinal juice	Disaccharides
	Disaccharidases	Intestinal juices	Simple sugars
Lipid	Lipase	Intestinal juice / Pancreatic juice	Fatty acids and glycerol

not be considered serious. Usually the person can control it simply by not drinking milk or eating dairy products and supplementing their diet with calcium and the other nutrients normally obtained from dairy products.

During the four to eight hours that the contents of the stomach are slowly emptied into the small intestine, a large amount of fluids is added to aid the digestive process. Along with the 2,000 ml of stomach fluids added, about 6,000 ml more

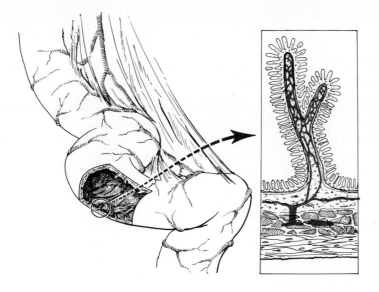

Figure 7.12
Villi of the Small Intestine.
The lining of the small intestine is not smooth but covered with microscopic projections called villi (see inset). These greatly increase the surface of the small intestine and allow better absorption of nutrients into the body.

fluids are added to the digestive tract each day. This makes the contents of the small intestine very watery. Dissolving the food materials in this solution helps these molecules be digested and absorbed into the cells that form the wall of the intestine. In fact, most of the foods you eat are taken into the body from the small intestine. Direct contact with this lining is possible since the pH of the intestine is not as acid as the stomach. The products of the pancreas change the pH of the small intestine to a more neutral condition, so a person is less likely to get an ulcer in the small intestine. On the surface of the wall of the small intestine are a great many small fingerlike projections called **villi** (singular, *villus*). These greatly increase the surface area of the small intestine (figure 7.12). Each villus contains many small blood vessels that take up the molecules of simple carbohydrates and amino acids. These blood vessels connect with larger and larger blood vessels, and the nutrients are circulated to the rest of the body. In addition, the villi contain a vessel called a lacteal (part of the lymphatic system) that absorbs fatty acids and glycerol. These lymph vessels eventually empty into the bloodstream so that the components of fats are available to cells throughout the body. The circulatory system distributes the digested nutrients to all of the cells. As a small capillary passes near a cell, the molecules of food diffuse from the blood vessel through the cell membrane into the cell. These digested food molecules, once inside the cell, can be used by the mitochondria for respiration. The process of cellular respiration was discussed in chapter 6.

Some nutrients are not easily absorbed through the villi, but are still necessary for the health of the animal. The fats are an example. Through the action of the bile, the fats can be altered in a way that allows them to pass through the villi. The bile salts break down fat into smaller droplets so they can be acted upon by the enzyme lipase, which is produced by the pancreas and intestine. The bile contains many types of molecules. One of these is the bile salts, and another is **cholesterol.** No specific function has been identified for the cholesterol in the bile, but we do know that the bile salts are produced from the cholesterol. In most people, the cholesterol is pumped from the gallbladder into the small intestine and does not build up. Some people, however, produce more cholesterol or absorb a great deal of water from the bile in the gallbladder. The cholesterol molecules can form crystals in the gallbladder, which are called **gallstones.** Gallstones may reduce the flow of bile into

Box 7.2
William Beaumont and Alexis St. Martin

On the morning of 6 June 1822, a nineteen-year-old French-Canadian fur trapper named Alexis St. Martin was shot in the stomach by the accidental discharge from a shotgun. The Army surgeon at Fort Mackinac, Michigan, Dr. William Beaumont, was called to attend the wounded man. Since part of the stomach and body wall had been shot away, Beaumont quickly dressed the wound but expected the patient to die. Finding St. Martin alive the next day, Beaumont was surprised and encouraged to do what he could to extend his life. In fact, Beaumont cared for St. Martin for two years, and the wound healed. However, the wound healed in such a way that the stomach formed an opening through the body wall. Beaumont found that he could look through the opening and observe the activities in the stomach. As a result, Beaumont was able to perform a number of experiments related to human digestion. He obtained pure gastric juice from the stomach and noted its effects on food outside of the body. He also suspended food by a string and noted the progress of digestion in St. Martin's stomach. St. Martin did not take kindly to these probings and twice ran away from Beaumont's care, but he did not die until the age of eighty-three, having lived over sixty years with a hole in his stomach.

Note: Photo from *Reveille Till Taps: Soldier Life at Fort Mackinac: 1780–1895* by Keith R. Widder. Mackinac Island State Park Commission, Mackinac Island, Michigan.

the small intestine by blocking the bile duct (figure 7.11). This can reduce the amount of fat digestion in the small intestine. Gallstones can be dissolved by medication or broken apart by ultrasound, but in other cases the entire gallbladder must be removed surgically. After removal of the gallbladder, the person can no longer store a supply of bile to aid in fat digestion. Therefore the diet must be changed to reduce the amount of fatty foods eaten.

The Colon

Very little digestion takes place in the next portion of the intestine, the large intestine, or **colon.** Any materials that were not absorbed in the small intestine are usually wastes and are changed in the colon in preparation for elimination from the body. To prevent an extensive loss of valuable water from the digestive tract, much of the water that was added to the food to aid in digestion is taken back into the body. This reabsorbing of water changes the contents from a thin, watery material to a thicker mass called **feces.** The digestion in the intestines usually takes about twelve hours.

The processes that prepare the fecal material for elimination from the body are controlled by the cells of the colon and a variety of bacteria that live in the colon. The total number of bacteria in the colon is great. In fact, some people have estimated that half of the fecal material is composed of bacteria. These bacteria degrade the materials not digested in the small intestine. The bacteria produce a variety of products, some of which benefit people. Vitamins are among these products. As the diet changes, the bacteria change in type and in the effects they have on the colon. A change in water, for example, may favor the bacterial production of gases. As the gas accumulates, it causes pain and bulging of the colon. To prevent such discomfort when you travel to another region or country, sample the water and foods of that area very cautiously. Drink small amounts of the water to begin with and make sure that the foods are well cleaned and cooked. Once the bacteria in your intestine have become accustomed to these new foods, they should not give you much trouble.

The bacteria in your colon may also be disrupted by antibiotics. These drugs help you defend yourself against bacterial infections. The antibiotic, however, may not recognize the difference between infectious bacteria and the normal bacteria in your colon. It could destroy many of the helpful bacteria and cause diarrhea. The loss of bacteria would also decrease the amount of vitamins you receive. In that case, your doctor may recommend that you take a vitamin supplement for a while and eat plenty of fresh fruit, since the bacteria on the fruit will help restock your colon with new bacteria.

The final portion of the tract is called the **rectum.** After this section of the digestive tract has been filled, the feces are eliminated from the body through an opening from the rectum called the **anus.**

Summary

For an organism composed of many cells to live in good health, it must receive nutrient molecules that can enter the cells and function in the metabolic processes. The digestive tract of humans is a tube that carries the food through a number of

Figure 7.13
Summary of Human Digestive Organs.

Name of structure

Function of structure

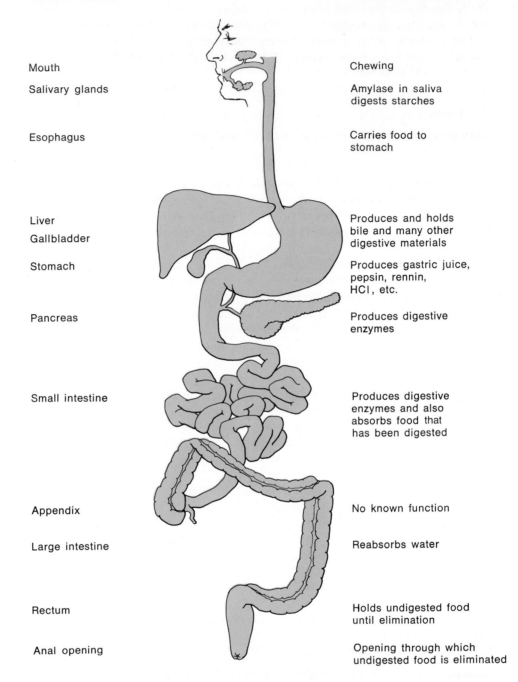

Name of structure	Function of structure
Mouth	Chewing
Salivary glands	Amylase in saliva digests starches
Esophagus	Carries food to stomach
Liver	Produces and holds bile and many other digestive materials
Gallbladder	
Stomach	Produces gastric juice, pepsin, rennin, HCl, etc.
Pancreas	Produces digestive enzymes
Small intestine	Produces digestive enzymes and also absorbs food that has been digested
Appendix	No known function
Large intestine	Reabsorbs water
Rectum	Holds undigested food until elimination
Anal opening	Opening through which undigested food is eliminated

parts: mouth, esophagus, stomach, small and large intestine, rectum, and anus (figure 7.13). Each of these sections has become specialized in function. Through absorption of the digested food into the villi of the small intestine, small nutrient molecules are carried to all the cells of the body for use in cellular metabolism. The proper quantity and quality of nutrients (proteins, carbohydrates, lipids, and growth factors) are essential to good health. Disruptions in the quantity or quality of the foods may result in metabolic problems such as obesity, starvation, kwashiorkor, vitamin deficiency diseases, and ulcers.

Fantastic Journey

This is the title of a new movie just released from the studios of Hospital Productions, Inc. The plot involves a trip through the digestive tract of a lactose-intolerant patient in an attempt to identify the exact nature of the patient's problem. To gain entry into the tract, the doctors are reduced in size and placed in a hamburger. They are mistakenly washed down with a glass of milk instead of water.

As the scriptwriter, it is your job to describe in detail the events of their journey. (P. S. The hamburger has everything on it!)

Keep a diary for one day. Write down the amount and kinds of foods you eat and drink. Get a book from the library and calculate the amount of calories you ingest. See how you stack up in comparison to the recommended daily intake for your sex, height, and body form.

1. Name two enzymes involved in the breakdown of proteins in the digestive tract of humans.
2. What role does the liver play in fat digestion?
3. Why is chewing important in digestion?
4. What is a villus, and what is its function?
5. To lose weight, what two things must a person do?
6. How do the bacteria in the intestine help maintain good health?
7. What is kwashiorkor? Why is it a serious health problem?
8. What factors aid the movement of food through the digestive tract?
9. What particular malfunctions of the digestive tract result in each of the following conditions: gallstones, ulcers, heartburn, lactose intolerance?
10. How do the small and large intestines differ in structure and function?

amylase (am′ĭ-lās) The enzyme in saliva that begins the hydrolysis of starch.

anorexia nervosa (an′′o-rek′se-ah ner-vo′sah) A nutritional deficiency disease characterized by severe, prolonged weight loss due to fear of becoming fat. This eating disorder is thought to stem from sociocultural factors.

anus (ān′us) The opening of the digestive system through which undigested food is eliminated.

basal metabolism (ba′sal mĕ-tab′o-lizm) The amount of energy required to maintain normal body activity while at rest.

bulimia (bu-lim′e-ah) A nutritional deficiency disease brought on by repeated food binging and purging. It is thought to stem from psychological disorders.

calorie (kal′o-re) The amount of heat necessary to raise the temperature of one gram of water one degree Celsius.

cholesterol (ko-les′ter-ol) A steroid molecule found in the bile that can crystallize to form gallstones.

colon (ko′lon) The large intestine. Water and salts are absorbed from the contents of the colon.

complete protein (kom-plēt pro′te-in) Protein molecules that provide all the essential amino acids.

digestive tract (di-jes′tiv trakt) The series of tubes and structures that break down complex food molecules and absorb nutrients into the body.

duodenum (du''o-dēn'um) The upper portion of the small intestine into which flow various digestive enzymes and materials from the stomach, liver, and pancreas.

esophagus (e-sof'a-gus) A tube that conducts food from the mouth to the stomach.

essential amino acids (e-sen'shal ah-mēn'o ass-ids) Those amino acids that must be part of the diet, such as lysine, tryptophan, and valine.

feces (fe'sēz) The undigested food material eliminated from the digestive tract through the anus.

gallbladder (gawl blad-der) A saclike structure that holds bile before it is released into the small intestine.

gallstones (gawl-stōns) Crystals of cholesterol formed in the gallbladder.

growth factor (grōth fak-tōr) A nutrient needed by the body for proper functioning, but only in very small amounts, such as vitamins and minerals.

kilocalorie (kil''o-kal'o-re) A measure of heat energy one thousand times larger than a calorie.

kwashiorkor (kwash-e-or'kor) A protein deficiency disease.

lactase (lak'tās) An enzyme produced by the cells lining the small intestine that breaks the complex sugar, lactose, into simple sugars.

lactose intolerance (lak-tōs in-tol'er-ans) A condition resulting from the inability to digest lactose.

minerals (min'er-als) Growth factors, usually inorganic salts such as calcium and magnesium.

mouth (mowth) The opening to the digestive tract.

mucus (miu-kus) A slimy material produced in various parts of the digestive tract that aids the movement of food through the system and protects the lining of the digestive tract from being harmed by acids and enzymes.

obese (o-bēs) Extremely overweight.

partial protein (par-shal pro'te-in) Protein molecules that do not provide all the essential amino acids.

pepsin (pep'sin) An enzyme produced by the cells lining the stomach that begins the breakdown of proteins.

peptic ulcer (pep'tik ul-ser) A cavity formed in the wall of the digestive tract caused by the enzyme, pepsin.

peristalsis (per''i-stal'sis) Wavelike contractions of the muscles of the digestive tract that move food through the tube.

rectum (rek'tum) The final portion of the digestive tract in which undigested food is stored before being eliminated through the anus.

saliva (sah-li'vah) A digestive juice produced by the salivary glands that aids in the digestion of starches and moistens the food.

small intestine (smal in-test'in) The portion of the digestive tract in which most of the digestion and absorption of food occurs.

sphincter muscle (sfingk'ter mus'el) A circular muscle that closes the digestive tube when contracted.

stomach (stum-ak) The portion of the digestive tract that holds the food while digestive enzymes are mixed with it.

ulcer (ul-ser) A wound that does not heal.

villus (vil-lus) A microscopic fingerlike projection from the lining of the small intestine that increases the surface area of the digestive tract.

vitamin deficiency disease (vi-tah-min de-fish'en-se di-zēz) Poor health caused by the lack of a certain vitamin in the diet.

vitamins (vi-tah-mins) Growth factors needed in the diet in very small amounts.

vomiting (vom-i-ting) The forceful ejection of food from the stomach through the mouth.

Information Systems

Chapter Outline

Purpose

In previous chapters we have considered a variety of biological structures and their functions. Organic molecules found in living cells are not haphazard arrangements of atoms, but are highly organized and can be classified into major groups. The group known as the nucleic acids has a unique structure and is the primary control molecule of the cell. This chapter considers how the structure of these complex molecules is converted into actions by living cells.

For Your Information

The genetic code is written in an alphabet of four letters used to make three-letter words. Select any four letters from our twenty-six letter alphabet and make up as many three-letter words as you can. Remember, you must only use four letters and you can only write three-letter words. Good luck!

Learning Objectives

- Recognize the structure of DNA and RNA.

- Distinguish between DNA, nucleoprotein, chromatin, and chromosome.

- Diagram the DNA replication process.

- Diagram the DNA transcription process.

- Diagram the process of translation.

- Give examples of mutagenic agents and how they might affect DNA.

- Describe the processes involved in recombinant DNA procedures.

The Structure of DNA and RNA

In the nucleus of a eukaryotic cell is a very important library of molecular information. This library contains all the directions for making structural and regulatory proteins required for life processes. It is like a library of how-to books that will not allow its books to circulate. It is a reference library only. One may copy information from books and remove the copy, but the original always stays in the nucleus. It may seem impossible that the directions for growth and development are stored in a place as small as the nucleus of a cell. The secret lies in the language these directions are written in. This language is deoxyribonucleic acid (DNA). DNA has four properties that enable it to function as genetic material. It is able to: (1) *replicate* by directing the manufacture of copies of itself; (2) *mutate* or chemically change and transmit these changes to future generations; (3) *store* information that determines the characteristics of cells and organisms; and (4) use this information to *direct* the synthesis of structural and regulatory proteins essential to the operation of the cell or organism.

Like the other groups of organic macromolecules, the nucleic acids are made up of subunits. The subunits of nucleic acids are called **nucleotides.** Each nucleotide is composed of a *sugar* molecule (S) containing five carbon atoms, a **phosphate group** (P)*, and a kind of molecule called a **nitrogenous base** (B) (figure 8.1).

There are eight common types of nucleotides available in a cell for building nucleic acids. Nucleotides differ from one another in the kind of sugar and nitrogenous base they contain. Because of these differences, it is possible to classify nucleic acids into two main groups, ribonucleic acid (RNA) and deoxyribonucleic acid (DNA). The name of each tells you about the structure of the molecules. For example, the prefix *ribo* in RNA tells you that the sugar part of this nucleic acid is **ribose.** Similarly, DNA contains a ribose sugar that has been deoxygenated (lost an oxygen atom) and is called **deoxyribose** (figure 8.2a). The nucleotide units contain nitrogenous bases of two sizes. The larger nitrogenous bases are **adenine** (A) and **guanine** (G), which differ in the kinds of atoms attached to their double-ring structure (figure 8.2b). The smaller nitrogenous bases are **cytosine** (C), **thymine** (T), and **uracil** (U). Each of these differs from the others by the atoms attached to its single-ring structure (figure 8.2c). These differences in size are important, as you will see later. Table 8.1 shows the differences between the makeup of RNA and DNA. **DNA** is a nucleic acid containing deoxyribose sugar, phosphates, and the nitrogenous bases

Figure 8.1
Nucleotide Structure.
All nucleotides are constructed in this basic way. The nucleotide is the basic structural unit of all nucleic acid molecules. Notice that the phosphate group is written in its "shorthand" form as a P inside a circle.

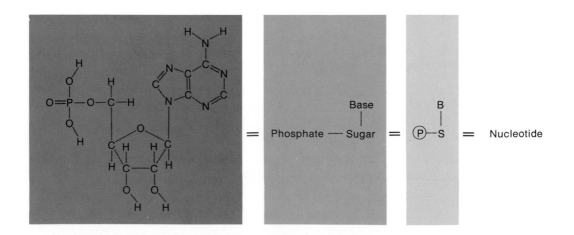

*In previous discussions, a phosphate has always been referred to with its complete chemical formula (PO_4). The phosphate group will now be represented in illustrations by a single capital letter P in a circle, ℗.

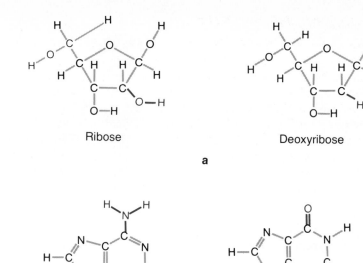

Ribose

Deoxyribose

a

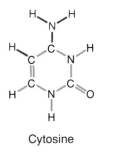

Adenine

Guanine

b

Cytosine

Uracil

Thymine

c

Figure 8.2
The Building Blocks of Nucleic Acids.
All nucleic acids are composed of two organic components; a five-carbon sugar molecule and a nitrogenous base. Notice the difference between the two sugar molecules highlighted in color (a). The nitrogenous bases are divided into two groups according to their size. The large purines (b) adenine and guanine molecules differ from each other in their attached groups (in color), as do the three smaller (c) pyrimidine nitrogenous bases: cytosine, thymine, and uracil.

Table 8.1 *Comparison of DNA and RNA*
The composition of the RNA molecule differs from that of DNA in both the base present and the sugar present. Both DNA and RNA contain phosphate.

Nucleic Acid	RNA	DNA
Base Present	Adenine	Adenine
	Guanine	Guanine
	Cytosine	Cytosine
	Uracil	Thymine
Acid Present	Phosphoric	Phosphoric
Sugar Present	Ribose	Deoxyribose

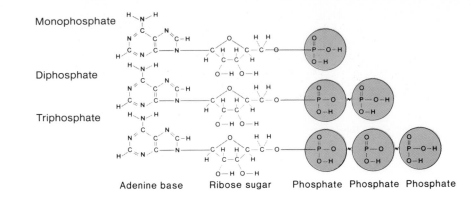

adenine, thymine, guanine, and cytosine that functions as the original blueprint for the synthesis of polypeptides. **RNA** is a type of nucleic acid containing ribose sugar, phosphates, and the four nitrogenous bases adenine, guanine, cytosine, and uracil (no thymine in RNA) that is directly involved in the synthesis of polypeptides at ribosomes.

The composition of the RNA molecule differs from DNA in both the base present and the sugar present. Both RNA and DNA contain phosphoric acid. The construction of a nucleotide involves the bonding of a nitrogenous base to a five-carbon sugar. For example, when adenine is chemically bonded to ribose sugar, the result is a molecule, *adenosine.* When a phosphate is added to the ribose of the adenosine, a new molecule is formed called *adenosine monophosphate,* AMP. This is a complete nucleotide. Monophosphate nucleotides are energetically stable and the building blocks of RNA. However, it is possible to increase the potential energy of a mono-phosphate nucleotide by adding a second phosphate group to form a *diphosphate* nucleotide. For example, when a second phosphate is bonded to AMP, it becomes *adenosine diphosphate,* ADP. This molecule contains potential energy in the bond formed between the two phosphate groups. Because this covalent bond contains chemical-bond energy, it is easily made available to power many types of chemical reactions in the cell. This special bond is represented by a curved line and called a *high-energy phosphate bond.* The maximum amount of potential chemical-bond energy is packed into nucleotides when a third and final phosphate group is added to diphosphates forming *triphosphate* nucleotides. ADP becomes *adenosine tri-phosphate,* ATP, the most widely used source of chemical-bond energy in all cells.

Chapter 6 dealt with the function of ATP as a source of chemical-bond energy. Keep in mind that all nucleotides exist in mono-, di-, and tri-phosphate forms (figure 8.3). As in ATP, the other triphosphates contain high-energy phosphate bonds used in the dehydration synthesis reaction that binds nucleotides together as RNA or DNA. The result of this nucleic acid synthesis is a polymer that can be compared to a rat-tail comb. The protruding "teeth" (different nitrogenous bases) are connected to a common "backbone" (sugar and monophosphate molecules). This is the basic structure of both RNA and DNA (figure 8.4).

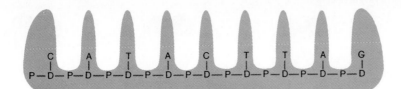

Figure 8.4
Single Strand of DNA.
A single strand of DNA resembles
a comb. The molecule is much
longer than pictured here and is
composed of a sequence of linked
nucleotides.

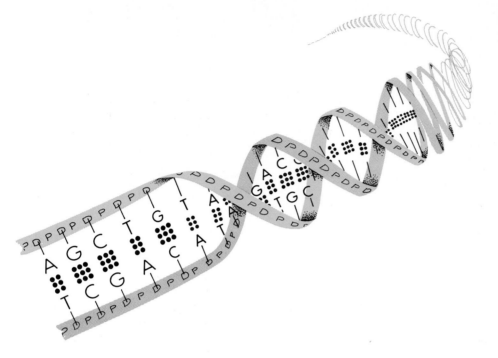

Figure 8.5
Duplex DNA.
Duplex DNA is found in the form of
a three-dimensional helix. One
strand is a chemical code that
contains the information necessary
to control and coordinate the
activities of the cell.

The Molecular Structure of DNA

DNA is actually a double molecule. It consists of two comblike strands held together between their protruding bases by hydrogen bonds. The two strands are twisted about each other in a double helix known as **duplex DNA** (figure 8.5). The four kinds of bases always pair in a definite way, adenine (A) with thymine (T), and guanine (G) with cytosine (C). The bases that pair are said to be **complementary bases.** Notice that the large bases (A and G) pair with the small ones (T and C), thus keeping the backbones of two complementary strands parallel. Three hydrogen bonds are formed between guanine and cytosine:

$$G \vdots C$$

and two between adenine and thymine:

$$A \vdots T$$

Notice in figure 8.4 that it is possible to make sense out of the sequence of nitrogenous bases. If you "read" them from left to right in groups of three, you can read three words—CAT, ACT, and TAG. You can "write" a message in the form of a stable DNA molecule by combining the four different DNA nucleotide units (A, T, C, G) in particular sequences. In this case, the four DNA nucleotides are being used as an alphabet to construct three-letter words. In order to make sense

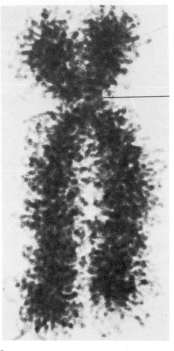

a

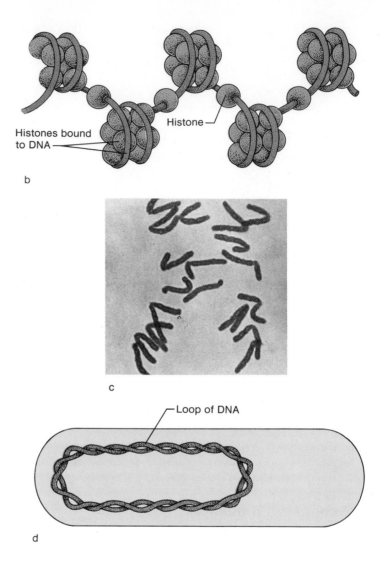

Histones bound to DNA

Histone

b

c

Loop of DNA

d

Figure 8.6 a–d
The Many Forms of Deoxyribonucleic Acid, DNA.

(a) The term nucleoprotein is used to describe the combination of DNA and its bonded protein of eukaryotic cells. When these giant nucleoprotein molecules are found loose inside of a cell, it is called chromatin. (b) Upon close examination of a portion of a eukaryotic nucleoprotein, it is possible to see how the DNA and protein are arranged. The protein component, histone, is found along the DNA in globular masses; they may be individual histones or arranged in groups. (c) During certain stages in the reproduction of a eukaryotic cell, the nucleoprotein coils and "super" coils forming tightly bound masses that are easily seen through a microscope when stained. In their super-coiled form they are called chromosomes, meaning colored bodies. (d) The nucleic acid of prokaryotic cells does not have histone proteins permanently affixed to the loop. The ends of the giant molecule overlap and bind with one another. (**b** from Avers, Charlotte J., *Molecular Cell Biology.* © 1986 Addison-Wesley Publishing Company, Inc., Menlo Park, CA. All Rights Reserved. Reprinted by permission.

out of such a code, it is necessary to read in a consistent direction. Reading the sequence in reverse does not always make sense, just as reading this paragraph in reverse would not make sense.

There are two different forms of DNA. The genetic material of eukaryotes is duplex DNA, but it has histone proteins attached along its length. The duplex DNA strands with attached proteins are called **nucleoproteins** or **chromatin fibers** (figure 8.6a). Histone and DNA are not arranged haphazardly, but come together in a highly organized pattern. The duplex DNA spirals around repeating clusters of eight histone spheres. Histone clusters with their encircling DNA are called **nucleosomes** (figure 8.6b). When eukaryotic chromatin coils into condensed, highly knotted bodies, they are seen easily through a microscope after staining with dye. Condensed like this, a chromatin fiber is referred to as a **chromosome** (figure 8.6c). The genetic material in prokaryotic cells is pure duplex DNA with the ends of the polymer connected to form a loop (figure 8.6d).

Each chromatin strand is different because each strand has a different chemical code. Coded DNA serves as a central cell library. Tens of thousands of messages are in this storehouse of information. This information tells the cell such things as: (1) what chemicals in the environment are nutrients; (2) how to produce enzymes for their ingestion; (3) how to manufacture enzymes that will metabolize the nutrients and eliminate harmful wastes; (4) how to repair and assemble cell parts; (5) how to reproduce healthy offspring; (6) when and how to react to favorable and unfavorable changes in the environment; and (7) how to coordinate and regulate all of life's essential functions. Should any of these functions fail to operate, the cell will die. The importance of maintaining essential DNA in a cell can be seen in cells that have lost it. For example, human red blood cells lose their nuclei as they become specialized in carrying oxygen and carbon dioxide throughout the body. Without DNA they are unable to manufacture the essential cell components needed to sustain themselves. They continue to exist for about 120 days, functioning only on enzymes manufactured earlier in their lives. When these enzymes are gone, the cells die. Since these specialized cells begin to die the moment they lose their DNA, they are more accurately called red blood corpuscles (RBCs), little dying red bodies.

DNA Replication

Since parent cells have a set of genetic information, there must be a doubling of DNA in order to have enough to pass on to the offspring. **DNA replication** is the process of duplicating the genetic material prior to its distribution to daughter cells. When a cell divides into two daughter cells, each new cell must receive a complete copy of the parent cell's genetic information or it will not function long. The accuracy of duplication is essential in order to guarantee the continued existence of that type of cell. Should the daughters not receive exact copies, they may be unable to manufacture structural and regulatory proteins essential for their survival. They may turn out to be so different that they are no longer members of the same cell family.

The DNA replication process begins when an enzyme breaks the hydrogen bonds between the bases of the two strands of DNA. It begins at a fixed point and, going in both directions at once, "unzips" the halves of the duplex DNA. As this enzyme proceeds down the length of the DNA, an enzyme known as **DNA polymerase** brings new DNA triphosphate nucleotides into position for bonding. The complementary bases pair (A ::: T, G ::: C) with the exposed nitrogenous bases of both DNA strands by forming new hydrogen bonds. Once properly aligned, a covalent bond is formed between sugar and phosphates of the newly positioned nucleotides using DNA polymerase. Energy is released when the last two phosphates of each incoming nucleotide are removed. A strong sugar and phosphate backbone is formed in the process (figure 8.7).

A new complementary strand of DNA forms on each of the old DNA strands, resulting in the formation of two double-stranded, duplex DNA molecules. In this way, the exposed nitrogenous bases of the original DNA serve as a **template,** or pattern, for the formation of the new DNA. As the new DNA is completed, it twists into its double helix shape.

The completion of the process yields two double helices identical in their nucleotide sequences. The DNA replication process is highly accurate. It has been estimated that there is only one error made for every 2×10^9 nucleotides. A human cell contains 46 chromosomes consisting of about 10,000,000,000 (ten billion) nucleotides! This averages to about five errors per cell. Don't forget that this figure is an estimate, and while some cells may have five errors per replication, others may have more and some no errors at all. It is also important to note that some errors may be major and deadly, while others are insignificant. Since this error rate is so

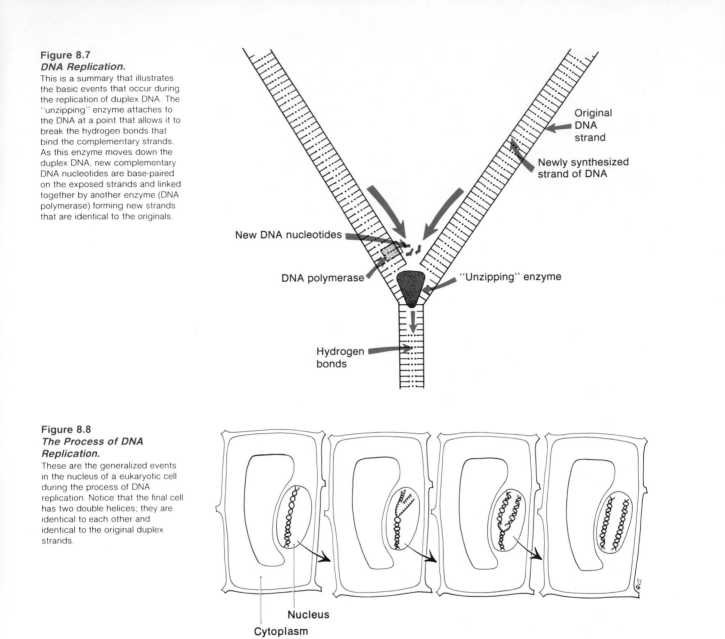

Figure 8.7
DNA Replication.
This is a summary that illustrates the basic events that occur during the replication of duplex DNA. The "unzipping" enzyme attaches to the DNA at a point that allows it to break the hydrogen bonds that bind the complementary strands. As this enzyme moves down the duplex DNA, new complementary DNA nucleotides are base-paired on the exposed strands and linked together by another enzyme (DNA polymerase) forming new strands that are identical to the originals.

Original DNA strand

Newly synthesized strand of DNA

New DNA nucleotides

DNA polymerase

"Unzipping" enzyme

Hydrogen bonds

Figure 8.8
The Process of DNA Replication.
These are the generalized events in the nucleus of a eukaryotic cell during the process of DNA replication. Notice that the final cell has two double helices; they are identical to each other and identical to the original duplex strands.

Nucleus

Cytoplasm

small, DNA replication is considered by most to be error free. Following DNA replication, the cell now contains twice the amount of genetic information and is ready to begin the process of distributing one set of genetic information to each of its two daughter cells.

Distribution of DNA involves splitting the cell and distributing a set of genetic information to the two new daughter cells. In this way, each new cell has the necessary information to control its activities. The mother cell ceases to exist when it divides its contents between the two smaller daughter cells (figure 8.8).

A cell does not really die when it reproduces itself; it merely starts over again. This is called the life cycle of a cell. A cell may divide and redistribute its genetic information to the next generation in a number of ways. These processes will be dealt with in detail in chapters 9 and 10.

DNA Transcription

As noted earlier, DNA is like a reference library that does not allow its books to circulate. Information from the originals must be copied. The second major function of DNA is the process of making a single-stranded, complementary RNA copy of DNA. This operation is called **transcription,** which means to transfer data from one form to another. In this case, the data is copied from DNA language to RNA language. The same base-pairing rules that control the accuracy of DNA replication apply to the process of transcription. Using this process, the genetic information stored as a DNA chemical code is carried in the form of an RNA copy to other parts of the cell. It is RNA that is used to guide the assembly of amino acids into structural and regulatory molecules, such as polypeptides and enzymes. Without the process of transcription, genetic information would be useless in directing cell functions.

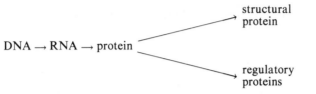

Transcription begins in a way that is similar to DNA replication. The duplex DNA is separated by an enzyme, exposing the nitrogenous base sequence of the two strands. However, unlike DNA replication, transcription only occurs on one of the two DNA strands that serves as a template or pattern for the synthesis of RNA. But which strand is copied? Where does it start and when does it stop? Where along the sequence of thousands of nitrogenous bases does the chemical code for the manufacture of a particular enzyme begin and where does it end? If transcription begins randomly, the resulting RNA may not be an accurate copy of the code and the enzyme product may be useless or deadly to the cell. To answer these questions, it is necessary to explore the nature of the genetic code itself.

The Genetic Code

We know that genetic information is in chemical code form in the DNA molecule. When the coded information is used or *expressed,* it guides the assembly of particular amino acids into structural and regulatory polypeptides and proteins. If DNA is molecular language, then each nucleotide in this language can be thought of as a letter within a four-letter alphabet. Each word or code is always three letters (nucleotides) long and only three-letter words can be written. A **DNA code** is a triplet nucleotide sequence that codes for one of the twenty common amino acids. The number of codes in this language is limited since there are only four different nucleotides, which are only used in groups of three. The order of these three letters is just as important in DNA language as it is in our language. We recognize that CAT is not the same as TAC. If all the possible three-letter codes were written using only the four DNA nucleotides for letters, there would be a total of sixty-four combinations. When codes are found at a particular place along a strand of DNA and the sequence has meaning, the sequence is called a **gene.** "Meaning" in this case means that the gene can be transcribed into an RNA molecule, which in turn may control the assembly of individual amino acids into a polypeptide.

All the genes in prokaryotic cells are attached end-to-end forming the typical bacterial loop of DNA. Each bacterial gene is made of attached nucleotides that

Figure 8.9
Transcription of mRNA in prokaryotic cells.
This is a summary that illustrates the basic events that occur during the transcription of one side of duplex DNA. The "unzipping" enzyme attaches to the DNA at a point that allows it to break the hydrogen bonds that bind the complementary strands. As this enzyme moves down the duplex DNA, new complementary RNA nucleotides are base-paired on one of the exposed strands and linked together by another enzyme (RNA polymerase) forming a new strand that is complementary to the nucleotide sequence of the DNA. The newly formed RNA is then separated from its DNA complement. Depending on the DNA segment that has been transcribed, this RNA may be a messenger (mRNA), a transfer (tRNA), or a ribosomal (rRNA) molecule.

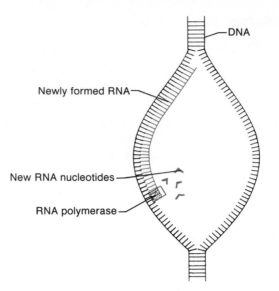

DNA

Newly formed RNA

New RNA nucleotides

RNA polymerase

are transcribed in order into a single strand of RNA. This RNA molecule is used to direct the assembly of a specific sequence of amino acids to form a polypeptide. This system follows the pattern of one DNA gene → one RNA → one polypeptide. The beginning of each gene on a DNA strand is identified by the presence of a region known as the **promoter** just ahead of an **initiation code,** which has the base sequence TAC. The gene ends with a terminator region just in back of one of three possible **termination codes,** ATT, ATC, or ACT. These are the "start reading here" and "stop reading here" signals. The actual genetic information is located between initiation and termination codes:

promoter::initiator code:::::gene:::::terminator code::terminator region

When a bacterial gene is transcribed into RNA, the duplex DNA is "unzipped" and an enzyme known as **RNA polymerase** attaches to the DNA at the promoter region (figure 8.9). It is from this region that the enzymes will begin to assemble RNA nucleotides into a complete, single-stranded copy of the gene, including initiation and termination codes. Triplet RNA nucleotide sequences complementary to DNA codes are called **codons.** Remember that there is no thymine in RNA molecules; it is replaced with uracil. Therefore, the initiation code in DNA (TAC) would be base paired by RNA polymerase to form the RNA codon AUG. When transcription is complete, the newly assembled RNA is separated from its DNA template and made available for use in the cell and the DNA recoils into its original double helix form.

The system is different in eukaryotic cells. A eukaryotic gene begins with a promoter region and an initiation code and ends with a termination code and region. However, the gene is split into sections composed of meaningful coding sequences called *exons* and meaningless sequences called *introns.* When such *split genes* are

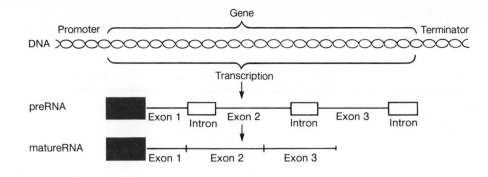

Figure 8.10
Transcription of mRNA in Eukaryotic Cells.
This is a summary of the events that occur during the manufacture of mRNA in a eukaryotic cell. Notice that the original nucleotide sequence is first transcribed into an RNA molecule that is later "clipped" and rebonded to form a shorter version of the original. It is during this time that the introns are removed.

transcribed, RNA polymerase synthesize a strand of preRNA that initially includes copies of both exons and introns. However, soon after its manufacture, this preRNA molecule has the meaningless introns clipped out and the exons spliced together into mature RNA. It is mature RNA that is used by the cell (figure 8.10).

Three general types of RNA are produced by transcription: messenger RNA, transfer RNA, and ribosomal RNA. Each kind of RNA is made from a specific gene and performs a specific funtion in the synthesis of polypeptides from individual amino acids at ribosomes. **Messenger RNA (mRNA)** is a mature, straight chain copy of a gene that describes the exact sequence in which amino acids should be bonded together to form a polypeptide.

Transfer RNA (tRNA) molecules are responsible for picking up particular amino acids and transferring them to the ribosome for assembly into the polypeptide. All tRNA molecules have a cloverleaf shape formed when they fold and some of the bases form hydrogen bonds that hold the molecule together. One end of the tRNA is able to attach to a specific amino acid. Toward the midsection of the molecule, a triplet nucleotide sequence can base pair with a codon on mRNA. This triplet nucleotide sequence on tRNA that is complementary to a codon of mRNA is called an **anticodon**. **Ribosomal RNA (rRNA)** is a highly coiled molecule and is used along with protein in the manufacture of all ribosomes. rRNA is involved in some unknown way in the synthesis of polypeptides.

Translation or Protein Synthesis

The mRNA molecule is a coded message written in nucleic acid language. The code is read in one direction starting at the initiator. The information is used to assemble amino acids into protein by a process called **translation.** The word "translation" refers to the fact that nucleic acid language is being changed to protein language. To translate mRNA language into protein language, a dictionary is necessary. Remember, the four letters in the nucleic acid alphabet yield sixty-four possible three-letter words. The protein language has twenty words in the form of twenty common

Table 8.2 Amino Acids
There are twenty common amino acids used in the protein synthesis operation of a cell.
Each has a known chemical structure.

Amino Acid	Abbreviation	Amino Acid	Abbreviation
Alanine	Ala	Leucine	Leu
Arginine	Arg	Lysine	Lys
Asparagine	AspN	Methionine	Met
Aspartic acid	Asp	Phenylalanine	Phe
Cysteine	Cys	Proline	Pro
Glutamic acid	Glu	Serine	Ser
Glutamine	GluN	Threonine	Thr
Glycine	Gly	Tryptophan	Try
Histidine	His	Tyrosine	Tyr
Isoleucine	Ileu	Valine	Val

amino acids. Thus, there are more than enough nucleotide words for the twenty amino acid molecules, because each nucleotide triplet codes for an amino acid (table 8.2).

Table 8.3 is a mRNA amino acid-nucleic acid dictionary. Notice that more than one codon may code for the same amino acid. Some have called that needless repetition, but such "synonyms" can have survival value. If, for example, the gene or the mRNA becomes damaged in a way that causes a particular nucleotide base to change to another type, the chances are still good that the proper amino acid will be read into its proper position. But not all such changes can be compensated for by the codon system and an altered protein may be produced (figure 8.11). Changes can occur that cause great harm. Some damage is so extensive that the entire strand of DNA is broken, resulting in improper **protein synthesis** or a total lack of synthesis. Such a change in DNA is called a **mutation.**

The construction site of the protein molecules (i.e., the translation site) is on the ribosome, a cellular organelle that serves as the meeting place for mRNA and the tRNA-carrying amino acid building blocks. The mRNA molecule is placed on the ribosome two codons (six nucleotides) at a time (figure 8.12). The tRNA, which is carrying an amino acid, forms hydrogen bonds with the mRNA (between the codon and anticodon) only long enough for certain reactions to occur. The ribosome tRNA-mRNA complex is formed. Both RNA molecules combine first with the smaller of the two ribosome units, and then the larger unit is added. The molecules of tRNA-carrying amino acids move in to combine with the mRNA on the ribosome.

Amino acids are transferred to the ribosome by tRNA molecules that are so specific they are only capable of transferring one particular amino acid. tRNA molecules are cloverleaf shaped. There are at least twenty different coding types of tRNA. Each tRNA transfers a specific amino acid to a ribosome for synthesis into a polypeptide. The tRNA properly aligns each amino acid so that it may be chemically bonded to another amino acid and form a long chain.

Each amino acid is bonded in sequence to form the new protein. As each is bonded in order, the unit is moved along the mRNA to allow the next molecule of tRNA and its amino acid to fit into position. Once the final amino acid is bonded into position, all the molecules are released from the ribosome. The termination codons signal this action. The intact ribosome is again free to become engaged in

Table 8.3   *Amino Acid-Nucleic Acid Dictionary*

A dictionary can come in handy for learning any new language. This one is used to translate nucleic acid language into protein language.

Amino Acid	mRNA	Amino Acid	mRNA Codons
Phenylalanine	UUU	Tyrosine	UAU
	UUC		UAC
Leucine	UUA	Histidine	CAU
	UUG		CAC
	CUU	Glutamine	CAA
	CUC		CAG
	CUA	Asparagine	AAU
	CUG		AAC
Isoleucine	AUU	Lysine	AAA
	AUC		AAG
	AUA	Aspartic acid	GAU
Methionine	AUG		GAC
Valine	GUU	Glutamic acid	GAA
	GUC		GAG
	GUA	Cysteine	UGU
	GUG		UGC
Serine	UCU	Tryptophan	UGG
	UCC	Arginine	CGU
	UCA		CGC
	UCG		CGA
	AGU		CGG
	AGC		AGA
Proline	CCU		AGG
	CCC	Glycine	GGU
	CCA		GGC
	CCG		GGA
Threonine	ACU		GGG
	ACC	Terminator	UAA
	ACA		UAG
	ACG		UGA
Alanine	GCU	Initiator	AUG
	GCC		
	GCA		
	GCG		

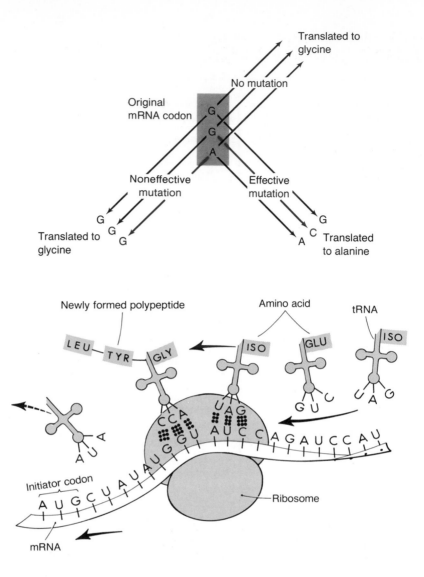

Figure 8.11
Noneffective and Effective Mutation.

A nucleotide substitution changes the genetic information only if the changed codon results in a different amino acid being substituted into a protein chain.

Figure 8.12
Translation.

Translation results in the formation of a specific sequence of amino acids, each covalently bonded to one another by a peptide bond. This amino acid aligning and bonding process takes place on the ribosomes. Each amino acid is brought into its proper position by a particular tRNA molecule.

another protein synthesis operation. In most cells the mRNA travels through more than one ribosome at a time. When viewed with the electron microscope, this appears as a long thread (the mRNA) with several dark knots (the ribosomes) along its length. This sequence of several translating ribosomes attached to the same mRNA is known as a **polysome** (figure 8.13). The newly synthesized chains of amino acids leave the ribosomes and fold into their typical three-dimensional structure for use in the cell.

In that way, the mRNA moves through the ribosome, its specific codon sequence allowing for the chemical bonding of a specific sequence of amino acids. Remember, the sequence was originally determined by the DNA. Figure 8.14 shows a possible result of protein synthesis. After transcribing the DNA code, the mRNA delivers its message to a ribosome, where a protein is made. In the case of figure 8.14, the protein contains the amino acid phenylalanine, serine, lysine, and arginine.

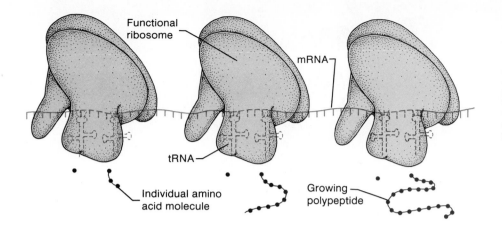

Figure 8.13
Polysomes.
Any single mRNA molecule can be translated on a series of ribosomes. The mRNA and its several translating ribosomes is called a polysome. The first ribosome will complete the translation of a polypeptide and the last ribosome to bind to the mRNA will synthesize the polypeptide last.

PROTEIN SYNTHESIS

	TRANSCRIPTION		TRANSLATION		
DNA →		RNA →		PROTEIN	
DNA: Cover (complementary to gene)	DNA: Gene	mRNA: Codons (complementary to gene)	tRNA: Anticodons (complementary to mRNA)	Amino acids specified	Protein
T	A	U	A		
T	A	U	A	Phe	Phenylalanine
T	A	U	A		
T	A	U	A		
C	G	C	G	Ser	Serine
T	A	U	A		
A	T	A	U		
A	T	A	U	Lys	Lysine
A	T	A	U		
C	G	C	G		
G	C	G	C	Arg	Arginine
T	A	U	A		

Each protein has a specific sequence of amino acids that determines its three-dimensional shape. This shape determines the activity of the protein molecule. The protein may be a structural component of a cell or a regulatory protein such as an enzyme. Any changes in amino acids or their order changes the action of the protein molecule. The protein insulin, for example, has a different amino acid sequence than the digestive enzyme trypsin. Both proteins are essential to human life and must be produced constantly and accurately. The amino acid sequence of each is determined by a different gene. Each gene is a particular sequence of DNA nucleotides. Any alteration of that sequence can directly alter the protein structure and, therefore, the survival of the organism.

Figure 8.14
Protein Synthesis.
There are several steps involved in protein synthesis. (1) Transcription involves the manufacture of mRNA from a DNA molecule. (2) The mRNA enters the cytoplasm and attaches to ribosomes. (3) Translation involves tRNA carrying amino acids to the ribosome and positioning them in the order specified by the mRNA. (4) The amino acids are combined chemically to form a protein.

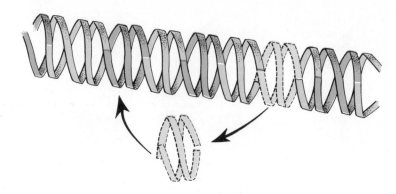

Figure 8.15
Jumping Genes.

Alteration in the sequence of DNA nucleotides occurs when one segment of DNA "jumps" out of its original position in the duplex DNA and inserts itself into a different position. This changes the original nucleotide sequence and, therefore, the encoded message.

Alterations of DNA

Several kinds of changes to DNA may result in mutations. Phenomena that are either known or suspected causes of DNA damage are called **mutagenic agents.** Two such agents known to cause damage to DNA are x-radiation (X rays) and the chemical nicotine found in tobacco. Both have been studied extensively and there is little doubt that they cause mutations. *"Jumping genes,"* yet another cause of mutation, are segments of DNA capable of moving from one position in a strand of DNA to another. When the jumping gene is spliced into its new location, it alters the normal nucleotide sequence, causing normally stable genes to be misread during transcription. The result is a mutant gene (figure 8.15).

Changes in the structure of DNA have harmful effects in the next generation if they occur in the sex cells. Some damage to DNA is so extensive that the entire strand of DNA is broken, resulting in improper position synthesis or a total lack of synthesis. This changes more than just a nucleotide base and is called a **chromosomal mutation.** In some cases the damage is so extensive that cells die. If enough cells are destroyed, the whole organism will die. A number of experiments indicate that many "street drugs," such as LSD (lysergic acid diethylamide), are mutagenic agents and cause DNA to break into smaller pieces.

Another example of the effects of altered DNA may be seen in human red blood corpuscles. Red corpuscles contain the oxygen-transport molecule hemoglobin. Normal hemoglobin molecules are composed of 150 amino acids in four chains, two alpha and two beta chains. The nucleotide sequence of the gene for the beta chain is known, as is the amino acid sequence for this chain. In normal individuals, the sequence begins . . .

Val-His-Leu-Thr-Pro-*Gly*-Gly-Lys. . . .

In some persons, a single nucleotide of the gene controlling synthesis of the beta chain changes. This type of mutation is called a **point mutation.** The result is a new amino acid sequence in all the red corpuscles:

Val-His-Leu-Thr-Pro-*Val*-Gly-Lys. . . .

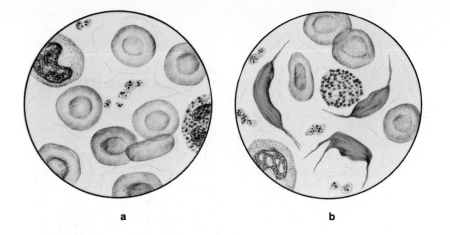

Figure 8.16
Normal and Sickled Red Blood Corpuscles.
Normal red blood corpuscles (a) are shown in comparison with corpuscles having the sickle shape (b). This sickling is the result of a single amino acid change in the molecule, hemoglobin.

a   b

This single nucleotide change, which causes a single amino acid to change, may seem minor. However, it is the cause of **sickle-cell anemia,** a disease that affects the red blood corpuscles by changing them from a circular to a sickle shape when oxygen levels are low (figure 8.16). When this sickling occurs, the red blood corpuscles do not flow smoothly through capillaries. Their irregular shapes cause them to clump, clogging the blood vessels. This prevents them from delivering their oxygen load to the oxygen-demanding tissues. A number of complicated physical disabilities may result including physical weakness, brain damage, pain and stiffness of the joints, kidney damage, rheumatism, and in severe cases, death.

There is no cure for this disease, since all of a person's cells contain the same wrong genetic information. However, a powerful new science of gene manipulation, **biotechnology,** suggests that in the future, genetic diseases may be controlled or cured. Since 1953, when the structure of the DNA molecule was first stated, there has been a rapid succession of advances in the field of genetics. It is now possible to transfer the DNA from one organism to another. This made possible the manufacture of human genes and gene products by bacteria.

Biotechnology includes the use of a method of splicing genes from one organism into another. The resulting new form of DNA is called **recombinant DNA** (box 8.1). This process is accomplished using enzymes that are naturally involved in the DNA replication process. Once new chunks of DNA have been inserted into another organism, they function naturally. This process is not as simple as described here. Today's techniques make it possible to determine the sequence of bases in a gene, manufacture genes, and splice these genes into different organisms. It is also possible to selectively remove genes from organisms. At first this was an extremely controversial issue. People were afraid that monsters would be created or that disease organisms for which there were no protection would be released. Most people, however, have now realized that these were overcautious reactions. Important biological compounds like insulin and interferon have been produced by recombinant

Box 8.1
Gene Splicing

In recent years, biologists have been able to control the natural biological processes of genetic recombination to develop microbes that contain genes from other organisms. Recombinant DNA is the result of splicing genes from different organisms into host cells. The host microbe replicates these new, "foreign" genes and synthesizes proteins encoded by them. Gene splicing begins with the laboratory isolation of DNA from an organism that contains the desired gene, for example, human cells that contain the gene for the manufacture of insulin. If the gene is short enough and its base sequence is known, it may be synthesized in the laboratory from separate nucleotides. If the gene is too long and complex, it is cut from the chromosome with enzymes called restriction endonucleases. They are given this name since these enzymes (-ases) only cut DNA (nucle-) at certain base sequences (restricted in their action) and work inside (endo-) the DNA. These enzymes do not cut the DNA straight across, but in a zig-zag pattern that leaves one strand slightly longer than its complement. The short nucleotide sequence that sticks out and remains unpaired is called a "sticky end" since it can be reattached to another complementary strand. DNA segments have successfully been cut from rats, frogs, bacteria, and humans.

This isolated gene with its "sticky ends" is spliced into microbial DNA. The host DNA is opened up with the proper restriction endonuclease and ligase (i.e., tie together) enzymes are used to attach the "sticky ends" into the host DNA. This gene-splicing procedure may be performed with small loops of bacterial DNA that are not part of the main chromosome. These small DNA loops are called plasmids. Once the splicing is completed, the plasmids can be inserted into the bacterial host by treating the cell with special chemicals that encourage it to take in these large chunks of DNA. A more efficient alternative is to splice the desired gene into the DNA of a bacterial virus so that it can carry the new gene into the bacterium as it infects the host cell. Once inside the host cell, the genes may be replicated along with the rest of the DNA to clone the "foreign" gene, or they may begin to synthesize the encoded protein.

As this highly sophisticated procedure has been refined, it has become possible to quickly and accurately splice genes from a variety of species into host bacteria, making possible the synthesis of large quantities of medically important products. For example, recombinant DNA procedures are responsible for the production of human insulin, used in the control of diabetes; interferon, currently being investigated as an antiviral agent; human growth hormone, used to stimulate growth in children lacking this hormone; and somatostatin, a brain hormone also implicated in growth.

DNA techniques. Insulin is a hormone that most of us produce in normal amounts but is needed by people with diabetes. Interferon is normally produced in the body as a reaction to and a defense against the invasion of a virus. Interferon is produced in such tiny amounts in an organism that it formerly was difficult to obtain a sufficient quantity for medical use. Today, with the gene for interferon inserted into bacterial DNA, quantities can be produced in large bacterial cultures at relatively low cost.

The possibilities that open up with the development of recombinant DNA methods are revolutionary. These methods enable cells to produce molecules that they would not normally make. Some research laboratories have even investigated the possibility of splicing the genes into human genetically deficient cells. Should such a venture prove to be practical, genetic diseases such as sickle-cell anemia could be controlled. The process of recombinant DNA gene splicing also enables cells to produce molecules they normally synthesize more efficiently. Some of the likely rewards are: (1) the production of additional, medically useful proteins; (2) mapping the location of genes on human chromosomes; (3) a more complete understanding of how genes are regulated; (4) the production of crop plants with increased yields; and (5) the development of new species of garden plants.

The discovery of the structure of DNA over thirty-five years ago seemed very far removed from the practical world. The importance of this "pure" research is just now being realized. Several companies are involved in recombinant DNA research with the aim of alleviating or curing disease.

Summary

The successful operation of a living cell depends on its ability to accurately reproduce genes and control chemical reactions. DNA replication results in an exact doubling of the genetic material. The process virtually guarantees that identical strands of DNA will be passed on to the next generation of cells.

The enzymes are responsible for the efficient control of a cell's metabolism. However, the production of protein molecules is under the control of the nucleic acids, the primary control molecules of the cell. The structure of the nucleic acids DNA and RNA determines the structure of the proteins, while the structure of the proteins determines their function in the cell's life cycle. Protein synthesis is decoding of the DNA into specific protein molecules and involves the use of the intermediate molecules mRNA and tRNA at the ribosome (figure 8.17). Errors in any of the codons of these molecules may produce observable changes in the cell's functioning and lead to the death of the cell.

Methods of recombining DNA have led to the controlled transfer of genes from one kind of organism to another. This has made it possible for bacteria to produce a number of human gene products.

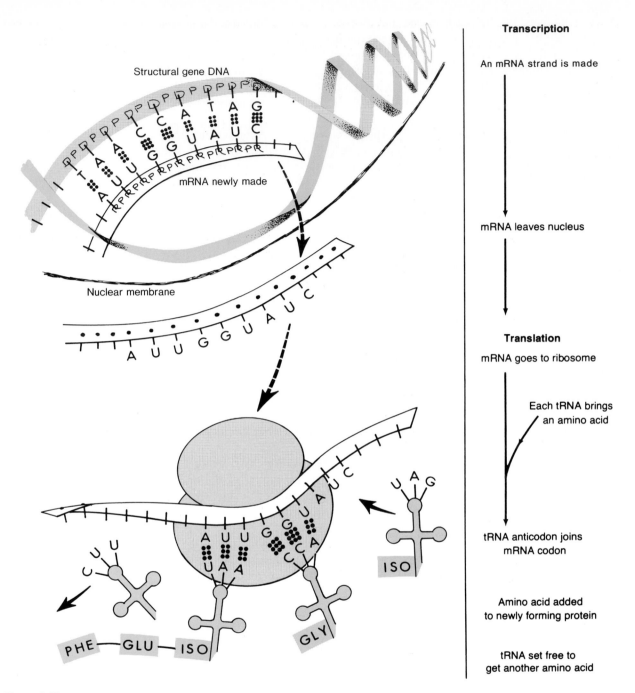

Structural gene DNA

mRNA newly made

Nuclear membrane

A U U G G U A U C

PHE — GLU — ISO

GLY

ISO

Transcription

An mRNA strand is made

↓

mRNA leaves nucleus

↓

Translation

mRNA goes to ribosome

↓

Each tRNA brings
an amino acid

↓

tRNA anticodon joins
mRNA codon

Amino acid added
to newly forming protein

tRNA set free to
get another amino acid

Figure 8.17
Summary of Protein
Synthesis Events.
(1) The DNA strand separates and
an mRNA is synthesized
(transcription). (2) mRNA leaves
the nucleus and enters the
cytoplasm where it attaches to a
ribosome. (3) The ribosome moves
along the mRNA codon by codon.
(4) tRNAs carry specific amino
acids to the mRNA and base-pair
with the mRNA. (5) The amino
acids combine to form protein.

Today Dr. Fritz Von Adelhoff, III, announced the discovery of a new cellular organelle. According to Von Adelhoff, this structure is composed entirely of protein. It is in the shape of a donut and is slightly smaller than a ribosome. As yet, no function is known for this organelle. However, Von Adelhoff speculates that it may be responsible for epoxy glue, which holds the cell together. Further information on the possible origins and function of this organelle will be looked into by Von Adelhoff's able assistant (*fill in your name*). S/he has extensive research experience in the areas of organic chemistry, cell structure and function, and exceptional knowledge of DNA and protein synthesis.

Go ahead doctor, write the follow-up article!

The method of DNA replication described in this chapter (and found to be the valid method) is called "semiconservative" replication. It is called this because the "new" strands of DNA are formed from the "old" strands that remain intact during the entire process. The discovery of this method came after considerable thought and "playing around" with other hypotheses. One discarded method was called the "conservative" replication because both original strands remained together during the process of replication. Diagram a short section of duplex DNA and use it to show the difference between semiconservative and conservative DNA replication.

1. What is the difference between a nucleotide, a nitrogenous base, and a codon?
2. What are the differences between DNA and RNA?
3. List the sequence of events when a DNA message is translated into protein.
4. Chromosomal and point mutation both occur in DNA, but they differ in what ways? How is this related to recombinant DNA?
5. Why is DNA replication necessary?
6. What is polymerase and how does it function?
7. How does DNA replication differ from the manufacture of an RNA molecule?
8. If a DNA nucleotide sequence is CATAAAGCA, what is the mRNA nucleotide sequence that would base pair with it?
9. What amino acids would occur in the protein chemically coded by the sequence of nucleotides in question eight?
10. How do tRNA, rRNA, and mRNA differ in function?

adenine (ad'ĕ-nēn) A double-ring nitrogenous base molecule in DNA and RNA. It is the complementary base of thymine or uracil.

anticodon (an''te-ko'don) A sequence of three nitrogenous bases on a tRNA molecule, capable of forming hydrogen bonds with three complementary bases on an mRNA codon during translation.

biotechnology (bi-o-tek-nol'uh-je) The science of gene manipulation.

chromosomal mutation (kro-mo-sōm'al miu-ta'shun) Change in the gene arrangement in a cell as a result of breaks in the DNA molecule.

chromosome (kro-mo-sōm) A duplex DNA molecule with attached histone protein (nucleoprotein) coiled into a short, compact unit.

codon (ko'don) A sequence of three nucleotides of an mRNA molecule that directs the placement of a particular amino acid during translation.

complementary base (kom''ple-men'tah-re) A base that can form hydrogen bonds with another base of a specific nucleotide.

cytosine (si'to-sēn) A single-ring nitrogenous base molecule in DNA and RNA. It is complementary to guanine.

deoxyribonucleic acid (DNA) (de-ok''se-ri-bo-nu-kle'ik ass-id) A polymer of nucleotides that serves as genetic information. In prokaryotic cells, it is a duplex DNA (double-stranded) loop and contains no permanently attached proteins. In eukaryotic cells, it is found in strands with attached histone proteins. When tightly coiled, it is known as a chromosome.

deoxyribose (de-ok''se-ri'bōs) A five-carbon sugar molecule; a component of DNA.

DNA code (D-N-A cōd) A sequence of three nucleotides of a DNA molecule.

DNA polymerase (po-lim'er-ās) An enzyme that brings new DNA triphosphate nucleotides into position for bonding on another DNA molecule.

DNA replication (rep''lĭ-ka'shun) The process by which the genetic material (DNA) of the cell reproduces itself prior to its distribution to the next generation of cells.

duplex DNA (du-pleks) DNA in a double helical shape

gene (jēn) Any molecule, usually a segment of DNA, that is able to: (1) *replicate* by directing the manufacture of copies of itself; (2) *mutate* or chemically change and transmit these changes to future generations; (3) *store* information that determines the characteristics of cells and organisms; and (4) use this information to *direct* the synthesis of structural and regulatory proteins.

guanine (gwah'nēn) A double-ring nitrogenous base molecule in DNA and RNA. It is the complementary base of cytosine.

initiation code (ĭ-nĭ'she-a''shun cōd) The code on DNA that begins the process of translation for an amino acid.

messenger RNA (mRNA) (mes'en-jer) A molecule composed of ribonucleotides that functions as a copy of the gene and is used in the cytoplasm of the cell during protein synthesis.

mutagenic agent (miu-tah-jen'ik a-jent) Anything that causes permanent change in DNA.

mutation (miu-ta'shun) Any change in the genetic information of a cell.

nitrogenous base (ni-trah'jen-us bās) A category of organic molecules found as components of the nucleic acids. There are five common types: thymine, guanine, cytosine, adenine, and uracil.

nucleoproteins (chromatin fibers) (nu-kle-o-pro'te-inz) The duplex DNA strands with attached proteins.

nucleosomes (nu'kle-o-somz) Histone clusters with their encircling DNA.

nucleotide (nu'kle-o-tĭd) The building block of the nucleic acids. Each is composed of a five-carbon sugar, a phosphate, and a nitrogenous base.

phosphate (fos-fāt) Part of a nucleotide, composed of phosphorus, oxygen, and hydrogen atoms.

point mutation (point miu-ta'shun) A change in the DNA of a cell as a result of a loss or change in a nitrogenous base sequence.

polysome (pah'le-sōm) A sequence of several translating ribosomes attached to the same mRNA.

promoter (pro-mo'ter) A region of DNA at the beginning of each gene just ahead of an initiator code.

protein synthesis (pro'te-in-sin'thĕ-sis) The process whereby the tRNA utilizes the mRNA as a guide to arrange the amino acids in their proper sequence according to the genetic information in the chemical code of DNA.

recombinant DNA (re-kom'bĭ-nant) DNA that has been constructed by inserting new pieces of DNA into the DNA of another organism, such as a bacterium.

ribonucleic acid (RNA) (ri-bo-nu-kle'ik ass-id) A polymer of nucleotides formed on the template surface of DNA by transcription. Three forms that have been identified include mRNA, rRNA, and tRNA.

ribose (ri′bōs) A five-carbon sugar molecule component of RNA.

ribosomal RNA (rRNA) (ri-bo-sōm′al) A globular form of RNA; a part of ribosomes.

RNA polymerase (po-lim′er-ās) An enzyme that attaches to the DNA at the promoter region of a gene when the genetic information is transcribed into RNA.

sickle-cell anemia (si̇-kul sel ah-ne′me-ah) A disease caused by a point mutation. This malfunction produces sickle-shaped red blood corpuscles.

template (tem′plāt) A model from which a new structure can be made. This term has special reference to DNA as a model for both DNA replication and transcription.

termination codes (ter-mi-na′shun cōdz) The DNA nucleotide sequence just in back of a gene with the code ATT, ATC, or ACT that signals ''stop here.''

thymine (thi′mēn) A single-ring nitrogenous base molecule of DNA but not RNA. It is complementary to adenine.

transcription (tran-skrip′shun) The process of manufacturing RNA from the template surface of DNA. Three forms of RNA that may be produced are mRNA, rRNA, and tRNA.

transfer RNA (tRNA) (trans-fur) A molecule composed of ribonucleic acid. It is responsible for transporting a specific amino acid into a ribosome for assembly into a protein.

translation (trans-la′shun) The assembly of individual amino acids into a polypeptide.

uracil (yu′rah-sil) A single-ring nitrogenous base molecule in RNA and not in DNA. It is complementary to adenine.

PART

3

Reproduction and Heredity

All living things are capable of reproducing themselves so that they can perpetuate their own kind. Several methods of reproduction have been identified. One method ensures the exact replication of a cell. When a cell divides by this method, it ceases to exist and becomes its own offspring, identical to the parent cell. Another method follows a series of events that results in the formation of cells that display different gene combinations that are found in the parent cell. These differences may better enable the offspring to survive in a changing environment. When multicellular organisms reproduce, they go through a very complex series of cellular events that lead to the formation of the next generation. These events are controlled by chemicals produced in the body and by outside factors. Exactly which characteristics are displayed in an individual depends on which genetically controlled traits are transmitted to the offspring. Characteristics are also affected by factors that influence how, when, or if these genes are expressed in the offspring. Should genetically controlled traits be found to enhance the survival and further reproduction of the organism, more individuals in the population may display these features as time goes by. An increase in the frequency of such genes in a population may change the population's appearance, behavior, or other characteristics. ■

Mitosis—The Cell Copying Process

▪ Purpose

In the previous chapter we saw how the molecule DNA replicates. Once this process is complete, doubled DNA is distributed to two new daughter cells. The way this cell-splitting process occurs assures that the daughter cells will have the same genetic message as the original DNA molecule. However, cells with identical genetic messages may differ in the way they are built and the specific roles they perform. The relationships among DNA, genes, and chromosomes are concepts for later consideration of genetics, evolution, and sex cell formation.

For Your Information

Many factors influence the growth of cells. Several years ago, investigators discovered that human mothers' milk contains a compound that encourages cell division in the cells lining the intestinal tract of a child.

Learning Objectives

■ List the purposes of cell division.

■ Explain the cell cycle.

■ State the processes occurring during interphase.

■ Name the stages of mitosis and explain what is happening during each stage.

■ Define differentiation.

■ Explain how cancer is caused and treated.

The Importance of Cell Division

Cells may die, but during the lifetime of an organism, the process of cell division assures that organisms can increase in size, dead cells can be replaced by new cells, and damaged tissues can be repaired. For example, you began as a single cell that resulted from the union of a sperm and an egg. One of the first activities of this single cell was to divide. As this process continued, the number of cells in your body increased, and as an adult your body consists of several trillion cells. The second function of cell division is to provide for the maintenance of the body. Certain cells in your body, like red blood cells, gut lining, and skin wear out. As they do, they must be replaced with new cells. Altogether you lose about one million cells per second; this means that at any time, cell division is happening a million times in your body. A third purpose of cell division is for repair. When a bone is broken, the break heals because cells divide, increasing the number of cells needed to knit the broken pieces together. If some skin cells are destroyed by a cut or abrasion, cell division produces new cells to repair the damage.

During cell division, two events occur. The replicated genetic information of a cell will be equally distributed to two daughter nuclei in a process called **mitosis.** Following the division of the nucleus, there is a division of the cytoplasm into two new cells called **cytokinesis,** cell splitting. Each new cell gets one of the two daughter nuclei so that both have a complete set of genetic information. Usually these two processes happen in sequence.

The Cell Cycle

All cells go through a basic life cycle, but they vary in the amount of time they spend in the different stages. A generalized picture of a cell's life cycle may help you to understand it better (figure 9.1). Once begun, cell division is a continuous process without a beginning or an end. It is a cycle in which cells continue to grow and divide. There are four stages to the life cycle of a eukaryotic cell: (1) G_1, gap phase one; (2) S, synthesis; (3) G_2, gap phase two; and (4) cell division (mitosis and cytokinesis).

The first three phases of the cell cycle, G_1, S, and G_2 occur during a period of time known as interphase. **Interphase** is the stage between cell divisions. During the G_1 stage, the cell grows in volume as it produces tRNA, mRNA, ribosomes, enzymes, and other cell components. During the S stage, DNA replication occurs in preparation for the distribution of genes to daughter cells. The G_2 stage follows as final preparations are made for mitosis with the synthesis of spindle fiber proteins.

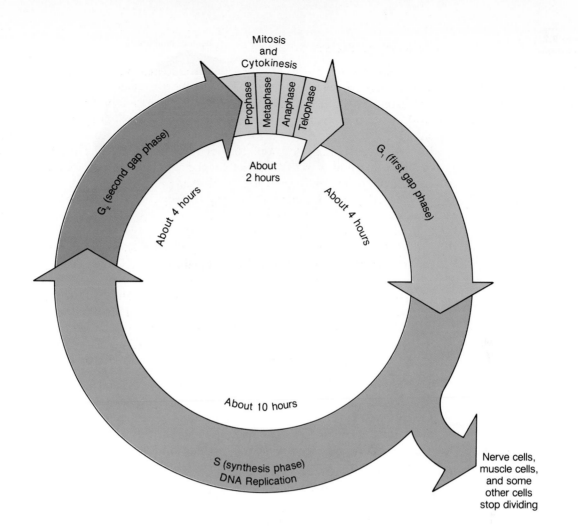

Mitosis
and
Cytokinesis

Prophase | Metaphase | Anaphase | Telophase

About
2 hours

G_2 (second gap phase)

About 4 hours

G_1 (first gap phase)

About 4 hours

About 10 hours

S (synthesis phase)
DNA Replication

Nerve cells,
muscle cells,
and some
other cells
stop dividing

Figure 9.1
Cell Cycle.
During the cell cycle, tRNA, mRNA,
ribosomes, and enzymes are
produced in the G_1 stage. DNA
replication occurs in the S stage.
Proteins required for the spindle
are synthesized in the G_2 stage.
The nucleus is replicated in mitosis
and two cells are formed by
cytokinesis. Some cells, such as
the nerve cell, remain in the G_1
stage. The time periods indicated
are relative and vary depending
upon the type of cell and age of
the organism.

Figure 9.2
Interphase.

DNA replication occurs during
interphase. The individual
chromosomes are not visible, but a
distinct nuclear membrane and
nucleolus are present.

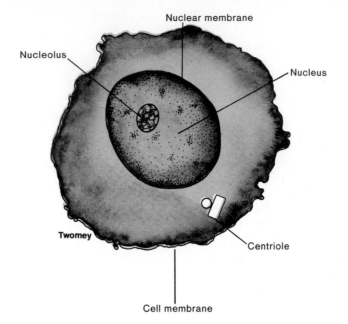

In this stage, the cell is not dividing but is engaged in metabolic activities, such as muscle cell contractions, nerve cell transmission, or glandular cell secretion. During interphase, the nuclear membrane is intact and the individual chromosomes are not visible (figure 9.2). The individual chromatin strands are too thin and tangled to be seen. Remember that **chromosomes** are composed of various kinds of histone proteins and DNA that contain the cell's genetic information. Basic to any chromosome is the chromatid, a double helix of DNA and the nucleosomes. It is these chromosomes that were replicated in the S stage and will be distributed during mitosis.

The Stages of Mitosis

All stages in the life cycle of a cell are continuous, there is no precise point when the G_1 stage ends and the S stage begins, or when the interphase period ends and mitosis begins. Likewise, in the individual stages of mitosis, there is a gradual transition from one stage to the next. During mitosis there are four recognizable stages known as prophase, metaphase, anaphase, and telophase.

Prophase

As the G_2 phase of interphase ends, prophase begins. **Prophase** is the first stage of mitosis. One of the first noticeable changes is that the individual chromosomes become visible (figure 9.3). The thin, tangled chromatin present during interphase gradually coils and thickens and becomes visible as separate chromosomes. The DNA portion of the chromosome carries genes that are arranged in a specific order. Each chromosome carries its own set of genes that is different from the sets of genes on other chromosomes.

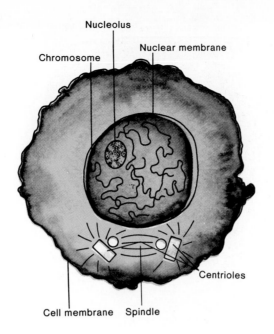

Nucleolus

Chromosome

Nuclear membrane

Centrioles

Cell membrane   Spindle

Figure 9.3
Early Prophase.
Chromosomes begin to appear as thin tangled threads and nucleolus and nuclear membrane are present.

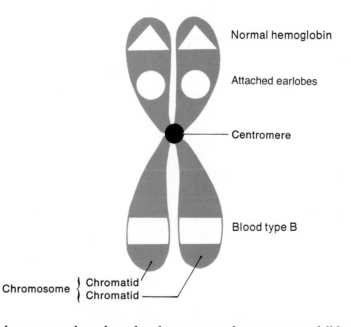

Normal hemoglobin

Attached earlobes

Centromere

Blood type B

Chromosome { Chromatid
              { Chromatid

Figure 9.4
Chromatids.
During interphase when chromosomal replication happens, the original double strand of DNA unzips and forms two identical double strands that are attached at the centromere. Each of these double strands is a chromatid. (We really don't know on which chromosomes most human genes are located. The examples presented here are for illustrative purposes only.)

As prophase proceeds and as the chromosomes become more visible, we recognize that each chromosome is made of two parallel, threadlike parts laying side by side. Each parallel thread is called a **chromatid** (figure 9.4). These chromatids were formed during the S stage of interphase when DNA synthesis occurred. The two identical chromatids are attached at a region called the **centromere.**

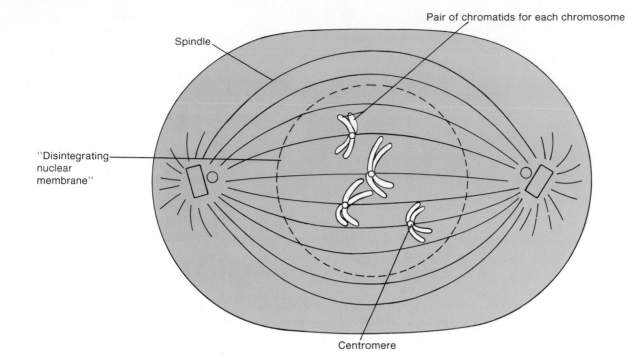

Spindle

Pair of chromatids for each chromosome

"Disintegrating nuclear membrane"

Centromere

Figure 9.5
Late Prophase.
In late prophase, the chromosomes appear as two chromatids connected by a centromere. The nucleolus and the nuclear membrane have disintegrated. The centromeres have moved farther apart and the spindle is longer.

In the diagrams in this text, a few genes are shown as they might occur on human chromosomes. The diagrams show fewer chromosomes and fewer genes on each chromosome than are actually present. Normal human cells have ten billion nucleotides arranged into forty-six chromosomes, each chromosome with thousands of genes. In this book, smaller numbers of genes and chromosomes are used to make it easier to follow the events that happen in mitosis.

> We really don't know on which chromosomes most human genes are located. Those shown here are for illustrative purposes only.

Several other events occur as the cell proceeds to the late prophase stage (figure 9.5). One of these events is the duplication of the **centrioles.** Remember that some cells contain centrioles that are microtubule-containing organelles located just outside the nucleus. As they duplicate, they move to the poles of the cells. As the centrioles move to the poles, the microtubules form the spindle. The **spindle** is an array of microtubules extending from pole to pole used in the movement of chromosomes.

As prophase is occurring, the nuclear membrane gradually disintegrates. Although it is present at the beginning of prophase, it disappears when this stage is finished. In addition to the nuclear membrane, the nucleoli within the nucleus dis-

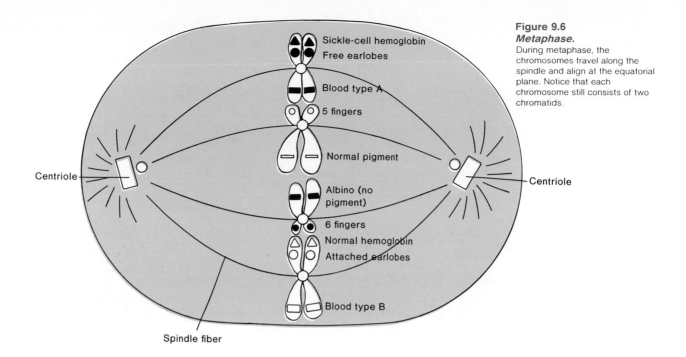

Figure 9.6
Metaphase.
During metaphase, the chromosomes travel along the spindle and align at the equatorial plane. Notice that each chromosome still consists of two chromatids.

Sickle-cell hemoglobin
Free earlobes
Blood type A
5 fingers
Normal pigment
Albino (no pigment)
6 fingers
Normal hemoglobin
Attached earlobes
Blood type B
Centriole
Centriole
Spindle fiber

appear. Because of the disintegration of the nuclear membrane, the chromosomes are free to move anywhere within the cytoplasm of the cell. As a result of this movement, the cell enters into the next stage of mitosis.

Metaphase

Metaphase is the second stage in mitosis when the chromosomes align at the equatorial plane. There is no nucleus present during metaphase and the spindle, which started to form during prophase, is completed. The centrioles are at the poles and the microtubules extend between them to form the spindle. At the beginning of metaphase, the chromosomes become attached to the spindle fibers at their centromeres. Initially they are distributed randomly throughout the cytoplasm. Then the chromosomes move until all their centromeres align themselves along the equatorial plane at the equator of the cell (figure 9.6). At this stage in mitosis, each chromosome still consists of two chromatids. In a human cell, there are forty-six chromosomes or ninety-two chromatids aligned at the cell's equatorial plane during metaphase.

If we view a cell in the metaphase stage from the side (figure 9.6), it is an equatorial view. In this view, the chromosomes appear as if they were in a line. If we view the cell from the pole, it is a polar view. The chromosomes are seen on the equatorial plane (figure 9.7). Chromosomes viewed from this direction appear as hot dogs scattered on a plate.

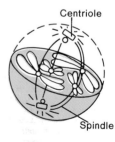

Centriole
Spindle

Figure 9.7
Polar View of Metaphase.
The polar view shows the chromosomes spread out on a plane.

Figure 9.8
Anaphase.
The pairs of chromatids separate
as the centromeres divide. The
chromatids, now called
chromosomes, are separating and
moving toward the poles.

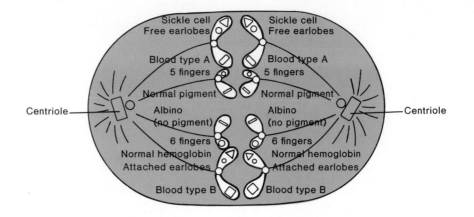

Anaphase

Anaphase is the third stage of mitosis. During this stage, the nuclear membrane is still absent and the spindle extends from pole to pole. In anaphase, the chromosome splits at the centromere and the two chromatids within the chromosome separate as they move along the spindle fibers toward the poles (figure 9.8). Though this movement of chromosomes has been observed repeatedly, no one knows the exact mechanism of its action. After this separation of chromatids occurs, the chromatids are called **daughter chromosomes.** Daughter chromosomes contain identical genetic information.

Examine figure 9.8 closely and notice that the four chromosomes moving to one pole have the identical genetic information as the four moving to the opposite pole. It is the alignment of the chromosomes in metaphase and their separation in anaphase that causes this type of distribution. At the end of anaphase, there are two identical groups of chromosomes, one group at each pole. The next stage completes the mitosis process.

Telophase

Telophase is the last stage in mitosis during which daughter nuclei are formed. Each set of chromosomes becomes enclosed by a nuclear membrane and the nucleoli reappear. Now the cell has two identical **daughter nuclei** (figure 9.9). In addition, the microtubules disintegrate so the spindle disappears. With the formation of the daughter nuclei, mitosis, the first process in cell division, is completed and the second process, cytokinesis, may happen. Cytokinesis splits the cytoplasm of the original cell and forms two smaller daughter cells. **Daughter cells** are two cells formed by

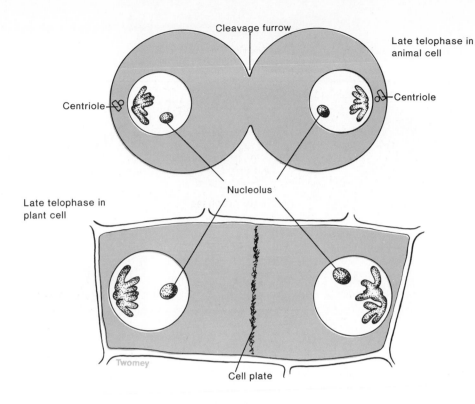

Cleavage furrow

Late telophase in animal cell

Centriole

Centriole

Nucleolus

Late telophase in plant cell

Twomey

Cell plate

Figure 9.9
Telophase.
During telophase, the spindle disintegrates and the nucleolus and nuclear membrane form.

cell division that have identical genetic information. Each of the newly formed daughter cells then enters the G_1 phase of interphase. These cells can grow, replicate their DNA, and enter another round of mitosis and cytokinesis to complete the cell cycle.

Plant and Animal Cell Differences

Cell division is similar in both plant and animal cells. However, there are some minor differences. One difference concerns the centrioles (figure 9.10). Centrioles are essential in animal cells, but they are absent in plant cells. However, by some process, plant cells do produce a spindle. There is also a difference in the process of cytokinesis (figure 9.11). In animal cells, cytokinesis results from a **cleavage furrow.** This is an indentation of the cell membrane of an animal cell that pinches the cytoplasm into two parts as if a string were tightened about its middle. In an animal cell, cytokinesis begins at the cell membrane and proceeds to the center. In plant cells, a **cell plate** begins at the center and proceeds to the cell membrane, resulting in a cell wall that separates the two daughter cells.

Figure 9.10
Comparison of Plant and Animal Mitosis.
(I) Drawings of mitosis in an animal cell. (II) Photographs of mitosis in a whitefish blastula. (III) Drawings of mitosis in a plant cell.
(IV) Photographs of mitosis in an onion root tip. (Carolina Biological Supply Company Photographs)

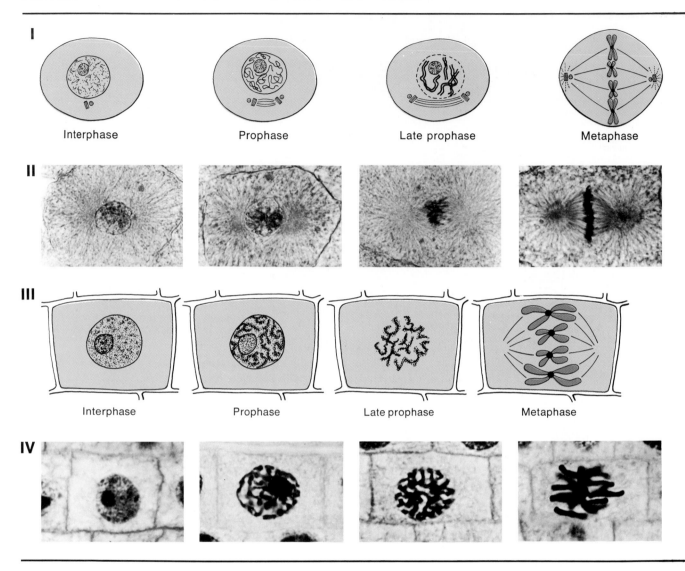

I

Interphase · Prophase · Late prophase · Metaphase

III

Interphase · Prophase · Late prophase · Metaphase

Figure 9.11
Cytokinesis.
In animal cells, there is a pinching in of the cytoplasm that eventually forms two daughter cells. Daughter cells in plants are formed when a cell plate separates the cell into two cells.

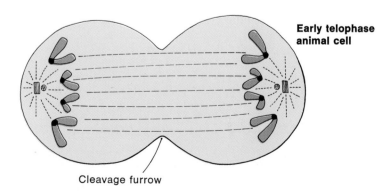

Early telophase animal cell

Cleavage furrow

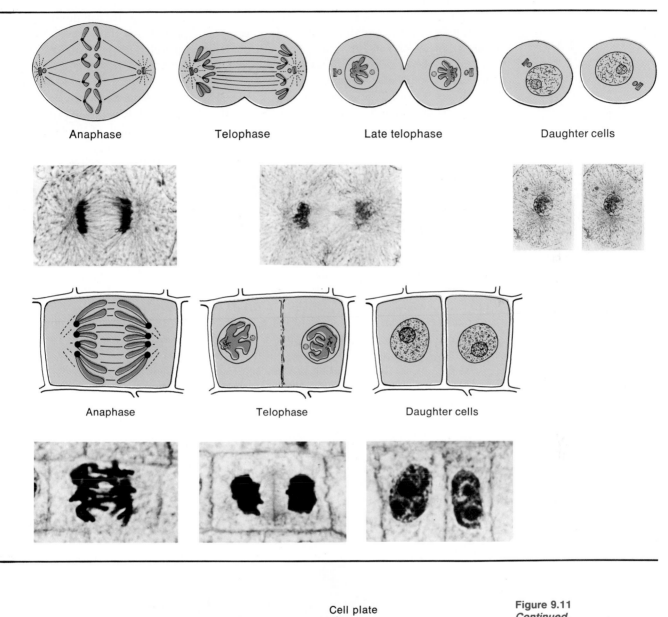

Anaphase Telophase Late telophase Daughter cells

Anaphase Telophase Daughter cells

Early telophase plant cell

Cell plate

Figure 9.11
Continued.

Differentiation

Because of the two processes in cell division, mitosis and cytokinesis, the daughter cells have the same genetic composition. You received a set of genes from your father in his sperm, and a set of genes from your mother in her egg. By cell division, this cell formed two daughter cells. This process was repeated, and there were four cells, all of which had the same genes. All the trillions of cells in your body were formed by the process of cell division. This means that, except for mutations, all the cells in your body have the same genes.

However, all the cells in your body are not the same. There are nerve cells, muscle cells, bone cells, skin cells, and many other types. How is it possible that cells with the same genes can produce different types of cells? Think of the genes in a cell being like the individual recipes in a cookbook. It is possible to give a copy of the same cookbook to one hundred people and, although they all have the same book, each person could prepare a different dish. If you use the recipe to make a chocolate cake, you ignore the directions for making salads, fried chicken, or soups, although these recipes are in the book.

It is the same with cells. Although some genes are used by all cells, some cells only activate certain selective genes. Muscle cells produce proteins capable of contraction. Most other cells do not use these genes. Pancreas cells use genes that result in the formation of digestive enzymes, but never produce contractile proteins. **Differentiation** is the process of forming specialized cells within a multicellular organism. Some cells lose their ability to divide and remain permanently in the interphase condition, such as muscle and nerve cells. Other cells retain their ability to divide as their form of specialization. Cells that line the digestive tract or form the surface of your skin are examples of dividing cells.

As part of the differentiation process, some cells lose their ability to divide. In growing organisms, such as infants, seedlings, or embryos, most cells are capable of division and divide at a rapid rate. In older organisms, many cells lose their ability to divide as a result of differentiation and the frequency of cell division decreases. As the organism ages, the lower frequency of cell division may affect many bodily processes, including healing. In some older people, there may be so few cells capable of dividing that a broken bone may never heal. It is also possible for a cell to undergo mitosis but not cytokinesis. In voluntary muscle cells, the cells undergo mitosis but not cytokinesis and this results in multinucleated cells (figure 9.12).

Figure 9.12
Skeletal Muscle.

In some cells, such as the skeletal muscles, the cell undergoes mitosis but not cytokinesis. The result is cells with many nuclei.

Nucleus

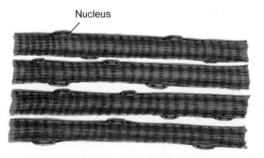

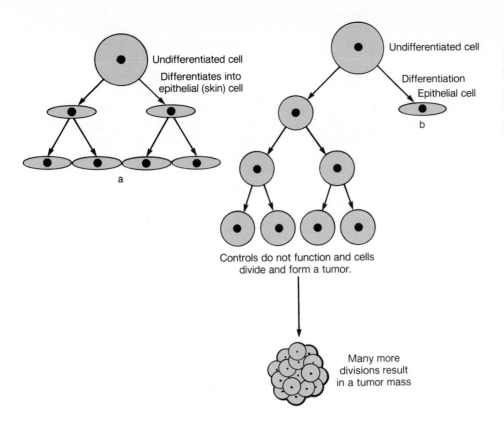

Figure 9.13
Tumor.

(a) In normal cell division, the mechanism that controls cell division and differentiation functions properly. The cells differentiate into normal skin cells. (b) In a tumor, differentiation does not occur and the mechanism to control cell division does not work. The result is a tumor in the skin.

Controls do not function and cells divide and form a tumor.

Many more divisions result in a tumor mass

Abnormal Cell Division

Understanding mitosis can help you understand certain biological problems and how to solve them. All cells do not divide at the same rate, but each kind of cell has a regulated division rhythm. Regulation of the cycle can come from inside or outside the cell. When human white blood cells are grown outside the body under special conditions, they develop a regular cell division cycle. The cycle is determined by the DNA of the cells. However, white blood cells in a person may increase their rate of mitosis as a result of outside influences. Disease organisms entering the body, tissue damage, and changes in cell DNA all may alter the rate at which white blood cells divide. An increase in white blood cells in response to the invasion of disease organisms is valuable since these white blood cells are capable of destroying the disease organisms.

On the other hand, an uncontrolled increase in the rate of mitosis of white blood cells causes a kind of cancer known as leukemia. This condition causes a general weakening of the body because the excess number of white blood cells diverts necessary nutrients from other cells of the body and interferes with their normal activities.

Normally cells differentiate and become specialized in their functioning. Each cell type has its cell division process regulated so that it does not interfere with the activities of other cells or the whole organism. However, some cells de-differentiate (i.e., revert to their embryonic state) and begin to divide in an uncontrolled fashion. When this happens, a group of cells forms that is known as a tumor (figure 9.13). A benign tumor is a cell mass that does not fragment and spread beyond its original area of growth. A benign tumor can become harmful by growing to a point that it interferes with normal body functions. Some tumors are malignant. Cells of these tumors move from the original site (metastasis) and establish new colonies in other regions of the body. **Cancer** is an abnormal growth of cells that has a malignant potential.

Once cancer has been detected, treatment requires the elimination of the tumor. There are three kinds of treatment. If the cancer is confined to a few specific locations, it may be possible to surgically remove it. However, there are cases where surgery is impractical. If the tumor is located where it can't be removed without destroying healthy, vital tissue, surgery may not be used. For example, the removal of certain brain cancers could severely damage the brain. In these cases, two other methods may be used to treat cancer.

Chemotherapy uses various types of chemicals to destroy cancer cells. This treatment may be used without knowing exactly where the cancer is located. However, it has negative effects on normal cells. It lowers the body's immune reaction because it decreases its ability to produce white blood cells. Other side effects include intestinal disorders and the loss of hair. It is damage to healthy, dividing cells in the scalp and the intestinal tract that causes these disorders.

Radiation therapy uses large amounts of X rays or gamma rays. Since this treatment will damage surrounding healthy cells, it is used only very cautiously when surgery is impractical. Radiation therapy can be effective because cells undergoing division are usually not in the G_1 stage. Nondividing cells pause in the G_1 phase and are less likely to be damaged by treatment. Cancer cells are usually in the S or G_2 stage. In these stages DNA replication or protein synthesis is occurring and the cell is more subject to damage. Therefore, while radiation therapy may destroy healthy cells, it usually destroys more cancer cells.

Chemotherapy and radiation treatment are often used to control leukemia by taking advantage of the fact that these cells are undergoing an unusual mitosis. Dividing cells are likely to be damaged by x-radiation because the radiation can more easily destroy essential molecules (DNA) of the cell. Since cancer cells spend more time dividing than normal cells, they have a greater chance of being killed by radiation. Physicians, therefore, often prescribe cobalt therapy for certain cancer patients with widely dispersed cancers such as leukemia. Cobalt is radioactive and releases radiation.

Radiation is dangerous for the same reasons that it is beneficial in cancer treatment. In cases of extreme exposure to radiation, people develop what is called radiation sickness. The symptoms of this disease include loss of hair, bloody vomiting and diarrhea, and a reduced white blood cell count. These symptoms occur in parts of the body where mitosis is common. The lining of the intestine is constantly being lost as food travels through and it must be replaced by the process of mitosis. Hair growth is the result of the continuous division of cells at the roots. White blood cells are also continuously reproduced in the bone marrow and lymph nodes. When radiation strikes these rapidly dividing cells and kills them, the lining of the intestine wears away and bleeds, hair falls out, and few new white blood cells are produced to defend the body against infection.

Summary

Cell division is necessary for growth, repair, and reproduction. Cells go through a cell cycle that includes cell division (mitosis and cytokinesis) and interphase. Interphase is the period of growth and preparation for division. Mitosis is divided into four stages: prophase, metaphase, anaphase, and telophase. During mitosis, two daughter nuclei are formed from one parent nucleus. These nuclei have identical sets of chromosomes and genes that are exact copies of those of the parent. Although the process of mitosis has been presented as a series of phases, you should realize that it is a continuous, flowing process from prophase through telophase. Following mitosis, cytokinesis divides the cytoplasm and the cell returns to interphase.

The regulation of mitosis is important if organisms are to remain healthy. Regular divisions are necessary to replace lost cells and allow for growth. However, uncontrolled cell division may result in cancer and disruption of the total organism's well-being.

Consider This

A chemical known as colchicine is extracted from the seeds of a small crocuslike plant. This chemical is used in biological laboratories because it can prevent the formation of the spindle. Which parts of the cell cycle would proceed normally and which parts would be altered if this chemical were used on cells? If you know that the cells are not killed by colchicine and begin mitosis normally, what changes might occur in the number of chromosomes of the next cell generation, and how might this change the metabolism of the cell?

Experience This

What is cancer?

Cancer is a large group of diseases characterized by uncontrolled growth and spread of abnormal cells. If the spread is not controlled or checked, it results in death. However, many cancers can be cured if detected and treated promptly.

Cancer's Seven Warning Signals

1. Change in bowel or bladder habits
2. A sore that does not heal
3. Unusual bleeding or discharge
4. Thickening or lump in breast or elsewhere
5. Indigestion or difficulty in swallowing
6. Obvious change in wart or mole
7. Nagging cough or hoarseness

Thirty percent of Americans will eventually have cancer. Cancer will strike three out of four families. If you notice one of the above symptoms, see your doctor.

To become better informed on cancer, contact your local Cancer Society office or write American Cancer Society, Inc., 90 Park Ave., New York, N.Y. 10016. Request a copy of the latest publication *Cancer Facts and Figures*.

Questions

1. Name the four stages of mitosis and describe what occurs in each stage.
2. What is meant by the phrase "cell cycle?"
3. During which phase of a cell's cycle does DNA replication occur?
4. At what stage of mitosis does the DNA become most visible?
5. What are the differences between plant and animal mitosis?
6. Why can X ray treatment be used to control cancer?
7. What is the purpose of mitosis?
8. What is the difference between a cell plate and a cell wall?
9. What type of activities occur during interphase?
10. List five differences between an interphase cell and a cell in mitosis.

Chapter Glossary

anaphase (an'a-fāz) The third stage of mitosis when the centromeres split and the chromosomes move to the poles.

cancer (kan'sur) A tumor that is malignant.

cell plate (sel plāt) A plant cell structure that begins to form in the center of the cell and proceeds to the cell membrane, resulting in cytokinesis.

centrioles (sen'tre-ōls) Organelles containing microtubules located just outside the nucleus.

centromere (sen'tro-mēr) Region where two chromatids are joined.

chromatid (kro'mah-tid) A replicated chromosome physically attached to an identical chromatid at the centromere.

chromosomes (kro-mo-sōms) Structure composed of various kinds of histone proteins and DNA that contain the cell's genetic information.

cleavage furrow (kle'vaj fur-ro) Indentation of the cell membrane of an animal cell that pinches the cytoplasm into two parts.

cytokinesis (si-to-kǐ-ne'sis) Division of the cytoplasm of one cell into two new cells.

daughter cells (daw'tur sels) Two cells formed by cell division.

daughter chromosomes (daw'tur kro'mo-sōms) Chromosomes produced by DNA replication and containing identical genetic information; formed after chromosome division in anaphase.

daughter nuclei (daw'tur nu'kle-i) Two nuclei formed by mitosis.

differentiation (dif''fur-ent-she-a'shun) The process of forming specialized cells within a multicellular organism.

interphase (in'tur-fāz) Stage between cell divisions in which the cell is engaged in metabolic activities.

metaphase (me'tah-fāz) Second stage in mitosis during which the chromosomes align at the equatorial plane.

mitosis (mi-to'sis) Process that results in equal and identical distribution of replicated chromosomes into two newly formed nuclei.

prophase (pro'fāz) First stage of mitosis during which individual chromosomes become visible.

spindle (spin'dul) An array of microtubules extending from pole to pole and used in the movement of chromosomes.

telophase (tel'uh-fāz) Last stage in mitosis during which daughter nuclei are formed.

Meiosis—Sex Cell Formation

■ Purpose

How can the chromosome number in humans remain 46 generation after generation if both parents contribute equally to the genetic information of the child? In this chapter, we will discuss the mechanics of the process of meiosis. Meiosis is a specialized cell division resulting in the formation of sex cells. Knowing the mechanics of this process is essential to understanding how genetic variety can occur in sex cells. This variety ultimately shows up as differences in offspring.

For Your Information

General appearance: round face
Personality: strong sense of curiosity, easy
  going, stable temperament
IQ: unknown
Music: proficient
Athletics: competent in softball and
  swimming
Manual dexterity: good
General health: excellent; three
  grandparents lived to 80s, the fourth
  dying young in a car accident

A summary sheet for a human sperm bank lists this information regarding one of its donors. Since the 1960s, some 250,000 women in the United States have conceived babies by artificial insemination.

Learning Objectives

■ Explain why sexually reproducing organisms must form cells with the haploid number of chromosomes.

■ Describe the stages in meiosis I.

■ Describe the stages in meiosis II.

■ Understand how genetic variety in offspring is generated by mutation, crossing-over, segregation, independent assortment, and fertilization.

■ Explain the similarities and differences between mitosis and meiosis.

Sexual Reproduction

The most successful kinds of plants and animals are those that have developed a method of shuffling and exchanging genetic information. This usually involves organisms that have two sets of genetic data, one inherited from each parent. **Sexual reproduction** is the formation of a new individual by the union of sex cells. Before sexual reproduction can occur, the two sets of genetic information must be reduced to one set. This is somewhat similar to shuffling a deck of cards and dealing out hands. This shuffling and dealing assures that each of the hands will be different. An organism with two sets of chromosomes can produce many combinations of chromosomes when it produces sex cells, just as many different hands can be dealt from one pack of cards. When one of these sex cells unites with another, a new organism containing two sets of genetic information is formed. This new organism's information might very well be superior to the information found in either parent; this is the value of sexual reproduction.

In chapter 9, we discussed the cell cycle and pointed out that it is a continuous process without a beginning or an end. The process of mitosis followed by growth is important in the life cycle of any organism. Thus, the *cell cycle* is part of an organism's *life cycle* (figure 10.1).

The sex cells produced by male organisms are called **sperm,** and those produced by females are called **eggs.** A general term sometimes used to refer to either eggs or sperm is the term **gamete.** The uniting of an egg and sperm (gametes) is known as **fertilization.** The **zygote,** which results from the union of an egg and a sperm, divides repeatedly by mitosis to form the complete organism. Notice that the zygote and its descendants have two sets of chromosomes. However, the male gamete and the female gamete each contain one set of chromosomes. These sex cells are said to be **haploid.** The haploid number of chromosomes is represented by the initial **N.**

Figure 10.1
Life Cycle.
The cells of this adult organism have eight chromosomes in their nuclei. In preparation for sexual reproduction, the number of chromosomes must be reduced by half so that fertilization will result in the original number of eight chromosomes in the new individual. The offspring will grow and produce new cells by mitosis.

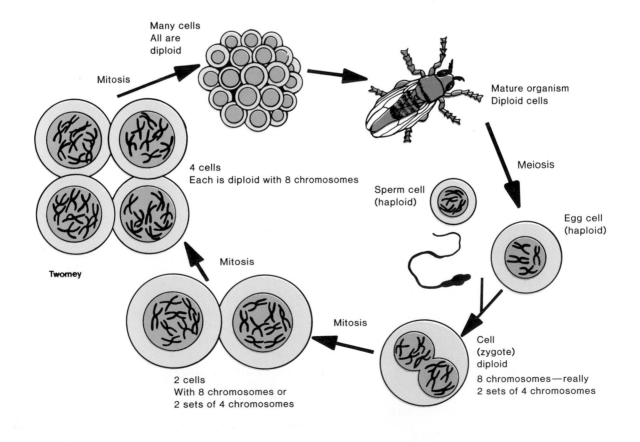

Many cells
All are
diploid

Mitosis

Mature organism
Diploid cells

Meiosis

4 cells
Each is diploid with 8 chromosomes

Sperm cell
(haploid)

Egg cell
(haploid)

Mitosis

Twomey

Mitosis

Cell
(zygote)
diploid

8 chromosomes—really
2 sets of 4 chromosomes

2 cells
With 8 chromosomes or
2 sets of 4 chromosomes

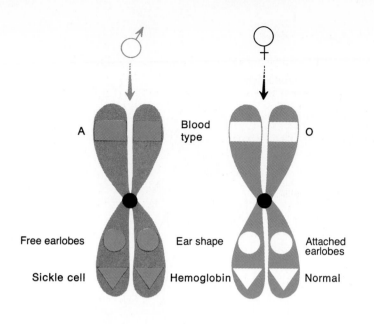

Figure 10.2
*Pair of Homologous
Chromosomes.*
A pair of chromosomes of similar
size and shape having genes for
the same traits are said to be
homologous. Notice that the genes
are not the same but are for the
same type of information.

Table 10.1 *Chromosome Numbers*

Organism	Haploid Number	Diploid Number
Mosquito	3	6
Fruit fly	4	8
Housefly	6	12
Toad	18	36
Cat	19	38
Human	23	46
Hedgehog	23	46
Chimpanzee	24	48
Horse	32	64
Dog	39	78
Onion	8	16
Kidney bean	11	22
Rice	12	24
Tomato	12	24
Potato	24	48
Tobacco	24	48
Cotton	26	52

A zygote contains two sets and is said to be **diploid.** The diploid number of chromosomes is represented by the number and initial *2N (N + N = 2N)*. Diploid cells have two sets of chromosomes, one set from each parent. Remember, a chromosome is composed of two chromatids, each containing duplex DNA. These two chromatids are attached to each other at a point called the centromere. In a diploid nucleus, the chromosomes occur as **homologous chromosomes**—a pair of chromosomes in a diploid cell that contain similar genes throughout their length. One of the chromosomes of a homologous pair was donated by the father, the other by the mother (figure 10.2). Different species of organisms vary in the number of chromosomes they contain. Table 10.1 lists several different organisms and their haploid and diploid chromosome numbers.

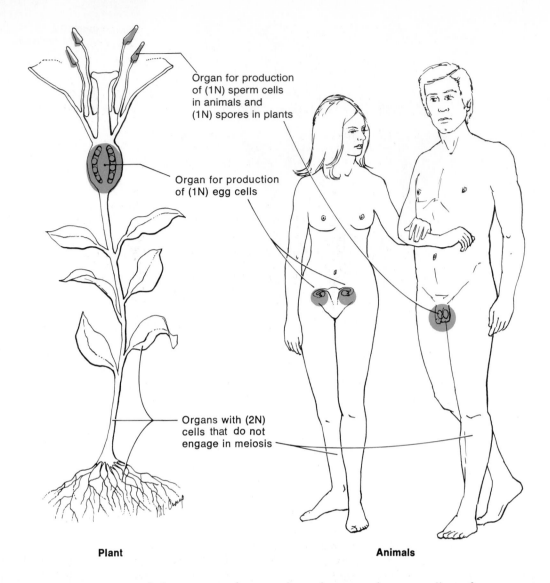

Organ for production
of (1N) sperm cells
in animals and
(1N) spores in plants

Organ for production
of (1N) egg cells

Organs with (2N)
cells that do not
engage in meiosis

Plant   Animals

Figure 10.3
Haploid and Diploid Cells.
Both plants and animals produce
cells with a haploid number of
chromosomes. The male anther in
plants and the testes in animals
produce haploid male cells. In both
plants and animals, the ovary
produces haploid female cells.

It is necessary for organisms that reproduce sexually to form gametes having only one set of chromosomes. If gametes contained two sets of chromosomes, when they combined to form a zygote, that new individual would have four sets of chromosomes. The number of chromosomes would continue to double with each new generation, which could result in death. However, this does not happen, the number of chromosomes remains constant generation after generation. Since cell division by mitosis and cytokinesis results in cells having the same number of chromosomes as the parent cell, two questions arise: how are sperm and egg cells formed, and how do they get only one-half of the chromosomes of the diploid cell? The answers lie in the process of **meiosis,** the specialized pair of cell divisions that reduce the chromosome number from diploid (2N) to haploid (N). The major function of meiosis is to produce cells that have one set of genetic information. Therefore, when fertilization occurs, the zygote will have two sets of chromosomes as did each parent.

Not every cell goes through the process of meiosis. Only specialized organs are capable of producing haploid cells (figure 10.3). In animals, the organs that undergo

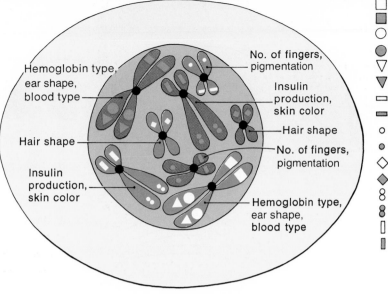

Legend:
- □ Blood type O
- ■ Blood type A
- ○ Attached earlobes
- ● Free earlobes
- ▽ Normal hemoglobin
- ▼ Sickle-cell hemoglobin
- ▭ Normal pigment
- ▬ Albino (no pigment)
- ○ 5 fingers
- ◉ 6 fingers
- ◇ Curly hair
- ◆ Straight hair
- 8 Light skin color
- 8 Dark skin color
- | | Normal insulin
- || Diabetes

Labels in figure: Hemoglobin type, ear shape, blood type; No. of fingers, pigmentation; Insulin production, skin color; Hair shape; Hair shape; No. of fingers, pigmentation; Insulin production, skin color; Hemoglobin type, ear shape, blood type

Figure 10.4
Chromosomes in a Cell.
In this diagram of a cell, the eight chromosomes are scattered in the nucleus. Even though they are not arranged in pairs, note that there are four pairs of homologous chromosomes.

meiosis are called **gonads.** The female gonads that produce eggs are called **ovaries.** The male gonads that produce sperm are called **testes.** Organs that produce gametes are also found in flowering plants. In plants, the **pistil** produces eggs or ova and the **anther** produces pollen, which contain sperm.

To illustrate meiosis in this chapter, we arbitrarily chose an organism whose cells have a diploid number of eight (figure 10.4).

The haploid number of chromosomes for this organism is four, and these haploid cells contain only one complete set of four chromosomes. You can see there are eight chromosomes in this cell, four from the mother and four from the father. A closer look at figure 10.4 shows you there are only four types of chromosomes, but two of each type:

1. long chromosomes consisting of chromatids attached near the center
2. long chromosomes consisting of chromatids attached near one end
3. short chromosomes consisting of chromatids attached near one end
4. short chromosomes consisting of chromatids attached near the center

We can therefore talk about the number of chromosomes in two ways. We can say that our hypothetical diploid cell has eight chromosomes, or we can say that it has four pairs of homologous chromosomes.

Haploid cells, on the other hand, do not have homologous chromosomes. They have only one of each type of chromosome. The whole point of meiosis is to distribute the chromosomes and the genes they carry so that each daughter cell gets one member of each homologous pair. In this way, each daughter cell gets one complete set of genetic information.

Figure 10.5
Meiosis I.
The stages in meiosis I result in reduction division. This reduces the number of chromosomes in the parent cell into two haploid daughter cells.

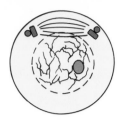
Prophase I

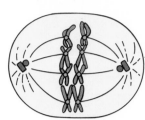
Metaphase I

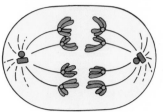
Anaphase I

Telophase I

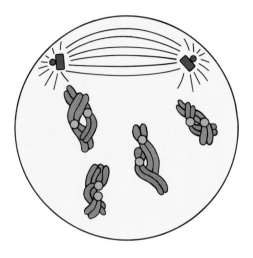

Figure 10.6
Prophase I.
During prophase I, the cell is preparing for division. A unique event that occurs in prophase I is that the chromosomes are synapsed.

Mechanics of Meiosis: Meiosis I

Meiosis is preceded by an interphase stage when DNA replication occurs. In a sequence of events called meiosis I, members of homologous pairs of chromosomes divide into two complete sets. This is sometimes called a **reduction division,** a type of cell in which daughter cells get only half the chromosomes from the parent cell. The division begins with chromosomes composed of two chromatids. The sequence of events in meiosis I is artificially divided into four phases: prophase I, metaphase I, anaphase I, and telophase I. Figure 10.5 shows the events in meiosis I.

Prophase I

During prophase I the cell is preparing itself for division (figure 10.6). During prophase I the chromatin material coils and thickens into chromosomes, the nucleoli disappear, the nuclear membrane disintegrates, and the spindle begins to form. The spindle is formed in animals when the centrioles move to the poles. There are no centrioles in plant cells, but the spindle does form. However, there is an important difference between the prophase stage of mitosis and prophase I of meiosis. During prophase I, homologous chromosomes come to lie next to each other in a process called **synapsis.** While the chromosomes are synapsed, a unique event called crossing-over can occur. Crossing-over is the exchange of equivalent sections of DNA on homologous chromosomes. We will fit crossing-over into the whole picture of meiosis later.

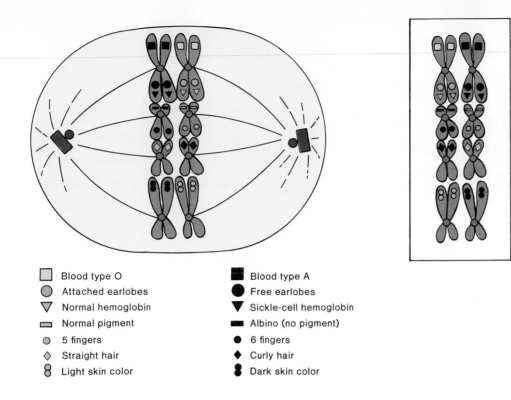

☐ Blood type O	■ Blood type A
◯ Attached earlobes	● Free earlobes
▽ Normal hemoglobin	▼ Sickle-cell hemoglobin
⊟ Normal pigment	▬ Albino (no pigment)
◌ 5 fingers	● 6 fingers
◇ Straight hair	◆ Curly hair
8 Light skin color	8 Dark skin color

Figure 10.7
Metaphase I.
Notice that the homologous chromosome pairs are arranged on the equatorial plane in the synapsed condition. The cell at the left shows one way the chromosomes could be lined up. The rectangle on the right shows a second arrangement. How many other ways can you diagram?

Metaphase I

The synapsed pair of homologous chromosomes move into position on the equatorial plane of the cell. In this stage, the centromeres of each chromosome attaches to the spindle. The synapsed homologous chromosomes move to the equator of the cell as single units. The way they are arranged on the equator (which one is on the left and which one is on the right) is determined by chance (figure 10.7). In the cell in figure 10.7, three "gray" chromosomes from the father and one "blue" chromosome from the mother are lined up on the left. Similarly, one "gray" chromosome from the father and three "blue" chromosomes from the mother are on the right. They could have aligned themselves in several other ways. For instance, they could have lined up as shown in the rectangular box at the right in figure 10.7.

Anaphase I

Anaphase I is the stage during which homologous chromosomes separate (figure 10.8). During this stage, the chromosome number is reduced from diploid to haploid. The two members of each pair of homologous chromosomes move away from each other toward opposite poles. The direction each takes is determined by how each pair was originally arranged on the spindle. Each chromosome is independently attached to a spindle fiber at its centromere. Unlike the anaphase stage of mitosis, the centromeres that hold the chromatids together *do not divide* during anaphase I of meiosis. Each chromosome still consists of two chromatids. Because the chromosomes and the genes they carry are being separated from one another, this process is called **segregation.** The way a single pair of homologous chromosomes segregates does not influence how other pairs of homologous chromosomes segregate. That is, each pair segregates independently of other pairs. This is known as **independent assortment** of chromosomes.

Figure 10.8
Anaphase I.
During this phase, one member of the homologous chromosome pair is segregated from the other member of the pair. Notice that the centromeres of the chromosomes do not split.

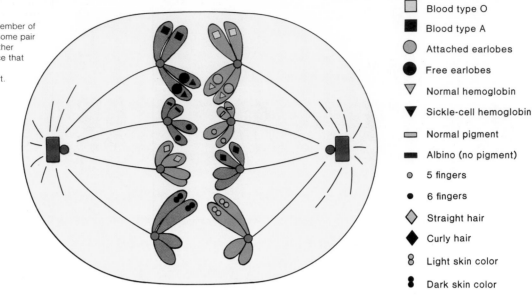

☐	Blood type O
■	Blood type A
⬤ (gray)	Attached earlobes
⬤ (black)	Free earlobes
▽	Normal hemoglobin
▼	Sickle-cell hemoglobin
▭ (outline)	Normal pigment
▬ (filled)	Albino (no pigment)
∘	5 fingers
•	6 fingers
◇	Straight hair
◆	Curly hair
8 (light)	Light skin color
8 (dark)	Dark skin color

Figure 10.9
Telophase I.
What activities would you expect during a telophase stage of cell division?

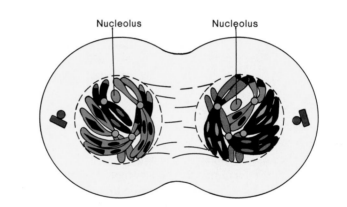

Nucleolus   Nucleolus

Prophase II   Metaphase II   Anaphase II   Telophase II

Figure 10.10
Meiosis II.
During meiosis II, the centromere splits and each chromosome divides into separate chromatids.

Telophase I

Telophase I consists of changes that return the cell to an interphase condition (figure 10.9). The chromosomes uncoil and become long, thin threads, the nuclear membrane re-forms around them, and nucleoli reappear. During this activity, cytokinesis divides the cytoplasm into two separate cells.

Because of meiosis I, the total number of chromosomes is divided equally, and each daughter cell has one member of each homologous chromosome pair. Each individual chromosome is still composed of two chromatids joined at the centro-

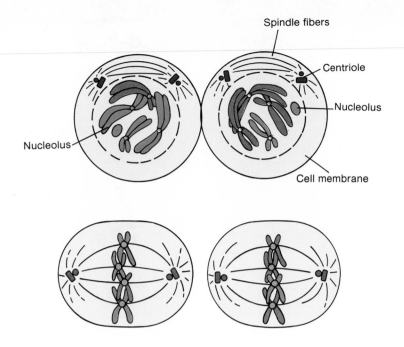

Figure 10.12
Metaphase II.
During this metaphase, each chromosome lines up on the equatorial plane. Each chromosome is composed of two chromatids and a centromere.

mere, and the chromosome number is reduced from diploid (2N) to haploid (N). In the cell we have been using as our example, the number of chromosomes is reduced from eight to four. The four pairs of chromosomes have been distributed to the two daughter cells.

Depending on the type of cell, there may be a time following telophase I when a cell engages in normal metabolic activity that corresponds to an interphase stage. However, the chromosomes do not replicate before the cell enters meiosis II.

Mechanics of Meiosis: Meiosis II

Meiosis II is composed of four stages: prophase II, metaphase II, anaphase II, and telophase II. The two daughter cells formed during meiosis I continue through meiosis II, so that usually four cells result from the two divisions. The events of meiosis II are summarized in figure 10.10.

Prophase II

Prophase II is similar to prophase in mitosis; the nuclear membrane disintegrates, nucleoli disappear, and the spindle apparatus begins to form. However, it differs from prophase I since these cells are haploid, not diploid (figure 10.11). Also, there is no synapsis, crossing-over, segregation, or independent assortment during prophase II.

Metaphase II

The metaphase II stage is typical of any metaphase stage because the chromosomes attach by their centromeres to the spindle at the equatorial plane of the cell. Since pairs of chromosomes are no longer together in the same cell, each chromosome moves as a separate unit (figure 10.12).

Figure 10.13
Anaphase II.

This anaphase stage is very similar to the anaphase of mitosis. The centromere of each chromosome divides, and one chromatid separates from the other. At this time, these chromatids are known as chromosomes.

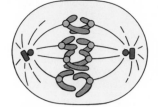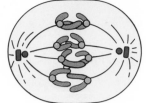

Figure 10.14
Telophase II.

During the telophase stage, what events would you expect?

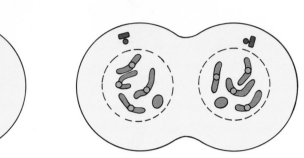

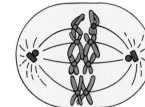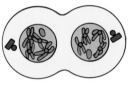

Prophase I   Metaphase I   Anaphase I   Telophase I

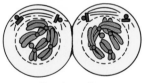

Prophase II   Metaphase II   Anaphase II   Telophase II

Figure 10.15
Meiosis I and II.

During the two divisions of meiosis, the diploid number (2N = 8) of chromosomes is reduced to the haploid number (N = 4) of chromosomes, and four haploid cells are produced from one diploid parent cell.

Anaphase II

Anaphase II differs from anaphase I because the centromere of each chromosome splits in two and the chromatids, now called daughter chromosomes, move to the poles (figure 10.13). Remember, there are no paired homologs in this stage and, therefore, segregation and independent assortment cannot occur.

Telophase II

During telophase II, the cell returns to a nondividing condition. As cytokinesis occurs, new nuclear membranes form, chromosomes uncoil, nucleoli re-form, and the spindles disappear (figure 10.14). This stage is followed by differentiation in which the four cells mature into gametes, either sperm or eggs. The two divisions of meiosis are summarized in figure 10.15 so that you can better see the flow of the entire process.

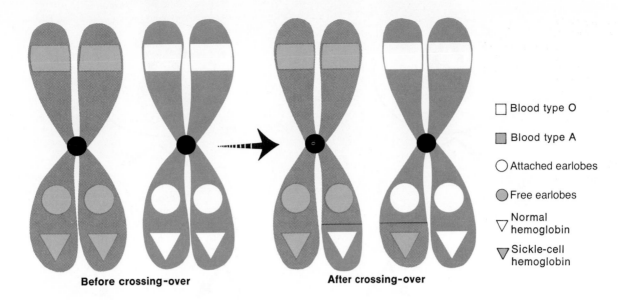

Before crossing-over   **After crossing-over**

□ Blood type O

■ Blood type A

○ Attached earlobes

● Free earlobes

▽ Normal hemoglobin

▼ Sickle-cell hemoglobin

In many organisms, egg cells are produced in such a manner that three of the four cells resulting from meiosis in a female disintegrate. However, since the one that survives is randomly chosen, the likelihood of any one particular combination of genes being formed is not affected. The whole point of learning the mechanism of meiosis is to see how variation happens. Now we can look at variation and how it comes about.

Figure 10.16
Synapsis Allows Crossing-Over to Occur.
While pairs of homologous chromosomes are in synapsis, one part of one chromatid can break off and be exchanged for an equivalent part of its homologous chromatid.

Sources of Variation

The process of forming a haploid cell by meiosis, and the combination of two haploid cells to form a diploid cell by sexual reproduction, results in variety in the offspring. There are five factors that influence genetic variation in offspring: mutations, crossing-over, segregation, independent assortment, and fertilization.

Two types of mutations were discussed in chapter 8. These were point mutations and chromosomal mutations. In point mutations, there is a change in a DNA nucleotide that results in the production of a different protein. In chromosomal mutations, genes are rearranged. By causing the production of different proteins, both types of mutations increase variation. The second source of variation is crossing-over.

Crossing-Over

Crossing-over is the exchange of a part of a chromatid from one homologous chromosome with an equivalent part of a chromatid from the other homologous chromosome. This exchange results in a new gene combination. Crossing-over occurs during meiosis I when homologous chromosomes are synapsed. Remember that a chromosome is a double strand of DNA. To break a chromosome, bonds between sugars and phosphates are broken. This is done at the same spot on both chromatids. The two pieces switch places. After switching places, the two pieces of DNA can be bonded together by re-forming the bonds between the sugar and the phosphate molecules. Examine figure 10.16 carefully to note precisely what occurs during crossing-over (figure 10.16).

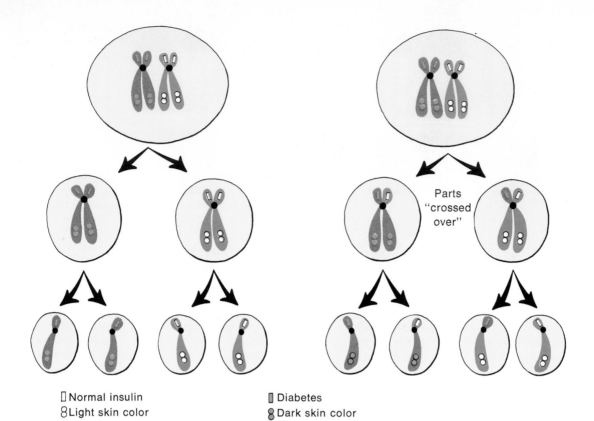

☐ Normal insulin
⊗ Light skin color

▮ Diabetes
⊗ Dark skin color

Figure 10.17
*Variations Resulting from
Crossing-Over.*

The cells on the left resulted from
meiosis without crossing-over.
Those cells on the right had one
cross-over. Compare the results of
meiosis in both cases.

Figure 10.16 shows a pair of homologous chromosomes close to each other. Notice that each gene occupies a specific place on the chromosome. This is the locus, a place on a chromosome where a gene is located. Homologous chromosomes contain an identical order of genes. For the sake of simplicity, only a few loci are labeled on the chromosomes used as examples. Actually, the chromosomes contain hundreds or possibly thousands of genes.

What does crossing-over have to do with the possible kinds of cells that result from meiosis? Consider figure 10.17. Notice that without crossing-over, only two kinds of genetically different gametes result. Two of the four gametes have one type of chromosome, while the other two have the other type of chromosome. With crossing-over, four genetically different gametes are formed.

With just one cross-over, we double the number of kinds of gametes possible from meiosis. Since crossing-over can occur at almost any point along the length of the chromosome, great variation is possible. In fact, crossing-over can occur at a number of different points on the same chromosome. That is, there can be more than one cross-over per chromosome pair (figure 10.18).

Crossing-over helps to explain why a child can be a mixture of family characteristics. If the violet chromosome was the chromosome that a mother received from her mother, the child could receive some genetic information not only from the mother's mother, but also from the mother's father (figure 10.19).

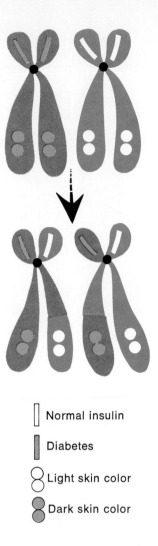

Figure 10.18
Multiple Cross-Overs.
Crossing-over can occur several
times between one pair of
homologous chromosomes.

▯ Normal insulin

▮ Diabetes

⬀ Light skin color

⬀ Dark skin color

Segregation

After crossing-over has taken place, segregation occurs. This involves the separation
and movement of homologous chromosomes to the poles. Let us assume a person
has a gene for insulin production on one chromosome and a gene for diabetes on
the homologous chromosome. Such a person would produce enough insulin to be
healthy. When this pair of chromosomes segregates during metaphase I, one daughter
cell receives a chromosome with a gene for insulin production and the second
daughter cell receives a chromosome with a gene for diabetes. The process of segregation causes genes to be separated from one another so that they have an equal
chance of being transmitted to the next generation. If the mate also has one gene
for insulin production and one gene for diabetes, that person also produces two kinds
of gametes.

Both of the parents have normal insulin production. If one or both of them
contributed a gene for normal insulin production during fertilization, the offspring
would produce enough insulin to be healthy. However, if, by chance, both parents
contributed the gamete with a gene for diabetes, the child would be a diabetic. This

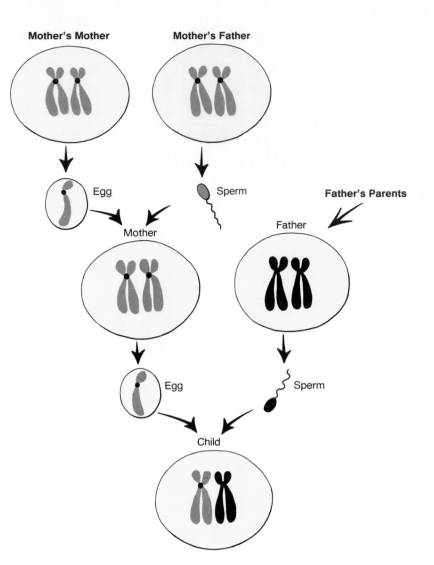

Figure 10.19
Mixing of Genetic Information Through Several Generations.

The mother of this child has information from both of her parents. The child receives a mixture of this information from its mother.

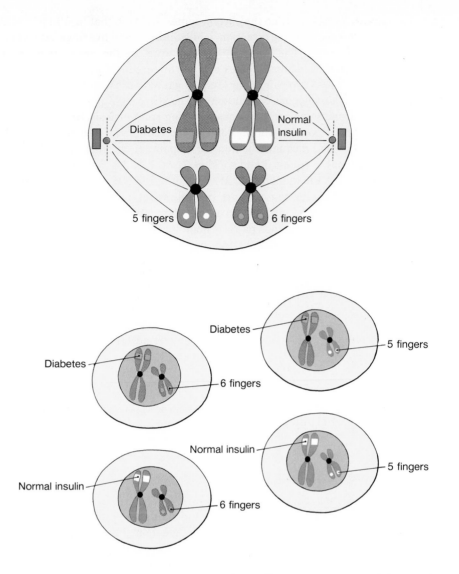

Figure 10.21
Variety Generated by Independent Assortment.
When a cell has two pairs of homologous chromosomes, four kinds of haploid cells can result from independent assortment.

is variation because the parents may produce offspring with traits different from themselves. No new genes were created but simply redistributed in a fashion that allowed for the new combination of genes in the offspring to be different from the parents' genes. This will be explored in greater detail in chapter 12.

Independent Assortment

So far in discussing variety, we have only dealt with one pair of chromosomes where it is possible to have two varieties of gametes. Let us see how variation increases when we add a second pair of chromosomes (figure 10.20).

In figure 10.21, chromosomes carrying insulin production information always separate from each other. The second pair of chromosomes with the information for the number of fingers also separates. Since the pole to which a chromosome moves is a chance event, half of the time the chromosomes divide so that insulin production and six fingeredness move in one direction, while diabetes and five fingeredness move in the opposite direction. The other half of the time, insulin production and five fingeredness go together, while diabetes and six fingeredness go to the other pole. With four chromosomes (two pair), four kinds of gametes are possible (figure 10.21). With three pairs of homologous chromosomes, there are eight possible kinds of cells with respect to chromosome combinations resulting from meiosis. See if you can list them. The number of possible chromosomal combinations of gametes is found by the expression 2^n, where n equals the number of pairs of chromosomes. With three pairs of chromosomes, n equals 3 and so $2^n = 2^3 = 2 \times 2 \times 2 = 8$. With 23 pairs of chromosomes, as in the human cell, $2^n = 2^{23} = 8,388,608$. More than eight million kinds of sperm cells or egg cells are possible from a single human parent organism. This huge variation is possible because each pair of homologous chromosomes assorts independently of the other pairs of homologous chromosomes (independent assortment). In addition to this variation, crossing-over creates new gene combinations and mutation can cause the formation of new genes, thereby increasing this number greatly.

Fertilization

Because of the large number of possible gametes resulting from independent assortment, segregation, mutation, and crossing-over, an incredibly large number of types of offspring can result. Since human males can produce millions of genetically different sperm and females can produce millions of genetically different eggs, the number of kinds of offspring possible is infinite for all practical purposes. With the possible exception of identical twins, every human that has ever been born is genetically unique.

Chromosomes and Sex Determination

You already know that there are several different kinds of chromosomes and that each chromosome carries genes unique to it and that these genes are found at specific places. Furthermore, diploid organisms have homologous pairs of chromosomes. Sexual characteristics are determined by genes in the same manner as other types of characteristics. In many organisms, sex-determining genes are located on specific chromosomes known as **sex chromosomes.** All other chromosomes not involved in determining the sex of an individual are known as **autosomes.** In humans, other mammals, and some other organisms such as fruit flies, the sex of an individual is determined by the presence of a certain chromosome combination. The genes that determine maleness are located on a small chromosome known as the Y chromosome. This Y chromosome behaves as if it were homologous with another larger chromosome known as the X chromosome. Males have one X and one Y chromosome. Females have two X chromosomes. Some animals like bees have their sex determined in a completely different way. The females are diploid and the males are haploid. Other plants and animals have still other chromosomal mechanisms for determining their sex.

Table 10.2 *A Comparison of Mitosis and Meiosis*

Mitosis	Meiosis
1. One division completes the process.	1. Two divisions are required to complete the process.
2. Chromosomes do not synapse.	2. Homologous chromosomes synapse in prophase I.
3. Homologous chromosomes do not cross-over.	3. Homologous chromosomes do cross-over.
4. Centromeres divide in anaphase.	4. Centromeres divide in anaphase II, but not in anaphase I.
5. Daughter cells have the same number of chromosomes as the parent cell ($2N \rightarrow 2N$ or $N \rightarrow N$).	5. Daughter cells have half the number of chromosomes as the parent cell ($2N \rightarrow N$).
6. Daughter cells have the same genetic information as the parent cell.	6. Daughter cells are genetically different from the parent cell.
7. Results in growth, replacement of worn out cells, and repair of damage.	7. Results in sex cells.

Comparison of Mitosis and Meiosis

As you read this chapter on meiosis, you may have seen some similarities and differences between mitosis and meiosis. Study table 10.2 and acquaint yourself with the differences between the processes of mitosis and meiosis.

Summary

Meiosis is a specialized process of cell division resulting in the production of four cells, each of which has the haploid number of chromosomes. The total process involves two sequential divisions during which one diploid cell reduces to four haploid cells. Since the chromosomes act as carriers for genetic information, genes separate into different sets during meiosis. Crossing-over and segregation allow hidden characteristics to be displayed, while independent assortment allows characteristics donated by the mother and the father to be mixed in new combinations.

Together, crossing-over, segregation, and independent assortment assure that all sex cells are unique. Therefore, when any two cells unite to form a zygote, the zygote will also be one of a kind. The sex of many kinds of organisms is determined by specific chromosome combinations. In humans, females have two X chromosomes, while males have an X and a Y chromosome.

Consider This

Assume that corn plants have a diploid number of only 2. Each plant's chromosomes are diagrammed below.

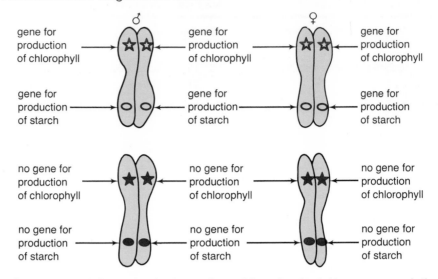

Show sex cell formation in the male and female plant. How many variations in sex cells can occur and what are they? What variations can occur in the production of chlorophyll and starch in the descendants of these parent plants? Note:

gene for production of chlorophyll	= green plant
no gene for chlorophyll	= white, dead plant
gene for production of starch	= regular corn
no gene for starch	= sweet corn

Experience This

Models can be very useful in helping us to understand complex biological events such as meiosis. You can do this at home very easily by using various lengths of colored strings, threads, or yarns to simulate the twenty-three pairs of homologous chromosomes in a human cell. Each homologous pair should be different from the other pairs, either in color, length, or both. Begin your modeling with each chromosome in its replicated form (i.e., two chromatids per chromosome). Attach the two chromatids with a loose twist. Using your twenty-three pairs of string model chromosomes, manipulate them through the stages of meiosis I and II. If you have performed the actions properly, you should end up with four cells, each haploid (N = 23).

Questions

1. List three differences between mitosis and meiosis.
2. How do haploid cells differ from diploid cells?
3. What are the major sources of variation that occur during meiosis?
4. Can a haploid cell undergo meiosis?
5. What is unique about prophase I?
6. Why is meiosis necessary in organisms that reproduce sexually?
7. Define the terms zygote, fertilization, and homologous chromosomes.

8. How much variation as a result of independent assortment can occur in cells with the following diploid numbers: 2, 4, 6, 8, and 22.
9. Diagram the metaphase I stage of a cell with the diploid number of 8.
10. Diagram fertilization as it would occur between a sperm and an egg with the haploid number of 3.

anther (an-ther) A sex organ in plants that produces the sperm.

autosome (aw′to-som) Chromosomes not involved in determining the sex of individuals.

crossing-over (kros-sing o-ver) The exchange of a part of a chromatid from one chromosome with an equivalent part of a chromatid from a homologous chromosome.

diploid (dip′loid) A cell that has two sets of chromosomes, one set from the maternal parent and one set from the paternal parent.

egg cells (eg sels) The haploid sex cell produced by sexually mature females.

fertilization (fer″ti-li-za′shun) The joining of haploid nuclei, usually from an egg and a sperm cell, resulting in a diploid cell called the zygote.

gamete (gam′ēt) A haploid sex cell.

gonad (go-nad) Animal organ that produces gametes.

haploid (hap′loid) A single set of chromosomes resulting from the reduction division of meiosis.

homologous chromosomes (ho-mol′o-gus kro′mo-sōms) A pair of chromosomes in a diploid cell that contains similar genes on corresponding loci throughout their length.

independent assortment (in″de-pen′dent ă-sort′ment) The segregation, or assortment, of one pair of homologous chromosomes is not dependent upon the segregation, or assortment, of any other pair of chromosomes.

meiosis (mi-o′sis) The specialized pair of cell divisions that reduce the chromosome number from diploid (2N) to haploid (N).

ovaries (o′var-ēz) The female sex organ that produces haploid sex cells, the eggs or ova.

pistil (pis′til) A sex organ in plants that produces eggs or ova.

reduction division (re-duk′shun di-vi′zhun) A type of cell division in which daughter cells get only half the chromosomes from the parent cell.

segregation (seg″re-ga′shun) The separation and movement of homologous chromosomes to the poles of the cell.

sex chromosomes (seks kro′mo-sōmz) Chromosomes that carry genes that determine the sex of the individual.

sexual reproduction (sek′shu-al re″pro-duk′shun) The propagation of organisms involving the union of gametes from two parents.

sperm cell (spurm sel) The haploid sex cell produced by sexually mature males.

synapsis (sin-ap′sis) The condition in which the two members of a pair of homologous chromosomes come to lie close to one another.

testes (tes′tēz) The male sex organ that produces haploid cells, the sperm.

zygote (zi′gōt) A diploid cell that results from the union of an egg and a sperm.

11

Reproduction and Development

Chapter Outline

Purpose

In the previous chapter on meiosis, we discussed how haploid cells are formed. Meiosis also results in the mixing of genes as they are repackaged into gametes. The uniting of two gametes brings two sets of genes together in a single cell. This chapter deals with reproduction and development—particularly human reproduction—and the various structures that are involved. We will follow the process from the production of gametes to the birth of a baby. Finally, we discuss some factors that can alter the developmental process.

For Your Information

People in developed countries have come to use various birth- and conception-control methods as a regular part of their lives. This is not the case in other parts of the world where the population continues to increase at a staggering rate. Many social scientists and biologists contend that one of the main factors that enables a "poor" country to "pull itself up by its own bootstraps" and advance is their acceptance and use of effective birth- and conception-control methods.

Learning Objectives

■ List the structures associated with the male and female human reproductive systems.

■ Describe how hormones function in the process of ovulation and pregnancy.

■ Recognize the role of the placenta in the development of mammals.

■ Understand the differences between spermatogenesis and oogenesis in human reproduction.

■ List the events necessary for conception and pregnancy to occur in humans.

■ Know the different hormones that regulate the functioning of the human reproductive system and describe their actions.

Reproductive Anatomy and Physiology

Organisms that reproduce sexually have specialized structures that are responsible for producing gametes. Many organisms, including mammals, have additional structures for storing and transporting gametes such as sperm ducts and oviducts. Mammals also have specialized structures that house and feed the developing embryo. The specific structures used in reproduction vary from one kind of organism to another, and the specific ways gametes form vary as well. In animals, the process of meiosis leads to the production of gametes and is called **gametogenesis** (gamete formation). Because gametogenesis and reproduction vary only slightly from animal to animal, the remainder of this chapter deals primarily with humans.

The term **spermatogenesis** is one form of gametogenesis and specifically refers to the production of sperm. The other form of gametogenesis that refers to the production of eggs is called **oogenesis.** Spermatogenesis takes place in the **testes,** which are located in a saclike structure called the **scrotum.** The two testes are bean-shaped and are composed of many tubes (**seminiferous tubules** and others) held together by a thin, covering membrane (figure 11.1). All of the tubes join to lead into a larger tube, which eventually leads out of the body through the **penis.** Before puberty, the seminiferous tubules are packed solid with diploid cells. At around age 11–13, these cells specialize and begin the process of meiosis. The seminiferous tubules become hollow and can transport the sperm. Each cell that begins meiosis results in four smaller, round haploid cells. Spermatogenesis involves several steps (figure 11.2).

The cells that make up the wall of the seminiferous tubules enlarge and become **primary spermatocytes.** These diploid cells undergo the first meiotic division, which produces two haploid **secondary spermatocytes.** The secondary spermatocytes go through the second meiotic division, resulting in four haploid **spermatids** that lose much of their cytoplasm and develop long tails. These cells are known as **sperm.** The sperm have only a small amount of food reserves. Therefore, once they are released and become active swimmers, they live no more than seventy-two hours.

Figure 11.1
The Human Male Reproductive System.
The male reproductive system consists of two testes that produce sperm, ducts that carry sperm, and glands. Muscular contractions propel the sperm through the vas deferens past the seminal vesicles, prostate gland, and bulbo-urethral gland, where most of the liquid of the semen is added. The semen passes through the urethra of the penis to the outside of the body.

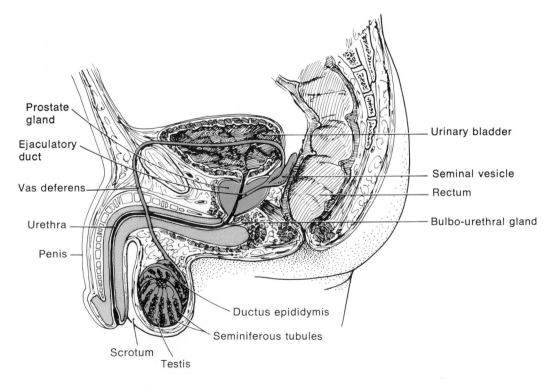

Prostate gland

Ejaculatory duct

Vas deferens

Urethra

Penis

Scrotum

Testis

Ductus epididymis

Seminiferous tubules

Urinary bladder

Seminal vesicle

Rectum

Bulbo-urethral gland

GAMETOGENESIS

Spermatogenesis

Oogenesis

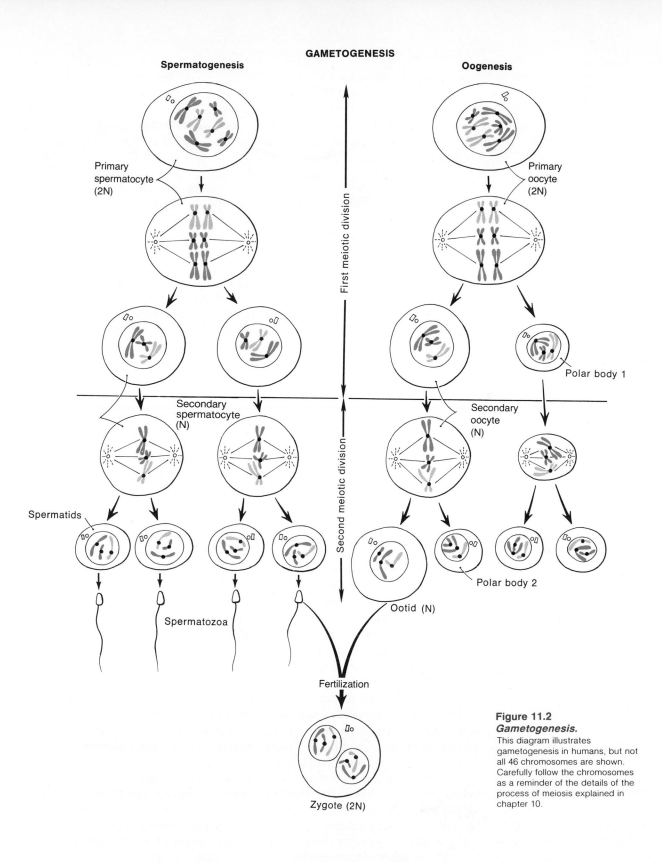

Primary spermatocyte (2N)

First meiotic division

Primary oocyte (2N)

Polar body 1

Secondary spermatocyte (N)

Second meiotic division

Secondary oocyte (N)

Spermatids

Polar body 2

Spermatozoa

Ootid (N)

Fertilization

Zygote (2N)

Figure 11.2
Gametogenesis.
This diagram illustrates gametogenesis in humans, but not all 46 chromosomes are shown. Carefully follow the chromosomes as a reminder of the details of the process of meiosis explained in chapter 10.

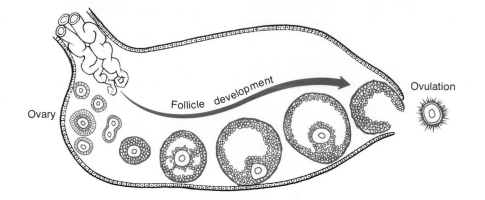

Figure 11.3
Ovulation.
The egg begins development within the ovary inside a sac of cells known as a follicle. Each month one of the follicles develops and releases its product. This release through the wall of the ovary is known as ovulation.

Ovary

Follicle development

Ovulation

However, the life of a sperm may be increased in some cases. If the temperature is lowered drastically by using special equipment, the sperm become deactivated and can live for years outside the testes. This has led to the development of sperm banks. Although human sperm banks have been formed, they have few depositors. The most successful use of sperm banks is in cattle breeding, where the sperm from selected bulls are stored and used to fertilize (inseminate) genetically superior cows. As a result, a particular animal can continue to reproduce his own kind for years after his death. Another reason for using artificial insemination in cattle is that the sperm are easier to transport than cattle; therefore, sperm from a particular bull can fertilize more cows. The method of fertilization has improved the quality of many animals.

Spermatogenesis in the human male takes place continuously; new gametes are produced throughout a male's reproductive life. Although the number of sperm produced decreases as a man ages, most men continue to produce sperm until death. A great number of sperm are produced. Sperm counts can be taken and used to determine the probability of successful fertilization. For reasons not really understood, a man must be able to release at least one hundred million sperm at one insemination to be fertile. A healthy male probably releases about one billion sperm during each act of **sexual intercourse,** also known as **coitus** or **copulation.**

While men produce many billions of gametes each year, women do not. Just below the surface of the **ovary,** a single diploid **primary oocyte** begins to undergo meiosis in the normal manner. But in telophase I, the two cells that form get unequal portions of cytoplasm. You might think of it as a lopsided division (figure 11.2). The smaller of the two cells is called a **polar body** and the larger is the **secondary oocyte.** During the second meiotic division, the secondary oocyte again divides unevenly so that another polar body forms. None of the polar bodies survive to become eggs; therefore, only one large egg is produced by oogenesis.

Oogenesis begins before the female is born. The DNA replicates and the first meiotic division begins in thousands of cells of the ovaries. Oogenesis halts after the first meiotic division after the formation of the secondary oocytes. These cells remain just under the surface of the ovary until puberty, which occurs between the ages of 11 and 14. Only one secondary oocyte is then released from the surface of the ovary every twenty-eight days during the process called **ovulation.** The other secondary oocytes remain in the ovary. Ovulation begins when the soon-to-be-released secondary oocyte becomes encased in a saclike structure, known as a **follicle,** near the surface of the ovary. Internal pressure increases, the covering of the ovary ruptures, and the secondary oocyte shoots off the surface (figure 11.3). The cell then travels through the tubes of the reproductive system (figure 11.4).

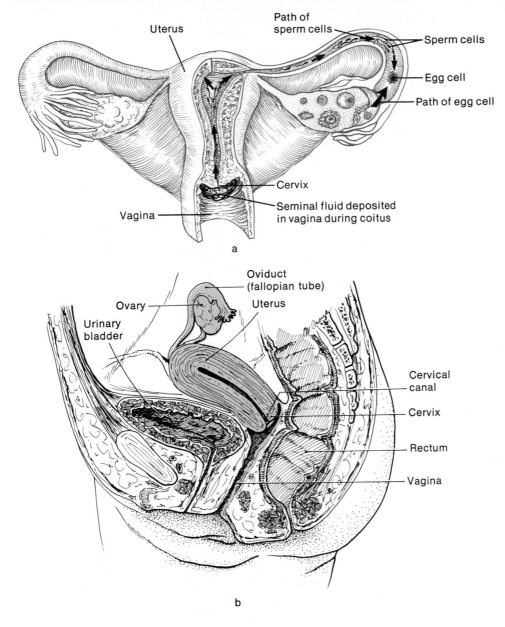

Uterus

Path of
sperm cells

Sperm cells

Egg cell

Path of egg cell

Cervix

Seminal fluid deposited
in vagina during coitus

Vagina

a

Oviduct
(fallopian tube)

Ovary

Uterus

Urinary
bladder

Cervical
canal

Cervix

Rectum

Vagina

b

Figure 11.4
*The Human Female
Reproductive System.*
(a) After ovulation, the cell travels
down the oviduct to the uterus. If it
is not fertilized, it is shed when the
uterine lining is lost during
menstruation. (b) The human
female reproductive system.

It is swept into the **oviduct** by ciliated cells and propelled toward the **uterus.** If the secondary oocyte is fertilized, the cell completes the meiotic division by proceeding through meiosis II with the sperm DNA inside. If the cell is not fertilized, the secondary oocyte passes through the **vagina** to the outside during menstruation. During her lifetime, a female releases about three hundred to five hundred secondary oocytes. Obviously, few of these cells are fertilized.

One of the most important differences to note here is the age of the cell. In males, meiosis is continuous, occurring each time a new sperm is manufactured. Sperm do not remain in the tubes of the male reproductive system for very long. They are either released shortly after they form or die and are harmlessly absorbed. In females, meiosis begins before birth, but the oogenesis process is not completed and the cell is not released for many years. A secondary oocyte released when a

Figure 11.5
*Nondisjunction in
Oogenesis.*
Notice that the secondary oocyte
has an extra chromosome as a
result of an uneven division of
homologous chromosomes in
meiosis I. Therefore, the ootid
(egg) also has an extra
chromosome.

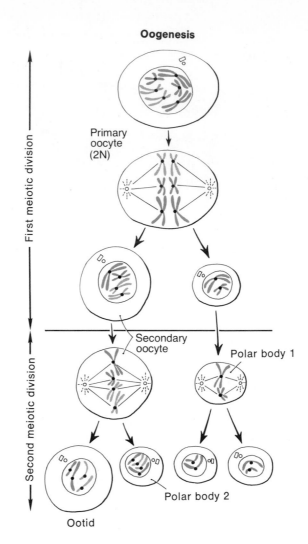

Oogenesis

Primary
oocyte
(2N)

First meiotic division

Secondary
oocyte

Polar body 1

Second meiotic division

Polar body 2

Ootid

woman is thirty-seven years old began meiosis thirty-seven years before! During that
time, the cell was exposed to many changes, a number of which may damage the
DNA or interfere with the meiotic process. Such alterations are less likely to occur
in males.

Nondisjunction

In chapter 10, we described the normal process of reducing diploid cells to haploid
cells. This involved segregating homologous chromosomes into separate cells during
the first meiotic division. Occasionally a pair of homologous chromosomes does not
divide properly during gametogenesis and both chromosomes of a pair end up in the
same gamete. This abnormal kind of division is known as **nondisjunction.** As you
can see in figure 11.5, one cell is missing a chromosome and the genes that were
carried on it. This usually results in the death of the cell. The other cell has a double
dose of one chromosome. Apparently the genes of an organism are balanced against
one another. A double dose of some genes and a single dose of others results in
abnormalities that may lead to the death of the cell. Some of these abnormal cells,

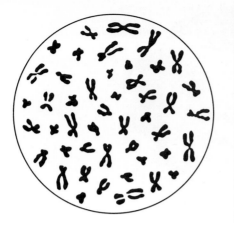
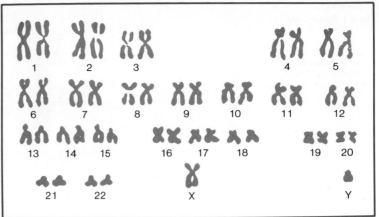

however, do live and develop into sperm or eggs. If an abnormal sperm or egg happens to result in a zygote, the offspring will have an abnormal number of chromosomes. All the cells that develop from the zygote will also be abnormal. It is possible to examine the cells of a person and count the chromosomes. The usual way to do this is to grow cells and take photographs of them during metaphase of mitosis. The pictures of the chromosomes can then be cut out and arranged to make identification easier (figure 11.6).

One example of the effects of nondisjunction is the condition known as **Down's syndrome** (mongolism). If an egg cell with two copies of chromosomes 21 had been fertilized by a normal sperm cell, the resulting zygote would have 47 chromosomes (figure 11.7). The child that developed from this fertilized egg would have the symptoms characteristic of Down's syndrome, such as thickened eyelids, a low level of intelligence, and faulty speech (figure 11.8). The genetic difference between downics and normal persons is the presence of an extra chromosome number 21.

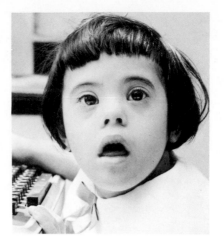

Figure 11.8
Down's Syndrome.
Every cell in this child's body has one extra chromosome. With special care and training, persons with this syndrome can be taught to do many things, but they will still be abnormal.

Figure 11.9
Age of Mother and Nondisjunction.
Notice that as the age of the female increases, the rate of nondisjunction increases only slightly until the age of thirty-seven. From that point on, the rate increases drastically.

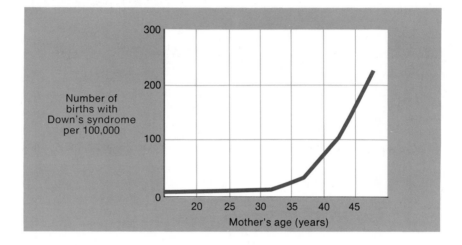

Considering the time of replication of DNA and the time of meiosis I, it seems likely that the chances of having a mongoloid child increases as the age of the mother increases. Figure 11.9 illustrates the frequency of occurrence of nondisjunction at different ages in women. Notice that the frequency of nondisjunction increases very rapidly after age thirty-seven. For this reason, many physicians encourage couples to have their children in their early to mid-twenties and not in their late thirties or early forties.

Another case of nondisjunction can be seen in persons with the condition known as **Turner's syndrome.** Nondisjunction in this case results in the loss of one of the sex-determining chromosomes. In humans, those persons who receive two X chromosomes develop into females. Those who receive one X and a smaller Y chromosome develop into males. (See chapter 12 for a more complete discussion.) Persons with Turner's syndrome have only one X chromosome and, therefore, a total of 45 chromosomes. These persons have difficulty in perceiving spacial relations and in arithmetic skills. They are also female, sterile, short, and have a "webbed neck" caused by abnormal development of neck muscles (figure 11.10).

A third case of nondisjunction involves persons who have an extra sex chromosome. They have two X chromosomes *and* a Y chromosome, making a total of 47 chromosomes (figure 11.11). Such persons exhibit **Klinefelter's syndrome,** developing into tall males who are sterile and often have a lowered mental ability.

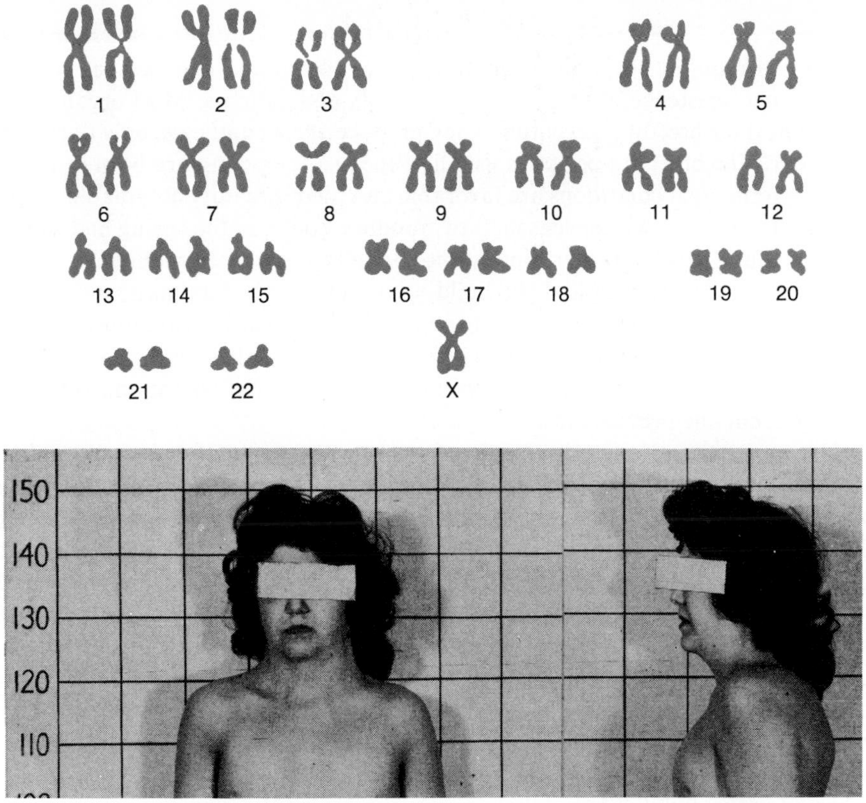

Figure 11.10
Turner's Syndrome.
This illustration shows one of the
typical symptoms of Turner's
syndrome. The person appears to
have a very wide or webbed neck.
Every cell of this person's body
lacks one of the sex chromosomes.

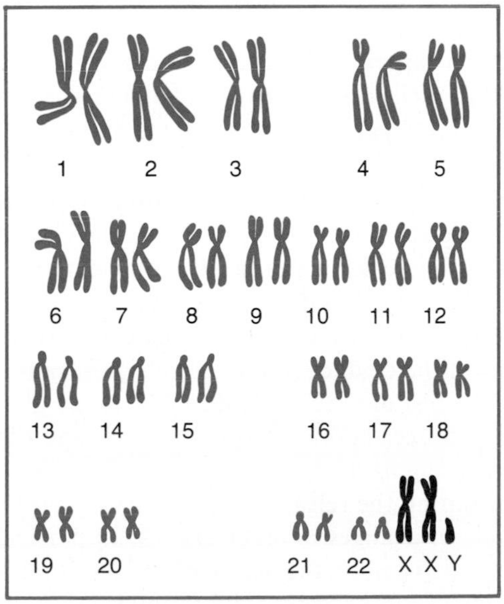

Figure 11.11
Klinefelter's Syndrome.
Persons with Klinefelter's
syndrome have 47 chromosomes.
They have an extra sex
chromosome.

When Does Gametogenesis Occur?

In terms of energy, the production of gametes is an expensive process. A good deal of energy is required to produce the millions of sperm and eggs needed to assure the continued existence of most species of plants and animals. Most organisms are seasonal in their breeding activities. They produce sperm and eggs only during part of the year. The breeding period is usually timed so the young are born or hatched when environmental conditions are favorable and food is readily obtainable. In parts of the world having a winter season, birds produce young in the spring and summer. Since the young develop rather rapidly, the eggs and sperm are produced in the early spring just prior to mating and the building of nests. Other animals, such as deer, also produce young in the spring when environmental conditions are most favorable. But since the gestation period (time when the embryo is developing in the uterus) for deer is about seven months, the production of eggs and sperm and the mating act must occur the previous fall.

Those plants and animals that live in areas without seasonal changes do not show the typical **seasonal reproductive patterns.** There may still be periods of time, however, when individuals are unable to produce sperm or eggs. Humans differ from most other animals in that both sexes continuously produce eggs and sperm. As with most organisms, humans produce many more sperm than eggs. This is probably related to the fact that the sperm must seek out the egg. Most of the sperm will never fertilize an egg. Releasing sperm during the mating act is similar to shooting at a bird with a shotgun. With a large number of pellets flying through the air, there is a greater chance of hitting the bird. If the production of gametes is seasonal, what controls the process? **Hormones** are involved in the control of the production of eggs and sperm. They are also involved in the control of other organs and the behavior associated with reproduction. A hormone is a chemical substance produced by one part of the body (usually a gland) that alters the activity of a different part of the body. Table 11.1 lists some hormones and their functions. Hormones are usually proteins or steroids. Each kind of hormone is produced by specific cells and is "broadcast" to all of the cells of the body through the circulatory system. Only certain target organs, however, are capable of receiving the chemical message and responding to it. When a target organ receives a message, it alters its activities in one of several ways. Many cells begin to grow and divide when stimulated by specific hormones, other cells secrete particular compounds and still others respond by increasing specific metabolic activities.

Certain hormones in women (follicle-stimulating hormone, FSH, and luteinizing hormone, LH) bring about the release of a secondary oocyte (ovulation) about every twenty-eight days. These hormones cause a follicle inside the ovary to enlarge. Inside the saclike follicle, one cell is developing into a secondary oocyte. When this maturation is complete, the follicle erupts and the cell is released. The secondary oocyte then travels down the oviduct (fallopian tube). Because of the action of the luteinizing hormone, the rest of the follicle develops into a glandlike structure, the **corpus luteum,** which produces hormones (progesterone and estrogen) that prevent the release of other eggs (figure 11.12).

In addition to regulating the release of secondary oocytes from the ovary, hormones control the cycle of changes in other organs. In particular, the breasts and lining of the uterus are changed by the action of progesterone and estrogen (figure 11.12). The lining of the uterus becomes thicker and is filled with blood prior to the release of the oocyte. This assures that if it becomes fertilized, the new embryo will be able to attach itself to the wall of the uterus and receive nourishment. If the cell

Table 11.1 *Human Reproductive Hormones*
Listed here are a few hormones, their production sites, target organs, and some of their functions.

Hormone	Production Site	Target Organ	Function
1. Prolactin (lactogenic or luteotropic hormone)	Pituitary gland	Breasts, ovary	Milk production; also helps maintain normal ovarian cycle
2. Follicle stimulating hormone	Pituitary gland	Ovary, testis	Stimulates egg production in females and sperm production in males
3. Luteinizing hormone	Pituitary gland	Ovary, testis	Stimulates ovulation and sex hormone (estrogen and testosterone) production in both males and females
4. Estrogens	Follicle of the ovary	Entire body	Development of female reproductive tract and secondary sexual characteristics
5. Testosterone	Testes	Entire body	Development of male reproductive tract and secondary sexual characteristics
6. Progesterone	Ovary	Uterus, breasts	Uterine thickening and maturation; maintains pregnancy
7. Oxytocin	Pituitary gland	Breasts, uterus	Causes uterus to contract and breasts to release milk

is not fertilized, the lining of the uterus is shed. This is known as the **menstrual flow, menses,** or **period.** Once the wall of the uterus has been shed, it begins to build up again. This continual building up and shedding of the wall of the uterus is known as the **menstrual cycle.**

During the time that the hormones are regulating the release of eggs and the menstrual cycle, some changes are taking place in the breasts. The same hormones that prepare the uterus to receive the embryo also prepare the breasts to produce milk. These changes in the breasts, however, are relatively minor unless **pregnancy** occurs.

Hormonal Control of Fertility

An understanding of how various hormones regulate the menstrual cycle, egg release, milk production, and sexual behavior has led to the medical use of certain hormones. Some women are unable to have children because they do not release eggs from their ovaries or they release them at the wrong time. Physicians can now regulate the release of eggs from the ovary using certain hormones (commonly called fertility drugs). These hormones can be used to stimulate the release of cells for capture and use in what is called *in vitro* fertilization (test tube fertilization), or to increase the probability of natural conception, *in vivo* fertilization (in life fertilization). Unfortunately, the use of these drugs all too often results in dangerous multiple implantations, since too many secondary oocytes are released at one time. The implantation of multiple embryos makes it difficult for one embryo to develop properly and be carried through the entire nine-month gestation period. When we better

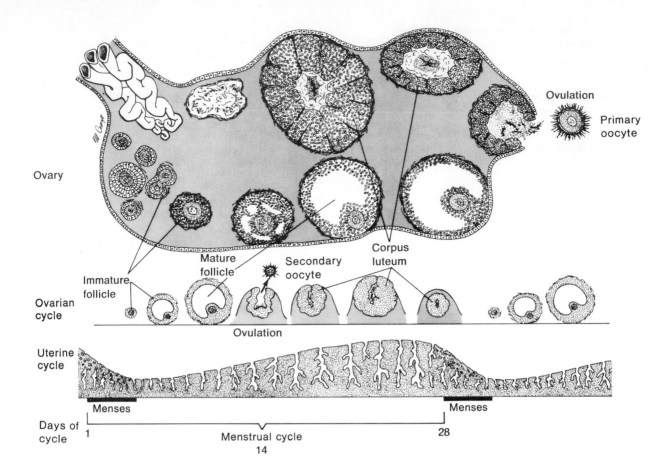

Ovary

Ovulation

Primary
oocyte

Ovarian
cycle

Immature
follicle

Mature
follicle

Secondary
oocyte

Corpus
luteum

Ovulation

Uterine
cycle

Menses

Menses

Days of
cycle

1

Menstrual cycle
14

28

Figure 11.12
*Ovarian Cycle in Human
Females.*

The release of secondary oocytes
(ovulation) is timed to match with
the thickening of the uterus. The
uterine cycle in mammals involves
the preparation of the uterine wall
to receive the embryo if fertilization
occurs. Knowing how these two
cycles compare, it is possible to
determine when pregnancy is most
likely to occur.

understand the action of hormones, we may be able to control the effects of fertility
drugs and eliminate the problem of multiple implantations. A second medical use
of hormones is the control of conception by the use of **birth-control pills (oral contraceptives).** Birth-control pills have the opposite effect of fertility drugs. They raise
the level of estrogen and progesterone, which suppresses the production of FSH and
LH, preventing the release of secondary oocytes from the ovary.

Hormonal control of fertility is not as easy to achieve in men because there is
no comparable cycle of gamete release. The major male hormone is **testosterone,**
which is produced in the testes. It controls such secondary sexual characteristics as
beard growth, deepness of the voice, and sexual drive. The growth of the testes and
the production of sperm and testosterone are under the control of other hormones.
Perhaps a better understanding of these male hormone interactions may someday
lead to the production of effective birth-control pills for men. One that was tested
had an unusual side effect that prevented it from becoming popular. It changed the
whites of a man's eyes to green.

Fertilization and Pregnancy

In most women, a secondary oocyte is released from the ovary at about the middle
of the menstrual cycle. The cycle is usually said to begin on the first day of menstruation. If a woman had a regular twenty-eight day menstrual cycle, the cell is
released approximately on day fourteen. Some women, however, have very irregular
menstrual cycles, and it is difficult to determine just when the cell will be released

Table 11.2 *Timing Fertilization*

Length of Shortest Cycle	First Unsafe Day After Start of Any Period	Length of Longest Cycle	Last Unsafe Day After Start of Any Period
22 Days	4th Day	22 Days	11th Day
23 Days	5th Day	23 Days	12th Day
24 Days	6th Day	24 Days	13th Day
25 Days	7th Day	25 Days	14th Day
26 Days	8th Day	26 Days	15th Day
27 Days	9th Day	27 Days	16th Day
28 Days	10th Day	28 Days	17th Day
29 Days	11th Day	29 Days	18th Day
30 Days	12th Day	30 Days	19th Day
31 Days	13th Day	31 Days	20th Day
32 Days	14th Day	32 Days	21st Day
33 Days	15th Day	33 Days	22nd Day
34 Days	16th Day	34 Days	23rd Day
35 Days	17th Day	35 Days	24th Day
36 Days	18th Day	36 Days	25th Day
37 Days	19th Day	37 Days	26th Day

Based on our understanding of the varying uterine and ovarian cycles, it is possible to figure a woman's "safe days" (least likely days on which to become pregnant) and her "unsafe days" (most likely days on which to become pregnant). For example, if a woman has a regular twenty-eight day menstrual cycle, she is most likely to become pregnant if she has intercourse between the tenth and seventeenth day of her cycle. However, if a woman has an irregular cycle that varies from twenty-six days to thirty-two days, she should not have intercourse between the eighth and twenty-first day following the beginning of her period if she wants to avoid pregnancy.

to become available for **fertilization.** Once the cell is released, it is swept into the oviduct and moved toward the uterus. If it does not meet the sperm, the secondary oocyte dies and is absorbed by the body or shed with the lining of the uterus during menstruation. If sperm are present as a result of sexual intercourse, they swim through the uterus and oviduct. Because of the possible irregularities in the time of ovulation and the fact that sperm can be active for up to three days, it is difficult to determine a specific period of time during which fertilization *cannot* occur (table 11.2).

If sperm are present, they swarm around the secondary oocyte as it passes down the oviduct, but only one sperm penetrates the outer layer to fertilize it. This cell, which now has two complete sets of chromosomes, undergoes meiosis II. During this division, a polar body is pinched off and the egg is formed. Since chromosomes from the sperm are already inside, they simply intermingle with those of the egg cell forming a **zygote** or fertilized egg. As the zygote continues to travel down the oviduct, it begins dividing by mitosis into smaller and smaller cells. This division process is called **cleavage.** Eventually a solid ball of cells is produced, known as the **morula** stage of embryological development (figure 11.13). Following the morula stage, the solid ball of cells becomes hollow in the middle and is then known as the **blastula stage.** During this stage, when the **embryo** is about six days old, it becomes embedded in the lining of the uterus. This is known as **implantation.** In mammals, the blastula has a region of cells, known as the inner cell mass, that develops into the embryo proper. The outer cells become structures associated with the embryo.

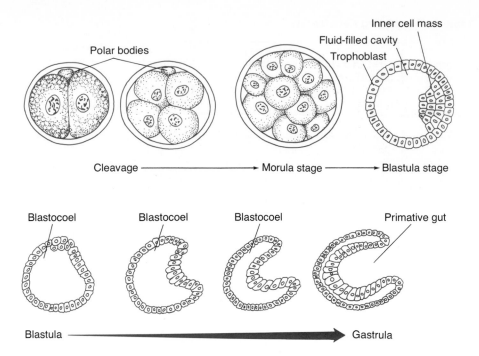

Figure 11.13
Early Embryonic Development.
Following fertilization, the zygote begins a series of divisions called cleavage, which eventually result in a solid ball of cells called the morula. Eventually the morula becomes hollow as the cells migrate. The hollow ball of cells is called a blastula or blastocyst in mammals.

Polar bodies

Inner cell mass
Fluid-filled cavity
Trophoblast

Cleavage ——————→ Morula stage ——————→ Blastula stage

Figure 11.14
Gastrula Formation.
Many animals form the gastrula stage by folding one side of the embryo into its cavity.

Blastocoel   Blastocoel   Blastocoel   Primative gut

Blastula ————————————————————→ Gastrula

The next stage in the development is known as the **gastrula stage,** since during this time the gut is formed. In many kinds of animals, the gastrula is formed by an infolding of one side of the blastula, a process similar to poking a finger into a balloon (figure 11.14).

Gastrula formation in mammals is a much more complicated process, but the result is the same. The embryo develops a tube that eventually becomes the gut. The formation of the primitive gut is just one of a series of changes that eventually result in an embryo recognizable as a miniature human being. Most of the time during its development, the embryo is enclosed in a water-filled membrane known as the **amnion,** which protects it from blows and keeps it moist. Two other membranes, the **chorion** and the **allantois,** fuse together with the lining of the wall of the mother's uterus to form the **placenta** (figure 11.15). A fourth sac, known as the **yolk sac,** is well developed in birds, fish, and reptiles, but poorly developed in mammals. The placenta produces hormones that prevent menstruation and ovulation during the nine months that the embryo is in the uterus. It also provides for the metabolic needs of the embryo.

As the embryo's cells divide and grow, some of them become specialized. These become nerve cells, bone cells, blood cells, or other specialized cells. In order to divide and grow, cells must receive nourishment. This is provided by the mother through the placenta. The placenta is an amazing organ in which the circulatory system of the mother and the circulatory system of the embryo come close enough to each other to exchange materials (figure 11.16). The materials diffusing across the placenta include oxygen, carbon dioxide, nutrients, and a variety of waste products. The materials entering the embryo travel through blood vessels in the **umbilical cord.** The embryo is contained within the mother's body and relies upon her body to provide all its needs. The major parts of the body develop by the tenth week of pregnancy (figure 11.17). After this time, the embryo increases in size and the structure of the body is refined.

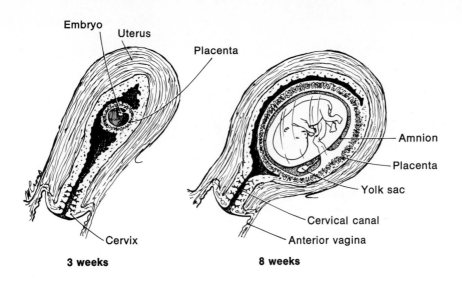

Embryo Uterus
Placenta

Figure 11.15
Placental Development.
A portion of the cells that result from mitosis of the zygote form part of the placenta. This structure is really an organ composed of cells of the uterus and the developing child.

Amnion
Placenta
Yolk sac
Cervical canal
Anterior vagina

Cervix

3 weeks   **8 weeks**

Figure 11.16
Placental Structure.
The blood vessels that supply the developing child with nutrients and remove metabolic wastes are separate from the blood vessels of the mother. Because of this separation, the placenta can selectively filter many types of incoming materials.

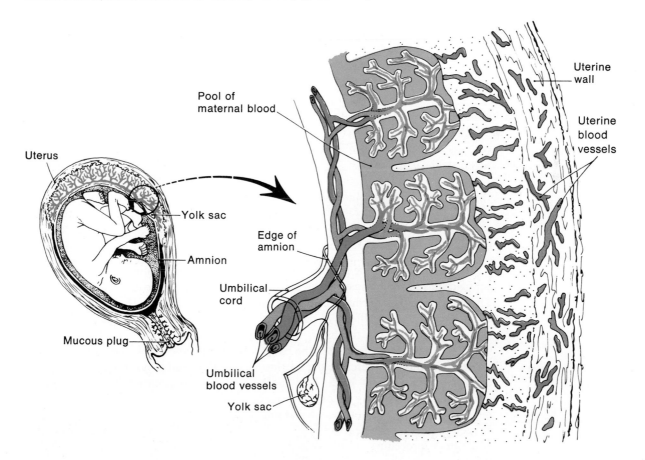

Uterus

Pool of maternal blood

Uterine wall

Uterine blood vessels

Yolk sac
Amnion

Edge of amnion

Umbilical cord

Mucous plug

Umbilical blood vessels
Yolk sac

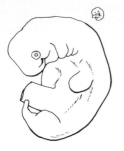

4 weeks. Limb buds showing, lungs and stomach being formed, heart forming, nerves and brain just forming.

6 weeks. Face showing jaws, external ear forming, lungs obvious, blood being formed, eyes showing pigment.

7 weeks. Fingers beginning to appear, tail shortening, tongue forming, stomach almost in final shape, eyelids forming.

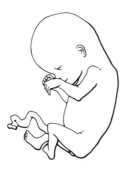

12 weeks. Head dominant, bridge of nose formed, cheeks formed, tooth buds formed, bile secreted, blood forming in bone marrow, bone becoming hard, brain cells becoming well defined.

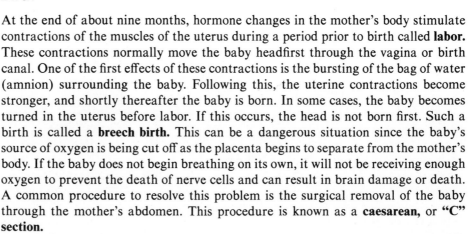

15 weeks. Face looks "human," hair forming, muscles contracting spontaneously, tonsils beginning to form, kidneys showing typical shape, heart beginning to contract, sweat glands developing.

Figure 11.17
Early Development
of the Embryo.

Birth

At the end of about nine months, hormone changes in the mother's body stimulate contractions of the muscles of the uterus during a period prior to birth called **labor.** These contractions normally move the baby headfirst through the vagina or birth canal. One of the first effects of these contractions is the bursting of the bag of water (amnion) surrounding the baby. Following this, the uterine contractions become stronger, and shortly thereafter the baby is born. In some cases, the baby becomes turned in the uterus before labor. If this occurs, the head is not born first. Such a birth is called a **breech birth.** This can be a dangerous situation since the baby's source of oxygen is being cut off as the placenta begins to separate from the mother's body. If the baby does not begin breathing on its own, it will not be receiving enough oxygen to prevent the death of nerve cells and can result in brain damage or death. A common procedure to resolve this problem is the surgical removal of the baby through the mother's abdomen. This procedure is known as a **caesarean,** or **"C"** **section.**

Following the birth of the baby, the placenta is born and is usually referred to as the afterbirth. Once born, the baby begins to function on its own. The umbilical cord collapses and the baby's lungs, kidneys, and digestive system must now support all bodily needs. This change is quite a shock, but the baby's loud protests fill the lungs with air and stimulate breathing.

Over the next few weeks, the mother's body returns to normal with one major exception. The breasts, which have also undergone changes during the period of pregnancy, are ready to produce milk to feed the baby. Following birth, hormones stimulate the production and release of this milk. If the baby is breast-fed, the stimulus of the baby's sucking will prolong the time during which milk is produced. In some cultures, breast-feeding continues for two to three years. The continued production of milk often delays the reestablishment of the normal cycles of ovulation and menstruation. Many people believe that a woman cannot become pregnant while she is nursing a baby, but because there is so much variation among women, relying on this as a natural conception-control method is unrealistic. Many women have been surprised to find themselves pregnant again a few months after delivery.

Contraception

Throughout history people have tried various methods of conception-control (figure 11.18). In ancient times, conception-control was encouraged during times of food shortage or when tribes were on the move from one area to another in search of a new home. Writings as early as 1500 B.C. indicate that the Egyptians used a form of tampon medicated with the ground powder of a shrub to prevent fertilization. This may sound primitive, but we use the same basic principle today to destroy sperm in the vagina. Contraceptive jellies and foams make the environment of the vagina more acidic, which diminishes the sperm's chances of survival. The spermicidal (sperm-killing) foam or jelly is placed in the vagina before intercourse. When the sperm make contact with the acidic environment, they stop and soon die. The aerosol foams are an effective method of conception-control, but the pill is more effective (table 11.3).

Death of the sperm is not the only method to prevent conception. Any method that prevents the sperm from reaching the egg prevents conception. One method is not to have intercourse during those times of the month when a secondary oocyte may be present. This is known as the **rhythm method** of conception-control. While at first glance it appears to be the simplest and least expensive, determining just when a secondary oocyte is likely to be present can be very difficult. You can see from table 11.2 that in the case of a woman with an irregular menstrual cycle, a couple might have only a few days each month to engage in intercourse without fear of pregnancy. In addition to counting days, a woman can better estimate the time of ovulation by keeping a record of changes in her body temperature and vaginal pH. Both of these changes are tied to the menstrual cycle and therefore can help a woman predict ovulation. In particular, at about the time of ovulation, a woman has a slight rise in body temperature (less than 1° C). Thus, one should use an extremely sensitive thermometer. (There is even a "digital" readout thermometer on the market that spells out the word "yes" or "no".)

Three other methods of conception-control that prevent the sperm from reaching the secondary oocyte include the **diaphragm, condom,** and **sponge.** The diaphragm is a specially fitted membranous shield that is inserted before intercourse into the vagina and positioned so that it covers the opening of the uterus. Because of anatomical differences among females, diaphragms must be fitted by a physician. The

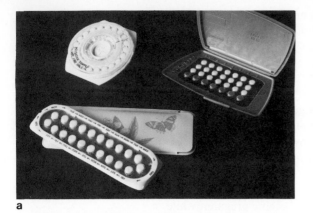

a

b

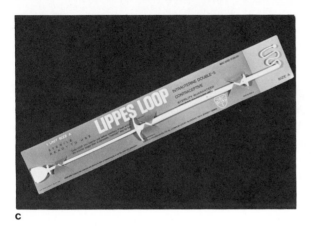

c

Figure 11.18
Contraceptive Methods.
These are the primary methods of conception control used today.
(a) Oral contraception (pills),
(b) diaphragm and spermicidal jelly,
(c) intrauterine device,
(d) spermicidal vaginal foam, and
(e) condom.

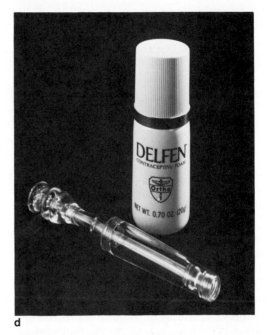

d

e

Table 11.3 *Effectiveness of Contraceptive Methods*

Method	Pregnancies Per 100 Women Per Year[a]	
	High[b]	Low
No contraceptive[c]	80	40
Coitus interruptus	23	15
Condom[d]	17	8
Douche	61	34
Chemicals (spermicides)[e]	40	9
Diaphragm and jelly	28	11
Rhythm[f]	58	14
Pill	2	0.03
IUD	8	3
Sterilization	0.003	0

a Data describe the number of women per 100 who will become pregnant in a one-year period while using a given method.
b High and low values represent best and worst estimates from various demographic and clinical studies.
c In the complete absence of contraceptive practice, 8 out of 10 women can expect to become pregnant within one year.
d Effectiveness increases if spermicidal jelly or cream is used in addition.
e Aerosol foam is considered to be the best of the chemical barriers.
f Use of a clinical thermometer to record daily temperatures increases effectiveness.

Note: From Volpe, E. Peter, *Biology and Human Concerns,* 3d. ed. © 1975, 1979, 1983 Wm. C. Brown Publishers, Dubuque, Iowa. All Rights Reserved. Reprinted by permission.

effectiveness of the diaphragm is increased if spermicidal foam or jelly is also used. The condom is probably the most popular male contraceptive device. It is a thin, rubber sheath that is placed over the erect penis before intercourse. This physical barrier not only prevents sperm from reaching the secondary oocyte, but can also prevent sexually transmitted diseases such as syphilis and gonorrhea from being passed from one person to another during sexual intercourse (box 11.1).

The **intrauterine device (IUD)** is not a physical barrier that prevents the gametes from uniting. How this contraceptive device works is not completely known. It may in some way interfere with the implantation of the embryo to the lining of the uterus. IUDs such as the "shield" must be fitted and inserted into the uterus by a physician. A physician can remove the IUD if pregnancy is desired. For all practical purposes, this method of birth-control is not available in the United States as a result of legal pressures on physicians that had inserted the devices in women who later became pregnant or developed uterine cancers. The device continues to be used successfully in many other countries.

Two contraceptive methods that require surgery are tubal ligation and vasectomy. Tubal ligation involves the cutting and tying off of the oviducts and can be done on an outpatient basis in most cases. Ovulation continues as usual, but the sperm and egg cannot unite. Vasectomy can also be performed in a physician's office and does not require hospitalization. A small opening is made in the scrotum, and the spermatic cord (**vas deferens**) is cut and tied. This prevents sperm from moving through the ducts to the outside. Because the majority of the sperm-carrying fluid (**semen**) is produced by the **seminal vesicles, prostate gland,** and **bulbo-urethral glands,**

Box 11.1
Sexually Transmitted Diseases

Diseases currently referred to as *sexually transmitted diseases* (STDs) were formerly called *venereal diseases* (VDs). The term venereal is derived from the name of the Roman goddess for love, Venus. Although these kinds of illnesses are most frequently transmitted by sexual activity, many can also be spread by other methods of direct contact or by objects such as hypodermic needles, blood transfusions, or blood contaminated materials. Currently the Centers for Disease Control in Atlanta, Georgia, recognize eighteen diseases as being sexually transmitted and a nineteenth, gay bowel syndrome, which is actually caused by a great variety of different microorganisms.

Some of the most important STDs are described on the following page because of their high prevalence in the population and our inability to bring some of them under control. For example, there is no known cure for the HTLV III virus that is responsible for the disease AIDS. There has also been a sharp rise in the number of gonorrhea cases in the United States caused by a form of the bacterium *Neisseria gonorrhoeae* that is able to resist the drug penicillin by producing an enzyme that actually destroys the antibiotic. Those troublemakers are known as PPNGs, penicillinase producing *Neisseria gonorrhoeae.* However, most of the infectious agents are able to be controlled if diagnosis occurs early and treatment programs are carefully followed by the patient. The spread of STDs during sexual intercourse is significantly diminished by the use of condoms. Other types of sexual contact (i.e., hand, oral, anal) and congenital transmission (i.e., from the

mother to the fetus during pregnancy) help maintain some of these diseases in the population at high enough levels to warrant attention by public health officials, the United States Public Health Service, Centers for Disease Control, and state and local public health agencies. All of these agencies are involved in the special area of study known as epidemiology, and each attempts to raise the general public health to a higher level. Their investigations have resulted in the successful control of many diseases and the identification of special epidemiological problems, such as those associated with the STDs.

Members of all public health agencies are responsible for telling the public of the things that are good for them and for warning them of things that are dangerous. In order to meet these obligations when dealing with sexually transmitted diseases, such as AIDS and syphilis, they encourage the use of one of their most potent weapons, sex education. Everyone must know about their own sexuality if they are to understand the transmission and nature of STDs. Then it will be possible to alter their behavior in ways that will prevent the spread of these diseases. The intent is to present the public with biological facts, not to scare people. Public health officials do not have the luxury of personal opinions when it comes to their job. The biological nature of sexual behavior is not a moral issue, but biological facts are needed if people are to make intelligent decisions relating to their sexual behavior. It is hoped that through education, people will alter their high risk sexual behavior so that they will not put themselves in situations where they

could become infected with one of the venereal diseases. As one health official stated, each person should be knowledgeable enough about their own sexuality and the STDs to answer the question "Is what I'm about to do worth dying for?"

DISEASE	AGENT
Genital herpes	Virus
Gonorrhea	Bacterium
Syphilis	Bacterium
Acquired Immune Deficiency Syndrome (AIDS)	Virus
Candidiasis	Yeast
Chancroid	Bacterium
Condyloma acuminatum (venereal warts)	Virus
Gardnerella vaginalis	Bacterium
Genital *Chlamydia* infection	Bacterium
Genital Cytomegalovirus infection	Virus
Genital *Mycoplasma* infection	Bacterium
Group B *Streptococcus* infection	Bacterium
Nongonococcal urethritis	Bacterium
Pelvic inflammatory disease (PID)	Bacterium
Reiter's syndrome	Bacterium
Scabies	Insect
Trichomoniasis	Protozoan
Viral hepatitis (HBV)	Virus
Gay bowel syndrome	Variety of agents

Note: From *Introductory Microbiology* by Frederick C. Ross. Copyright © 1986, 1983 by Scott, Foresman and Company. Reprinted by permission.

Sexually Transmitted Diseases

Disease	Causative Agent	Symptoms	Incubation Period
Chancroid	*Hemophilus ducreyi* (bacterium)	Symptoms include a discharge of pus containing blood, small sores with irregular soft edges, and some pain.	1–3 days
Gonorrhea (clap, strain)	*Neisseria gonorrhoeae* (bacterium)	In male: a mucous and pus-containing discharge; painful and frequent urination. In females, no definite symptoms.	3–9 days or up to 2 weeks
Granuloma inguinale	*Calymmatobacterium granulomatis* (bacterium)	No discharge; a moist ''pimple'' on or in the area of the external genitalia.	Unknown
Lymphogranuloma venereum (LGV)	*Chlamydia trachomatis* (bacterium)	Ulcers in the urogenital system; fever, body ache, swelling.	3–21 days
Syphilis	*Treponema pallidum* (bacterium)	*Primary Stage:* chancre: painless ulcer formed at site of infection. *Secondary Stage:* rash, brain infection, bone infections, and infections of other body organs and tissues. *Tertiary Stage:* soft ulcerations of various body tissues and organs. *Congenital:* affects tooth development, central nervous system, and may result in blindness, deafness, or death.	9–90 days
Yeast infection (vaginitis)	*Candida albicans* (fungus)	Thick, yellow discharge with severe itching.	Variable
Trich	*Trichomonas vaginalis* (protozoan)	Green-whitish, foamy foul-smelling discharge and irritation.	Variable
NGU (nongonococcal urethritis)	*Chlamydia* or other bacterial types	Similar to gonorrhea but milder.	Variable
Herpes	Herpes simplex virus	Similar to ''cold sores''; clusters of blisterlike ''pimples'' that break open and release fluid containing virus.	Variable
AIDS (acquired immune deficiency syndrome)	HTLV III virus (human T-cell lymphotrophic virus group III)	The virus interferes with the T-helper lymphocytes of the immune system preventing the body from defending itself against various infectious diseases and cancer. Victims may experience secondary infections that may become fatal such as pneumonia and meningitis, or they develop malignant cancers that are unable to be brought under control.	Variable, possibly years

Figure 11.19
Tubal Ligation and Vasectomy.
Two very effective contraceptive methods require surgery. Tubal ligation (a) involves severing the oviducts and sealing the cut ends. This prevents the sperm and primary oocyte from meeting. This procedure usually requires a short hospitalization period. Vasectomy (b) requires minor surgery usually in a clinic under local anesthesia. A man may experience minor discomfort only for several days following the surgery. The cutting and sealing of the vas deferens prevents the sperm cells from being released from the body.

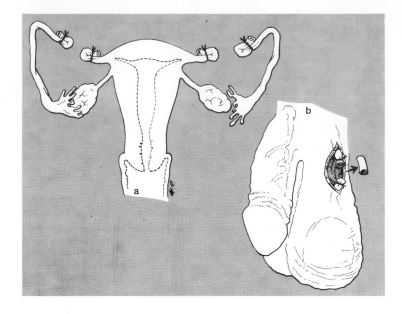

a vasectomy does not interfere with normal ejaculation. The sperm that are still being produced die and are reabsorbed in the testes. Neither tubal ligation nor vasectomy interferes with normal sex drive. However, these methods of contraception should be considered permanent, and only persons who do not wish to have more children should consider tubal ligation and vasectomy (figure 11.19).

Abortion

Another medical procedure often associated with birth-control is **abortion;** it has been used throughout history as a method of birth-control. Abortion involves various medical procedures that cause the death and removal of the developing embryo. Abortion is obviously not a method of conception-control, rather, it prevents the normal development of the embryo and causes its death. Abortion is a very emotional subject. Some people feel that abortion should be prohibited by law in all cases. Others feel that abortion should be allowed in certain situations, such as in pregnancies that endanger the mother's life or in pregnancies that are the result of rape or incest. Still others feel that abortion should be available to any woman under any circumstances. Regardless of the moral and ethical issues that surround abortion, it is still a common method of terminating unwanted pregnancies.

The abortion techniques used today all involve the possibility of infections, particularly if done by poorly trained personnel. The three most common techniques are scraping the inside of the uterus with special instruments (called a D and C or dilation and curettage), injecting a saline solution into the uterine cavity, and using a suction device to remove the embryo from the uterus.

Problems, Surprises, and Changing Attitudes

What we have discussed so far is a description of what usually occurs during reproduction in humans. One common side effect of pregnancy is known as **morning sickness.** The dizziness, nausea, and vomiting may be the result of hormonal and other chemical changes taking place in the body. Once these changes are made, the morning sickness usually disappears. Changes occurring during pregnancy also affect the metabolic balance of the mother. She requires a greater amount of water, nutrients, and growth factors. Both the developing embryo and the mother suffer if the mother's diet is poor. Proteins are extremely important during pregnancy, since they are required by the mother and the embryo for the production of additional cells and enzymes. Without adequate protein, the baby will suffer both mentally and physically. In countries with high birthrates and a low-protein diet, the population is caught in a cycle of mental and physical poverty, which can be broken only by increasing the protein available or reducing the population so that each person receives enough of the proper foods (refer to chapter 7).

Other things may also damage the baby mentally and physically. For instance, drugs taken by the mother or diseases such as German measles and syphilis can influence the development of the baby during pregnancy. A number of years ago in Europe, the drug thalidomide was given to pregnant women as a sedative. Unfortunately, the drug was able to cross the placenta and interfere with the normal development of arms and legs. As a result of birth abnormalities throughout the world, thalidomide was removed from the market. Mothers addicted to heroin have babies who are also addicted to the drug. After these babies are born and no longer have a constant supply of the drug coming to them through the placenta, they go through withdrawal symptoms. These are extreme cases in which drugs taken by a pregnant woman have severe effects on her unborn child. All medication should be carefully controlled by a physician during pregnancy. Most women should restrict their intake of caffeine, alcohol, and nicotine since these substances have been demonstrated to increase the incidence of birth abnormalities.

Occasionally the mother is surprised to find out that she will have twins. Twins are produced in two ways. **Identical twins** are of the same sex and have the same genes. When the zygote divides by cleavage, it may split into two separate groups of cells. Each will develop into an independent embryo. Since they came from the same single fertilized egg, they are genetically identical (figure 11.20).

Fraternal twins do not contain the same genetic information and may be of different sexes. They result from the fertilization of two separate eggs by different sperm. Therefore they are no more identical to each other than ordinary brothers and sisters.

As attitudes toward sex have changed and people have become less inhibited, attitudes toward childbirth have changed. In the past, fathers were prohibited from seeing the birth of their children. Today, an increasing number of physicians and hospitals are encouraging fathers to be present during the birth of their children. Many fathers want to see the birth of their child and provide emotional support to the mother during childbirth.

For years the use of anesthetics during childbirth was routine. But today, many physicians encourage mothers to have their children with a minimum of drugs for relief of pain. Any drug given to the mother reaches the child through the placenta and may slow reactions at birth. A limited use of anesthetics also allows for a speedier recovery by the mother. Along with this trend toward natural childbirth, there has been an increase in the number of mothers who wish to breast-feed their babies. For many mothers, breast-feeding is less expensive, easier, and more fulfilling than bottle-feeding. In addition, human milk is designed to meet the needs of the human baby.

Summary

Sexual reproduction involves the production of gametes by meiosis in the ovaries and testes. The production and release of these gametes is controlled by the interaction of hormones. In males, each cell that undergoes spermatogenesis results in four active sperm, while in females, each cell that undergoes oogenesis results in one egg and polar bodies. Nondisjunction is an abnormal kind of meiosis that causes

Figure 11.20
Twinning.
Identical twins are the result of an
early separation of the dividing
cells of a zygote. These separated
cells are from the same parent cell
and are genetically identical.

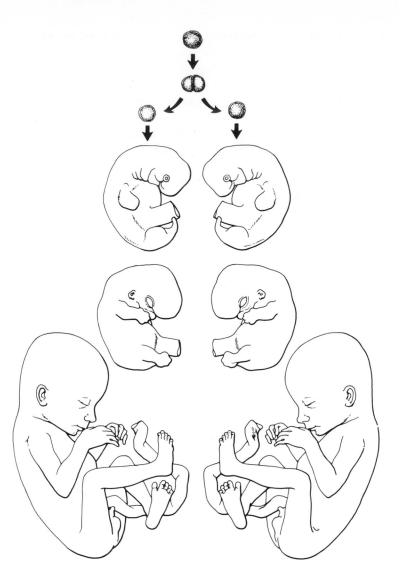

abnormal embryos, as found in Down's, Turner's, and Klinefelter's syndromes. Many organisms, including humans, have specialized structures for the support of the developing embryo. Many factors influence the development of the embryo in the uterus. Successful sexual reproduction depends on proper hormone balance, proper meiotic division, fertilization, placenta formation, proper diet of the mother, and birth. Hormones control ovulation and menstruation and may also be used to encourage or discourage ovulation. Fertility drugs and "the pill," for example, involve hormonal control. In addition to the pill, a number of conception-control methods have been developed including the diaphragm, condom, IUD, spermicidal jellies and foams, the sponge, tubal ligation, and vasectomy. During pregnancy, it is important that the mother receive proper nutrition and avoid using unnecessary drugs, since these things influence the developing embryo. During labor, the contractions of the uterus move the baby through the birth canal headfirst. The placenta is expelled following the baby. Even though the processes of sexual reproduction and birth have remained the same, many attitudes toward breast-feeding, natural childbirth, and the father's participation during birth have changed over the last few years.

A great world adventurer discovered a tribe of women in the jungles of Brazil. After many years of very close study and experimentation, he found that sexual reproduction was not possible. He also noticed that the female children resembled their mothers to a great degree and found that all the women had a gene that prevented meiosis. Ovulation occurred as usual, and pregnancy lasted nine months. The mothers nursed their children for three months after birth and became pregnant the next month. This cycle was repeated in all the women of the tribe.

Consider the topics of meiosis, mitosis, sexual reproduction, and regular hormonal cycles in women and explain in detail what may be happening in this tribe.

The continuation of the human species is based on sexual reproduction. Yet countless millions know more about how an automobile engine works than how their own bodies function. Because of this fact, many school systems have either expressed a desire to introduce or have made commitments to introduce sex education programs into their curricula. Where does your school system and community stand with regard to sex education? What do they see as the pros and cons of such a program? At what age is the topic introduced to students?

1. What structures are associated with the human female reproductive system? What are their functions?
2. What structures are associated with the human male reproductive system? What are their functions?
3. What are the differences between oogenesis and spermatogenesis in humans?
4. What is nondisjunction? Give an example.
5. How are ovulation and menses related to each other?
6. What changes occur in ovulation and menstruation during pregnancy?
7. What are the functions of the placenta?
8. If a woman begins to menstruate on the first of June and has a regular twenty-eight day menstrual cycle, on what dates would sexual intercourse most likely result in pregnancy? What if the cycle lasted twenty-two days?
9. In which part of the female body is fertilization most likely to occur?
10. What are some advantages of breast-feeding?

abortion (ă-bor′shun) A medical procedure that causes the death and removal of the embryo from the uterus.

allantois (al″lan′to-is) A saclike extension of the embryo that forms part of the placenta.

amnion (am′ne-on) A fluid-filled sac that surrounds the embryo.

birth-control pill (burth kon-trol′ pil) Pills containing hormones that prevent the release of eggs from the ovary.

blastula stage (blas′chu-lah stāj) An early embryological stage that follows the morula; it consists of a hollow ball of cells.

breech birth (brēch birth) A birth in which the head is not born first.

bulbo-urethral gland (bul″bo-yu-re′thral gland) A part of the male reproductive system that produces a portion of the semen.

caesarean (se-zār'e-an) The surgical removal of a baby through the mother's abdominal wall.

chorion (ko're-on) A saclike extension of the embryo that forms part of the placenta.

cleavage (kle'vuj) An early embryological stage during which the zygote divides by mitosis into smaller cells.

condom (kon'dum) A thin, rubber sheath placed over an erect penis before sexual intercourse.

corpus luteum (kōr'pus lu'te-um) The follicle after the primary oocyte has erupted that becomes a glandular structure.

diaphragm (di'uh-fram) A mechanical conception-control device that covers the entrance to the uterus.

Down's syndrome (downs sin'drōm) A genetic disorder resulting from the presence of an extra chromosome number 21. Symptoms include thickened eyelids, a low level of intelligence, and faulty speech. Sometimes called mongolism.

embryo (em'bre-o) The early stage in the development of a sexually reproduced organism.

fertilization (fur''ti-li-za'shun) The uniting of a male gamete (sperm) with a female gamete (egg) to form a zygote.

follicle (fol'li-kul) The saclike structure in the ovary that contains the developing egg and develops into the corpus luteum.

fraternal twins (fra-tur'nal twins) Two offspring formed at the same time; results from the fertilization of two separate eggs.

gametogenesis (ga-me''to-jen'e-sis) The generating of gametes. The meiotic cell division process that produces sex cells; oogenesis and spermatogenesis.

gastrula stage (gas'tru-lah stāj) The embryological stage following the blastula stage in which the primitive gut is formed.

hormone (hōr'mōn) Chemical substance that is released from glands in the body to regulate other parts of the body.

identical twins (i-den'ti-cal twins) Two offspring born at the same time as a result of the fertilization of one egg that separates into two individuals.

implantation (im-plan-ta'shun) The embedding of the early embryo in the lining of the uterus.

intrauterine device (IUD) (in-trah-yu'tur-in de-vis) A mechanical contraception-control device placed in the uterus to prevent embryo implantation.

Klinefelter's syndrome (Klin'fel-turs sin'drōm) A genetic disorder caused by having two X chromosomes as well as a Y chromosome. Symptoms include tallness, sterility, and mental impairment.

labor (la'bor) The contractions of the uterus that result in the birth of the baby.

menses (period, menstrual flow) (men'sēz) The shedding of the lining of the uterus.

menstrual cycle (men'stru-al si'kul) The repeated building up and shedding of the lining of the uterus.

morning sickness (mor'ning sik'nes) One of the symptoms of pregnancy characterized by nausea, vomiting, and dizziness.

morula (mōr'yu-lah) An early embryological stage consisting of a solid ball of cells.

nondisjunction (non''dis-junk'shun) An abnormal meiotic division that results in sex cells having too many or too few chromosomes.

oogenesis (o''o-jen'e-sis) The specific name given to the gametogenesis process that leads to the formation of eggs.

oral contraceptive (ōr-al kon-trah-sep'tiv) A pill containing hormones that prevent the release of primary oocytes from the ovary.

ovary (o'vah-re) The female sex organ responsible for the production of the haploid egg cells.

oviduct (o'vi-dukt) The tube (fallopian tube) that carries the primary oocyte to the uterus.

ovulation (ov-yu-la'shun) The release of a secondary oocyte from the ovary.

penis (pe-nis) The portion of the male reproductive system that deposits sperm in the female reproductive tract.

placenta (plah-sen′tah) An organ made up of tissues from the embryo and the uterus of the mother that allows for the exchange of materials between the mother's bloodstream and the embryo's bloodstream. It also produces hormones.

polar body (po′lar bod-e) The smaller cell formed by an unequal meiotic division during oogenesis.

pregnancy (preg′nan-se) In mammals, the period of time when the embryo is developing in the uterus of the female.

primary oocyte (pri′ma-re o′o-sīt) A diploid cell found in the ovary that undergoes the first meiotic division after the sperm has entered this cell.

primary spermatocyte (pri′ma-re spur-mat′o-sīt) A diploid cell in the testes that undergoes the first meiotic division in the process of spermatogenesis.

prostate gland (pros-tāt gland) A part of the male reproductive system that produces a portion of the semen.

rhythm method (rith′m meth′od) A method of conception-control in which couples avoid sexual intercourse when the egg is most likely to be present.

scrotum (skro′tum) The sac that contains the testes.

seasonal reproductive pattern (se′zon-al re′′pro-duk-tiv pat′ern) A behavior typical of most plants and animals in which breeding activities take place only during a particular time of the year.

secondary oocyte (sĕ-kon-da-re o′o-sīt) A haploid cell that goes through the second meiotic division to produce an egg.

secondary spermatocyte (sĕ-kon-da-re spur-mat′o-sīt) A haploid cell found in the testes that goes through the second meiotic division to provide spermatids.

semen (se-men) The fluid produced by the seminal vesicle, prostate gland, and bulbo-urethral gland of males that carries sperm.

seminal vesicle (sem-in-al ves′ĭ-kul) A part of the male reproductive system that produces a portion of the semen.

seminiferous tubules (sem′′in-if′ur-us tūb-yūls) Sperm-producing tubes in the testes.

sexual intercourse (copulation, coitus) (seks′u-al in′ter-kors) (kop-yu-la′shun, ko′ĭ-tus) The mating of male and female. The action of the male depositing sperm in the reproductive tract of the female.

sperm (′spurm) A male gamete.

spermatids (spurm′ah-tids) Haploid cells produced by spermatogenesis that change into sperm.

spermatogenesis (spur-mat-o-jen′uh-sis) The specific name given to the gametogenesis process that leads to the formation of sperm.

sponge (′spunj) A vaginal contraceptive device that operates by preventing sperm from reaching the egg.

testes (tes-tēz) The male organs that produce sperm.

testosterone (tes-tos′tur-ōn) The male sex hormone produced in the testes that controls the secondary sex characteristics.

Turner's syndrome (tur′nurz sin′drōm) A genetic disorder caused by the lack of one sex chromosome. Symptoms include difficulty with spacial relations, webbed neck, sterility, and usually shortness.

umbilical cord (um-bil′ĭ-cal cord) The cord containing the blood vessels that transports materials between the placenta and the embryo.

uterus (yu′tur-us) An organ found in mammals in which the embryo develops.

vagina (vuh-ji′nah) The passageway between the uterus and outside of the body; the birth canal.

vas deferens (vas def′ur-ens) The portion of the sperm duct that is cut and tied during a vasectomy.

yolk sac (yōk sak) A small sac that is present as a rudiment in mammalian embryos.

zygote (zi′gōt) The fertilized egg.

Mendelian Genetics

■ Purpose

This chapter considers the fundamentals of inheritance. Previous chapters have introduced you to the concepts and importance of DNA as a molecule for storing genetic information used to manufacture proteins, mitosis, meiosis, and sexual reproduction. Throughout this chapter we will discuss how characteristics are passed from one generation to the next using many human characteristics to illustrate these patterns of inheritance.

For Your Information

As the buying habits of the American public change from beef to more lean sources of protein, such as fish and chicken, beef producers are seeing their share of the market decline. In response to this change in consumer demand, scientists involved in cattle genetics are developing new strains of beef cattle that have less fat and, therefore, fewer calories per pound. Since the bison (American Buffalo) has lean meat, some animal geneticists have crossbred standard beef animals with the buffalo. In this way, the buffalo genes for lean meat have been introduced into beef cattle and a new "breed" has been developed, beefalo. Such an animal produces meat that has 80% less fat and 55% fewer calories than regular beef. Some of the problems encountered with these attempts to create lean beef have been differences in flavor, color, and texture from the standard fatty beef currently being consumed by the American public. Further work by geneticists will be needed to develop new strains of animals that provide meat with acceptable flavor, color and texture, as well as low fat content.

Learning Objectives

■ Be able to work monohybrid and dihybrid genetic problems dealing with traits that show dominance, recessiveness, and lack of dominance.

■ Be able to work genetic problems dealing with multiple alleles, polygenic inheritance, and X-linked characteristics.

■ Explain how environmental conditions influence an organism's phenotype.

Genetics, Meiosis, and Cells

Why do you have a particular blood type or hair color? Why do some people have the same skin color as their parents, while others have a skin color different from their parents? These questions can be better answered if you understand how genes work. A **gene** is a portion of DNA that determines characteristics. Through meiosis and reproduction these genes can be transmitted from one generation to another. The study of genes, how genes produce characteristics, and how the characteristics are inherited is the field of biology called **genetics.** The first person to systematically study inheritance and formulate laws about how characteristics are passed from one generation to the next was an Augustinian monk named Johann Gregor Mendel (1822–84) (figure 12.1). However, his work was not generally accepted until 1900 when three men working independently rediscovered some of the ideas that Mendel had formulated over thirty years earlier. Because of his early work, the study of the pattern of inheritance that follows the laws formulated by Johann Gregor Mendel is often called **Mendelian genetics.**

To understand this chapter, you need to know some basic terminology. One term that you encountered earlier is gene. Mendel thought of a gene as a "particle" that could be passed from the parents to the **offspring** (children, descendants, progeny). Today we know that genes are actually composed of specific sequences of DNA nucleotides. The "particle" concept is not entirely inaccurate, since genes are located on specific portions of chromosomes; however, they are not like beads on a string.

Another important idea to remember is that all sexually reproducing organisms are diploid. Since gametes are haploid and most organisms are diploid, the conversion of diploid to haploid cells during meiosis is an important process. The diploid cells have two sets of chromosomes, one set inherited from each parent. Therefore, they have two chromosomes of each kind and have two genes for each characteristic. When sex cells are produced by meiosis, reduction division occurs and the diploid number is reduced to haploid. Therefore, the sex cells produced by meiosis have one

Figure 12.1
Gregor Mendel.
Mendel raised peas in the monastery garden to help him determine how characteristics are inherited in plants.

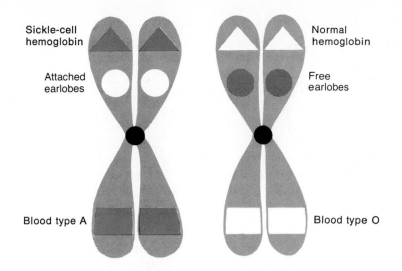

Sickle-cell hemoglobin

Attached earlobes

Blood type A

Normal hemoglobin

Free earlobes

Blood type O

Figure 12.2
A Pair of Homologous Chromosomes.
Homologous chromosomes contain genes for the same characteristics at the same place. Note that the attached earlobe allele is located at the ear-shape locus on one chromosome, and the free earlobe allele is located at the ear-shape locus on the other member of the homologous pair of chromosomes. (We really don't know on which chromosomes most human genes are located. The examples presented here are for illustrative purposes only.)

chromosome of each of the pairs that was in the diploid cell that began meiosis. Diploid organisms usually result from the fertilization of a haploid egg by a haploid sperm. Therefore, they inherit one gene of each type from each parent. For example, each of us has two genes for earlobe shape: one came with our father's sperm, the other in our mother's egg.

Genes and Characteristics

Although each diploid organism has two genes for each characteristic, there may be several alternative forms of each gene within the population. Alternative forms of genes for the same trait at the same region on homologous chromosomes are called **alleles.** In people, there are alleles for earlobe shape. One allele produces an earlobe that is fleshy and hangs free, while the other allele produces a lobe that is attached to the side of the face and does not hang free. The type of earlobe that is present is determined by the type of allele (gene) received from each parent and how these alleles interact with one another. Alleles are always located on the pair of homologous chromosomes, one allele on each chromosome. These alleles are also always at the same specific location on a chromosome, **locus,** on the same type of chromosome in all individuals of a species (figure 12.2).

The **genome** is a set of all the genes necessary to specify an organism's complete list of characteristics. A diploid (2N) cell has two genomes and a haploid cell (N) has one genome. The **genotype** of an organism is a listing of the genes present in that organism. It consists of the cell's DNA; therefore, you cannot see the genotype of an organism. It is impossible to know the complete genotype of most organisms, but it is often possible to figure out the genes present that determine a particular characteristic. For example, in humans there are two alleles for earlobe shape. There are three possible genotypic combinations of these two alleles. A person's genotype could be: (1) two genes for attached earlobes; (2) one gene for attached and one gene for free earlobes; or (3) two genes for free earlobes.

How would individuals with each of these three genotypes appear? The way each combination of genes expresses itself is known as the **phenotype** of the organism. A person with two alleles for attached earlobes will have earlobes that are

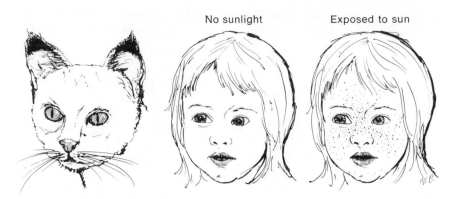

No sunlight   Exposed to sun

Figure 12.3
*Environment and Gene
Expression.*

The expression of many genes is
influenced by the environment. The
gene for dark hair in the cat is
sensitive to temperature and
expresses itself only in the parts of
the body that stay cool. The gene
for freckles expresses itself more
fully when the person is exposed
to sunlight.

attached. A person with one allele for attached earlobes and one allele for free ear-lobes will have a phenotype that appears as free hanging lobes. An individual with two alleles for free earlobes will also have free hanging lobes. Notice that there are three genotypes, but only two phenotypes. The individuals with the free earlobe phenotype have different genotypes.

For various reasons, certain genes may not express themselves. Sometimes, the physical environment determines if certain genes function. For example, some cats have coat color genes that do not reveal themselves unless the temperature of the skin is below a certain point. Often, the only parts of a cat that become cool enough to allow the genes to express themselves are the tips of the ears and the feet. Consequently these areas will differ in color from the rest of the cat's body. Another example in humans is the presence of genes for freckles that do not show themselves fully unless the person's skin is exposed to sunlight (figure 12.3).

The expression of some genes is directly influenced by the presence of other genes in the organism. For any particular pair of alleles in an individual organism, the two alleles are either identical or not identical. An individual is **homozygous** when a diploid organism has two identical alleles for a particular characteristic. A person with two alleles for freckles is homozygous for that trait. Also, a person with two alleles for no freckles is homozygous.

An individual is **heterozygous** when a diploid organism has two different allelic forms of a particular gene. A person with one allele for freckles and one allele for no freckles is heterozygous. If an organism is homozygous, the characteristic expresses itself in a specific manner. A person homozygous for free earlobes has free earlobes, and a person homozygous for attached earlobes has attached earlobes. However, if an organism is heterozygous, how do these two different genes interact?

Often, one gene expresses itself and the other does not. A **dominant allele** expresses itself and masks the effect of other alleles for the trait. For example, if a person has one allele for free earlobes and one allele for attached earlobes, that person has a phenotype of free earlobes. The allele for free earlobes is dominant. A **recessive allele** is one that, when present with another allele, does not express itself and is masked by the effect of the other allele. Attached earlobes is a recessive characteristic. A person with one allele for free earlobes and one allele for attached earlobes has the phenotype, free earlobes. Recessive traits only express themselves when the organism is homozygous for the recessive alleles. If you have attached earlobes, you have two alleles for that trait. Recessive genes are not necessarily bad. The term recessive has nothing to do with the significance of the gene. It simply describes how it can be expressed. They are not less likely to be inherited, but must be present in a homozygous condition to express themselves.

Mendelian genetics involves the study of passing genes from one generation to another and how the genes received from the parents influence the traits of the offspring. Before you go on, be certain that you understand the principles of meiosis and how the gametes formed from this process are combined by fertilization. It is in meiosis that the two alleles in a pair of genes segregate. Although we will talk about the segregation of genes, remember, in this chapter it is the chromosomes and not the individual genes that actually segregate.

Notice that the chromosomes in figure 12.2 have two different alleles for hemoglobin, one for sickle-cell hemoglobin and one for normal hemoglobin. By meiosis, the parent may contribute a gamete containing a gene for sickle-cell hemoglobin or a gene for normal hemoglobin. By fertilization, the offspring will receive two genes for hemoglobin production, but only one from each parent. For most of the remainder of this chapter, we will discuss genetic problems dealing with how you may determine which genes are passed on by the parents, and how you go about determining the genetic makeup of the offspring resulting from fertilization.

Mendel's Laws of Heredity

Heredity problems are concerned with determining which genes are passed from the parents to the offspring, and the probability of producing various types of offspring. The first person to develop a method of predicting the outcome of inheritance patterns was Mendel. This was done in his experiments concerning the inheritance of certain characteristics in sweet pea plants. From his work, Mendel concluded which traits were dominant and which were recessive. Some of his results were:

Characteristic	Alleles	Dominant	Recessive
plant height	tall and dwarf	tall	dwarf
pod shape	full and constricted	full	constricted
pod color	green and yellow	green	yellow
seed surface	round and wrinkled	round	wrinkled
seed color	yellow and green	yellow	green
flower color	violet and white	violet	white

As a result of analyzing his data, Mendel formulated several genetic laws to describe how characteristics are passed from one generation to the next and how they are expressed in an individual.

Mendel's law of dominance When an organism has two different alleles for a trait, the allele that is expressed and over-shadows the expression of the other allele is said to be dominant. The gene whose expression is over-shadowed is said to be recessive.

Mendel's law of segregation When gametes are formed by a diploid organism, the alleles that control a trait separate from one another into different gametes, retaining their individuality.

Mendel's law of independent assortment Members of one gene pair separate from each other independently of the members of other gene pairs.

When Mendel was doing his research, biologists knew nothing of chromosomes or DNA or of the process of mitosis or meiosis. Mendel assumed that each gene was separate from other genes. It was fortunate for him that each characteristic he picked to study was found on a separate chromosome. If two or more of these genes had been located on the same chromosome (linked genes), he probably would not have been able to formulate his laws. The discovery of chromosomes and DNA have caused modifications to be made in Mendel's laws. However, it was Mendel's work that formed the foundation for the science of genetics.

Probability Versus Possibility

In order to solve heredity problems you must have an understanding of probability. **Probability** is the chance that an event will happen, and is often expressed as a percent or a fraction. *Probability* is not the same as *possibility*. It is possible to toss a coin and have it come up heads. But the probability of getting a head is more precise than just saying it is possible to get a head. The probability of getting a head is one out of two (1/2 or 0.5 or 50%) because there are two sides to the coin, only one of which is a head. Probability can be expressed as a fraction:

$$\text{Probability} = \frac{\text{the number of events that can produce a given outcome}}{\text{the total number of possible outcomes}}$$

What is the probability of cutting a deck of cards and getting the ace of hearts? The number of times that the ace of hearts can occur is one. The total number of possible outcomes (number of cards in the deck) is fifty-two. Therefore, the probability of cutting an ace of hearts is 1/52.

What is the probability of cutting an ace? The total number of aces in the deck is four, and the total number of cards is fifty-two. Therefore, the probability of cutting an ace is 4/52 or 1/13.

It is also possible to determine the probability of two independent events occurring together. *The probability of two or more events occurring simultaneously is the product of their individual probabilities.* If you throw a pair of dice, it is possible that both will be a four. What is the probability that they both will be a four? The probability of one die being a four is 1/6. The probability of the other die being a four is also 1/6. Therefore, the probability of throwing two fours is

$$1/6 \times 1/6 = 1/36.$$

Steps in Solving Heredity Problems—Monohybrid Crosses

The first type of problem we will work is the easiest type, a monohybrid cross. A **monohybrid cross** is a genetic cross or mating in which a single characteristic is followed from one generation to the next.

In humans, the allele for free earlobes is dominant and the allele for attached earlobes is recessive. If both parents are heterozygous (have one allele for free earlobes and one allele for attached earlobes), what is the probability that they can have a child with free earlobes? with attached earlobes?

In solving a heredity problem there are five basic steps:

Step 1: Assign a symbol for each allele.

Usually, a capital letter is used for a dominant allele and a small letter for a recessive allele. Use the symbol E for free earlobes and e for attached earlobes.

E = free earlobes
e = attached earlobes

Step 2: Determine the genotype of each parent and indicate a mating.

Since both parents are heterozygous, the male genotype is *Ee*. The female genotype is also *Ee*. The X between them is used to indicate a mating.

$$Ee \times Ee$$

Step 3: Determine all the possible kinds of gametes each parent can produce.

Remember that gametes are haploid; therefore, they can only have one allele instead of the two present in the diploid cell. Since the male has both the free earlobe allele and the attached earlobe allele, half of his gametes will contain the free earlobe allele and the other half will contain the attached earlobe allele. Since the female has the same genotype, the genotype of her gametes will be the same as his.

For problems dealing with one pair of alleles, a Punnett Square is used. A **Punnett Square** is a box figure that allows you to determine the probability of genotypes and phenotypes of the offspring of a particular cross. Remember, because of the process of meiosis, each gamete receives only one allele for each characteristic listed. Therefore, the male will give either an *E* or *e;* the female will also give either an *E* or *e*. The possible gametes produced by the male parent are listed on the left side of the square, while the female gametes are listed on the top. In our example, the Punnett Square would show a single dominant allele and a single recessive allele from the male on the left side. The alleles from the female would appear on the top.

female genotype
Ee
possible female gametes

E & e

male genotype *Ee*   possible male gametes   = *E & e*

	E	e
E		
e		

Step 4: Determine all the gene combinations that can result when these gametes unite.

To determine the possible combinations of alleles that could occur as a result of this mating, simply fill in each of the empty squares with the alleles that can be donated from each parent. Determine all the gene combinations that can result when these gametes unite.

	E	e
E	EE	Ee
e	Ee	ee

Step 5: Determine the phenotype of each possible gene combination.

In this problem, three of the offspring, *EE, Ee,* and *Ee,* have free earlobes. One offspring, *ee,* has attached earlobes. Therefore, the answer to the problem is that the probability of having offspring with free earlobes is 3/4 and for attached earlobes is 1/4.

Take the time to learn these five steps. All monohybrid problems can be solved using this method; the only variation in the problems will be the types of alleles and the number of possible types of gametes the parents can produce. Let's work a problem with one parent heterozygous and the other homozygous for a trait.

Some people are unable to convert the amino acid phenylalanine into the amino acid tyrosine. Such individuals suffer from phenylketonuria (PKU) and may become mentally retarded. The normal condition is to convert phenylalanine to tyrosine. It is dominant over the condition for PKU. If one parent is heterozygous and the other parent is homozygous for PKU, what is the probability that they can have a child who is normal? with PKU?

Step 1:

Use the symbol N for normal and n for PKU.

$$N = \text{normal}$$
$$n = \text{PKU}$$

Step 2:

$$Nn \times nn$$

Step 3:

	n	*n*
N		
n		

Step 4:

	n	*n*
N	*Nn*	*Nn*
n	*nn*	*nn*

Step 5:

In this problem, one-half of the offspring will be normal and one-half will have PKU.

The Dihybrid Cross

A **dihybrid cross** is a genetic study in which two pairs of alleles are followed from the parental generation to the offspring. This problem is basically worked the same as a monohybrid cross. The main difference is that in a dihybrid cross, you are working with two different characteristics from each parent.

It is necessary to use Mendel's *law of independent assortment* when working dihybrid problems. This law states that members of one allelic pair separate from each other independently of the members of other pairs of alleles. This happens during meiosis when the chromosomes segregate. (Mendel's law of independent assortment only applies if the two pairs of alleles are located on separate chromosomes. This is an assumption we will use in dihybrid crosses.)

In humans, the allele for free earlobes dominates the allele for attached earlobes. The allele for dark hair dominates the allele for light hair. If both parents are heterozygous for earlobe shape and hair color, what types of offspring can they produce and what is the probability for each type?

Step 1:

Use the symbol E for free earlobes and e for attached earlobes. Use the symbol D for dark hair and d for light hair.

E = free earlobes
e = attached earlobes
D = dark hair
d = light hair

Step 2:

Determine the genotype for each parent and show a mating. The male genotype is *EeDd,* the female genotype is *EeDd,* and the $\times$ between them indicates a mating.

$$EeDd \times EeDd$$

Step 3:

Determine all the possible gametes each parent can produce and write the symbols for the alleles in a Punnett Square. Since there are two pair of alleles in a dihybrid cross, each gamete must contain one allele from each pair; one from the E pair (either E or e) and one from the D pair (either D or d). In this example, each parent can produce four different kinds of gametes. The four squares on the left indicate the gametes produced by the male, the four on the top indicate the gametes produced by the female.

To determine the possible gene combinations in the gametes, select one gene from one of the pairs of genes and match it with one gene from the other pair of genes. Then match the second gene from the first pair of genes with each of the genes from the second pair. This may be done as follows:

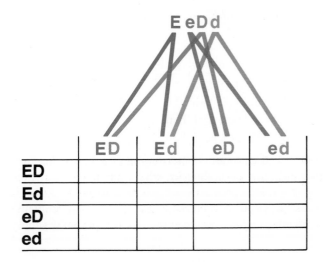

	ED	Ed	eD	ed
ED				
Ed				
eD				
ed				

Determine all the gene combinations that can result when these gametes unite. Fill in the Punnett Square.

	ED	Ed	eD	ed
ED	EEDD	EEDd	EeDD	EeDd
Ed	EEDd	EEdd	EeDd	Eedd
eD	EeDD	EeDd	eeDD	eeDd
ed	EeDd	Eedd	eeDd	eedd

Step 5:

Determine the phenotype of each possible gene combination. In this dihybrid problem there are sixteen possible ways in which gametes could combine to produce off-spring. There are four possible phenotypes in this cross. They are represented as:

Genotype	Phenotype	Symbol
EEDD or *EEDd* or *EeDD*	free earlobes and dark hair	*
EEdd or *Eedd*	free earlobes and light hair	∧
eeDD or *eeDd*	attached earlobes and dark hair	''
eedd	attached earlobes and light hair	+

	ED	Ed	eD	ed
ED	EEDD *	EEDd *	EeDD *	EeDd *
Ed	EEDd *	EEdd ∧	EeDd *	Eedd ∧
eD	EeDD *	EeDd *	eeDD ''	eeDd ''
ed	EeDd *	Eedd ∧	eeDd ''	eedd +

The probability of having a given phenotype is:

> 9/16 free earlobes, dark hair
> 3/16 free earlobes, light hair
> 3/16 attached earlobes, dark hair
> 1/16 attached earlobes, light hair

If a man with attached earlobes is heterozygous for hair color and his wife is homozygous for free earlobes and light hair, what can they expect their offspring to be like?

This problem has the same characteristics as the previous problem. Following the same steps, the symbols would be the same but the parental genotypes would be:

$$eeDd \times EEdd$$

The next step is to determine the possible gametes that each parent could produce and place them in a Punnett Square. The male parent can produce two different kinds of gametes, *eD* and *ed*. The female parent can only produce one kind of gamete, *Ed*.

	Ed
eD	
ed	

If you combine the gametes, only two kinds of offspring can be produced:

	Ed
eD	*EeDd*
ed	*Eedd*

They should expect either a child with free earlobes and dark hair or a child with free earlobes and light hair.

Real World Situations

So far we have considered a few simple cases in which a characteristic is determined by simple dominance and recessiveness between two alleles. Other situations, however, do not fit these patterns because some genetic characteristics are determined by more than two alleles. Some traits are influenced by gene interactions, and some traits are inherited differently depending on the sex of the offspring.

Lack of Dominance

All the cases that we have considered so far involved pairs of alleles in which one was clearly dominant over the other. Although this is common, it is not always true. In some combinations of alleles, there is a **lack of dominance.** This is a situation in which two unlike alleles both express themselves, neither being dominant. A classic example involves the color of the petals of snapdragons. There are two alleles for the color of these flowers. Because neither allele is recessive, we cannot use the traditional capital and small letters as symbols for these alleles. Instead, the allele for white petals is given the symbol F^W, and one for red petals is given the symbol F^R. There are three possible combinations of these two alleles:

Genotype	Phenotype
$F^W F^W$	White flowers
$F^R F^R$	Red flowers
$F^R F^W$	Pink flowers

Notice there are only two different alleles, red and white, but there are three phenotypes, red, white, and pink. Both the red flower allele and the white flower allele partially express themselves when both are present, and this results in pink.

Heredity problems dealing with lack of dominance are worked with the same five steps used in other problems. In the remainder of this chapter, we will list the steps to solve the problems, but will not state what is done in those steps. Next, we will work a lack of dominance problem. Since this problem concerns only one trait, it is worked as a monohybrid problem.

If a pink snapdragon is crossed with a white snapdragon, what phenotypes can result and what is the probability of each phenotype?

Step 1:

Genotype	Phenotype
$F^W F^W$	White flowers
$F^R F^R$	Red flowers
$F^R F^W$	Pink flowers

Step 2:

$$F^R F^W \times F^W F^W$$

Step 3:

	F^W	F^W
F^R		
F^W		

Step 4:

	F^W	F^W
F^R	$F^W F^R$ pink	$F^W F^R$ pink
F^W	$F^W F^W$ white	$F^W F^W$ white

Step 5:

This cross results in two different phenotypes, pink and white. No red flowers can result since it would require that both parents be able to contribute at least one red allele. The white flowers are homozygous for white, and the pink flowers are heterozygous.

Multiple Alleles

So far we have only discussed traits that are determined by two alleles. However, there can be more than two different alleles for a trait. The fact that some characteristics are determined by three or more different alleles is called **multiple alleles.** However, an individual can only have a maximum of two of the alleles for the characteristic. A good example of a characteristic that is determined by multiple alleles is the ABO blood type. There are three alleles for blood type:

$$I^A = \text{blood type A}$$
$$I^B = \text{blood type B}$$
$$i = \text{blood type O}$$

Box 12.1
The Inheritance of Eye Color

It is commonly thought that eye color is inherited in a simple dominant/recessive manner. Brown eyes are considered to be dominant over blue eyes. The real pattern of inheritance, however, is considerably more complicated than this. Eye color is determined by the amount of a brown pigment, known as melanin, present in the iris of the eye. If there is a large quantity of melanin present on the anterior surface of the iris, the eyes are dark. Black eyes have a greater quantity of melanin than brown eyes.

If there is not a large amount of melanin present on the anterior surface of the iris, the eyes will appear as blue, not because of a blue pigment but because blue light is returned from the iris. The iris appears blue for the same reason that deep bodies of water tend to appear blue. There is no blue pigment in the water, but blue wavelengths of light are returned to the eye from the water. People appear to have blue eyes because the blue wavelengths of light are reflected from the iris.

Just as black and brown eyes are determined by the amount of pigment present, colors such as green, gray, and hazel are produced by the various amounts of melanin in the iris. If a very small amount of brown melanin is present in the iris, the eye tends to appear green, whereas relatively large amounts of melanin produce hazel eyes.

Several different genes are probably involved in determining the quantity and placement of the melanin and, therefore, in determining eye color. These genes interact in such a way that a wide range of eye color is possible. Eye color is probably determined by polygenic inheritance, just as skin color and height are. (Some newborn babies have blue eyes that later become brown because at the time of birth they have not yet begun to produce melanin in their irises.)

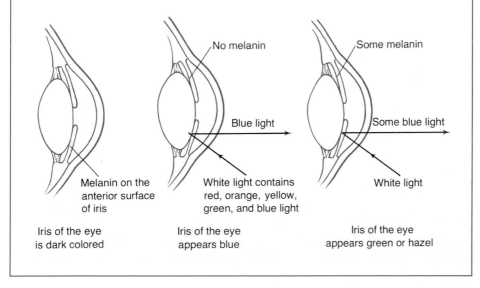

Melanin on the anterior surface of iris

Iris of the eye is dark colored

No melanin

Blue light

White light contains red, orange, yellow, green, and blue light

Iris of the eye appears blue

Some melanin

Some blue light

White light

Iris of the eye appears green or hazel

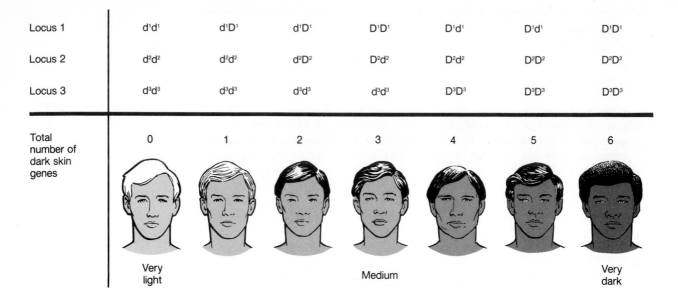

Locus 1	d¹d¹	d¹D¹	d¹D¹	D¹D¹	D¹d¹	D¹d¹	D¹D¹
Locus 2	d²d²	d²d²	d²D²	D²d²	D²d²	D²D²	D²D²
Locus 3	d³d³	d³d³	d³d³	d³d³	D³D³	D³D³	D³D³

Total number of dark skin genes	0	1	2	3	4	5	6

Very light   Medium   Very dark

Figure 12.4
Polygenic Inheritance.
Skin color in humans is an example of polygenic inheritance. The darkness of the skin is determined by the number of dark skin genes a person inherits from their parents.

A and B show a lack of dominance when they are together in the same individual, but both are dominant to the O allele. These three alleles can be combined as pairs in six different ways resulting in four different phenotypes:

Genotype	Phenotype
$I^A I^A$	blood type A
$I^A i$	blood type A
$I^B I^B$	blood type B
$I^B i$	blood type B
$I^A I^B$	blood type AB
ii	blood type O

Multiple allele problems are worked as a monohybrid problem. Some examples are in the practice problems at the end of this chapter.

Polygenic Inheritance

Thus far we have considered phenotypic characteristics that are determined by alleles at a specific single place on homologous chromosomes. There are, however, some characteristics that are determined by genes at several different loci (on different chromosomes or different places on a single chromosome). This is called **polygenic inheritance.** A number of different pairs of alleles may combine their efforts to determine a characteristic. Skin color in humans is a good example of this inheritance pattern. According to some experts, genes for skin color are located at a minimum of three different loci. At each of these loci, the allele for dark skin is dominant over the allele for light skin. Therefore, a wide variety of skin colors is possible depending on how many dark skin alleles are present (figure 12.4). Polygenic inheritance is very common. Just as in the skin color example, many other characteristics cannot be categorized in terms of "either . . . or. . . ." because there can be such great variation in phenotypes that are the result of polygenic inheritance. People show great variations in height. There are not just tall and short people. There is a wide range with some people being as short as one meter and others being taller than two meters. In addition, intelligence varies significantly from those who are severely retarded to those who are geniuses. In addition, many of these traits are influenced by outside, environmental factors such as diet, disease, accidents, and social factors. These are just some examples of polygenic inheritance patterns.

Pleiotropy

Often a gene can have a variety of effects on the phenotype of the organism. In fact, every gene probably affects or modifies the expression of many different characteristics shown by the organism. This is called pleiotropy. **Pleiotropy** is a term used to describe the multiple effects that a gene may have on the phenotype of an organism. For example, the gene for sickle-cell hemoglobin has two major effects. One is good and one is bad. Having the allele for sickle-cell hemoglobin can result in abnormally shaped red blood corpuscles. This occurs because the hemoglobin molecules are synthesized with the wrong amino acid sequence. These abnormal hemoglobin molecules tend to attach to one another in long, rodlike chains when oxygen is in short supply. These rodlike chains distort the shape of the red blood corpuscle into a sickle shape. When these abnormal red blood corpuscles change shape, they clog small blood vessels. The sickled red corpuscles are destroyed more rapidly than normal corpuscles. This results in a shortage of red blood corpuscles causing anemia and an oxygen deficiency in the tissues that have become clogged. This results in pain, swelling, and damage to organs such as the heart, lungs, brain, and kidneys.

Although sickle-cell anemia is usually lethal in the homozygous condition, it can be beneficial in the heterozygous state. Three genotypes can exist (N = normal hemoglobin, n = sickle-cell hemoglobin):

Genotype	Phenotype
NN	normal hemoglobin and nonresistance to malaria
Nn	normal hemoglobin and resistance to malaria
nn	resistance to malaria but die of sickle-cell anemia

A person with a single sickle-cell allele is more resistant to malaria than a person without this gene. A heterozygous person only has ill effects when there is an increased demand for oxygen. This is likely to occur at high altitudes, during strenuous physical exercise, or under stress.

Sickle-cell anemia was originally found throughout the world in places where malaria was common. Today, however, this genetic disease can be found anywhere in the world. In the United States, it is most common among black populations whose ancestors came from equatorial Africa.

Let's look at another example of pleiotropy. In this example, a single gene affects many different chemical reactions in the way a cell metabolizes the amino acid phenylalanine (figure 12.5). People normally have a gene for the production of an enzyme that converts the amino acid phenylalanine to tyrosine. If this gene is functioning properly, phenylalanine will be converted to tyrosine, which will be available to be converted into thyroxin and melanin by other enzymes. If the enzyme that normally converts phenylalanine to tyrosine is absent, toxic materials can accumulate and result in a loss of nerve cells causing mental retardation. Because less tyrosine is produced, there is also less of the growth hormone, thyroxin; therefore, abnormal body growth results. Because tyrosine is necessary to form the pigment melanin, people who have this condition will be lighter in skin color because of an absence of this pigment. The one abnormal allele produces three different phenotypic effects: mental retardation, abnormal growth, and light skin.

Figure 12.5
Pleiotrophy.

Pleiotrophy is a condition in which a single gene has more than one effect on the phenotype. This diagram shows how the normal pathways work (these are shown in black). If the enzyme phenylalanine hydroxylase is not produced because of an abnormal gene, there are three major results: (1) mental retardation, because phenylpyruvic acid kills nerve cells; (2) abnormal body growth, since less of the growth hormone thyroxine is produced; and (3) pale skin pigmentation, because less melanin is produced (abnormalities are shown in color).

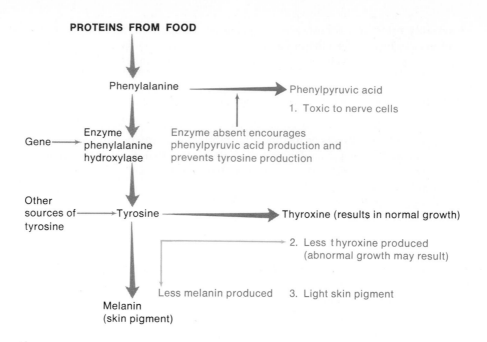

PROTEINS FROM FOOD

Phenylalanine → Phenylpyruvic acid

1. Toxic to nerve cells

Gene → Enzyme phenylalanine hydroxylase

Enzyme absent encourages phenylpyruvic acid production and prevents tyrosine production

Other sources of tyrosine → Tyrosine → Thyroxine (results in normal growth)

2. Less thyroxine produced (abnormal growth may result)

Less melanin produced   3. Light skin pigment

Melanin (skin pigment)

Linkage

Pairs of alleles located on nonhomologous chromosomes separate independently of one another during meiosis when the chromosomes separate into sex cells. Since each chromosome has many genes on it, these genes tend to be inherited as a group (see page 216). Genes located on the same chromosome that tend to be inherited together are called a **linkage group.** The closer two genes are to each other on a chromosome, the more probable it is that they will be inherited together. The process of crossing-over, which occurs during prophase I of meiosis, may split up these linkage groups. Crossing-over happens between the chromosomes donated by the mother and the chromosomes donated by the father, and results in a mixing of genes.

X-Linked Genes

Many organisms have two types of chromosomes (figure 12.6). **Autosomes** are not involved in sex determination and have the same loci on both members of the homologous pair of chromosomes. **Sex chromosomes** are a pair of chromosomes in mammals and some other animals that determine the sex of an organism. In humans and some other animals there are two types of sex chromosomes, the X chromosome and the Y chromosome. Y chromosomes are much shorter than the X chromosomes and probably have no genes for traits found on the X chromosome. One portion of the Y chromosome contains the male-determining genes. Females are produced when two X chromosomes are present. Males are produced when one X chromosome and one Y chromosome are present.

Genes found on the X chromosome are said to be **X-linked.** Because the Y chromosome is shorter than the X chromosome, it does not have any of the alleles that are found on the comparable portion of the X chromosome. Therefore, in men, the presence of a single allele on his only X chromosome will be expressed regardless of whether it is dominant or recessive. In humans, only a few genes have been identified as being found only on the Y chromosomes. One of these is a condition known as "porcupine men." This is a trait found only in males and causes their skin to be

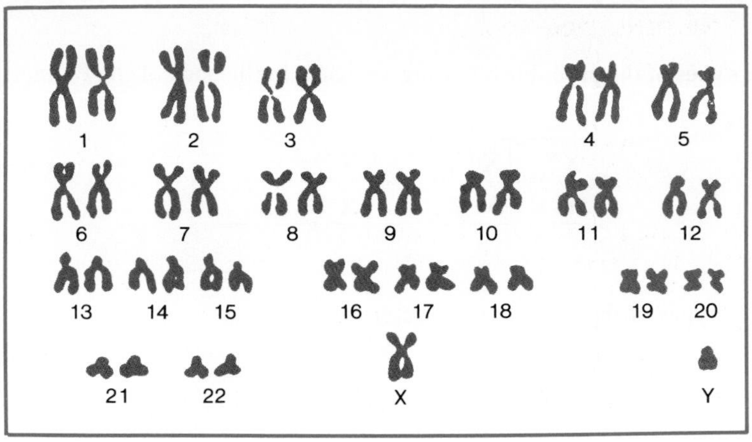

Figure 12.6
Human Chromosomes.

Notice in each of the chromosome pairs from 1 to 22 that both members of the pairs have the same shape; these are autosomes. However, in the X and Y pair of chromosomes, the X chromosome is longer than the Y; these are sex chromosomes.

covered with rough, bristly scales, and inch-long bristlelike outgrowths. Two other genes have been found on the Y chromosome. One is known to cause a web to develop between the second and third toes, while the other causes tufts of hair to grow on the outer edges of the ears. There are over one hundred genes that are linked to the X chromosome. Some of these X-linked genes are color blindness, hemophilia, brown teeth, and a form of muscular dystrophy. To better understand an X-linked gene, let us use the five steps of solving genetics problems and work an X-linked problem.

In humans, the gene for normal color vision is dominant and the gene for color blindness is recessive. Both genes are X-linked. A male who has normal vision mates with a female who is heterozygous for normal color vision. What type of children can they have for these traits, and what is the probability for each type?

Step 1:

Because this condition is linked to the X chromosome, it has become traditional to symbolize the allele as a superscript on the letter X. Since the Y chromosome does not contain a homologous allele, only the initial Y is used.

$$X^N = \text{normal color vision}$$
$$X^n = \text{color-blind}$$
$$Y \ \ = \text{male (no gene present)}$$

Step 2:

Male's genotype $= X^N Y$ (normal color vision)
Female's genotype $= X^N X^n$ (normal color vision)

$$X^N Y \times X^N X^n$$

Step 3:

The genotype of the gametes are listed in the Punnett Square:

	X^N	X^n
X^N		
Y		

Step 4:

The genotypes of the probable offspring are listed in the body of the Punnett Square:

	X^N	X^n
X^N	$X^N X^N$	$X^N X^n$
Y	$X^N Y$	$X^n Y$

Step 5:

The phenotypes of the offspring are determined:

normal female	carrier female
normal male	color-blind male

1/4 normal female
1/4 carrier female
1/4 normal male
1/4 color-blind male

A **carrier** is any individual who is heterozygous for a trait. In this situation, the recessive allele is hidden. In X-linked situations, only the female can be heterozygous since the males lack one of the X chromosomes. Heterozygous females will exhibit the dominant trait (normal vision in this problem), but have the recessive allele hidden. If a male has a recessive allele on the X chromosome, it will be expressed because there is no other allele on the Y chromosome to dominate it. If a heterozygous carrier (must be female) has sons, she should expect one half of them to be color-blind and one half to have normal color vision. For these reasons, there are many more color-blind males than there are color-blind females.

Environmental Influences on Gene Expression

The specific phenotype an organism exhibits is determined by the interplay of the genotype of the individual and the conditions the organism encounters as it develops. Therefore, it is possible for two organisms with identical genotypes (identical twins) to differ in their phenotypes. This is possible because all genes must express themselves through the manufacture of proteins. These proteins may be structural or enzymatic, and the enzymes may be more or less effective depending on the specific biochemical conditions when the enzyme is in operation. The expression of the genes will vary depending on the environmental conditions while the gene is operating.

Maybe you assumed that the dominant gene would always be expressed in a heterozygous individual. It is not so simple! Here, as in other areas of biology, there are exceptions. For example, the allele for six fingers is dominant over the allele for five fingers in humans. Some people who have received the gene for six fingers have a fairly complete sixth finger, while it may vary in others and appear as a little stub. In another case, a dominant gene causes the formation of a little finger that cannot be bent as a normal little finger. However, not all people who are believed to have inherited that allele will have a stiff little finger. In some cases, this dominant characteristic does not show or perhaps only shows on one hand. Therefore, there may be a variation in the degree to which a dominant gene expresses itself, or in some cases it may not even be expressed. Other genes may be interacting with these dominant genes, causing the variation in expression. It is important to recognize that the environment affects the expression of our genes in many ways.

Both internal and external environmental factors can influence the expression of genes. For example, at conception, a male receives genes that will eventually determine the pitch of his voice. However, these genes are expressed differently after puberty. At puberty, male sex hormones are released. This internal environmental change results in the deeper male voice. A male who does not produce these hormones retains a higher-pitched voice in later life. A comparable situation in females occurs when an abnormally functioning adrenal gland causes the release of large amounts of male hormones. This results in a female with a deeper voice.

Many external environmental factors can influence the phenotype of an individual. One such factor is diet. Diabetes mellitus, a metabolic disorder in which glucose is not properly metabolized and is passed out of the body in the urine, has a genetic basis. Some people who have a family history of diabetes are thought to have inherited the trait for this disease. Evidence indicates that they can delay the onset of the disease by reducing the amount of sugar in their diet. This change in the external environment influences gene expression much the same way that temperature influences the expression of color production in cats or sunlight affects the expression of freckles in humans (figure 12.3).

Summary

Genes are units of heredity composed of specific lengths of DNA that determine the characteristics an organism displays. Specific genes are at specific loci on specific chromosomes. The phenotype displayed by an organism is the result of the effect of the environment on the ability of the genes to express themselves. Diploid organisms have two genes for each characteristic. The alternative genes for a characteristic are called alleles. There may be many different alleles for a particular characteristic. Those organisms with two identical alleles are homozygous for a characteristic; those with different alleles are heterozygous for a characteristic. Some alleles are dominant over other alleles that are said to be recessive.

Sometimes two alleles will both express themselves, and often a gene may have more than one recognizable effect on the phenotype of the organism. Some characteristics may be determined by several different pairs of alleles. In humans and some other animals, males have an X chromosome with a normal number of genes and a Y chromosome with fewer genes. Though they are not identical, they behave as a pair of homologous chromosomes. Since the Y chromosome is shorter than the X chromosome and has fewer genes, many of the recessive characteristics present on the X chromosome appear more frequently in males than in females, who have two X chromosomes.

Experience This

Some humans inherit the ability to taste the chemical phenylthiocarbamide (PTC), and others are unable to taste PTC. Ask your instructor to furnish you with a supply of paper impregnated with PTC. Place a piece of this paper on your tongue. If you experience a bitter taste, you are a taster. If you do not, you are a nontaster. Take enough strips to test those members of your family you can readily contact. These may include your grandparents, parents, siblings, children, aunts, and uncles. After you have tested the members of your family, construct a pedigree. Is it possible for you to determine the genotypes of your family members? Is tasting PTC a dominant or a recessive trait? Could it be a case of lack of dominance? Might it be linked to another trait?

In a pedigree, circles represent females and squares represent males. Symbols of parents are connected by a horizontal mating line, and the offspring are shown on a horizontal line below the parents. Individuals who are tasters are represented by a solid symbol, nontasters by an open symbol.

A typical pedigree might look something like this one, which shows three generations.

Fill in your pedigree. You may need to modify this pedigree to fit your particular family situation.

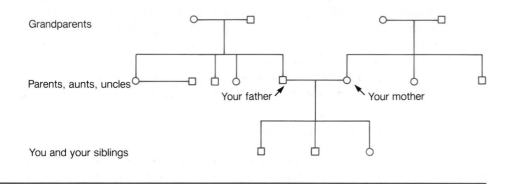

1. How many kinds of gametes are possible with each of the following genotypes?
    a. Aa
    b. AaBB
    c. AaBb
    d. AaBbCc
2. What is the probability of getting the gamete ab from each of the following genotypes?
    a. aabb
    b. Aabb
    c. AaBb
    d. AABb
3. What is the probability of each of the following sets of parents producing the given genotypes in their offspring?

Parents	Offspring Genotype
a. AA × aa	Aa
b. Aa × Aa	Aa
c. Aa × Aa	aa
d. AaBb × AaBB	AABB
e. AaBb × AaBB	AaBb
f. AaBb × AaBb	AABB

4. If an offspring has the genotype Aa, what possible combinations of parental genotypes can exist?
5. In humans, the gene for albinism is recessive to the gene for normal skin pigmentation.
    a. If two heterozygous persons have children, what is the probability that a child will be an albino?
    b. If a child is normal, what is the probability that it is a carrier of the recessive albino gene?

6. In certain pea plants, the gene *T* for tallness is dominant over *t* for shortness.
   a. If a homozygous tall and homozygous short plant are crossed, what will be the phenotype and genotype of the offspring?
   b. If both individuals are heterozygous, what will be the phenotypic and genotypic ratios of the offspring?
7. Smoos are strange animals with one of three shapes: round, cuboidal, or pyramidal. If two cuboidal smoos mate, they always have cuboidal offspring. If two pyramidal smoos mate, they always produce pyramidal offspring. If two round smoos mate, they produce all three kinds of offspring. Assuming only one locus is involved, answer the following questions:
   a. How is smoo shape determined?
   b. What would be the phenotypic ratio if a round and cuboidal smoo were to mate?
8. What is the probability of a child having type AB blood if one of the parents is heterozygous for A blood and the other is heterozygous for B? What other genotypes are possible?
9. A color-blind woman marries a normal vision man. They have ten children, six boys and four girls.
   a. How many are normal?
   b. How many are color-blind?
10. A light-haired man has blood type O, his wife has dark hair and blood type AB, but her father had light hair.
    a. What is the probability that they will have a child with dark hair and blood type A?
    b. What is the probability that they will have a light-haired child with blood type B?
    c. How many different phenotypes could their children show?
11. Certain kinds of cattle have two genes for coat color: *R* = red, and *r* = white. When an individual cow is heterozygous, it is spotted with red and white (roan). When two red genes are present, it is red. When two white genes are present, it is white. The gene H, for lack of horns, is dominant over h, for the presence of horns.
    a. If a bull and a cow both have the genotype RrHh, how many possible phenotypes of offspring can they have?
    b. How probable is each phenotype?
12. Hemophilia is a disease that prevents the blood from clotting normally. It is caused by a recessive gene located on the X chromosome. A boy has the disease; neither his parents nor his grandparents have the disease. What are the genotypes of his parents and grandparents?

Answers

1. a. 2—*A, a*
   b. 2—*AB, aB*
   c. 4—*AB, Ab, aB, ab*
   d. 8—*ABC, ABc, Abc, AbC, aBC, aBc, abC, abc*
2. a. 100%—only *ab* is possible
   b. 50%—*Ab* and *ab* are equally possible
   c. 25%—*AB, Ab, aB*, and *ab* are equally possible
   d. 0%—*ab* not possible
3. a. 100%
   b. 1/2 or 50%
   c. 1/4 or 25%
   d. 1/8 or 12.5%
   e. 1/4 or 25%
   f. 1/16 or 6.25%

4. AA $\times$ aa
   AA $\times$ Aa
   Aa $\times$ Aa
   Aa $\times$ aa
5. a. 1/4 or 25%
   b. 2/3 or 67%
6. a. Tall, Tt
   b. Phenotypic ratio—3 tall to 1 short
      Genotypic ratio—1 homozygous tall, 2
      heterozygous tall, 1 homozygous short
7. a. This is a case of lack of dominance.
   b. Fifty percent or 1/2 round, 50% or 1/2 cuboidal.
8. 1/4 or 25%
   $I^A i$, $I^B i$, ii
9. a. (4) All the girls have the normal phenotype but are carriers.
   b. (6) All the boys are color-blind.
10. a. 1/4 or 25%
    b. 1/4 or 25%
    c. 4
11. a. Six possible phenotypes
    b. Red with horns—RRhh = 1/16 or 6.25%
       Roan with horns—Rrhh = 2/16 or 12.5%
       White with horns—rrhh = 1/16 or 6.25%
       Red, hornless—RrHH or RrHh = 3/16 or 18.75%
       Roan, hornless—RrHH or RrHh = 6/16 or 37.5%
       White, hornless—rrHH or rrHh = 3/16 or 18.75%
12. Father $X^N Y$
    Mother $X^H X^N$
    Mother's father $X^N Y$
    Mother's mother $X^H X^N$
    Father's father $X^N Y$
    Father's mother $X^N X$—

Chapter Glossary

alleles (al′lēls) Alternative forms of a gene for a particular characteristic (e.g., attached earlobe genes and free earlobe genes are alternative alleles for ear shape).

autosomes (aw′to-sōmz) Chromosomes that are not involved in determining the sex of an organism.

carrier (kar′re-er) Any individual having a hidden recessive gene.

dihybrid cross (di-hi′brid kros) A genetic study in which two pairs of alleles are followed from the parental generation to the offspring.

dominant allele (dom′in-ant al′ēl) A gene that expresses itself and masks the effect of other alleles for the trait.

gene (jēn) A unit of heredity located on a chromosome and composed of a sequence of DNA nucleotides.

genetics (jĕ-net-iks) The study of genes, how genes produce characteristics, and how the characteristics are inherited.

genome (je′nōm) A set of all the genes necessary to specify an organism's complete list of characteristics.

genotype (je′no-tīp) The catalog of genes of an organism, whether or not these genes are expressed.

heterozygous (hĕ''ter-o-zi'gus) A diploid organism that has two different allelic forms of a particular gene.

homozygous (ho''mo-zi'gus) A diploid organism that has two identical genes for a particular characteristic.

lack of dominance (lak uv dom'in-ans) The condition of two unlike alleles both expressing themselves, neither being dominant.

law of dominance (law uv dom'in-ans) When an organism has two different alleles for a trait, the allele that is expressed and over-shadows the expression of the other allele is said to be dominant. The gene whose expression is over-shadowed is said to be recessive.

law of independent assortment (law uv in''de-pen'dent ă-sort'ment) Members of one gene pair will separate from each other independently of the members of other gene pairs.

law of segregation (law of seg''re-ga'shun) When gametes are formed by a diploid organism, the alleles that control a trait separate from one another into different gametes, retaining their individuality.

linkage group (lingk'ij grūp) Genes located on the same chromosome that tend to be inherited together.

locus (loci) (lo'kus) (lo'si) The spot on a chromosome where an allele is located.

Mendelian genetics (Men-de'le-an jĕ-net'iks) The pattern of inheriting characteristics that follows the laws formulated by Johann Gregor Mendel.

monohybrid cross (mon''o-hi'brid kros) A genetic study in which a single characteristic is followed from the parental generation to the offspring.

multiple alleles (mul'tĭ-p'l al'lēls) The concept that there are several different forms of genes for a particular characteristic.

offspring (of'spring) Descendants of a set of parents.

phenotype (fēn'o-tip) The physical, chemical, and psychological expression of the genes possessed by an organism.

pleiotropy (pli-ot'ro-pe) The multiple effects that a gene may have in the phenotype of an organism.

polygenic inheritance (pol''e-jen'ik in-her'ĭ-tans) The concept that a number of different pairs of alleles may combine their efforts to determine a characteristic.

probability (prob''a-bil'ĭ-te) The chance that an event will happen expressed as a percent or fraction.

Punnett Square (pun'net sqwār) A method to determine probabilities of gene combinations in a zygote.

recessive allele (re-sĕ'siv al'lēl) An allele that when present with its homologue does not express itself and is masked by the effect of the other allele.

sex chromosomes (seks kro'mo-sōmz) A pair of chromosomes that determine the sex of an organism.

X-linked gene (eks lingt jēn) A gene located on one of the sex-determining chromosomes.

Population Genetics

Chapter Outline

Purpose

This chapter is designed to help you understand why plants and animals of the same kind vary slightly in different parts of the world, and how we artificially maintain certain groups of characteristics in domesticated species. Later chapters on evolution will build on this information. You will notice that there is some overlap between this chapter and later chapters.

For Your Information

One of the forms that prejudice takes is the desire to keep your "ancestral genes clean." Koreans living in Japan are considered "less able" and "unfit" for many higher level jobs in Japan. The South African government attempts to rigidly segregate Indians, blacks, whites, and colored from one another. Many Native Central Americans are being eliminated by having their political and social freedoms denied. In the United States during WW II, Japanese-Americans were "detained" in relocation camps. Blacks and Native Americans have also been discriminated against for decades.

Learning Objectives

- Know the difference between species and population.

- Be able to distinguish between gene pool and deme.

- Describe the occurrence rate of a gene in a population in terms of gene frequency units.

- Relate the concepts of cloning and hybridization to asexual and sexual reproduction.

- Recognize the role of mutation, sexual reproduction, population size, and migration to gene frequency.

- Describe the importance and potential danger in the practice of monoculture.

- Describe the role of a genetic counselor.

Introduction

To understand the principles of genetics in chapter 12, we concerned ourselves with small numbers of organisms having specific genotypes. Plants and animals, however, don't usually exist as isolated individuals, but as members of populations. Before we go any further, we need to define two concepts that are used throughout this chapter, species and population.

A **species** is a group of organisms of the same kind that have the ability to interbreed to produce offspring that are also capable of reproducing. Usually the members of a species look quite similar, although there are some exceptions. For example, all the dogs in the world are of the same species, but a Saint Bernard does not look very much like a Pekingese. Mating can occur between two quite different (in appearance) organisms (figure 13.1). If you examine the chromosomes of reproducing organisms, you find that they are identical in number and size and usually carry very similar groups of genes. In the final analysis, the species concept concerns the genetic similarity of organisms.

The concepts of population and species are interwoven, since a **population** is considered to be all the organisms of the same species found within a specific geographic region. Population, however, is primarily concerned with numbers of organisms. This chapter mixes these two ideas. It deals with populations and why differences in **gene frequencies** (how often a gene occurs in a population) happen. For example, why there are more blue-eyed people in Scandinavia than in Spain, or why some populations have a high frequency of blood type A.

Figure 13.1
Phenotypic Differences in Breeds of Dogs.
Although these dogs look quite different, they all have the same number of chromosomes and have very similar arrangements of genes on the chromosomes.

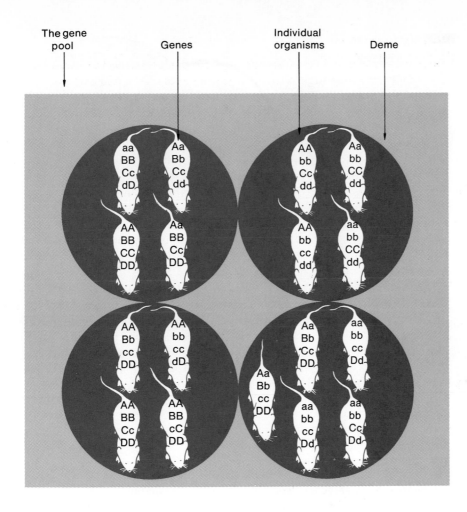

The gene pool | Genes | Individual organisms | Deme

Figure 13.2
Gene Pool.
The gene pool consists of all the genes of all the organisms in the species. Notice that the four demes have different gene frequencies but the same genes (a, b, c, and d).

In the introduction, we related the species concept to genetic similarity; however, the two are not identical. You know that any one organism has a specific genotype consisting of all the genes that organism has in its DNA. It can have a maximum of two alleles for a characteristic at one locus on the homologous chromosomes. In a group of organisms of the same kind, there may be more than two kinds of alleles (multiple alleles) for a specific characteristic. Since, theoretically, all organisms of a species are able to exchange genes, we can think of all the genes of all the individuals of the same species as a giant **gene pool.** Since each individual organism is like a container of a set of these genes, the gene pool contains many more kinds of genes than any one of the individuals. The gene pool is like a refrigerator full of cartons of different kinds of milk—chocolate, regular, skimmed, buttermilk, low fat, and so on. If you were blindfolded and reached in with both hands and grabbed two cartons, you might end up with two chocolate, a skimmed and a regular, or one of the many other possible combinations. The cartons of milk represent different alleles, and the refrigerator (gene pool) contains a greater variety than could be determined by randomly selecting two cartons of milk at a time. Figure 13.2 illustrates how the gene pool and individuals are related.

Figure 13.3a
**Range of Common Water
Snake and Island Water
Snake.**
The common water snake is found
throughout the northeastern part of
the United States. The island water
snake is limited to the islands in
the western section of Lake Erie.

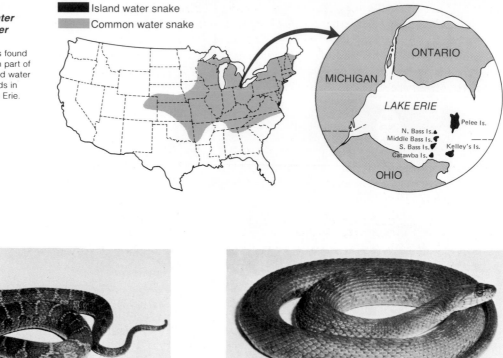

Island water snake
Common water snake

ONTARIO

MICHIGAN

LAKE ERIE

Pelee Is.

N. Bass Is.
Middle Bass Is.
S. Bass Is.
Catawba Is.

Kelley's Is.

OHIO

Common water snake

Island water snake

Figure 13.3b
**Common Water Snake and
Island Water Snake.**

Individuals are usually found in clusters in the pool as a result of several different factors, such as geographic barriers or the availability of resources. Gene clusters, called **demes,** may differ quite a bit from one place to another. There may be differences in the kinds of genes and the numbers of each kind of gene in each of the clusters. Figure 13.2 also indicates the relationship of demes to individuals and the total gene pool concept. Note, for example, that while all the demes contain the same kinds of genes, the frequency of genes *A* and *a* differ from one deme to another.

Since organisms tend to interbreed with other organisms located close by, local collections of genes tend to remain the same unless, in some way, genes are added to or subtracted from this local population (deme). The common water snake is found throughout the eastern portion of the United States (figure 13.3). The island water snake is one of the several demes within this species. This snake is confined to the islands in western Lake Erie. Most common water snakes of the mainland have light and dark bands. The island populations do not have this banded coloration; most of them have genes for solid coloration with very few genes for banded coloration. The island snakes are geographically isolated from the main gene pool, and mate only with one another. Thus, the different colors shown by island snakes and mainland snakes are a result of a difference in the numbers of genes for banded coloration in the different demes.

Genes are repackaged into new individuals from one generation to the next. Often there will be very little adding of new genes or subtracting of other genes from a deme, and a species will consist of a number of more or less separate groups (demes) that are known as **subspecies, races, breeds, strains,** or **varieties.** All these terms are used to describe different forms of organisms that are all members of the same species. However, certain terms are used more frequently than others depending of your field of interest. For example, dog breeders use the term breed, horticulturalists use the term variety, and anthropologists use the term race. The most general and widely accepted term is subspecies.

Gene Frequency

Gene frequency is typically stated in terms of a percentage or decimal fraction and is a mathematical statement of how frequently a particular gene shows up in the sex cells of a population (for example, 10% or 0.1, 50% or 0.5). It is possible for two demes to have all the same genes, but with very different frequencies.

As an example, all humans are of the same species and, therefore, constitute one large gene pool. There are, however, many distinct local populations scattered across the surface of the earth. These more localized populations (races or demes) show many distinguishing characteristics that have been perpetuated from generation to generation. In Africa, dark skin genes, tightly curled hair genes, and flat nose genes have very high frequencies. In Europe, the frequencies of light skin genes, straight hair genes, and narrow nose genes are the highest. Chinese tend to have moderately colored skin, straight hair, and broad noses. All three of these populations have genes for dark skin and light skin, straight hair and curly hair, narrow noses and broad noses. The three differ, however, in the frequencies of these genes (figure 13.4). Many other genes show differences in frequency from one race to

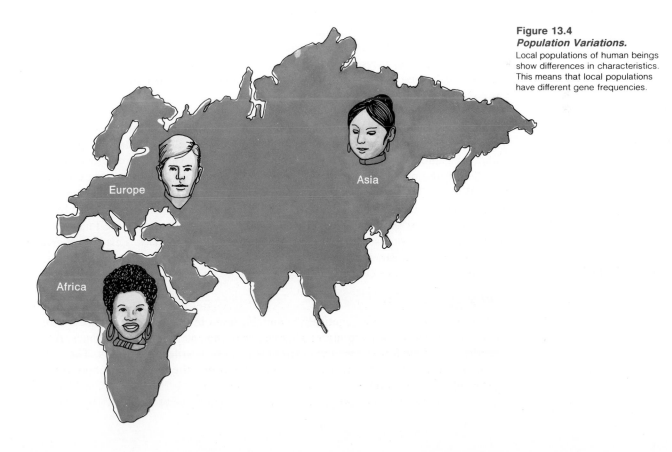

Figure 13.4
Population Variations.
Local populations of human beings show differences in characteristics. This means that local populations have different gene frequencies.

Table 13.1 *Recessive Traits with a High Frequency of Expression*
Many recessive characteristics are extremely common in some human populations. The corresponding dominant characteristic is also shown here.

Recessive	Dominant
Light skin color	Dark skin color
Straight hair	Curly hair
Five fingers	Six fingers
Type O blood	Type A or B blood
Normal hip joints	Dislocated hip birth defect
Blue eyes	Brown eyes
Normal eyelids	Drooping eyelids
No tumor of the retina	Tumor of the retina
Normal fingers	Short fingers
Normal thumb	Extra joint in the thumb
Normal fingers	Webbed fingers
Ability to smell	Inability to smell
Normal tooth number	Extra teeth
Presence of molars	Absence of molars
Normal palate	Cleft palate

another, but these three characteristics are easy to see. Once a particular mixture of genes is present in a population, that mixture tends to maintain itself unless something is operating to change the frequencies. In other words, gene frequencies are not going to change without reason. As transportation has developed, more persons have moved from one geographic area to another, and human gene frequencies have begun to change. Ultimately, as barriers to intermarriage (both geographic and sociological) between races are leveled, the human gene pool will show fewer and fewer racial differences. However, it may be thousands of years before significant changes are seen.

For some reason, people tend to think that gene frequency has something to do with the dominance or recessiveness of genes. This is not true. Often in a population, recessive genes are more frequent than their dominant counterparts. Straight hair, blue eyes, and light skin are all recessive characteristics, yet they are quite common in the population of certain European countries. See table 13.1 for other examples.

What really determines the frequency of a gene is the value that the gene has to the person possessing it. The dark skin genes are very valuable to a person living under the bright sun in tropical Africa. These genes are less valuable to someone living in the less intense sunlight of the cooler European countries. This idea of gene value and what it can do to gene frequencies will be dealt with more fully later.

Why Demes Exist

Since organisms are not genetically identical, some individuals may possess combinations of genes that are more valuable than other combinations of genes. As a result, some individuals find the environment more hostile than others. The individuals with unfavorable combinations of genes leave the population more often, either by death or migration.

Figure 13.5
Blind Cave Fish.
This fish lives in caves where there is no light. The eyes do not function, and the animal has very little color in its skin.

Different environments may select different combinations of genes as being good or bad, and so create separate demes. For example, a blind fish living in a lake is at a severe disadvantage. A blind fish living in a cave where there is no light, however, is not at the same disadvantage. Thus, these two environments might allow or encourage different genes to be present in the two populations (figure 13.5).

A second mechanism, which tends to create small demes with gene frequencies different from other demes, involves the founding of a new population. The collection of genes from a small founding population is likely to be different from the genes present in the larger parent population. After all, they are leaving because they find that environment hostile for one reason or another. Once a small founding population establishes itself, it tends to maintain its collection of genes, since the organisms mate only among themselves. This results in a reshuffling of genes from generation to generation, and does not allow new genes to be introduced into the population. Many species of plants and animals are divided up into quite distinct demes by barriers that prevent movement. Animals and plants that live in lakes tend to be divided into small separate populations by barriers of land. Whenever such barriers exist, there will very likely be differences in the gene frequencies from lake to lake, since each lake was colonized separately and their environments are not identical. The local group of organisms may be called a subspecies, race, variety, or deme. Other species of organisms experience few barriers and, therefore, subspecies are quite rare.

A population of organisms in which there is little genetic variety is likely to display an exceptionally high frequency of unfavorable characteristics or be on the verge of extinction. For example, some dogs have become isolated from the rest of the gene pool as a result of breeders using dogs only within a highly select group. This is usually done in an attempt to increase the frequency of genetically controlled traits that are highly prized by the breeder. However, this inbreeding can result in an increase in the frequency of unfavorable genes. As a result, some breeds have come to display such problems as decreased litter size, hip development problems, and aggressive behavior. In addition, such a stagnant gene pool does not provide new combinations of genes to prevent the death of the species as the environment changes. A large gene pool with a great variety of genes is more likely to contain some genes that will better adapt the organisms to a new environment. A number of mechanisms introduce this necessary variety into a population.

How Variety is Generated

Mutations

Mutations introduce new genes into a population. Gene frequency changes occur rather slowly. All alleles have originated as a result of mutation sometime in the past, and these changed genes have been maintained within the gene pool of the species. If a mutation produces a bad gene, that gene will remain uncommon in the population. While most mutations are bad, very rarely one will occur that is valuable to the organism. For example, at some time in the past, mutations occurred in the populations of certain insects that made them tolerant to the insecticide DDT, even though the chemical had not yet been invented. These alleles were very rare in these insect populations until DDT was invented. Then, these genes became very valuable to the insects that carried them. Since insects that lacked this gene died when they came in contact with DDT, more of the DDT-tolerant insects were left to reproduce the species and, therefore, the DDT-tolerant gene became much more common within these populations.

Sexual Reproduction

The process of *sexual reproduction* also tends to generate new genotypes (collections of genes) when the genes from two individuals mix during fertilization to generate a unique individual. This doesn't directly change the gene pool, but the new member may have a new combination of characteristics so superior to other members of the population that the new member will be much more successful in producing offspring. In a corn population, there may be genes for resistance to corn blight (a fungal disease) and resistance to attack by insects. Corn plants that possess both of these genes are going to be more successful than corn plants that have only one of these genes. They will probably produce more offspring (corn seeds) than the others because they will survive fungal and insect attacks, and will tend to pass on this same combination of resistant genes to their offspring (figure 13.6).

Migration

The *migration* of individuals from one deme to another is also an important way for genes to be added to or subtracted from a population. The extent of migration need not be great. As long as genes are entering or leaving a population, the gene pool will change. Plant breeders have to weed out "wild strains" that accidentally crop up in their fields of select breeds of corn, wheat, and other cereal grains. This is necessary in order to maintain the genetic uniqueness of these special hybrids.

Size of Population

The *size of the population* has a lot to do with how effective any of these mechanisms are at generating variety in a gene pool. The smaller the population, the less genetic variety it can contain. Therefore, migrations, mutations, and accidental death can have great effects on the genetic makeup of a small population. For example, if a town had a population of twenty people and only two had brown eyes and the rest had blue eyes, what happens to those brown-eyed people would be more critical than if the town had twenty thousand people and two thousand had brown eyes. While the ratio of brown eyes to blue eyes is the same in both cases, even a small change in a population of twenty could significantly change the gene frequency of the brown-eyed gene.

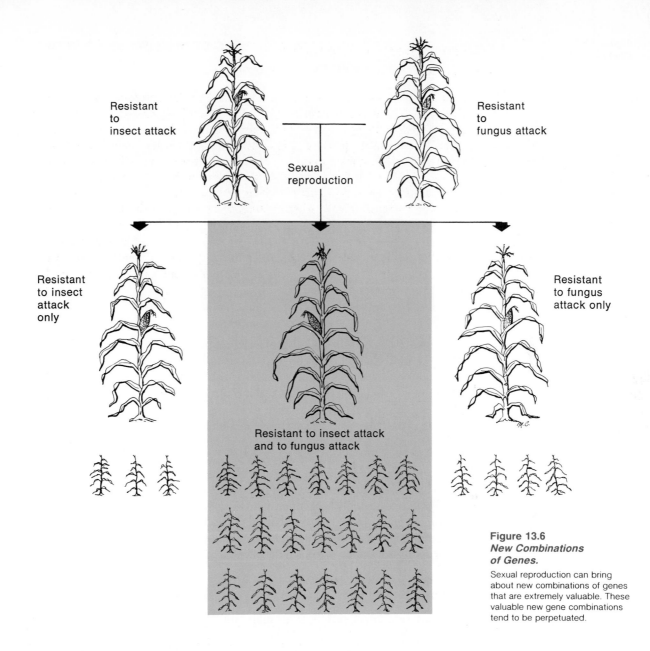

Resistant to insect attack

Resistant to fungus attack

Sexual reproduction

Resistant to insect attack only

Resistant to fungus attack only

Resistant to insect attack and to fungus attack

Figure 13.6
New Combinations of Genes.

Sexual reproduction can bring about new combinations of genes that are extremely valuable. These valuable new gene combinations tend to be perpetuated.

Domesticated Plants and Animals

Humans often work with small select populations of plants and animals in order to artificially construct specific gene combinations that are useful or desirable. This is particularly true of the plants and animals used for food. If we can collect domesticated animals and plants with genes for rapid growth, high reproductive capacity, resistance to disease and other desirable characteristics, we will be better able to supply ourselves with energy in the form of food. Plants are particularly easy to work with in this manner since we can often increase the numbers of specific organisms by asexual reproduction (without sex). Potatoes, apple trees, strawberries, and many other plants can be reproduced by simply cutting the original plant into a number of parts and allowing these parts to sprout roots, stems, and leaves. If a

Figure 13.7
Clones.
The rooting of cuttings is one of
the methods used to make clones
of plants. All of the individuals in a
clone have the same genotype.

single potato has certain desirable characteristics, it may be reproduced asexually. All of the individual plants reproduced asexually have exactly the same genes and are usually referred to as **clones.** Figure 13.7 shows how a clone is developed.

Humans can also bring together specific combinations of genes by selective breeding. This is usually not as easy as cloning. Sexual reproduction tends to mix up genes rather than preserve desirable combinations of genes. If, however, two different demes of the same species—each with particular desirable characteristics—are found, they can be crossed to produce a heterozygous **hybrid** that has the desirable characteristics of both demes. If hybridization is to be of any value, the desirable characteristics in each of the two demes should have homozygous genotypes. It is possible to sexually reproduce a small population until specific characteristics are homozygous. To make two characteristics homozygous in the same individual is more difficult. Therefore, hybrids are developed by crossing two different populations to collect in one organism all of the desirable characteristics (figure 13.8).

Monoculture

A serious side effect results from maintaining specific gene combinations in domesticated plants and animals. The genetic variety within the population is reduced. Most agriculture in the world is based on extensive plantings of the same species over large expanses of land (figure 13.9). This agricultural practice is called **monoculture.** In many cases the plants have been extremely specialized through selective breeding to possess just the qualities that people want. It is certainly easier to manage fields in which there is only one kind of plant growing. This is particularly true today when herbicides, pesticides, and fertilizers are tailored to meet the needs of specific crop species. However, with monoculture comes a significant risk.

Our primary food plants are derived from wild ancestors that possessed combinations of genes that allowed them to compete successfully with other organisms in their environment. However, when humans use selective breeding to increase the frequency of certain desirable genes, certain other genes are lost from the gene pool of our food plants. In effect, these plants are able to live only under conditions that

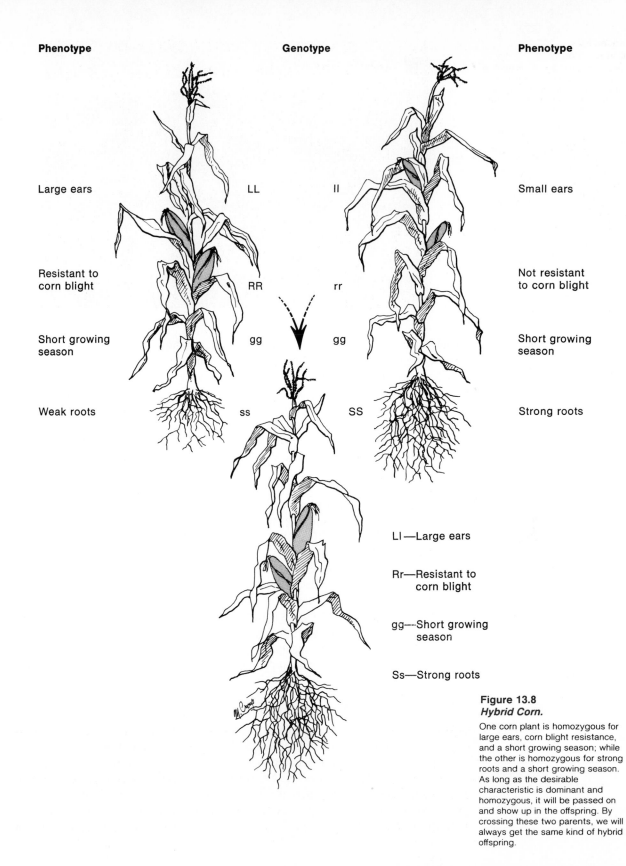

Phenotype	Genotype		Phenotype

Phenotype

Large ears

Resistant to
corn blight

Short growing
season

Weak roots

Genotype

LL   ll

RR   rr

gg   gg

ss   SS

Phenotype

Small ears

Not resistant
to corn blight

Short growing
season

Strong roots

Ll—Large ears

Rr—Resistant to
corn blight

gg—-Short growing
season

Ss—Strong roots

Figure 13.8
Hybrid Corn.

One corn plant is homozygous for
large ears, corn blight resistance,
and a short growing season; while
the other is homozygous for strong
roots and a short growing season.
As long as the desirable
characteristic is dominant and
homozygous, it will be passed on
and show up in the offspring. By
crossing these two parents, we will
always get the same kind of hybrid
offspring.

Figure 13.9
Monoculture.
These large, irrigated fields have all been planted with the same crop. For many reasons this is the most economic way to farm, but it also increases the chance of widespread crop failure.

people carefully maintain. Since these plants are adapted to live under highly controlled conditions and since we plant vast expanses of the same plant, there is tremendous potential for extensive crop loss from diseases.

Whether we are talking about a clone or a hybrid population, there is the danger of the environment changing and affecting the population. Since these organisms are so similar, most of them will be affected in the same way. If the environmental change is a new variety of disease to which the organism is susceptible, the whole population may be killed or severely damaged. Since new diseases do come along, plant and animal breeders are constantly developing new clones, strains, or hybrids that are resistant to the new diseases.

When we select specific good characteristics, we often get bad ones along with them. Therefore, these "special" plants and animals require constant attention. Insecticides, herbicides, cultivation, and irrigation practices are all used to aid the plants and animals that we need to maintain our dominant position in the world.

Another related problem in plant and animal breeding is the tendency of heterozygous organisms to mate and reassemble new combinations of genes by chance from the original heterozygotes. Thus, hybrid organisms must be carefully managed to prevent the formation of gene combinations that would be unacceptable. Since most economically important animals cannot be propagated asexually, the development and maintenance of specific gene combinations in animals is a more difficult undertaking.

Human Genetics

At the beginning of this chapter, we pointed out that the human gene pool consists of a number of categories called races. The particular characteristics that set one race apart from another originated many thousands of years ago before travel was as common as it is today. And yet we still associate certain racial types with certain geographic areas. Though there is much more movement of people and a mixing of racial types, people still tend to have children with others of the same social, racial, and economic background. This nonrandom mate selection can sometimes bring

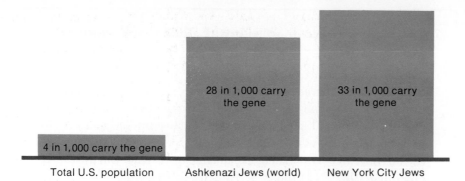

28 in 1,000 carry
the gene

33 in 1,000 carry
the gene

4 in 1,000 carry the gene

Total U.S. population   Ashkenazi Jews (world)   New York City Jews

Frequency of Tay-Sachs gene in three populations

Figure 13.10
Frequency of Tay-Sachs Gene.
The frequency of a gene can vary from one population to another. Genetic counselors use this information to advise people of their chances of having specific genes and of passing them on to their children.

together genes that are relatively rare because they have been common in the ancestral deme from which the specific racial groups of today are descended. The gene frequencies within specific human subpopulations can be very important to anyone who wishes to know the probability of having children with particular combinations of genes, particularly if the gene combination is bad. Tay-Sachs disease causes degeneration of the nervous system and the early death of children. Since it is caused by a recessive gene, both parents must pass the gene to their child in order for the child to have the disease. By knowing the frequency of the gene in the background of both parents, we can determine the probability of their having a child with this disease. Persons of the Ashkenazi Jewish background have a higher frequency of this recessive gene than do persons of any other group of racial or social origin (figure 13.10). Therefore, people of this particular background should be aware of the probability that they can have children who will develop Tay-Sachs disease. Likewise, sickle-cell anemia is more common in people of specific African ancestry than in any other human deme. These people should be aware that they might be carrying the gene for this type of defective hemoglobin. If they are, they should ask themselves, how likely is it that I will have a child with this disease? (figure 13.11). These and other cases make it very important that trained **genetic counselors** have information about the frequencies of genes in specific human demes, so they can help couples with genetic questions.

Genetics, Evolution, and Public Concern

Many current issues related to genetics and evolution are matters of public concern. These are often not easy concepts for a layperson to understand. When there is misunderstanding, there is often mistrust.

Some History

Modern genetics had its start in 1900 with the rediscovery of the fundamental laws of inheritance proposed by Mendel. For the next forty or fifty years, this rather simple understanding of genetics resulted in unreasonable expectations on the part of both scientists and laypeople. People generally assumed that much of what a person was in terms of structure, intelligence, and behavior was inherited. This led to the passage of **eugenics laws.** Their basic purpose was to eliminate "bad" genes from the human gene pool and encourage "good" gene combinations. These laws took a form that prevented the marriage of or permitted the sterilization of persons who were "known" to have "bad" genes (figure 13.12).

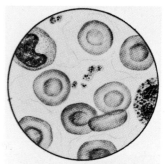

a

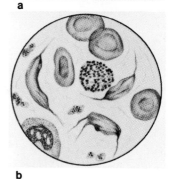

b

Figure 13.11
Normal and Abnormal Red Blood Corpuscles.
Sickle-cell anemia is caused by a recessive allele that slightly changes the hemoglobin molecules of the red blood corpuscles. As a result, the red blood corpuscles (a) will become sickle shaped (b) if they are deprived of oxygen.

> ### 720.301   Sterilization of mental defectives; statement of policy
>
> Sec. 1. It is hereby declared to be the policy of the state to prevent the procreation and increase in number of feeble-minded and insane persons, idiots, imbeciles, moral degenerates and sexual perverts, likely to become a menace to society or wards of the state. The provisions of this act are to be liberally construed to accomplish this purpose. As amended 1962, No. 106, § 1, Eff. March 28, 1963.

These laws overlooked the fact that many genetic abnormalities are caused by recessive genes. In most cases, the symptoms of these "bad" genes can be recognized only in homozygous individuals. Removing only the homozygous individuals from the gene pool would have little influence on the frequency of the "bad" genes in the population. There would be many "bad" genes masked by dominant alleles in heterozygous individuals, and these genes would continue to show up in future generations.

Often these laws were thought to save money, since sterilization would prevent the birth of future "defectives" and, therefore, reduce the need for expensive mental institutions or prisons. These laws were also used by people to legitimize racism and promote prejudice.

Genetic Disorders

Today, genetic diseases and the degree to which behavioral characteristics and intelligence are inherited are still important social and political issues. The emphasis, however, is on determining the specific method of inheritance or the specific biochemical pathways that result in what we currently label as insanity, lack of intelligence, or antisocial behavior. Although progress is rather slow, several genetic abnormalities have been "cured," or at least made tolerable, by medicines or control of the diet (figure 13.13).

Today, effective genetic counseling has become the preferred method of dealing with genetic abnormalities. A person known to be a carrier of a "bad" gene can be told the likelihood of passing that characteristic on to the next generation before deciding whether or not to have children. In addition, **amniocentesis** (a medical procedure that samples amniotic fluid) and other tests make it possible to diagnose some genetic abnormalities early in pregnancy. If an abnormality is diagnosed, an abortion can be performed. Since abortion is unacceptable to some people, the counseling process must include a discussion of the acceptability of abortion or alternatives.

Figure 13.13
Phenylketonuria (PKU).

This genetic disease is caused by an abnormal biochemical pathway. If children with this condition are allowed to eat foods containing the amino acid phenylalanine, they will become mentally retarded. However, if the amino acid phenylalanine is excluded from the diet and certain other dietary adjustments are made, the person will develop normally. Since NutraSweet is a phenylalanine based sweetener, people with this genetic disorder must use caution when buying such products. This abnormality can be diagnosed very easily with a simple test of the urine of newborn infants.

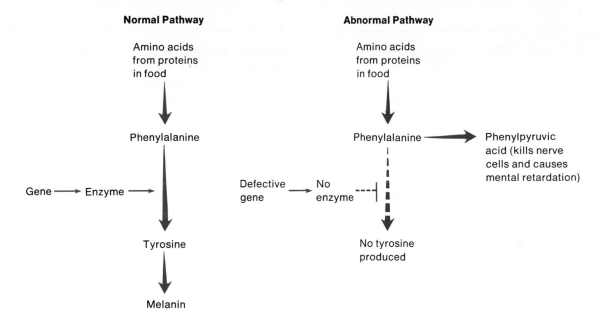

Summary

The gene pool is the collection of all the genes of a population. Organisms that can produce offspring that can also reproduce are members of the same species. Sub-populations (demes) may have different gene frequencies even though they are members of the same gene pool. The gene frequencies of a deme will stay the same if mutation does not occur, if the population is large, if random mating occurs, and if no organism migrates into or out of the deme. New variety in the gene pool can be generated by mutation, sexual reproduction, and migration of organisms into or out of the population. Knowledge of population genetics is useful for plant and animal breeders as well as for people who specialize in genetic counseling. Breeders can maintain certain desirable gene frequencies in a population if they understand the factors that control changes in gene frequencies. The development of clones and hybrid organisms are two examples of how population genetics is used. The genetic counselor cannot control human populations the way plant and animal breeders do, but can use gene frequency information to determine the probability of abnormal children being born. This helps couples make decisions about whether or not to have children.

Albinism is a trait caused by a recessive gene. Could this gene be eliminated from the human population? Include the following in your consideration: causes of mutations, how genes result in characteristics, gene frequency, and inheritance patterns.

Experience This

The next time you visit your local library, ask the reference librarian to refer you to the section that deals with the law. Take some time and see if you can locate laws or local ordinances that have a eugenic component to them. Look up topics that might deal with sterilization, conception control, fertility drugs, medicines that help select genetically ''unfit'' to survive and therefore reproduce, and so on.

Questions

1. How does the size of a population affect the gene pool?
2. List three factors that change gene frequencies in a population.
3. Why do races or subspecies develop?
4. Give an example of a gene pool containing a number of separate demes.
5. How is a clone developed? What are its benefits?
6. How is a hybrid formed? What are its benefits?
7. What forces maintain racial differences in the human gene pool?
8. How do the concepts of species, deme, and population differ?
9. What is meant by the term *gene frequency?*
10. How do the gene frequencies in clones and normal reproducing populations differ?

Chapter Glossary

amniocentesis (am''ne-o-sen-te'sis) A medical procedure that samples amniotic fluid. Fetal cells are examined to see if they are normal.

clones (klōnz) All of the individuals reproduced asexually that have exactly the same genes.

deme (dēm) A local recognizable population that differs in gene frequencies from other local populations of the same species. (Also see subspecies.)

eugenics laws (yu-jen'iks laws) Laws designed to eliminate ''bad'' genes from the human gene pool and encourage ''good'' gene combinations.

gene frequency (jēn fre'kwen-se) The measure of the number of times that a gene occurs in a population. The percentage of sex cells that contain a particular gene.

gene pool (jēn pool) All the genes of all the individuals of the same species.

genetic counselor (jĕ-ne'tik kown'sel-or) A professional biologist with specific training in human genetics.

hybrid (hi'brid) The offspring of two different genetic lines produced by sexual reproduction.

monoculture (mon''o-kul'chur) The agricultural practice of planting the same species over large expanses of land.

population (pop''u-la'shun) All the organisms of the same species found within a specified geographic region.

species (spe-shēz) A group of organisms of the same kind that have the ability to interbreed to produce offspring that are also capable of reproducing.

subspecies (**races, breeds, strains,** or **varieties**) (sub'spe-shēs) A number of more or less separate groups (demes) within the same gene pool. These groups differ from one another in gene frequency.

Evolution and Ecology

rganisms do not live alone. All organisms are members of populations that interact with other populations and their nonliving environment. Exactly when, how, and to what extent they interact influences their ability to survive. Organisms channel energy and materials through the environment, and interact with members of their own and other populations. In an attempt to maintain the species in a changing world, all populations experience genetic change over generations of time. These changes may be small or they may be extensive enough to result in the production of organisms that are so different from their ancestors that they form a new species. Understanding what factors play roles in this type of long-term population change enables biologists to predict and control them. Such efforts, we hope, will enhance the survival of organisms that are on the verge of extinction and possibly have value for our own species. ■

Chapter Outline

Purpose

Previous chapters have presented background in the areas of chemistry, information systems (DNA), sexual reproduction, heredity, and population genetics. These are all closely related to one another and are also related to the surroundings of organisms. Since the surroundings are always changing, the survival of living things depends on their ability to adjust their processes to these changes. The changes that assure survival can occur to any individual in the population, but unless they are genetically (DNA) determined and transmitted to the next generation, they will be of little value to the survival of the species. It is the purpose of this chapter to identify the ways differences come about and how these differences may change a sexually reproducing species over thousands of generations.

For Your Information

News item, 16 May 1986
Gonorrhea Resists Drugs

A new drug-resistant gonorrhea strain is rapidly spreading from coast to coast. Federal health researchers warn that the future could bring a strain resistant to every antibiotic on the market.

Since tetracycline-resistant gonorrhea was identified fourteen months ago, there have been seventy-nine cases of this antibiotic-resistant strain of the disease reported, according to the National Centers for Disease Control, Atlanta, Georgia.

Three of the cases were resistant not only to tetracycline, the second line defense against gonorrhea, but also to penicillin, the drug of choice for the sexually transmitted disease.

Those cases were treated with the drug ceftizoxime, but it is possible that the nation could later see a strain of gonorrhea that cannot be treated with any of the antibiotics available, said an official of the Centers for Disease Control.

Learning Objectives

- Understand the role of sexual reproduction in the process of natural selection.

- Recognize that gene information can be hidden if the gene is recessive.

- Recognize that acquired characteristics cannot be inherited.

- State events that generate variety in the gene pool.

- Understand the conditions necessary for the Hardy-Weinberg law to be valid.

- Know that differential reproduction is the key to natural selection.

- Understand why small subpopulations may have gene frequencies that differ from the total population.

Sexual Reproduction and Evolution

Each individual entering a population carries a unique set of genetic information. It is unique for two reasons. First, **spontaneous mutations,** which are natural changes in the DNA caused by unidentified environmental factors, have occurred in DNA. The causes of these spontaneous mutations have not been identified specifically, but it is thought that naturally occurring cosmic radiation and some naturally occurring chemical mutation-causing agents may be the cause. It is known that subjecting organisms to radiation or certain chemicals increases the mutation rate. Therefore, special safeguards are taken to protect persons who work in close contact with mutagenic agents. The safeguards include radiation badges to record exposure to radiation so that persons can regulate their exposure levels and special training and equipment for those people working with mutagenic chemicals (figure 14.1).

Figure 14.1
Protection From Mutagenic Agents.
Because radiation and certain chemicals increase the likelihood of mutations, those people who work with such things receive special training and use protective measures to reduce their exposure to mutagenic agents.

The cells in which mutations have the greatest importance are the sex cells, since they will provide the genetic information for the next generation. Spontaneous mutations in humans occur at an estimated rate of about one gene change in one hundred thousand genes. The total number of genes within the 46 chromosomes of a human can be reasonably estimated at two hundred thousand. This means that persons are likely to be carrying two mutated genes that were generated within their bodies, and were not transmitted to them by their parents. These new genes, however, may "show themselves" in the offspring as new characteristics or a minor change in the frequency of a characteristic. Most of these mutations will be bad for the organism, but occasionally a mutation occurs that is beneficial and gives the organism some improved ability or structure.

The second reason for the genetic uniqueness of each individual has to do with the processes of meiosis and fertilization. You will recall from chapter 10 that during meiosis, variety may be generated in gametes through the crossing-over and the independent assortment of chromosomes. Both of these processes bring about a reshuffling of genes prior to their being dealt to sperm or eggs. The exact combination of genes going into the offspring (next generation) at fertilization differs from that found in the parent. Thousands of unique gene packages in the form of gametes can be produced by any individual during meiosis as a result of crossing-over and independent assortment.

When fertilization occurs, genes donated by each of the parents are recombined into a single cell called the zygote. This zygote is unique since the gametes have new genes because of mutation and new gene combinations due to crossing-over and independent assortment. Furthermore, the zygote is the product of these processes occurring in two different parent organisms. The sum total of all the gene mixing that occurs during sexual reproduction is known as **genetic recombination.** This new individual has a **genome**—a complete set of genes—that is different from any other organism that has ever existed.

The Nature of Populations

Observations made by scientists and laypeople throughout history have confirmed the uniqueness of each individual. Another basic truth is that sexually reproducing organisms tend to overreproduce. They have the ability to reproduce far more offspring than are necessary to just replace the parents. For example, geese have a life span of about ten years, and on the average, a single pair raises a brood of about eight young each year. If these two parent birds and all their offspring were to survive and reproduce at this same rate for a ten-year period, there would be a total of 19,531,250 birds in the family (figure 14.2)!

But the size of most populations remains relatively constant over long periods of time. Minor changes in number may occur, but if the species is living in harmony with its environment, it does not experience such dramatic increases in population size. The population reproduces at the same rate at which death occurs. But don't think of this as a "static population." Although the total number of organisms of the species remains relatively constant, the makeup of the population still changes. In fact, to maintain itself in an ever-changing environment, the species must change in ways that better suit it to the new environment. For this to occur, members of the population must be eliminated in a nonrandom manner. Those that survive are those that are, for the most part, better suited to the environment. They reproduce more of their kind and transmit more of their genes to the next generation than do individuals who have genes that do not allow them to be as well adapted to the environment in which they live.

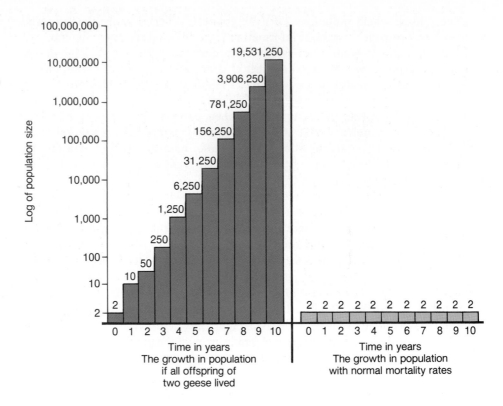

Figure 14.2
Potential Reproductive Ability and Realized Population.

All organisms have a reproductive capacity that is much greater than that needed just to replace the parents. Mortality in the young is often high; therefore, most populations remain relatively stable.

Log of population size

19,531,250
3,906,250
781,250
156,250
31,250
6,250
1,250
250
50
10
2

Time in years
The growth in population if all offspring of two geese lived

2 2 2 2 2 2 2 2 2 2 2

Time in years
The growth in population with normal mortality rates

Gene Concealment

Since the environment influences the survival and reproduction of the individual organisms within a population, it also causes changes in the gene pool of the species. Not all surviving organisms contribute equally to this creative change in the gene pool. An organism with a "good" gene won't necessarily survive to reproduce, since that organism may also contain "bad" genes and since genes do not always show themselves in the phenotype of the individual. Genes may remain concealed because they do not express themselves during the life of the individual or because their effects are overshadowed by another more powerful gene (figure 14.3).

Genes may not express themselves for many different reasons. Some genes only express themselves during a specific period in the life of the organism, so if the organism dies before the gene expresses itself, we do not know if it was there. In addition, many genes need an environmental trigger to initiate their expression. If the trigger is not encountered, the gene does not express itself.

Many genes are recessive and must be present in a homozygous condition before they will express themselves. If a recessive gene is combined with a dominant gene, only the dominant gene expresses itself and you do not know that the recessive gene is present. During sex cell formation, this recessive gene separates from its dominant allele and can recombine with another of its type at fertilization. This newly formed homozygous offspring may then show the recessive characteristic, and the new feature may be one that better adapts the organism to its environment. Thus, many genes are carried unseen within individuals of a population. These genes show up in later generations, allowing certain individuals to survive and reproduce and result in a population better able to cope with an environment that is in a constant state of change. Finally, "good" genes may occur in the population and show themselves to the environment, but are lost through random, accidental death.

Genotype	CC	Cc	cc
Phenotype	Pigmented	Pigmented	No pigment

Acquired Characteristics

Many organisms survive because they have characteristics that are not genetically determined. These **acquired characteristics** are gained during the life of the organism, are not caused by genes and, therefore, cannot be passed on to future generations during reproduction. Consider an excellent tennis player's skill. This ability is acquired through practice, not genes. An excellent tennis player's offspring will not "inherently" be excellent tennis players. Although they may inherit some of the genetically determined physical characteristics necessary to become an excellent tennis player, the skills are still acquired through practice (figure 14.4).

We often desire a specific set of characteristics in our domesticated animals. For example, the breed of dog known as boxers are "supposed" to have short tails. However, the genes for short tails are not present in the gene pool. Consequently, the tails of these dogs are amputated, a process called docking. Similarly the tails of lambs are also usually amputated. These acquired characteristics are not passed on to the next generation. Removing the tails of these animals does not remove the genes for tail production from their genomes.

Theory of Natural Selection

In a species population, we see that not all individuals are identical and some have advantages over others. Those individuals with the advantageous characteristics are better able to survive. They are also likely to produce greater numbers of offspring than those without the favorable characteristics. This idea—that the individuals who have the genes that make them better adapted to their surroundings are more

Figure 14.3
Gene Expression.
Genes must be expressed to allow the environment to select for or against them. The recessive gene *c* for albinism only shows itself in those individuals who are homozygous for the recessive characteristic. The characteristic is absent in those who are homozygous dominant and is hidden in those who are heterozygous.

Figure 14.4
Acquired Characteristics.

The ability of a person to play an outstanding game of tennis is learned by long hours of practice. The tennis skills this person acquires by practice cannot be passed on to her offspring.

likely to have higher survival rates and produce more offspring—is known as the **theory of natural selection.** It was first proposed by Charles Darwin and Alfred Wallace and was clearly set forth in 1859 by Darwin in his book *On the Origin of Species by Means of Natural Selection, or the Preservation of Favored Races in the Struggle for Life.* This title needs clarification on two points. First, **natural selection** and **differential reproduction** both mean that those individuals who have the genes that best fit them for their environment will have more offspring than other individuals. Therefore, the favorable gene will become increasingly common in future generations of the species.

Second, the phrase *struggle for life* does not necessarily refer to open conflict and fighting. It can be much more subtle than that. When a resource is in short supply—such as nesting materials, water, sunlight, or food—those organisms best able to deal with the stress will be most successful. The struggle is not obvious, but some survive the stress and some do not. Some individuals have few offspring and others have more. As an example of natural selection in action, consider what has happened to many insect populations as we have subjected them to a variety of kinds of insecticides. Since there is genetic variety within all species of insects, when an insecticide is used for the first time on a particular species, it kills all the individuals that are genetically susceptible. However, there are individuals with slightly different genetic compositions that may not be killed by the insecticide.

Suppose that in a population of a particular species of insect, 5% of the individuals possess genes that make them resistant to a specific insecticide. The first application of the insecticide could, therefore, kill 95% of the population. However, the tolerant individuals would then constitute the majority of the breeding population that has survived. This would mean that many insects in the second generation would be tolerant. The second use of the insecticide on this population would not be as effective as the first. Many species of insects will produce a new generation each month. If the insecticide continues to be used, each generation will become

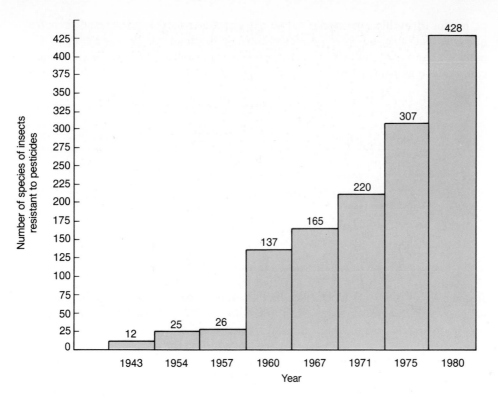

Figure 14.5
Resistance to Insecticides.
The continued use of insecticides has constantly selected for genes that give the insect resistance to a particular insecticide. As a result, many species of insects are now resistant to many kinds of insecticides and the number continues to increase.

more tolerant of the insecticide. In organisms with a short generation time, 99% of the population can become resistant to the insecticide in as little as five years. As a result, the insecticide is no longer useful as a control of the species. As a new factor (the insecticide) was introduced into the environment of the insect, natural selection occurred resulting in a population that was tolerant of the insecticide. Figure 14.5 indicates that over four hundred species of insects have populations that are resistant to many kinds of insecticides.

Gene Frequency Studies

The process of natural selection results in a slow but steady change in the frequency of occurrence of certain genes in a population. Thus, over generations, some genes increase in number while others decrease. This can be seen if we return to the gene pool concept presented in chapter 13 and deal with one pair of alleles as an example. By applying the Punnett Square method to the entire gene pool—not simply two mating individuals—we can determine the percentage of the population that is homozygous or heterozygous for a characteristic. Consider a gene pool composed of two alleles, *A* and *a*. In this hypothetical gene pool, we cannot know which individuals are male or female, nor can we know their genotypes. Therefore, we must guess which individuals will mate. For example, suppose that the *A* gene frequency were 60% (0.6), and the *a* gene frequency were 40% (0.4). There are only two alleles, and they make up 100% of the gene population. What possible genotypes can be produced in this gene pool? To find the answer, treat these genes and their frequencies as if they were individual genes being distributed into sperm or eggs. The males of the population can produce gametes with either A or a and the females can produce gametes with either A or a. Set this up in a Punnett Square as follows:

Possible female gametes

		$A = 0.60$	$a = 0.40$
	$A = 0.60$	Genotype of Offspring $AA = 0.60 \times 0.60$ $= 0.36 = 36\%$	Genotype of Offspring $Aa = 0.60 \times 0.40$ $= 0.24 = 24\%$
Possible Male Gametes	$a = 0.40$	Genotype of Offspring $Aa = 0.40 \times 0.60$ $= 0.24 = 24\%$	Genotype of Offspring $aa = 0.40 \times 0.40$ $= 0.16 = 16\%$

Inside each of the boxes is the frequency of occurrence of each of the three possible genotypes in this population: $AA = 36\%$, $Aa = 48\%$, and $aa = 16\%$. Sexual reproduction of this group results in what frequency of A and a in the population? To determine this, use the frequency figures for each of the genotypes in this parent generation. Organisms with the homozygous dominant genotype, AA, make up 36% of the total gene pool. They will contribute 36% of the A genes to the next generation in their gametes. Organisms with the homozygous recessive genotype, aa, make up 16% of the total gene pool. They will contribute 16% of the a genes to the next generation in their gametes. Those with the heterozygous genotype Aa make up 48% of the population and can contribute either the A or the a gene to the next generation. This means that half of their gametes (24% of all the gametes produced by all the members of the gene pool) will contain the A gene, while the remaining 24% will contain the a gene. The total frequency of each type of gene being donated to the next generation is as follows:

$$36\%A + 24\%A = 60\%A$$

and

$$16\%a + 24\%a = 40\%a.$$

If fertilization occurs in the four possible ways indicated above, *the resultant genotypes in the next generation will occur in the exact same frequencies as in the parent generation.*

At least on paper, gene frequencies do not change generation after generation provided certain assumptions are true. On paper it is possible to: (1) mate all organisms totally at random; (2) eliminate changes in gene frequency due to mutation; (3) eliminate the gain or loss of genes due to migration in or out of the population; and finally (4) deal with large enough numbers of individuals so that gene frequencies do not change because of accidental losses of a few organisms. These four conditions must exist for gene frequencies to remain unchanged. Now let us look at these four statements more closely. These four conditions and the resulting stability in gene frequencies are known as the **Hardy-Weinberg law.**

Box 14.1
The Voyage of HMS *Beagle*, 1831–1836

Probably the most significant event in Charles Darwin's life was his appointment in 1831 as naturalist on the British survey ship *HMS Beagle*. Surveys were common at this time, helping to refine maps and chart hazards to shipping. Darwin was just twenty-two years old at the time and probably would not have gotten the appointment had his uncle not persuaded Darwin's father to allow him to go. The post of naturalist was not a paid position.

The voyage of the *Beagle* lasted nearly five years. During the trip, the ship visited South America, the Galápagos Islands, Australia, and many Pacific Islands. Although he suffered greatly from seasickness, maybe because of it, he took a great interest in all forms of natural history and made extensive journeys by mule and on foot some distance inland from wherever the *Beagle* happened to be at anchor. His experience was unique for a man so young and very difficult to duplicate because of the slow methods of travel. Although many people had seen the places that Darwin visited, he was a student of nature and collected volumes of information on these places.

Most other people who had visited these far away places were not trained to recognize the significance of what they saw. Darwin's notebooks included information on plants, animals, rocks, geography, climate, and the native peoples he encountered on his many travels. The natural history notes he took during the voyage served as a vast storehouse of information, which he used in his writings during the rest of his life.

Since he was wealthy, he did not need to work to earn a living and could devote a good deal of his time to the further study of natural history and the analysis of his notes. He was a semi-invalid during much of his later life. Many people feel this was caused by a tropical disease he contracted during the voyage of the *Beagle*. As a result of his experiences, he wrote books on the formation of coral reefs, how volcanoes might have been involved in their formation, and, finally, the *Origin of Species*. This last book, written twenty-three years after his return from the voyage, changed biological thinking for all time.

The Voyage of HMS *Beagle*, 1831–1836

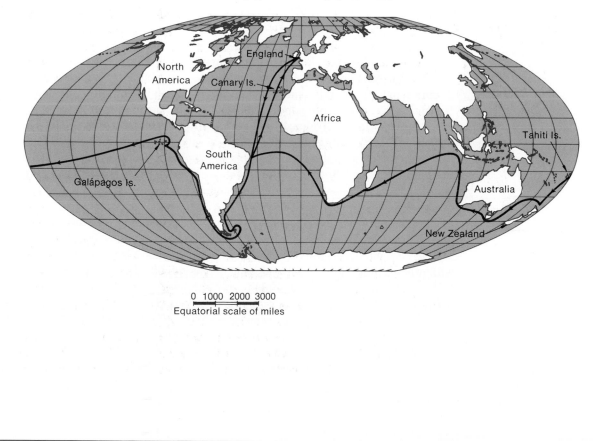

0 1000 2000 3000
Equatorial scale of miles

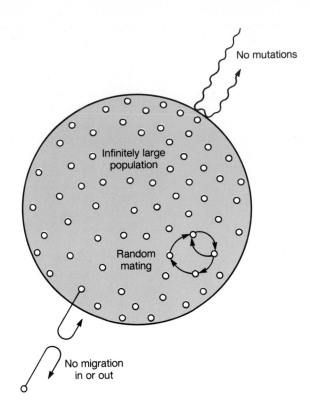

Figure 14.6
The Hardy-Weinberg Law.
The assumptions of the Hardy-Weinberg Law are that the population is infinitely large, has random mating, has no mutations, and has no migration.

No mutations

Infinitely large population

Random mating

No migration in or out

Hardy-Weinberg Law

This law states that populations of organisms will maintain the same gene frequencies from generation to generation if mating is random, if the population is large, if mutation does not occur, and if there is no migration into or out of the population. It appears that changes in populations would *not* occur (contrary to what was proposed by Darwin and Wallace) if these conditions held true in real life. This law holds comparable status to Mendel's laws of independent assortment and segregation (see chapter 12). But what happens on paper and what happens in real life are two different things! Both Hardy and Weinberg realized this and, therefore, this law is used only as a base from which to understand the reasons for gene frequency changes in gene pools (figure 14.6).

Return to the gene pool concept and you can understand why the four assumptions of the Hardy-Weinberg law do not apply to real populations. First of all, random mating does not occur. Many segments of a gene pool are isolated and do not mate with other segments during the lifetime of the individuals. In human populations, these isolations may be geographic, political, or social. Therefore, the Hardy-Weinberg law is not valid because nonrandom mating is a factor that leads to differential reproduction and natural selection (figure 14.7a). Second, you will recall that DNA is constantly being changed (mutated) spontaneously. The *A* and *a* genes will likely undergo other changes. They may change to totally new kinds, such as *a'* or *a''*, or *a* may change back to *A*. All of these mutations would automatically change the frequency of genes in the gene pool (figure 14.7b).

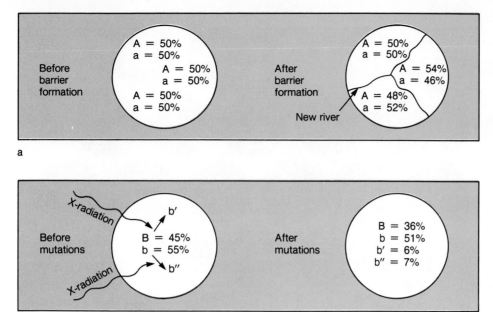

a

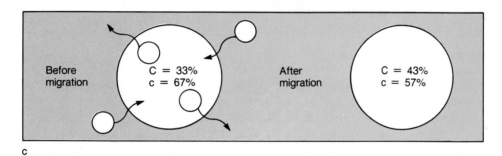

b

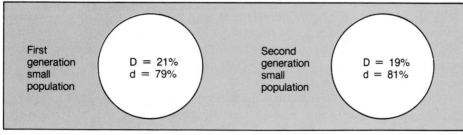

c

Figure 14.7
The Causes of Gene Frequency Changes.
When barriers prohibit totally random mating, gene frequencies may change in the isolated populations. Mutations result in the conversion of one gene into another and, therefore, result in gene frequency changes. When individuals enter or leave a population, they carry their individual genes with them and consequently can change the frequency of genes in a population. Small populations can have random gene frequency changes.

d

Third, environmental conditions may encourage immigration or emigration of individual organisms; thus, changing the frequency of genes within a particular deme. In many parts of the world, severe weather disturbances have lifted animals and plants and moved them over great distances, isolating them from their original gene pool (figure 14.7c). The island of Krakatoa was blown to bits in 1883 by a volcanic explosion. For two months it remained so hot that the rain that fell on the remaining island turned into steam. Essentially, all life was eliminated from the island. The

Figure 14.8
Probability.

If you flip two coins at the same
time, there are three possible ways
they might turn up. However, if you
flip four or five coins, there are
several different combinations that
are possible and it normally
requires large numbers of tosses
to approach the expected 50/50
ratio.

nearest possible source of new organisms was the island of Java forty kilometers
away. Yet in only one year following this disaster, plants were found growing on
Krakatoa, and by 1908, two hundred species of animals had established new populations on the island as they migrated from neighboring islands.

The final assumption of the Hardy-Weinberg law is that the population is infinitely large. If numbers are small, random events might give results that are quite
different from the expected statistical results. Take coin flipping as an analogy. If
you flip a coin once, there is a fifty-fifty (50:50) chance that the coin will turn up
heads. If you flip two coins, you may come up with two heads, two tails, or one head
and one tail. To come closer to the statistical probability of flipping 50% heads, 50%
tails, you would need to flip many coins at the same time. The more coins you flip,
the more likely you will end up with 50% of all coins showing heads and the other
50% showing tails (figure 14.8). The same is true of gene frequencies.

Gene frequency differences that result from chance are more likely to occur in
small populations rather than in large populations (fig. 14.7d). An example of this
kind of frequency difference occurs in a Pennsylvania settlement of the German
Baptist Brethren or Dunkers.

Because this group and other similar groups are socially and reproductively
isolated from the rest of the American population, many of their gene frequencies
can be maintained differently from the American population as a whole. For example, "hitchhiker's thumb" is the ability to bend the thumb backwards so that it
points toward the elbow (figure 14.9). The frequency of this gene in the general
population is 0.496, while the gene frequency in the Dunker population is 0.410.
This means that one-half of the individuals in the population at large will carry the
gene and about one-quarter will exhibit this recessive trait, while only about 40%
of the Dunker population carry the gene and only about 17% show it.

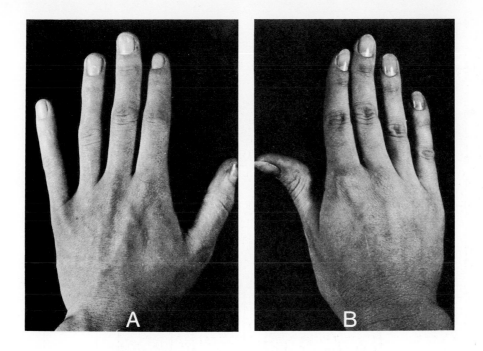

Table 14.1 *Differential Mortality*

The percentage of each genotype in the offspring differs from the genotype in the
original population as a result of differential reproduction.

Original Genes Frequencies and Genotypes	Total Number of Persons within Population 100,000	Number Lost Due to 50% Death	Total of Each Genotype in Reproducing Population of 58,000	New Percentage of Each Genotype in Population
AA = 36%	36,000	36,000 $-\underline{18,000}$ 18,000	18,000	$\dfrac{18,000}{58,000} = 32\%AA$
Aa = 48%	48,000	48,000 $-\underline{24,000}$ 24,000	24,000	$\dfrac{24,000}{58,000} = 41\%Aa$
aa = 16%	16,000	16,000 $-\underline{0}$ 16,000	16,000	$\dfrac{16,000}{58,000} = 27\%aa$

Differential Reproduction

Now we return to our original example of genes *A* and *a* to show how natural selection based on nonrandom mating can result in gene frequency changes in only
one generation. Again, assume that the parent generation has the following genotype frequencies: *AA* = 36%, *Aa* = 48%, and *aa* = 16%, with a total population
of 100,000 individuals. Suppose that 50% of all the organisms having at least one
A gene do not reproduce because they are more susceptible to disease. The parent

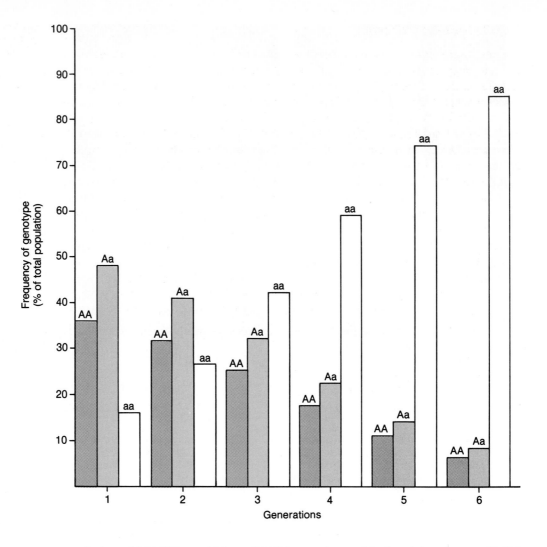

Frequency of genotype
(% of total population)

Generations

Figure 14.10
Changing Gene Frequency.
If the same selecting agent was in action for several generations and 50% of all individuals with the genotypes *AA* and *Aa* die, the frequency of the *aa* genotype will increase as the other two genotypes decrease in frequency.

population of 100,000 would have 36,000 organisms with the *AA* genotype, 48,000 with the *Aa* genotype, and 16,000 with the *aa* genotype. Because of the 50% loss, only 18,000 *AA* organisms and 24,000 *Aa* organisms will reproduce. All of the *aa* organisms will reproduce, however. Thus, there is a total reproducing population of only 58,000 individuals out of the entire original population of 100,000. What percentage of *A* and *a* will go into the gametes produced by these 58,000 individuals?

The percentage of *A*-containing gametes produced by the reproducing population will be 32% from the *AA* parents and 20.5% from the *Aa* parents (table 14.1). The frequency of the *A* gene in the gametes is 52.5%. The percentage of *a*-containing gametes is 47.5%; 20.5% from the *Aa* parents and 27% from the *aa* parents. The original parental gene frequencies were *A* = 60% and *a* = 40%. These have changed to *A* = 52.5% and *a* = 47.5%. More organisms in the population will have the *aa* genotype and fewer will have the *AA* and *Aa* genotypes.

If this process continued for several generations, the gene frequency would continue to shift until the *A* gene became rare in the population (figure 14.10). This is natural selection in action. Differential reproduction has changed the frequency of characteristics in this population of organisms.

The Selection Process

Any environmental factor that affects the probability that a gene will be passed on to the next generation is known as a **selecting agent.** Throughout this chapter we have stated that as the environment changes, it acts as a selecting agent on the gene pool of a species. As a result of natural selection, gene frequencies change and so do the frequencies of the phenotypes displayed by organisms. A classic example of this is in the English peppered moth (figure 14.11). Two color types are found in the species. One form is light colored and one is dark colored. These moths normally rest on the bark of trees during the day, where they may be spotted by birds and be eaten. About 150 years ago, the light-colored moths were most common. However, with the growth of the industrial revolution in England, which involved an increase in the use of coal, air pollution increased. The fly ash in the air settled on the trees, changing the bark to a darker color. Because the light moths were more easily seen against a dark background, the birds ate them. The darker ones were less conspicuous and, therefore, were less frequently eaten and more likely to reproduce successfully. The light-colored moth, which was originally the more common type, became much less common. This change in gene frequency occurred within the short span of fifty years. Scientists who have studied this situation have estimated that the dark-colored moths had a 20% better chance of reproducing than did the light-colored moths. This study is continuing today. As England solves some of her air pollution problems and the trees become lighter in color, the light-colored form of the moth is increasing in frequency.

Scientists have also studied changes in the frequency of the genes that control the height of clover plants (figure 14.12). Two identical fields of clover were planted,

Figure 14.11
Peppered Moth.
This photo of the two variations of the peppered moth shows that the light-colored moth is much more conspicuous against the dark tree trunk. The trees are dark because of an accumulation of pollutants from the burning of coal. The more conspicuous light-colored moths are more likely to be eaten by bird predators and the genes for light color should become more rare in the population.

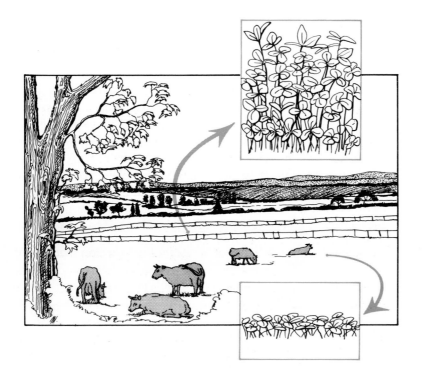

Figure 14.12
Selection for Tallness in Clover.
The clover field is undergoing natural selection by the grazing cattle who eat the tall plants and cause them to reproduce less than the short plants. The other field is not subjected to this selection pressure, and the clover population in it has more genes for tallness.

and cows were allowed to graze in only one of them. Cows acted as a selecting agent by eating the taller plants first. These tall plants never got a chance to reproduce. Only the shorter plants flowered and produced seeds. After some time, seeds were collected from both the grazed and ungrazed fields and grown in a greenhouse under identical conditions. The average height of the plants from the ungrazed field were compared to those of the grazed field. The plants from the ungrazed field had some tall, some short, but mostly medium sized plants. However, the plants from the grazed field had many more shorter plants than medium or tall individuals. The cows had selectively eaten the plants that had the genes for tallness, thus removing them from the reproducing population.

Summary

All sexually reproduced organisms naturally possess genetic variety among the individuals in the population as a result of mutations, meiosis, and genetic recombination resulting from fertilization. These genetic differences are important for the survival of the species, since natural selection must have genetic variety to select from. Natural selection by the environment results in better-suited individual organisms that have greater numbers of offspring than those less well off genetically. Not all genes are equally exposed, since some do not always express themselves and some may be recessive genes that only show themselves when in the homozygous state. Characteristics that are acquired during the life of the individual and not determined by genes can not be raw material for natural selection.

Selecting agents act to change the gene frequencies of the population if the conditions of the Hardy-Weinberg law are violated. The conditions of the Hardy-Weinberg law are random mating, no mutations, no migration, and large population size. After generations of time, the genes of the more favored individuals will make up a greater proportion of the gene pool. The process of natural selection allows the maintenance of a species in its environment, even as the environment changes.

Consider This

Penicillin was first introduced as an antibiotic in the early 1940s. Since that time, it has been found to be effective against the bacteria that causes the venereal disease gonorrhea. The drug acts on the dividing bacterial cell by preventing the formation of a new, protective cell wall. Without the wall, the bacteria can be killed by normal body defenses. Recently a new strain of this disease-causing bacterium has been found. This particular bacterium produces an enzyme that metabolizes penicillin. How can gonorrhea be controlled now that this organism is resistant to penicillin? How did a resistant strain develop? Include the following in your consideration: DNA, enzymes, selecting agents, gene pool, and the Hardy-Weinberg law.

Experience This

You can demonstrate variability in a population in the following manner. As members of the class enter the room ask them how tall they are. Plot height against the number of people at each height. Plot men and women separately. Perhaps you could do it on the blackboard and use pink and blue chalk to distinguish males from females. Calculate the average height for males and females. What are the extremes for both sexes?

Number
of
individuals
at
each
height

Height

Questions

1. Why are acquired characteristics of little interest to evolutionary biologists?
2. What factors can contribute to variety in the gene pool?
3. Why is overreproduction necessary for evolution?
4. What is natural selection? How does it work?
5. The Hardy-Weinberg law is only theoretical. What factors do not allow it to operate in a natural gene pool?
6. A gene pool has equal numbers of genes B and b. Half of the B genes mutate to b genes in this original generation. What will the gene frequencies be in the next generation?
7. The original gene frequencies in a gene pool are:

   $R = 50\%, (0.5)$

   $r = 50\%, (0.5)$

   If all the homozygous recessive individuals (rr) migrate from the population to another area, what will the gene frequencies be in the next generation?
8. How might a bad gene remain in a gene pool for generations without being eliminated by natural selection?
9. Give two examples of selecting agents and how they operate.
10. The smaller the population, the more likely it is that random changes will influence gene frequencies. Why is this true?

Chapter Glossary

acquired characteristic (ă-kwĭrd kăr''ak-ter-iss'tik) A characteristic of an organism gained during its lifetime, not caused by its genes and, therefore, not transmitted to the offspring.

differential reproduction (dif-fur-ent'shul re-pro-duk'shun) A process by which those organisms that have better genetic information for a particular environment reproduce more than individuals with poorer quality genetic information.

genetic recombination (jĕ-net'ik re-kom-bĭ-na'shun) The sum total of all the gene mixing that occurs during sexual reproduction.

genome (je-nōm) The complete set of genes of an individual.

Hardy-Weinberg law (har-de wĭn'burg law) The law that states populations of organisms will maintain constant gene frequencies from generation to generation as long as mating is random, the population is large, mutation does not occur, and no migration occurs.

selecting agent (se-lek'ting a-jent) Any factor that affects the probability that a gene will be passed to the next generation.

spontaneous mutation (spon-ta'ne-us miu-ta'shun) Natural changes in the DNA caused by unidentified environmental factors.

theory of natural selection (the'o-re uv nat'chu-ral se-lek'shun) This theory states: in a species of genetically differing organisms, the organisms with the genes that enable them to better survive in the environment and, thus, reproduce more offspring than others will transmit more of their genes to the next generation.

15

Speciation

Chapter Outline

Purpose

Most scientists accept the theory that plant and animal species have changed from their first appearance on earth and continue to change today. Some important questions can be raised. (1) How do species change? (2) What causes new species to be formed? (3) What evidence exists that new species are being produced? This chapter introduces the process of speciation and presents evidence that supports this theory.

For Your Information

Humans have invented several kinds of plants and animals for their own purposes. Most cereal grains are special plants that rely on human activity for their survival. Most would not live without fertilizer, cultivation, and other helps. These grains are the descendants of wild plants but in many cases, the wild ancestors have been lost and may be extinct. Thousands of generations of human selection for genes that gave characteristics that humans wanted have, in effect, caused the development of new species.

Learning Objectives

- Understand the importance of interrupting gene flow to the process of speciation.

- Recognize the role of reproductive isolation in maintaining species distinct from one another.

- Know that many plant species originate as a result of polyploidy.

- Appreciate that the concept of evolution changed as more information was gained.

- Recognize the steps necessary for speciation to occur.

- Understand that the basic pattern of evolution is divergence, but that there are several evolutionary patterns that can occur above the level of speciation.

- Know that the rate of evolutionary change differs with different organisms and at different times.

Review of Population Genetics

The previous two chapters have discussed the topics of population genetics and the theory of natural selection. Every species is well adapted to its environment because the environment influences which individuals reproduce the most offspring. The genes that are most favorable for survival and reproduction in a specific local environment are the ones that are most likely to be common in the gene pool of the species. The process of natural selection acts on all organisms over many generations resulting in altered gene frequencies. These altered gene frequencies will show up in the species as local differences in the behavior, physiology, and structure of the organisms. Thus, this change in the gene pool, brought about by the environment, results in local adaptations to the environment.

Over time, the genes of a species are passed from one generation to another. They may also spread from one geographical region to another. This movement of genes from one generation to another and from one place to another is called **gene flow.**

In 1957 a researcher in Sao Paulo, Brazil accidentally released an African strain of honeybees into the local population. The African type of honeybee is of the same species as honeybees that are most commonly used for honey production, the Italian honeybee. It has a relatively gentle disposition and is a better honey producer than the African strain. The African type of honeybee has two characteristics that make it much more difficult to work with; it is much more aggressive, tending to attack in large numbers when anything disturbs the hive and it often abandons old hives and establishes new ones. Both of these characteristics are negative from the point of view of the beekeeper. This aggressive type of honeybee has been steadily moving north in Central America toward the United States (figure 15.1). Some of these honeybees were found in southern California in June, 1985, when they were brought accidentally into the state with oil drilling equipment from Central America. These colonies were eliminated.

Since the African honeybees are the same species as the more commonly kept honeybees, it is possible that the genes for aggressiveness could be introduced into the gene pool of the commercially valuable types. This has caused much concern on the part of the public and some concern among commercial beekeepers in the southern United States. It is thought that the aggressive genes would be diluted in the much larger nonaggressive honeybee population and not lead to major disruptions in honey production. It is known that the African type of honeybee can not stand cold temperatures and, therefore, would not be successful in most of the United States.

We have just seen that a species disperses into new geographic areas. It can do this by traveling under its own power, by being carried by storms and wind, or by being attached to other organisms or human-transported objects. If the new home is suitable, a new colony starts, and the species has a wider geographic distribution. Examples of this abound. Most of the plants we consider to be weeds are introduced species that have arrived as a result of human activities. Kudzu (a vine imported from China), *Melaleuca* (a shrub imported from Australia), and dandelion and wild carrot (both accidental imports from Europe), are all introduced species that have

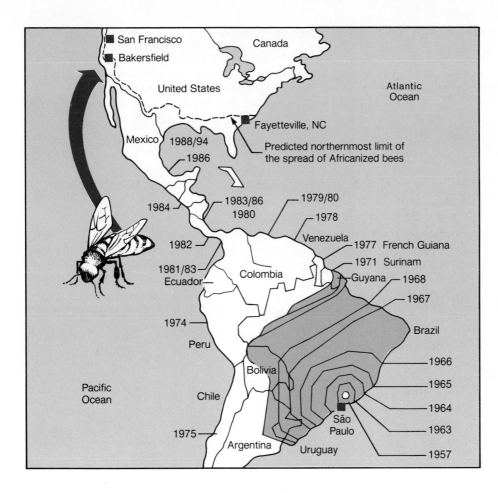

Figure 15.1
Gene Flow in the Honeybee Species.

The honeybee species consists of several races or subspecies. The African type was accidentally released in Brazil in 1957. This map shows the spread of this type of honeybee. It is thought that they will be in the southern United States by 1990.

spread over extensive new areas after their introduction (figure 15.2). The introduction of the European rabbit into Australia resulted in its rapid spread throughout the continent. Similarly, the house sparrow, starling, and gypsy moth have claimed major parts of the United States and Canada as new territory.

The geographical distribution of a species is known as its **range.** As a species expands its range, portions of the population can become separated from one another. This separation means that the new colony will have less frequent gene exchange (mating) with their geographically distant relatives. As you recall from chapter 13, each of these partially isolated populations is known as a deme. This geographic separation may become so complete that there is no gene flow between the original population and the newly established isolated group. If we define a species as a group of organisms that share the same gene pool, are these separate isolated groups really of the same species? This question has long concerned biologists and is at the heart of understanding the evolutionary process. The fundamental questions are, Can one species give rise to another? and What factors contribute to the production of new species?

Figure 15.2
Spread of Introduced Species.

Many species have been introduced into the United States either by accident or on purpose. When this happens, the species tends to spread from its point of introduction and increase the area that it lives in. Dandelion and Queen Anne's lace were probably accidental imports from Europe, while kudzu was purposely introduced to control erosion on steep slopes.

Species Definition

Before considering the question of how new species are produced, let us consider how one species is distinguished from another. A **species** is commonly defined as a group of organisms that can interbreed naturally to produce fertile offspring. The concept of a species, however, is a human invention that is applied to the living world. It is artificial and does not always fit every situation as nicely as we would like. Let us look at some examples to see how this definition may be applied. Coyotes have been known to mate with dogs to produce pups. However, even though the offspring are fertile, this mating does not commonly happen in nature; thus, coyotes and dogs are considered separate species. The mating of a male donkey and a female horse produces young that grow to adult mules, incapable of reproduction (figure 15.3). Since mules are sterile, this means that their parents (horses and donkeys) do not belong to the same species. Gene flow is not possible between these two kinds of animals. Similarly, lions and tigers can be mated in zoos to produce offspring. However, this does not happen in nature and the offspring are not likely to be fertile, so they are considered separate species.

Figure 15.3
Hybrid Sterility.
Even though they don't do so in
nature, donkeys and horses can be
mated. The offspring is called a
mule and is sterile. Because the
mule is sterile, the donkey and the
horse are considered to be of
different species.

How Species Originate

When a portion of the gene pool (part of the species) becomes separated from the
rest of the gene pool by some geographic change, such as a mountain range, river
valley, desert or ocean, we say that the local population shows **geographic isolation**
from the rest of the species. The geographic features that keep the different portions
of the species from exchanging genes are called **geographic barriers.** The uplifting
of mountains, the rerouting of rivers, and the formation of deserts all may separate

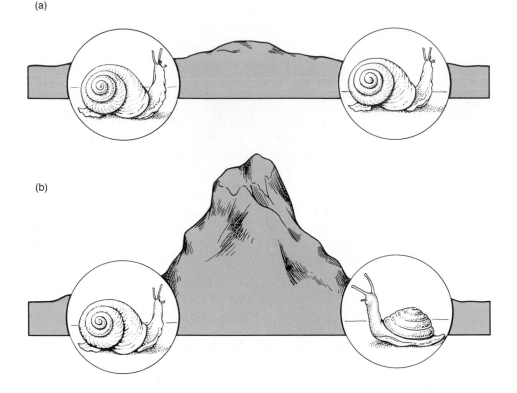

(a)

(b)

Figure 15.4
Effect of Geographic Isolation.

If a single species of snails (a) were divided into two populations by the uplifting of a mountain range, the two populations might eventually evolve enough differences to become different species (b).

one portion of a gene pool from another. Two kinds of squirrels are found on opposite sides of the Grand Canyon. Some people consider them to be separate species, while others consider them to be different isolated demes of the same species. Even small changes may cause geographic isolation in species that have little ability to move. A fallen tree, a plowed field, or even a new freeway may effectively isolate populations within such species. Snails in two valleys separated by a high ridge have been found to be closely related, but different species. The snails cannot get from one valley to the next because of the height and climatic differences presented by the ridge (figure 15.4).

The separation of a gene pool into several parts is not enough to generate new species. Even after many generations of geographic isolation, these separate groups may still be able to exchange genes (mate and produce fertile offspring) if they overcome the geographic barrier because they have not accumulated enough genetic differences to prevent reproductive success. Natural selection and differences in environments play a very important role in the process of forming new species. Following separation from the main portion of the gene pool by geographic isolation, the organisms within the small, local population will likely experience different environmental conditions. If, for example, a mountain range has separated a species into two populations, one of the two populations may receive more rain or more sunlight than the other (figure 15.5). These environmental differences act as natural selecting agents on the two gene pools and account for different genetic combinations in the two places. Furthermore, it is possible that different mutations happened in the two populations as well as different random combinations of genes as a result

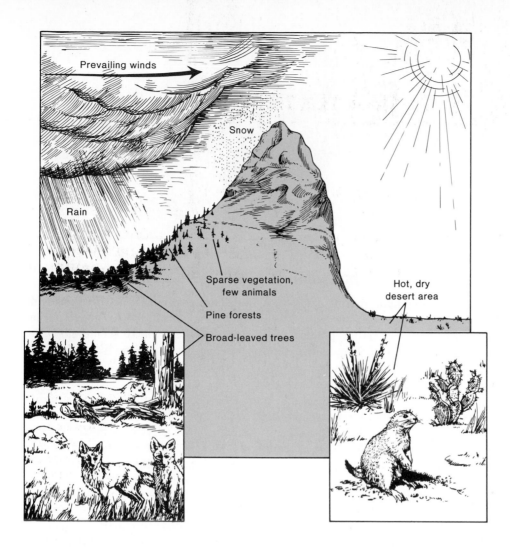

Prevailing winds

Snow

Rain

Sparse vegetation, few animals

Pine forests

Broad-leaved trees

Hot, dry desert area

of sexual reproduction. This would be particularly true if one of the demes was small. As a result, the two populations may show differences in color, height, enzyme production, time of seed germination or many other characteristics.

Over a long period of time, the genetic differences that accumulate may result in regional populations called **subspecies** that are significantly modified structurally, physiologically, or behaviorally. The differences among some subspecies many be so great that there is reduced reproductive success when subspecies mate. **Speciation** is the process of generating new species. This process only occurs if gene flow between isolated populations does not occur, even after barriers are removed. In other words, the process of speciation can begin with the geographic isolation of a portion of the species, but new species are generated only when they become separate from one another genetically. Speciation is really a three-step process. It begins with geographic isolation, is followed by selective agents choosing specific genetic combinations as being valuable, and ends with the genetic differences becoming so great that reproduction between the two groups is impossible.

Figure 15.5
Effect of Mountain Range on Plants and Animals.

More rain falls on the western side of the Rocky Mountains than on the eastern side. Consequently, there is abundant vegetation on the western slopes and only scattered, desert-type plants on the east. This, of course, affects the types of animals living in the two areas.

Maintaining Genetic Isolation

Organisms that allow mating across species lines are not going to be very successful since most cross-species matings result in no offspring or offspring that are sterile. As a part of the speciation process, it is typical that there will be the development of mechanisms called **reproductive isolating mechanisms** or **genetic isolating mechanisms,** which prevent cross-species matings. A great many different types of these genetic isolating mechanisms can be recognized.

In central Mexico, two species of robin-sized birds called towhees live in different environmental settings. The collard towhee lives on the mountainside in the pine forests, while the spotted towhee is found at lower elevations in oak forests. Geography presents no barriers to these birds. They are perfectly capable of flying to each other's habitat, but they do not. Because of their **habitat preference** or **ecological isolation,** mating between these two similar species does not occur.

Some plants flower only in the spring of the year, while other, closely related species flower in mid-summer or in the fall; therefore, they are not very likely to pollinate one another. Among many species of insects there is a similar spacing of the reproductive periods of closely related species so that they do not overlap. Thus, **seasonal isolation** (differences in the time of the year that reproduction takes place) is an effective genetic isolating mechanism.

Inborn behavior patterns that prevent breeding between species are referred to as **behavioral isolation.** The mating calls of frogs and crickets are highly specific. The sound pattern produced by the males is species specific and invites only females of the same species to come. The females have a built-in response to only the species-specific call.

The courtship behavior of birds involves both sound and visual signals that are species specific. For example, groups of male prairie chickens gather on meadows shortly before dawn in the early summer and begin their dances. The air sacs on either side of the neck are inflated so that the bright red skin is exposed. Their feet move up and down very fast while their wings are spread out and quiver slightly (figure 15.6). This combination of sight and sound attracts females. When they arrive, the males compete for the opportunity to mate with them. Other related species of birds conduct their own similar, but different, courtship displays. The differences among the dances are great enough so the females of one species can recognize the dance of a male of her own species.

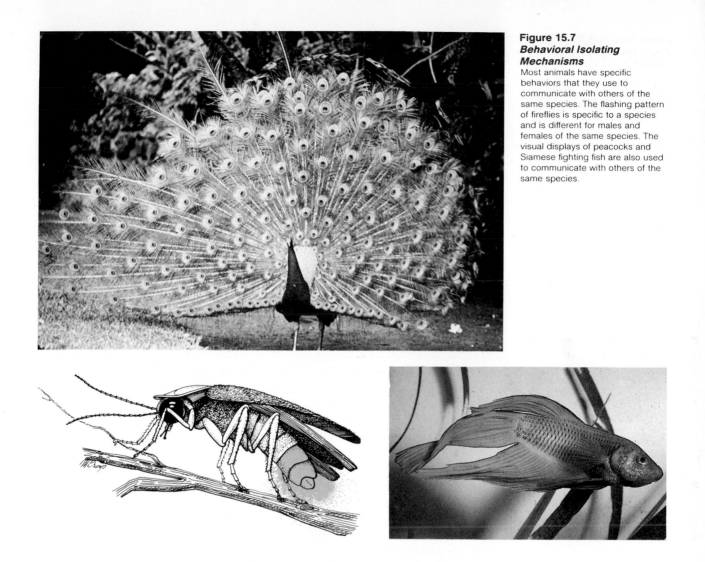

Figure 15.7
Behavioral Isolating Mechanisms
Most animals have specific behaviors that they use to communicate with others of the same species. The flashing pattern of fireflies is specific to a species and is different for males and females of the same species. The visual displays of peacocks and Siamese fighting fish are also used to communicate with others of the same species.

Behavioral isolating mechanisms such as these occur among other types of animals as well. The strutting of a peacock, the fin display of Siamese fighting fish, and the light-flashing patterns of different species of fireflies are all examples of behaviors that help to prevent different species from interbreeding and producing hybrids (figure 15.7).

Polyploidy

So far, we have considered only those hereditary changes that take a long time to add up to enough differences to make a new species. In plants, there are many examples of a special kind of speciation called **polyploidy,** in which the number of sets of chromosomes present is increased. This increase in chromosome number comes about as a result of abnormal cell divisions in which the chromosomes do not separate properly during mitosis or meiosis. For example, if a cell had the normal diploid chromosome number of 6 (2N=6), and the cell went through mitosis but did not divide into two cells, it would then contain 12 chromosomes. Similarly, it would

Figure 15.8
Polyploidy.

Many plants have been created by increasing the chromosome number. Many of the large flowered varieties of plants have been produced artificially by this technique. The smaller flower shown here is a normal diploid, the larger one is a tetraploid variety.

be possible to get 24 chromosomes, and by a slightly more complicated method to get 18 or 36 chromosomes. It is important to understand that plants do not do this on purpose, but when it happens, plants have a greater likelihood of surviving than do animals.

If an organism increases its chromosome number, it may be possible for that individual to form normal gametes to exchange with the parental gene pool. Many of these polyploids, however, can reproduce asexually. This can result in a population of polyploids that have the same chromosome number and are capable of sexual reproduction among themselves, even though they cannot reproduce with the original parent organism. In other words, a new gene pool has been formed that is automatically isolated from the parental species. Many of the plants we use are polyploids. Cotton, potatoes, sugar cane, wheat, and many types of garden flowers are examples (figure 15.8).

The Development of Evolutionary Thought

The natural selection process discussed in chapter 14 and the mechanism of speciation presented in this chapter are based on the fact that organisms differ from one another genetically and that populations can change over time. **Evolution** is the adaptation of a population of organisms to its environment by way of a change in the gene frequency over time. Natural selection is the process that causes evolution to occur and speciation is a primary step that can lead to greater evolutionary change.

Although most scientists accept evolutionary processes as central to an understanding of how various life forms arose and continue to change today, this was not always the case. For centuries people believed that the various species of plants and animals were fixed and unchanging—that is, they were thought to have remained unchanged from their creation. This was a reasonable assumption since these people knew nothing about DNA, meiosis, or population genetics. Furthermore, the process of evolution is so slow that they could not see the accumulation of changes we call evolution. It is even difficult for modern scientists to see this slow change in many kinds of organisms. In the mid-1700s, George Buffon, a French naturalist, expressed some concerns about the possibilities of change (evolution) in animals, but he did not suggest any mechanism that would result in evolution.

In 1809, Jean Baptiste de Lamarck, a student of Buffon's, suggested a process by which evolution could occur. He proposed that acquired characteristics could be transmitted to offspring. For example, he postulated that giraffes originally had short necks. Since giraffes constantly stretched their necks to obtain food, their necks got slightly longer. This slightly longer neck could be passed to the offspring who were themselves stretching their necks and over time, the necks of giraffes would get longer and longer. Although we now know Lamarck's theory was wrong (because acquired characteristics are not inherited), Lamarck did present an early theory for how evolution could occur. All during this period, from the mid-1700s to the mid-1800s, there continued to be lively arguments about the possibility of evolutionary change. Some, like Lamarck and others, thought that change did take place, while many others said that it was not possible. It was the thinking of two English scientists that finally provided a mechanism that could explain how evolution could occur.

In 1858, Charles Darwin and Alfred Wallace suggested the theory of natural selection as a mechanism for evolution. They based their theory on the following assumptions about the nature of living things:

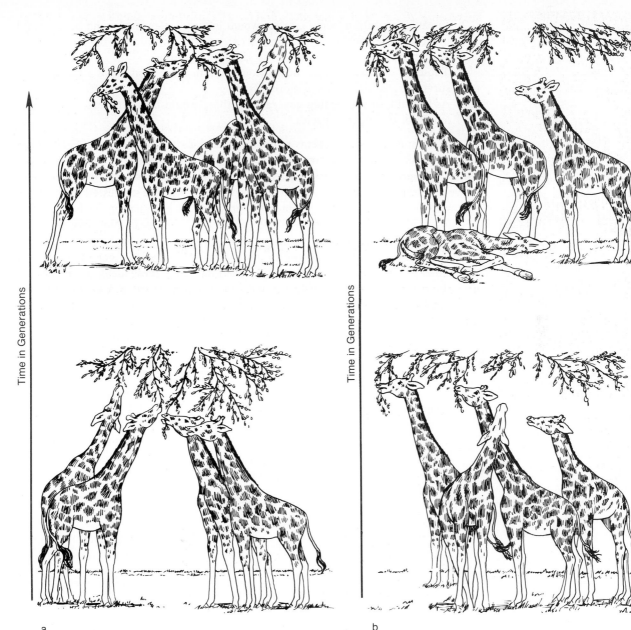

a

b

1. All organisms produce more offspring than can survive.
2. No two organisms are exactly alike.
3. Among organisms, there is a constant struggle for survival.
4. Those individuals that possess favorable characteristics have a higher rate of survival and produce more offspring.
5. Favorable characteristics become more common in the species and unfavorable characteristics are lost.

With these assumptions, the Darwin-Wallace theory of evolution by natural selection would have a different explanation for the development of long necks in giraffes (figure 15.9). Their theory was based on the following assumptions:

1. In each generation, more giraffes would be born than the food supply could support.

Figure 15.9
Two Theories of How Evolution Occurs.
Lamarck thought that acquired characteristics could be passed on to the next generation. Therefore, he postulated that as giraffes stretched their necks to get food, their necks got slightly longer. This characteristic was passed on to the next generation, which would have longer necks (a). The Darwin-Wallace theory states that there is variation within the population and that those with longer necks would be more likely to survive and reproduce and pass their genes for long necks on to the next generation (b).

2. In each generation, some giraffes would inherit longer necks and some would inherit shorter necks.
3. All giraffes would compete for the same food source.
4. Those giraffes with longer necks would obtain more food, have a higher survival rate, and produce more offspring.
5. As a result, in succeeding generations, there would be an increase in the neck length of the giraffe species.

This thought process seems simple and obvious today, but remember, at the time Darwin and Wallace proposed their theory, the processes of meiosis and fertilization were poorly understood and the concept of the gene was only beginning to be discussed. Nearly fifty years after Darwin and Wallace suggested their theory, the work of Gregor Mendel (chapter 12) provided an explanation for how characteristics could be transmitted from one generation to the next. Not only did Mendel's idea of the gene provide a means of passing traits from one generation to the next, but it provided the first step in understanding mutations, gene flow, genetic drift, and the significance of reproductive isolation. All of these ideas are interwoven into the modern concept of evolution and how it occurs. If we look at the same five ideas and update them with modern information, they might look something like this:

1. An organisms's capacity to reproduce results in surplus organisms.
2. Because of mutation and sexual reproduction involving meiosis and fertilization, new combinations of genes are present in every generation. These processes are so great that each individual in a population is genetically unique. The genes present are expressed as the phenotype of the organism.
3. Resources such as food, water, mates, and nest materials are in short supply, so some individuals will need to do without. Other environmental factors, like disease organisms, predators, and defense mechanisms affect survival. These are called selecting agents.
4. Those individuals with the best combination of genes will be more likely to survive and reproduce, passing more of their genes on to the next generation. An organism can be selected against if it has fewer offspring than its better-adapted species members. It does not need to die to be selected against.
5. Therefore, genes that produce characteristics favorable to survival will become more common and the species will become better adapted to its environment.

Evolution Above the Species Level

Tracing the evolutionary history of an organism back to its origins is a very difficult task, since many of its ancestors no longer exist. It is important to understand that the fossil record is incomplete and only provides limited information about the biology of the organism represented in that record. We may know a lot about the structure of the bones and teeth of an extinct ancestor, but know almost nothing about its behavior, physiology, and natural history. Biologists must use a great deal of indirect evidence to piece together the series of evolutionary steps that lead to a present day species. Figure 15.10 is typical of evolutionary trees that help us understand how time and structural changes are related in the evolution of birds,

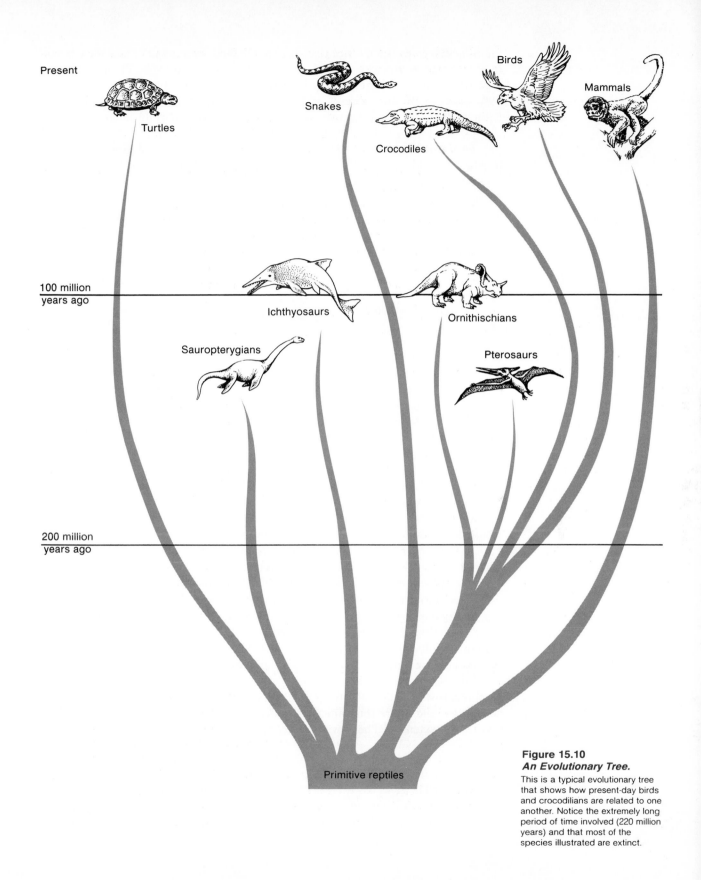

Figure 15.10
An Evolutionary Tree.

This is a typical evolutionary tree that shows how present-day birds and crocodilians are related to one another. Notice the extremely long period of time involved (220 million years) and that most of the species illustrated are extinct.

mammals, and reptiles. Notice that most of the kinds of animals shown on this evolutionary tree do not survive to the present. This is typical. Most of the species of organisms that have ever existed are extinct. Most estimates of extinction are around 99%—that is, 99% or more of all the species of organisms that ever existed are extinct. Extinction is a common happening in the long-term evolutionary picture of things.

The basic pattern seen in most evolutionary trees is one of **divergent evolution,** in which individual speciation events cause successive branches in the evolution of a group of organisms. This basic pattern is well illustrated by the evolution of the horse shown in figure 15.11. The modern horse, with its large size, single toe on each foot, and teeth designed for grinding grasses, is thought to have resulted from the accumulation of many changes from a small, dog-sized animal with four toes on its front feet, three toes on its hind feet, and teeth that were designed for chewing up leaves and small twigs. Even though we know a lot about the evolution of the horse, there are still many gaps that need to be filled before we have a complete evolutionary history.

Although divergence is the basic pattern in evolution, it is possible to superimpose several other patterns on it. One special evolutionary pattern, characterized by a rapid increase in the number of kinds of closely related species, is known as **adaptive radiation.** Adaptive radiation results in an evolutionary explosion of new species from a common ancestor. There are basically two kinds of situations that are thought to favor adaptive radiation. One is a condition in which an organism invades a previously unexploited environment. For example, at one time there were no animals on the land masses of the earth. The amphibians were the first vertebrate animals able to spend part of their life on land. A variety of different kinds of amphibians evolved rapidly that exploited several different kinds of life-styles. In the famous case of the finches of the Galápagos Islands, Ecuador, first studied by Charles Darwin, it is thought that one kind of finch arrived on these relatively isolated volcanic islands. Originally, there would have been no finches, or other land-based birds, on these islands. Adaptive radiation from the common ancestor is thought to have resulted in the many different kinds of finches found on the islands today (figure 15.12). Some of these finches took roles normally filled by other kinds of birds elsewhere in the world. Some even became warblerlike and one uses a cactus spine as a tool to probe for insects in a manner similar to a woodpecker.

A second set of conditions that can favor adaptive radiation is one in which a type of organism evolves a new set of characteristics that makes them able to displace organisms that previously filled roles in the environment. For example, although amphibians were the first vertebrates to occupy land, they were replaced by reptiles because reptiles possessed characteristics, such as dry skin and an egg, that could develop on land. These characteristics allowed them to replace most of the amphibians who could only live in relatively moist surroundings where they would not dry out and where their aquatic eggs could develop. The adaptive radiation of reptiles was extensive. They invaded most terrestrial settings and even evolved forms

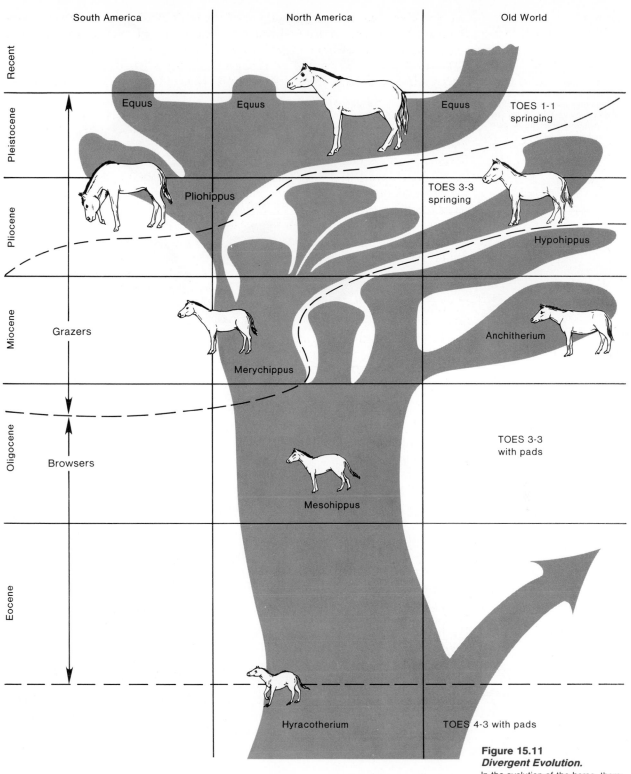

Figure 15.11
Divergent Evolution.

In the evolution of the horse, there have been many speciation events that have been added on top of one another. What began as a small, leaf-eating, five-toed animal of the forest has evolved into a large, grass-eating, single-toed animal of the plains. There are many related animals today but early ancestral types are extinct.

Figure 15.12
Adaptive Radiation.

When Darwin discovered the finches of the Galápagos Islands, he thought they might all have derived from one ancestor that arrived on these relatively isolated islands. If they were the only birds to inhabit the islands, they could have evolved very rapidly into the many different types shown here. The drawings show the specializations of beaks for different kinds of food.

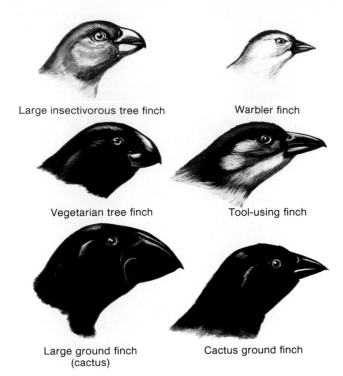

Large insectivorous tree finch

Warbler finch

Vegetarian tree finch

Tool-using finch

Large ground finch
(cactus)

Cactus ground finch

that flew and lived in the sea. Subsequently, the reptiles were replaced by the mammals, who went through a similar radiation. Figure 15.13 shows the sequence of radiations that occurred within the vertebrate group. The number of species of amphibians and reptiles has declined, while the number of species of birds and mammals has increased.

When organisms of widely different backgrounds develop similar characteristics, we see an evolutionary pattern known as **convergent evolution.** This particular pattern often leads people to misinterpret the evolutionary history of organisms. For example, many kinds of plants that live in desert situations have thorns and lack leaves during much of the year. Superficially they appear similar, but are often quite different from one another. They have not become one species, although they may resemble one another to a remarkable degree. The presence of thorns and the absence of leaves are adaptations to a desert type of environment, since the thorns discourage herbivores and the absence of leaves reduces water loss. Another example involves animals that make a living by catching insects while flying. Bats, swallows, and dragonflies all obtain food in this manner. They all have wings that provide flight, but they are derived from the modification of different structures (figure 15.14). At first glance, they may appear to be very similar and perhaps closely related, but detailed study of their wings and other structures shows that they are quite different kinds of animals. They have simply converged in structure, type of food they eat, and method of obtaining food. Likewise, whales, sharks, and tuna appear to be similar, but are different kinds of animals that all happen to live in the open ocean.

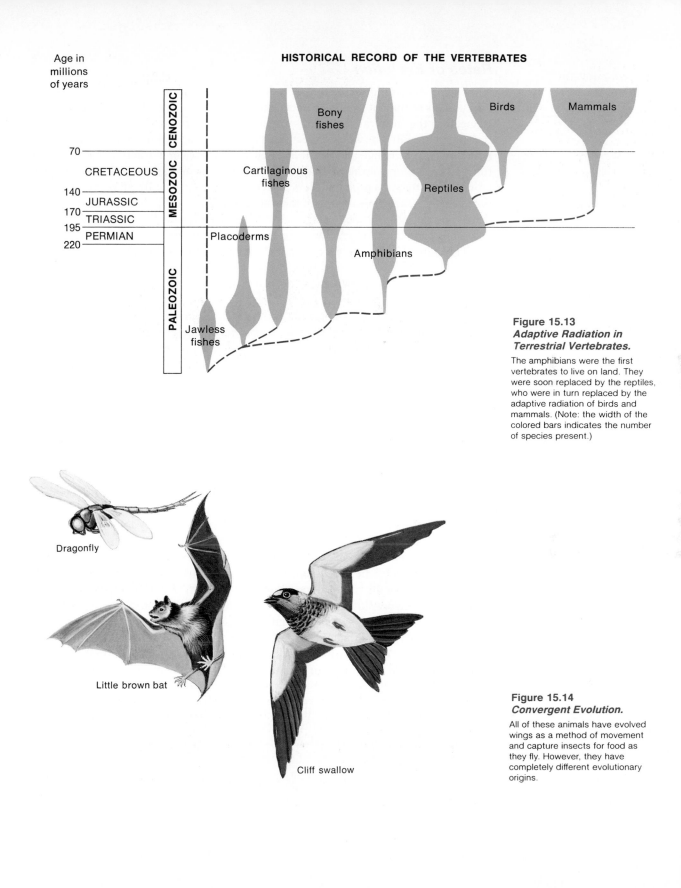

HISTORICAL RECORD OF THE VERTEBRATES

Age in millions of years

CENOZOIC

70

CRETACEOUS

MESOZOIC

140

JURASSIC

170

TRIASSIC

195

PERMIAN

220

PALEOZOIC

Jawless fishes

Placoderms

Cartilaginous fishes

Bony fishes

Amphibians

Reptiles

Birds

Mammals

Figure 15.13
Adaptive Radiation in Terrestrial Vertebrates.

The amphibians were the first vertebrates to live on land. They were soon replaced by the reptiles, who were in turn replaced by the adaptive radiation of birds and mammals. (Note: the width of the colored bars indicates the number of species present.)

Dragonfly

Little brown bat

Cliff swallow

Figure 15.14
Convergent Evolution.

All of these animals have evolved wings as a method of movement and capture insects for food as they fly. However, they have completely different evolutionary origins.

Rates of Evolution

Although it is commonly thought that evolutionary change is something that takes long periods of time, it should be understood that rates of evolution can vary greatly. Remember that natural selection is driven by the environment. If the environment is changing rapidly, one would expect rapid changes in the organisms that are present. Periods of rapid change also result in extensive episodes of extinction. During other times in the history of the earth, little change was taking place and the rate of evolutionary change was probably slow. Nevertheless, when we talk about evolutionary time, we are generally thinking in thousands to millions of years. Although both of these time periods are long compared to human life spans, the difference between thousands or millions of years in the evolutionary time scale is still very significant.

The view of evolution described by Charles Darwin was one of slow, gradual, progressive change, which has been called **microevolution.** This has been the generally accepted view of evolution for years, but recently some evolutionists have suggested different conditions and mechanisms that could have resulted in whole groups of characteristics changing at the same time. This kind of evolutionary pattern has been termed **macroevolution.** This concept suggests that species remained relatively unchanged for millions of years and then, within a few thousand years, they evolved into a number of different species. One theory that helps to account for this rapid change is that most characteristics displayed by an organism are not caused by a single gene, but caused by groups of genes interacting with each other. For example, flight in birds or insects is not the result of a single gene, but the development of flight seemed to have been a relatively rapid process. It has been suggested that, at times, there may have been a great rearrangement of genes on chromosomes that could have resulted in the sudden emergence of new species. The fossil record certainly shows long periods of time in which very little change occurred to the organisms present or the number of kinds present. This static condition is often followed by a period of great increase in the number of new species, extensive extinction, and rapid change in the structures possessed by the organisms that survive the period of change. This pattern of relatively slow evolutionary change followed by rapid change has been called **punctuated equilibrium.**

At the present time, the scientific community has no uniform concept about how such rapid bursts of evolution take place or what conditions favor this rapid change. The microevolutionists point to the fossil record as proof that evolution is a slow, steady process. The macroevolutionists point to the gaps in the fossil record as evidence that rapid change occurs. As with most controversies of this nature, more information is required to resolve the question. It will take decades to collect and even then, the differences of opinion may not be reconciled.

Summary

Populations are usually genetically diverse. Mutations, meiosis, and sexual reproduction tend to introduce genetic variety into a population. Organisms with wide geographic distributions often show different gene frequencies in different parts of their range. A species is a group of organisms that can interbreed to produce fertile offspring. The process of speciation usually involves the geographic separation of the species into two or more isolated populations. While they are separated, natural selection operates to adapt each population to its environment. If this generates enough change, the two populations may become so different that they cannot interbreed. Similar organisms that have recently evolved into separate species normally have mechanisms to prevent interbreeding. Some of these are habitat preference, seasonal isolation, and behavioral isolation. Plants have a special way of generating new species by increasing chromosome number as a result of abnormal mitosis or meiosis.

At one time people thought that all organisms had remained unchanged from the time of their creation. Lamarck suggested that change did occur and thought that acquired characteristics could be passed from generation to generation. Darwin and Wallace proposed the theory of natural selection as the mechanism that drives evolution. Evolution is basically a divergent process upon which other patterns can be superimposed. Adaptive radiation is a very rapid divergent evolution, while convergent evolution involves the development of superficial similarities among widely different organisms. The rate at which evolution has occurred probably varies. The fossil record shows periods of rapid change interspersed with periods of little change. This has caused some to look for mechanisms that could cause the sudden appearance of large numbers of new species in the fossil record and challenge the traditional idea of slow, steady change accumulating to enough differences to cause a new species to be formed.

Explain how all of the following are related to the process of speciation: mutation, natural selection, meiosis, Hardy-Weinberg law, geographic isolation, changes in the earth, gene pool, and competition.

Consider This

Spend a few minutes in a greenhouse, a park, or your back yard. How many examples of reproductive isolation can you discover?

Experience This

---

1. Why is geographic isolation important in the process of speciation?
2. How does speciation differ from the formation of subspecies or races?
3. Why aren't mules considered a species?
4. Describe three kinds of genetic isolating mechanisms that prevent interbreeding between different species.
5. How does a polyploid organism differ from a haploid or diploid organism?
6. Can you always tell by looking at two organisms whether or not they belong to the same species?
7. What is the difference between divergent evolution and adaptive radiation?
8. Describe two differences between convergent evolution and adaptive radiation.
9. Give an example of seasonal isolation, ecological isolation, and behavioral isolation.
10. List the series of events necessary for speciation to occur.

Questions

adaptive radiation (uh-dap'tiv ra-de-a'shun) A specific evolutionary pattern in which there is a rapid increase in the number of kinds of closely related species.
behavioral isolation (be-hav'yu-ral i-so-la'shun) A genetic isolating mechanism that prevents interbreeding between species because of differences in behavior.
convergent evolution (kon-vur'jent ev-o-lu'shun) An evolutionary pattern in which widely different organisms show similar characteristics.
divergent evolution (di-vur'jent ev-o-lu'shun) A basic evolutionary pattern in which individual speciation events cause many branches in the evolution of a group of organisms.

Chapter Glossary

ecological isolation (e-kŏ-loj′ĭ-kal i-so-la′shun) A genetic isolating mechanism that prevents interbreeding between species because they live in different areas.

evolution (ē-vo-lu′shun) The adaptation of a population of organisms to its environment by way of change in the gene frequencies over generations.

gene flow (jēn flo) The movement of genes from one generation to another or from one place to another.

genetic isolating mechanism (jĕ-net′ic i-so-la′ting mek′an-izm) A mechanism that prevents interbreeding between species.

geographic barrier (je-o-graf′ik bār′yur) Geographic features that keep different portions of a species from exchanging genes.

geographic isolation (je-o-graf′ik i-so-la′shun) A condition in which part of the gene pool is separated by geographic barriers from the rest of the population.

habitat preference (hab-i-tat pref′ur-ents) A genetic isolating mechanism that prevents interbreeding between species because they live in different areas.

macroevolution (ma″kro-ev-o-lu′shun) The concept that large numbers of characteristics can change in a very short period of time to produce rapid evolutionary change.

microevolution (mi″kro-ev-o-lu′shun) The concept that evolution is a slow, gradual, progressive process.

polyploidy (pah″lĭ-ploy′de) A condition in which cells contain multiple sets of chromosomes.

punctuated equilibrium (pung′chu-a-ted e-kwĭ-lib′re-um) The theory that evolution has consisted of intermittent periods of rapid change interspersed with long periods of relative stability.

range (rānj) The geographical distribution of a species.

reproductive isolating mechanism (re-pro-duk′tiv i-so-la-ting me′kan-ism) A mechanism that prevents interbreeding between species.

seasonal isolation (se′zun-al i-so-la′shun) A genetic isolating mechanism that prevents interbreeding between species because reproductive periods differ.

speciation (spe-she-a′shun) The process of generating new species.

species (spe-shēz) A group of organisms that can interbreed naturally to produce fertile offspring.

subspecies (sub′spe-shēz) Regional groups within a species that are significantly different structurally, physiologically, or behaviorally, yet are capable of exchanging genes by interbreeding.

Chapter Outline

Purpose

All living things require a continuous
source of energy. This energy is used for
growth, movement, reproduction, and
many other activities. There are certain
physical laws that describe how energy
changes occur. The second law of
thermodynamics states that during the
process of converting energy from one
form to another, some useful energy is lost
as useless heat. Many of the world's
problems result from our failure to
recognize the limits imposed by the laws
of thermodynamics. The purpose of this
chapter is to show how energy is used
and converted within groups of interacting
organisms, and how the laws of
thermodynamics apply to living systems.

For Your Information

News Item 26 July 1986
By the year 2000, Third World countries will need nearly one-third more food than they did in 1980, according to a study by the International Food Policy Research Institute. There are nineteen countries where many of the people are receiving less than 90% of the food energy considered necessary for good health. The countries are found in Asia (Bangladesh, Vietnam, Cambodia, Afghanistan and Nepal), the Caribbean (Haiti), and Africa (Kenya, Mozambique, Senegal, Madagascar, Burkina Faso, Central African Republic, Chad, Guinea, Guinea-Bissau, Mali, Mauretania, Niger, and Somalia). These countries will have a shortage of food of nearly thirty-one million tons. The major source of food for these people is grain.

Learning Objectives

■ Recognize the relationships organisms have to each other in an ecosystem.

■ Understand that useful energy is lost as energy passes from one trophic level to the next.

■ Appreciate that it is difficult to quantify energy flow through an ecosystem.

■ List characteristics of several different biomes.

■ Understand that humans have converted natural ecosystems to human use.

■ Recognize that the developed world uses far more resources per person than the majority of people in the world.

Ecology and Environment

Today we hear many different kinds of people using the terms ecology and environment. Students, homemakers, politicians, planners, and union leaders speak of "environmental issues" and "ecological concerns." Although these phrases are common, are they really understood by both the speaker and the listener? Often both speaker and listener interpret the same language in different ways, so we need to establish some basic definitions.

Ecology is a branch of biology that studies the relationships between organisms and their environment. This is a very simple definition for a very complex branch of science. Most ecologists define the word **environment** very broadly as anything that impacts on an organism during its lifetime. These environmental influences can be divided into two different categories. Those things that impact on an organism and are living are called **biotic factors** and those that are not alive are called **abiotic factors** (figure 16.1). If we consider a trout in a stream, we can identify many different environmental factors that are important to the life of the fish. The temperature of the water is extremely important as an abiotic factor, but it may be influenced by the presence of trees (biotic factor) along the stream bank that shade the stream and prevent the sun from heating it. Obviously, the kind and number of food organisms in the stream are very important biotic factors as well. The type of material that makes up the stream bottom and the amount of oxygen dissolved in the water are other important abiotic factors, both of which are related to how rapidly the water is flowing.

As you can see, the environment of an organism is complex and interrelated, everything seems to be influenced or modified by other factors. A plant is influenced by many different factors during its lifetime: the type and amount of minerals in the soil, the amount of sunlight hitting the plant, the animals that eat the plants, and the wind, water and temperature. Each item on this list can be further subdivided into other areas of study. For instance, water is important in the life of plants, and so rainfall is studied in plant ecology—not just how much rain, but also the time of year the rain falls. Though rainfall seems to be an easily understood portion of an ecological study, it really has many significant aspects to it. Is it a hard, driving rain or a soft, gentle rain? Does the water soak into the ground for later use, or does it quickly run off into the rivers?

Temperature is also very important to the life of a plant. For example, two areas of the world can have the same average daily temperature of 10° C,* but not have the same plants because of different temperature extremes. In one area, the temperature may be 13° C during the day and 7° C at night for a 10° C average. In

Figure 16.1
Biotic and Abiotic Environmental Factors.
Both living and nonliving things have an impact on an organism. The living influences are called biotic factors and the nonliving are called abiotic factors.

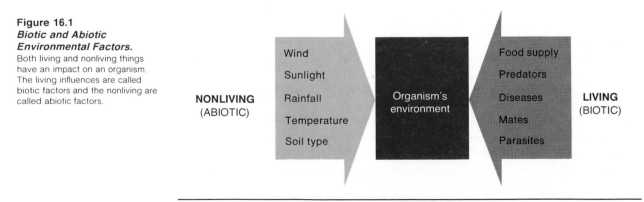

NONLIVING
(ABIOTIC)

Wind
Sunlight
Rainfall
Temperature
Soil type

Organism's environment

Food supply
Predators
Diseases
Mates
Parasites

LIVING
(BIOTIC)

---

*See the metric conversion chart inside the back cover for conversion to Fahrenheit.

another area, the temperature may be 20° C in the day and only 0° C at night, for a 10° C average. Plants react to extremes in temperature as well as to the daily average. Furthermore, different parts of the plant may respond differently to temperature. Tomato plants will grow at a temperature below 13° C, but will not begin to develop fruit at that temperature.

The animals in an area are influenced as much by abiotic factors as are the plants. If these nonliving factors do not favor the growth of plants, there will be little food and few hiding places for the animal life. Two types of areas that support only small numbers of living animals are deserts and polar regions. Near the polar regions of the earth, the low temperature and short growing season inhibits plant growth and, therefore, there are relatively few species of animals with relatively small numbers of individuals. Deserts receive little rainfall and, therefore, have poor plant growth and low concentrations of animals. On the other hand, tropical rain forests have high rates of plant growth and large numbers of animals of many kinds.

As you can see, living things are themselves part of the environment of other living things. If there are too many animals in an area, they could demand such large amounts of food that they would destroy the plant life and the animals themselves would die.

So far we have discussed how organisms interact with their environment in rather general terms. Ecologists have developed several concepts that help us to understand how biotic and abiotic factors interrelate to form a complex system.

The Organization of Living Systems

All living things require a continuous supply of energy to maintain life. Therefore, many people like to organize living systems by the energy relationships displayed among the different kinds of organisms present. An **ecosystem** is an interacting collection of organisms and the abiotic factors that affect them. Within an ecosystem, several different kinds of organisms can be identified. Organisms that trap sunlight for photosynthesis, resulting in the production of organic material from inorganic material, are called **producers.** Green plants and other photosynthetic organisms are, in effect, converting sunlight energy into the energy contained within the chemical bonds of organic compounds. There is a flow of energy from the sun into the living matter of plants.

The energy that plants trap can be transferred through a number of other organisms in the ecosystem. Since all of these other organisms must obtain energy in the form of organic matter, they are called **consumers.** Consumers cannot capture energy from the sun as plants do. Consumers eat plants directly or feed on plants indirectly by eating organic matter of other living things. Each time the energy enters a different consumer organism, it is said to enter a different **trophic level,** which is a step or stage in the flow of energy through an ecosystem (figure 16.2). The plants (producers) receive their energy directly from the sun and are said to occupy the *first trophic level.* Various kinds of consumers can be divided into several categories depending on how they fit into the flow of energy through an ecosystem. Those animals that feed directly on plants are called **herbivores** or **primary consumers** and occupy the *second trophic level.* Animals that eat other animals are called **carnivores** or **secondary consumers** and can be subdivided into different trophic levels depending on what specific animal they eat. Those animals that feed on herbivores occupy the *third trophic level* and are known as **primary carnivores.** Animals that feed on the primary carnivores are known as **secondary carnivores** and occupy

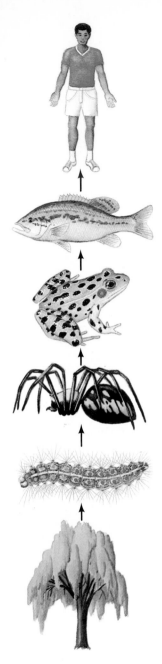

Figure 16.3
Trophic Levels.
As one organism feeds upon another organism, there is a flow of energy from one trophic level to the next. There are six trophic levels shown in this illustration.

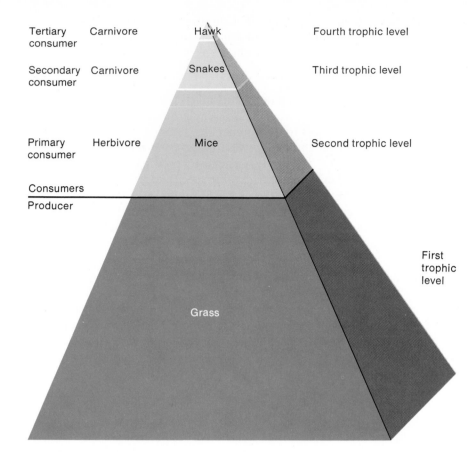

Figure 16.2
Organization of an Ecosystem.
Organisms within ecosystems can be divided into several different trophic levels based on how they obtain energy. There are several different sets of terminology used to identify these different roles. This illustration shows how the different sets of terminology are related to one another.

the *fourth trophic level*. For example, a human may eat a fish that ate a frog that ate a spider that ate an insect that consumed plants for food (figure 16.3). In this illustration there are six different trophic levels. Obviously there can be higher categories and, also, some organisms that don't fit neatly into this theoretical scheme. Some animals are carnivores at some times and herbivores at others and are called **omnivores.** They are classified into different trophic levels depending on what they happen to be eating at the moment.

If an organism dies, the energy contained within the organic compounds of its body is finally released to the environment as heat by organisms that decompose the dead body into carbon dioxide, water, ammonia, and other simple inorganic molecules. These organisms of decay, called **decomposers,** are things such as bacteria, fungi, and other organisms that use the dead organism as a source of energy. This group of organisms efficiently converts nonliving organic matter into simple inorganic molecules that can be used by producers in the process of trapping energy. Thus, decomposers are very important components of ecosystems since they cause materials to be recycled. As long as the sun supplies the energy, elements are cycled through ecosystems repeatedly. Table 16.1 summarizes the various categories of organisms within an ecosystem.

Table 16.1   *Roles in an Ecosystem*

Classification	Description	Examples
Producers	Plants that convert simple inorganic compounds into complex organic compounds by photosynthesis.	Trees, flowers, grasses, ferns, mosses, algae
Consumers	Organisms that rely on other organisms as food. Animals that eat plants or other animals.	
Herbivore	Eats plants directly.	Deer, duck, cricket, vegetarian human
Carnivore	Eats meat.	Wolf, pike, dragonfly
Omnivore	Eats plants and meat.	Rats, most humans
Scavenger	Eats food left by others.	Coyotes, skunks, vultures, crayfish
Parasite	Lives in or on another organism, using it for food.	Tick, tapeworm, many insects
Decomposer	Organism that returns organic compounds to inorganic compounds. Important component in recycling.	Bacteria, fungi

Figures 16.4 and 16.5 illustrate two different kinds of ecosystems. Can you identify producers, herbivores, carnivores, scavengers, and decomposers in each of these drawings? Now that we have a better idea of how ecosystems are organized, we can look more closely at the significance of the second law of thermodynamics to our understanding of how ecosystems work.

Figure 16.4
Forest Ecosystem.
This illustration shows many of the organisms in a forest ecosystem. Can you identify the trophic level of each organism within the ecosystem?

Figure 16.5
Aquatic Ecosystem.

This illustration shows many of the organisms in an aquatic ecosystem. Can you identify the trophic level each of the organisms occupies?

The Great Pyramids: Energy, Number, Biomass

The ancient Egyptians constructed elaborate tombs we call the pyramids. The broad base of the pyramid is necessary to support the upper levels of the structure, and it narrows to a point at the top. This same kind of relationship exists when we look at how the various trophic levels of ecosystems are related to one another. At the base of the pyramid is the producer trophic level, which contains the largest amount of energy of any of the trophic levels within an ecosystem. In an ecosystem, the total energy can be measured in several ways. The total producer trophic level can be harvested and burned. The number of calories of heat energy produced by burning is equivalent to the energy content of the organic material of the plants. Another way of determining the energy present is to measure the rate of photosynthesis and respiration and calculate the amount of energy being trapped in the living material of the plants.

Since only the plants in the producer trophic level are capable of capturing energy from the sun, all other organisms are directly or indirectly dependent on the producer trophic level. The second trophic level consists of herbivores that eat plants. This trophic level has significantly less energy in it for several reasons. In general, there is about a 90% loss of energy as we proceed from one trophic level to the next higher level. Actual measurements will vary from one ecosystem to another, but 90% is a good rule of thumb. One of the major reasons for this loss in energy content at the second and subsequent trophic levels is the second law of thermodynamics. This law states that whenever energy is converted from one form to another, some of the energy is converted to useless heat. Think of any energy-converting machine; it probably releases a great deal of energy. For example, an automobile engine must have a cooling system to get rid of the heat energy produced. An incandescent light bulb also produces large amounts of heat. Living systems are somewhat different, but they must follow the same energy rules.

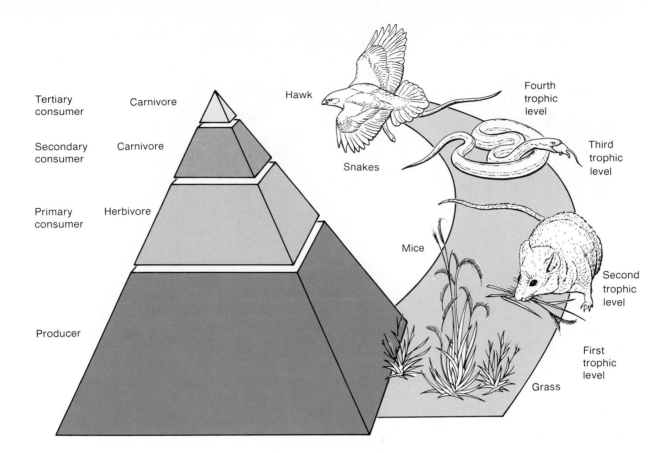

Tertiary consumer — Carnivore
Secondary consumer — Carnivore
Primary consumer — Herbivore
Producer

Hawk — Fourth trophic level
Snakes — Third trophic level
Mice — Second trophic level
Grass — First trophic level

In addition to the loss of energy in compliance with the second law of thermodynamics, there is an additional loss involved in the capture and processing of food material by herbivores. Although herbivores don't need to chase their food, they do need to travel to where food is available, then gather, chew, digest, and metabolize it. All these processes require energy.

Just as the herbivore trophic level experiences a 90% loss in energy content, the higher trophic levels of primary carnivores, secondary carnivores, and tertiary carnivores also have a reduction in the energy available to them. Figure 16.6 shows an energy pyramid in which the energy content decreases by 90% as we pass from one trophic level to the next. Since it is often difficult to measure the amount of energy in any one trophic level of an ecosystem, people often use other methods to quantify the different trophic levels. Often it is easy to simply count the number of organisms at each trophic level. This generally gives the same pyramid relationship and is called a *pyramid of numbers*. Obviously this is not a very good method to use if the organisms at the different trophic levels are of greatly differing size. For example, if you count all the small insects feeding on the leaves of one large tree, you would actually get an inverted pyramid (figure 16.7). Because of this difficulty, many people like to use biomass as a way of measuring ecosystems. **Biomass** is usually determined by collecting all the organisms at one trophic level and measuring their dry weight. This eliminates the size difference problem because all the organisms at each trophic level are weighed. This *pyramid of biomass* also shows the typical 90% loss at each trophic level. Although a biomass pyramid is better than a pyramid of numbers in measuring some ecosystems, it has some shortcomings.

Figure 16.6
Energy Flow Through an Ecosystem.
As energy flows from one trophic level to the next, approximately 90% of the energy is lost. This means that the amount of energy at the producer level must be ten times larger than the amount of energy at the herbivore level.

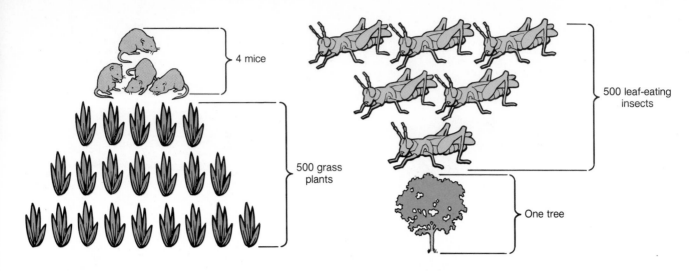

Figure 16.7
Pyramid of Numbers.

One of the easiest ways to quantify the various trophic levels in an ecosystem is to count the number of individuals in a small portion of the ecosystem. As long as all the organisms are of similar size and live about the same length of time, it gives a good picture of how different trophic levels are related. However, if the organisms at one trophic level are much larger or live much longer than the others, our picture of the relationship may be distorted. This is what happens when we look at the relationship between forest trees and the insects that feed on them. A pyramid of numbers becomes inverted in this instance.

Some organisms tend to accumulate biomass over long periods of time, while others do not. Many trees live for hundreds of years, while their primary consumers, insects, generally only live one year. Likewise, a whale is a long-lived animal, while its food organisms are relatively short-lived. Figure 16.8 show two biomass pyramids.

Although measuring differences in trophic levels within ecosystems is difficult, we should not forget that regardless of how the measurement is taken, there is a loss of energy that occurs each time energy passes from one trophic level to the next and that loss is large.

Ecological Communities

The way energy flows through an ecosystem involves many interchanges between organisms. We can distinguish between the ecosystem and the interacting organisms that are a part of it. An ecosystem is a unit that consists of the physical environment and all the interacting organisms within that area. The collection of interacting organisms within an ecosystem is called a **community.** The community consists of many kinds of organisms. The number of individuals of a particular species in an area is called a **population.** Therefore, we can look at the same organism from several points of view. We can look at it as an individual, as a part of a population of similar individuals, as a part of a community that includes other populations, and as a part of an ecosystem, which includes abiotic factors as well as living organisms.

As you know from the discussion in the previous section, one of the ways that organisms interact is by feeding on one another. The sequence of organisms that feed on one another is called a **food chain** (figure 16.9). Typical food chains contain a producer organism and various levels of consumers, including decomposers. Many factors determine an organism's position in a food chain. One such factor is the ability of the organism to digest cellulose, which is a complex carbohydrate typically found in the cell walls of plants. Humans, along with many other animals, do not have the enzymes necessary to break down the cellulose to simple sugars. However, many herbivorous animals are able to digest cellulose and derive nourishment from it. Carnivores have digestive systems that are designed to handle meat as a major food item. Plant eaters have teeth suitable for grinding plant material and long

Figure 16.8
Pyramid of Biomass.

Biomass is determined by collecting and weighing all the organisms in a small portion of an ecosystem. This method of quantifying trophic levels eliminates the problem of different-sized organisms at different trophic levels. However, it does not always give a clear picture of the relationship between trophic levels if the organisms have widely different lengths of life. For example, in aquatic ecosystems, many of the small producers may divide several times per day. The zooplankton that feed on them live much longer and tend to accumulate biomass over time. The single-celled algae produce much more living material but it is eaten as fast as it is produced, so it is not allowed to accumulate.

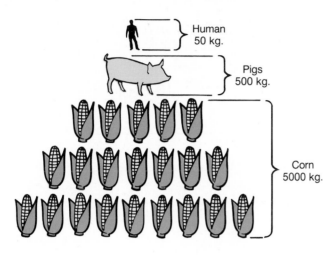

Human
50 kg.

Pigs
500 kg.

Corn
5000 kg.

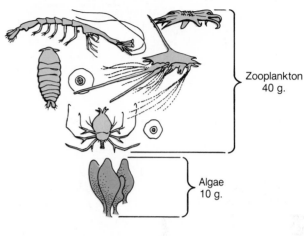

Zooplankton
40 g.

Algae
10 g.

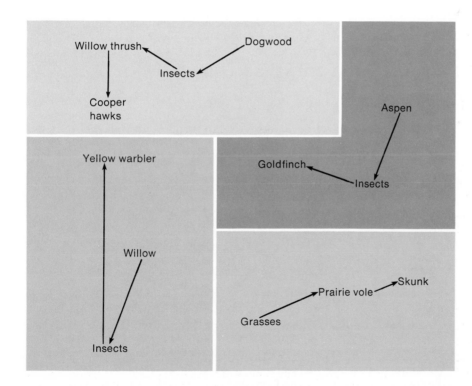

Willow thrush

Dogwood

Insects

Cooper hawks

Yellow warbler

Willow

Insects

Aspen

Goldfinch

Insects

Prairie vole

Skunk

Grasses

Figure 16.9
Food Chains.

There are a large number of different food chains in any community. Four food chains are illustrated here. They are a part of a larger food web seen in figure 16.10.

digestive tracts necessary to capture the relatively dilute nutrient content of their food. Carnivores usually have teeth that are useful for cutting and tearing food into chunks that can then be swallowed whole. Their digestive tracts are relatively short since the nutrient content of their animal food is high.

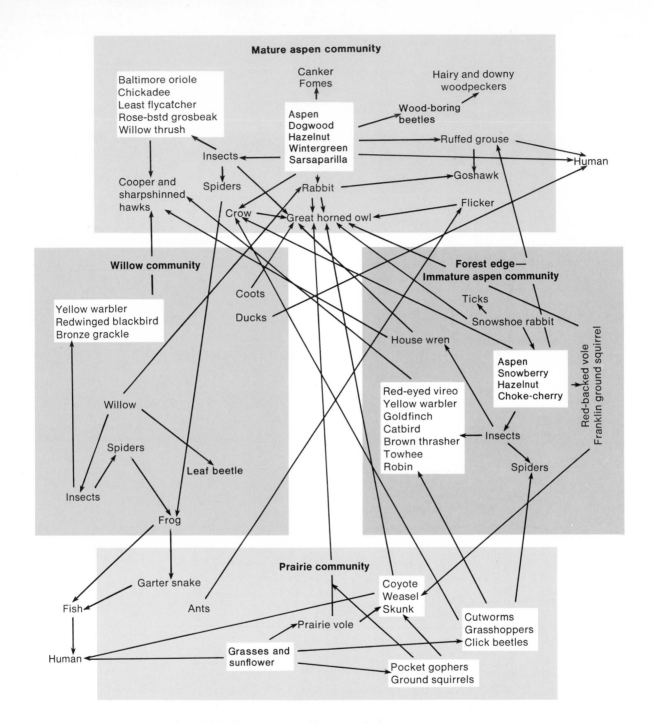

Figure 16.10
Food Web.

When many different food chains are interlocked with one another, a food web results. Notice that some organisms are a part of several food chains. Notice the great horned owl in particular. Because of the interlocking nature of the food web, as conditions change there can be shifts in the way in which food flows through this system.

Within a community there are many different food chains. Some organisms may be involved in several of the food chains at the same time, so the food chains become interwoven into a **food web** (figure 16.10). In a community, the interacting food chains usually result in a relatively stable mixture of populations. If a particular kind of organism is removed from a community, there is usually some adjustment that occurs in the populations of other organisms within the community. For example, humans have used insecticides to control the populations of many kinds of insects. This often disrupts the natural community and may result in lower numbers of insect-eating birds. Often the use of insecticides actually increases the insect

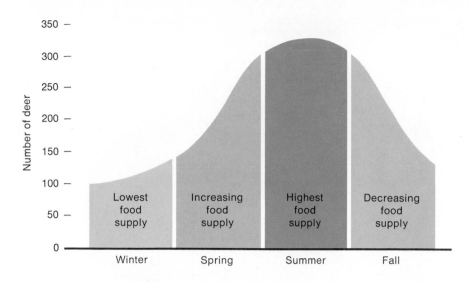

Figure 16.11
Annual Changes in Population Size.
The number of organisms living in an area varies during the year. The availability of food is the primary factor in determining the size of the population of deer in this illustration, but water availability, availability of soil nutrients, or other factors could also be important.

problem because insecticides kill many different kinds of insects rather than the one or two target species. Since many insects use other insects for food, herbivore insects may develop increased populations because there are no carnivore insects to eat them.

Communities are dynamic collections of organisms. As one population increases, another decreases. This might occur over several years, or even in the period of one year. Most ecosystems are not constant. There may be differences in rainfall throughout the year or changes in the amount of sunlight and average temperature. Because of this, you should expect populations to fluctuate as abiotic factors change. The change in the size of one population will trigger changes in other populations as well. Figure 16.11 shows what happens to the size of a population of deer as the seasons change. The area can support one hundred deer from January through February, when plant food for deer is least available. As spring arrives, plant growth increases, and the number of deer increases. It is no accident that deer breed in the fall and give birth in the spring, since it is during the spring that the producers are increasing and the area has more food. It is also no accident that wolves and other carnivores that feed on deer give birth in the spring. The increased available energy of producers means more food for deer (herbivores), which, in turn, means more energy for the wolves (carnivores) at the next trophic level.

Types of Communities

Since communities are complex and interrelated, it helps us to set artificial boundaries. This allows us to focus our study on a definite collection of organisms. An example of a community that has natural boundaries that are easy to determine is a small pond. The water's edge naturally defines the limits of this community. You would expect to find certain animals and plants living in the lake, such as fish, frogs, snails, insects, algae, pondweeds, bacteria, and fungi. But you might ask at this point, What about the plants and animals that live right at the water's edge? That leads us to think about the animals that only spend part of their lives in the water. That awkward looking, long-legged bird wading in the shallow and darting its long beak down to spear a fish has its nest atop some tall trees away from the water. Should it be considered part of the pond community? Should we also include the deer that comes to drink at dusk and then wanders away? Small parasites could enter the body of the deer as it drinks. The immature parasite would develop into

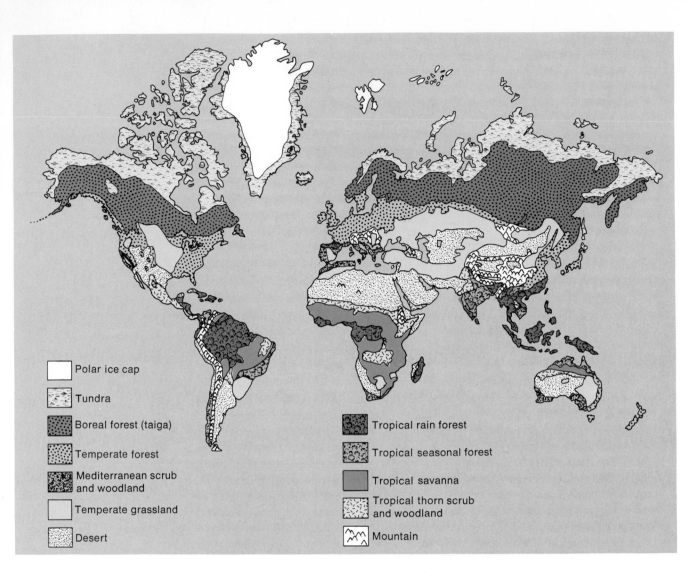

Polar ice cap

Tundra

Boreal forest (taiga)

Temperate forest

Mediterranean scrub and woodland

Temperate grassland

Desert

Tropical rain forest

Tropical seasonal forest

Tropical savanna

Tropical thorn scrub and woodland

Mountain

Figure 16.12
Biomes of the World.

Major climate differences
determine the kind of vegetation
that can live in a region of the
world. Associated with specialized
groups of plants are particular
communities of animals. These
major land types of ecosystems
are called biomes.

an adult within the deer's body. That same parasite must spend part of its life cycle in the body of certain snails. Are these parasites part of the pond community? Several animals are members of more than one community. What originally seemed to be a clear example of a community has become less clear-cut. Although the general outlines of a community can be arbitrarily set for the purposes of a study, we must realize that the boundaries of a community, or any ecosystem for that matter, must be considered somewhat artificial.

A community can be small, such as a tiny puddle or the organisms that live in and on a single leaf of a tree. It all depends on where the boundaries are set. The pond and the other small communities can all be collected into large regional communities called **biomes.** One of the large, land-based biomes in the eastern part of North America is the *temperate deciduous forest*. This biome, and other land-based biomes, is named for the major features of the ecosystem, which in this case happens to be the dominant vegetation. The predominant plants are large trees that lose their leaves more or less completely during the fall of the year and, therefore, are called deciduous. Most of the trees require a considerable amount of rainfall. This naming system works fairly well, since the major type of plant determines the other kinds

Figure 16.13
Prairie Biome.
This typical short-grass prairie of the western United States is associated with an annual rainfall of 38–50 centimeters. This community contains a unique grouping of plant and animal species.

of plants and animals that can occur. Of course, since the region is so large and has different climatic conditions in different areas, we can find some differences in the particular species of trees (and other organisms) found. For instance, in Maryland the tulip tree is one of the common large trees, while in Michigan it is so unusual that people plant it in lawns and parks as a decorative tree. Aspen, birch, cottonwood, oak, hickory, beech, and maple are typical trees found in this geographic region. Typical animals of this biome are skunks, porcupines, deer, frogs, opossums, owls, mosquitoes, and beetles.

Refer to the map of the various biomes (figure 16.12). The temperate deciduous forest covers a large area from the Mississippi River to the Atlantic Coast, and from Florida to southern Canada. This type of biome is also found in parts of Europe and Asia. Many local spots within this biome are quite different from one another. Many of them have no trees at all. For example, the tops of some of the mountains along the Appalachian Trail, the sand dunes of Lake Michigan, and the scattered grassy areas in Illinois are natural areas within this biome that lack trees. In much of this region the natural vegetation has been removed to allow for agriculture, so the original character of the biome is gone except where farming is not practical or the original forest has been preserved.

The biome located to the west of the temperate deciduous forest in North America is the *prairie biome*. This kind of biome is also common in parts of Eurasia, Africa, Australia, and South America. The dominant vegetation in this region is made up of various species of grasses. The rainfall in this grassland is not enough to support the growth of trees (figure 16.13). Trees are common on this biome only along streams, which provide larger amounts of water. The plants that are common in this area are those that can grow in drier conditions. Animals found in this area include the prairie dog, pronghorn antelope, prairie chicken, grasshopper, rattlesnake, and meadowlark. Most of the original grasslands, like the temperate deciduous forest, have been converted to agricultural uses. This required the breaking of a thick layer of sod formed by the original species of grasses that grew on the plains.

Figure 16.14
Savanna Biome.
A savanna is likely to develop in areas that have a rainy season and a dry season. During the dry season, fires are frequent. These fires kill tree seedlings and prevent the establishment of forests.

Figure 16.15
Desert Biome.
The desert gets less than 25 centimeters of precipitation per year, but it teems with life. Cactus, sagebrush, lichens, snakes, small mammals, birds, and insects inhabit the desert. Because daytime temperatures are high, most animals are only active at night when the air temperature drops significantly.

Sod is a thick spongy layer of roots that helps retain water and hold the soil in place. Today very little of the original prairie biome still exists. It has been converted to agricultural uses by breaking the sod so that wheat, corn, and other grains could be grown. This exposes the soil to the wind, which may cause excess drying and result in soil erosion that depletes the fertility of the soil.

A biome that is similar to a prairie is a *savanna* (figure 16.14). Savannas are typical of central Africa and parts of South America. Typically the area consists of grasses with scattered trees. The area generally has a wet and dry season and typically has fires during the dry part of the year.

Very dry areas are known as *deserts*. This kind of biome is found throughout the world wherever rainfall is low and irregular. Some deserts are extremely hot, while others can be quite cool during much of the year. The distinguishing characteristic of deserts is low rainfall, not high temperature. Furthermore, deserts show large daily fluctuations in air temperature. When the sun goes down at night, the land begins to cool off very rapidly. There is no insulating blanket of clouds to keep the heat from radiating into outer space. This biome is characterized by scattered, thorny plants that lack leaves or have reduced leaves. Many of the plants, like cacti, are capable of storing water in their fleshy stems (figure 16.15). Although this is a very harsh environment, there are many kinds of flowering plants, insects, reptiles, and mammals that are capable of living in this biome. Typically the animals avoid the hottest part of the day by staying in burrows or other shaded, cool areas.

Through parts of southern Canada, extending southward along the mountains of the United States, and in a major part of the Soviet Union we find communities that are dominated by evergreen trees. This is the *coniferous forest biome* (figure 16.16). The evergreen trees are especially adapted to withstand long, cold winters with abundant snowfall. Most of the trees in the wetter, colder areas are spruces and firs, but some drier, warmer areas have pines. The wetter areas generally have dense stands of small trees intermingled with many other kinds of vegetation and broken up by many small lakes and bogs. In the mountains of the western United States, the pines are often very large with few branches near the ground. They are widely scattered, giving a parklike appearance, since there is very little vegetation on the forest floor. Typical animals in this biome are mice, wolves, squirrels, moose, midges, and flies.

Figure 16.16
Coniferous Forest Biome.
Conifers are the dominant vegetation in most of Canada and at high altitudes in sections of western North America. It is characterized by cold winters with abundant snowfall.

North of the coniferous forest biome is an area known as the *tundra* (figure 16.17). It is characterized by extremely long, severe winters and short, cool summers. The deeper layers of the soil remain permanently frozen; this is known as the permafrost. Because of these conditions, very few kinds of animals and plants can survive. No trees can live in this region. Typical plants and animals of the area are dwarf willow and some other shrubs, reindeer moss, some flowering plants, caribou, wolves, musk oxen, fox, snowy owls, mice, and many kinds of insects. Many kinds of birds are summer residents only. The tundra community is relatively simple, so any changes may have drastic and long-lasting effects. The tundra is easy to injure and slow to heal; therefore, we must treat it gently. The construction of the Alaskan pipeline has left scars that could still be there a hundred years from now.

The climate opposite that found in the tundra is found in the *tropical rain forest*. Tropical rain forests are found primarily near the equator in Central and South America, Africa, parts of southern Asia, and some Pacific Islands (figure 16.18). The temperature is high, rain falls nearly every day, and there are thousands of species of plants in a small area. Balsa (a very light wood), teak (used in furniture), and ferns the size of trees are examples of trees from the tropical rain forest. Typically every plant has other plants growing on it. Tree trunks are likely to be covered with orchids, many kinds of vines, and mosses. Termites, tree frogs, bats, lizards, monkeys, and an almost infinite variety of birds and insects inhabit the rain forest. These forests are very dense and little sunlight reaches the forest floor. When the forest is opened up (by a hurricane or the death of a large tree) and sunlight reaches the forest floor, the opened area is rapidly overgrown with vegetation. Since plants grow so quickly in these forests, many attempts have been made to bring this land under cultivation. North American agricultural methods require the clearing of large areas and the planting of a single species of crop, such as corn. The constant rain falling on these fields quickly removes the soil's nutrients and heavy applications of

Figure 16.18
Tropical Rain Forest Biome.
The tropical rain forest is a moist, warm region of the world located near the equator. The growth of vegetation is extremely rapid. There is more variety in kinds of plants and animals in this biome than in any other.

fertilizer are required. Often these soils become hardened when exposed in this way. Although most of these forests are not suitable for agriculture, large expanses of tropical rain forest are being cleared yearly because of the pressure for more farmland in the highly populated tropical countries and the desire for high quality lumber from many of the forest trees.

Human Use of Ecosystems

How much humans use an ecosystem is often tied to the productivity of the ecosystem. **Productivity** is the rate at which an ecosystem can accumulate new organic matter. Since the plants are the producers, it is their activities that are most important. Those situations that provide the best circumstances for plant growth are the most productive. Warm, moist, sunny areas with high levels of nutrients in the soil are ideal. Some areas will have low productivity because one of the essential factors is missing. Deserts have low productivity because water is scarce, arctic areas because temperature is low, and the open ocean because nutrients are in short supply. Some communities, such as coral reefs and tropical rain forests, have high productivity. Marshes and estuaries are especially productive since the waters running into them are rich in the nutrients aquatic photosynthesizers need. Furthermore, these aquatic systems are usually shallow so that light can penetrate through most of the water column.

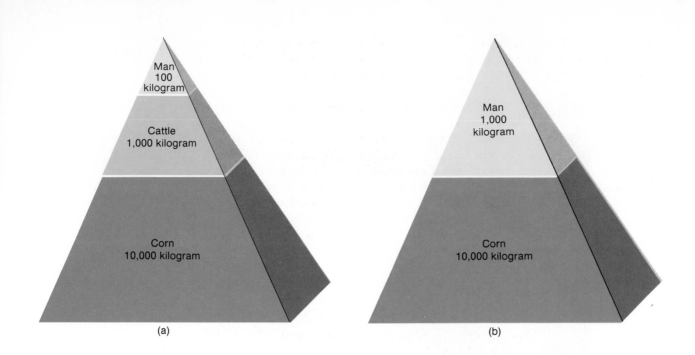

(a)

(b)

Figure 16.19
Human Biomass Pyramids.
Since approximately 90% of the
energy is lost as energy passes
from one trophic level to the next,
more people can be supported if
they eat producers directly than if
they feed on herbivores. Much of
the less developed world is in this
position today. Rice, corn, wheat,
and other producers provide the
majority of food for the world's
people.

Humans have been able to make use of naturally productive ecosystems by harvesting the food from them. However, in most cases, we have altered certain ecosystems substantially to increase productivity for our own purposes. In so doing, we have destroyed the original ecosystem and replaced it with an agricultural ecosystem. For example, the Native Americans living in the Great Plains area of the United States used buffalo as a source of food. There was much grass, many buffalo, and few humans. Therefore, in the Native Americans' pyramid of energy, the base was more than ample. However, with the discovery and settling of America, the human population in North America increased at a rapid rate. The top of the pyramid became larger. The food chain (prairie grass—buffalo—human) could no longer supply the food needs of the growing population. As the top of the pyramid grew, it became necessary for the producer base to grow larger. Since wheat and corn yield more biomass for humans than the original prairie grasses could, the settlers' domestic grain and cattle replaced the prairie grass and buffalo. This is fine if you are settlers, but unfortunate if you are buffalos or Native Americans.

In our pursuit of more productivity for our own purposes, we have often overlooked the alterations we have been making in the worldwide ecosystem known as the **biosphere.** In many parts of the world, the human demand for food is so large that it can only be met by having humans occupy the herbivore trophic level rather than the carnivore trophic level. Humans are omnivores that can eat both plants and animals as food, so they have a choice. However, as the size of the human population increases, it cannot afford the 90% loss that occurs when plants are fed to animals that are in turn eaten by humans. In much of the less developed world, the primary food is grain; therefore, the people are at the herbivore level. It is only in the developed countries that people can afford to eat meat. This is true from both an energy point of view and a monetary point of view. Figure 16.19a shows a pyramid of biomass having a producer base of 10,000 kilograms of corn. The second trophic level only has 1000 kilograms of cattle because of the 90% loss typical when energy is transferred from one trophic level to the next. The consumers at the third trophic level, humans in this case, experience a similar 90% loss. Therefore, only 100 kilograms of humans could be sustained by the two-step energy transfer. There has been a 99% loss in energy. 10,000 kilograms of corn are necessary to sustain 100 kilograms of humans.

Humans do not need to be carnivores at the third trophic level but can switch most of their food consumption to the second trophic level. There would then only be a 90% loss rather than a 99% loss and the 10,000 kilograms of corn could support 1000 kilograms of humans (figure 16.19b). By eliminating the cattle in the humans' food chain, ten times as much human life can be supported by the same amount of plant material. In parts of the world where food is scarce, people cannot afford the energy loss involved in passing food through the herbivore trophic level. Consequently, most of the people of the world are consumers at the second trophic level and rely on corn, wheat, rice, and other first trophic level organisms as food. Because much of the world's population is already feeding at the second trophic level, we cannot expect food production to increase to the extent that we could feed ten times more people than exist today.

It is unlikely that most people will be able to fulfill all of their nutritional needs by just eating grains. In addition to calories, people need a certain amount of protein in their diet. One of the best sources of protein is meat. Although protein is available from plants, the concentration is greater from animal sources. Figure 16.20 shows the quality of the diet experienced by people around the world. Major parts of Africa, Asia, and Latin America have diets that are deficient in both calories and protein. These people have very little food, and what food they do have is mainly from plant sources. These are also the parts of the world where human population growth is most rapid. In other words, these people are poorly nourished and as the population increases, they will probably experience greater calorie and protein deficiency. This example reveals that even when people live as consumers at the second trophic level, they may still not get enough food, and if they do, it may not have the necessary protein for good health.

Figure 16.20
Geography of Hunger.
This map illustrates the amount of calories the average person receives and the adequacy of the protein intake. (From *Human Ecology: Problems and Solutions* by Paul R. Ehrlich, Anne H. Ehrlich and John P. Holdren. W. H. Freeman and Company. Copyright © 1973.)

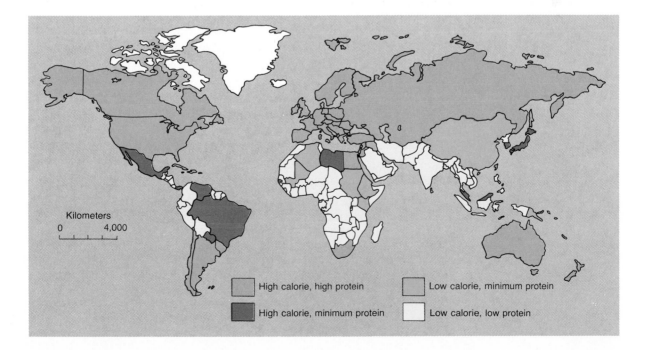

Kilometers
0   4,000

High calorie, high protein   Low calorie, minimum protein

High calorie, minimum protein   Low calorie, low protein

Box 16.1
Integrated Pest Management

Since the invention and use of DDT as a pesticide in the mid-1900s, pesticide use has become a major method of controlling unwanted insect pests. Insects are our major competitors for food, so any method that will control their populations will pay off in greater amounts of food for people. Insects and other pests destroy approximately 35% of the crops produced worldwide, so the problem is great. However, several kinds of problems are associated with pesticide use. Some pesticides are extremely toxic to humans and present a danger to human health. Some accumulate in food chains so that nontarget organisms at higher trophic levels are killed. Most pesticides are nonspecific and kill a variety of insects and other kinds of animals that are not target organisms. Finally, most insect populations develop resistance to pesticides after repeated long-term exposure to the pesticide. Because of these problems with pesticides, people have been looking for ways to reduce the dependence of agriculture on pesticides.

Integrated pest management is an approach to pest control that seeks to use a wide variety of techniques in an organized way to control specific pest species. It depends upon a complete understanding of all ecological aspects of the crop and the particular pests to which it is susceptible. Information about the time the pest hatches, what its natural enemies are, what conditions favor survival, and what conditions encourage reproduction are combined to plan an attack on the specific pest. Several different techniques have been successful. In some cases, a complete understanding of the ecological relations of the pest can suggest methods of applying the pesticide or allow timing of pesticide use so that it has its maximum impact on the pest and minimizes the impact on other beneficial insects. For example, the use of extremely tiny droplets that hang in the air in the evening is very effective against night flying insects like mosquitoes, but is less likely to kill large insects or those that are day flying like honeybees.

Many of the integrated pest control methods humans use are not pesticides at all. The screwworm fly weakens or kills large numbers of cattle, goats, deer, and other grazing animals when it lays its eggs in open wounds on the animal. The larvae feed and grow, causing illness in the animal. However, it was discovered that the females of this species only mate once. When large numbers of male flies were raised, made infertile by radiation, and released into the normal population, many of the wild females mated with these sterile males and the population was brought under control. This technique has been used effectively in parts of the southern United States when the screwworm fly was introduced from Mexico.

Grape growers in California's San Joaquin Valley had a serious problem with a grape leaf hopper that injured their grape plants. It was discovered that a parasitic wasp would cause the death of large numbers of leaf hoppers when it laid its eggs on the leaf hopper and the larva used the leaf hopper for food. However, the wasp only spent the summer in the vineyard and moved to blackberry plants during the winter. Most of the blackberry plants had been removed from the area as large numbers of grapes were planted. The removal of blackberry plants had actually made the pest problem worse.

There are now companies that raise various kinds of parasites, bacteria, and viruses that can be used to control the populations of specific target species. Although integrated pest management is one way of controlling pests, it does have some drawbacks. Much more detailed knowledge about the life history of the pest is needed, which is often difficult or expensive to obtain. The time needed to develop effective controls is often long as well, and we want quick, inexpensive control.

Resource Use

In our haste to improve our lot, we have often failed to look at what we have been doing to the total ecosystem and, in the final analysis, to ourselves. Many of the minerals obtained from ore deposits are the result of the accumulations from past ecosystems. When they are used up, they will be gone forever. Coal, oil, natural gas, and iron ore are examples of materials that are being used up much faster than they are being produced. These natural resources were built up over millions of years. We have been using them for only the past two hundred years but have already used up millions of years of accumulation.

As has been pointed out, survival is the name of the game. It is a fact of life that in order to live, we must change ecosystems. The question is, Must we change it as much or as fast as we have in the past?

The United States makes up approximately 5% of the world's population, yet we consume about 30% of the world's resources annually. Many of these resources must be imported. As other countries improve their standard of living, they will demand more of the world's energy and raw materials. Such a demand will cause our standard of living to change as well. We will need to lower our consumption and pay more for what we use. The shortage of some materials has come as a shock to many Americans. It shouldn't. Remember, we are only a small part of a large, complex ecosystem. The actions of other people in the world (remember they make up 95% of the world's population) will have a significant impact on our way of life because we are all interrelated within the biosphere.

Humans have a complex mixture of needs and wants. Some of our needs are basic to all animals, but some are unique cultural wants. "Human-made wants" determine our standard of living and this standard of living is part of the human ecosystem. Like all ecosystems, the human ecosystem requires energy to convert raw materials into items we can use. For all practical purposes, our present forms of energy and our raw materials are nonrenewable resources. We often lose sight of this fact and live as if there were an endless supply of energy and natural resources. In any ecosystem, the decomposers are essential. If we are to maintain our present standard of living, we must take a lesson from the decomposers and learn to recycle raw materials.

Summary

Ecology is the study of how organisms interact with their environment. The environment consists of biotic and abiotic components that are interrelated into an ecosystem. All ecosystems must have a constant input of energy from the sun. Producer organisms are capable of trapping the sun's energy and converting it into biomass. Herbivores feed on producers and are in turn eaten by carnivores, which may be eaten by other carnivores. Each level in the food chain is known as a trophic level. Other kinds of organisms involved in food chains are omnivores, which eat both plant and animal food, and decomposers that break down dead organic matter and waste products. All ecosystems have a large producer base with successively smaller amounts at the herbivore, primary carnivore, and secondary carnivore trophic levels because each time energy passes from one trophic level to the next, about 90% of the energy is lost to the ecosystem. A community consists of the interacting populations of organisms in an area. The organisms are interrelated in many ways in food chains that interlock to create food webs. Because of this interlocking, changes in one part of the community can have effects elsewhere.

Major land-based regional ecosystems are known as biomes. The temperate deciduous forest, coniferous forest, tropical rain forest, desert, savanna, and tundra are examples of biomes.

Humans use ecosystems to provide themselves with necessary food and raw materials. As the human population increases, most people will be living as herbivores at the second trophic level, since we cannot afford to lose 90% of the energy by first feeding it to an herbivore which we then eat. Humans have converted most productive ecosystems to agricultural production and continue to seek more agricultural land as population increases. In addition to food, we use ecosystems to provide other raw materials. We draw on the accumulated materials from past ecosystems for fuel and other materials.

Consider This   Describe a world in which there are no decomposers. List ten ways in which it would differ from the present world.

Experience This   The next time you are in the grocery store, determine the price per gram of the following kinds of items:

dry beans
whole wheat bread
flour
sugar
rice
hamburger
frozen fish
ham
chicken

Why are there differences in prices?

Questions
1. Why are rainfall and temperature important in an ecosystem?
2. Describe the flow of energy through an ecosystem.
3. What is the difference between *ecosystem* and *environment*?
4. What role does each of the following play in an ecosystem: sunlight, plants, second law of thermodynamics, consumers, decomposers, herbivores, carnivores, and omnivores?
5. Give an example of a food chain.
6. What is meant by the term *trophic level*?
7. Why is there usually a larger herbivore biomass than carnivore biomass?
8. List a predominant abiotic factor in each of the following biomes: temperate deciduous forest, coniferous forest, desert, tundra, tropical rain forest, and savanna.
9. Can energy be recycled through an ecosystem?
10. What is the difference between an ecosystem and a community?

Chapter Glossary   **abiotic factors** (a-bi-ot′ik fak′tōrz) Nonliving parts of an organism's environment.
biomass (bi-o-mas) The dry weight of a collection of designated organisms.
biomes (bi-ōmz) Large, regional communities.
biosphere (bi-o-sfēr) A worldwide ecosystem.
biotic factors (bi-ot′ik fak′tōrz) Living parts of an organism's environment.
carnivores (kar-nĭ-vōrz) Those animals that eat other animals.
community (kom-miu′nĭ-te) A collection of interacting organisms within an ecosystem.
consumers (kon-soom′urs) Organisms that must obtain energy in the form of organic matter.
decomposers (de-kom-po′zurs) Organisms that use dead organic matter as a source of energy.
ecology (e-kol′o-je) A branch of biology that studies the relationships between organisms and their environment.
ecosystem (e″ko-sis-tum″) An interacting collection of organisms and the abiotic factors that affect them.
environment (en-vi′ron-ment) Anything that impacts on an organism during its lifetime.

food chain (food chān) A sequence of organisms that feed on one another resulting in the flow of energy from a producer through a series of consumers.

food web (food web) A system of interlocking food chains.

herbivores (her′-bĭ-vōrz) Those animals that feed directly on plants.

omnivores (om′nĭ-vōrz) Those animals that are carnivores at some times and herbivores at others.

population (pop′′u-la′shun) The number of individuals of a specific species in an area.

primary carnivores (pri′ma-re kar′nĭ-vōrz) Those carnivores that eat herbivores and, therefore, are on the third trophic level.

primary consumers (pri′ma-re kon-su′merz) Those organisms that feed directly on plants—herbivores.

producers (pro-du′surz) Organisms that produce new organic material from inorganic material with the aid of sunlight.

productivity (pro-duk-tiv′ĭ-te) The rate at which an ecosystem can accumulate new organic matter.

secondary carnivores (sek′un-da-re kar′nĭ-vorz) Those carnivores that feed on primary carnivores and, therefore, are in the fourth trophic level.

secondary consumers (sek′un-da-re kon-su′murz) Those animals that eat other animals—carnivores.

trophic level (tro′fik le-vel) A step in the flow of energy through an ecosystem.

Community Interactions

■ Purpose

Within ecosystems organisms influence one another in many ways. Even organisms of the same species affect one another in the course of their normal daily activities. It is the purpose of this chapter to consider some of the kinds of interactions that occur within ecosystems and recognize the variety of ways that organisms within communities impact each other in the cycling of matter.

CHERNOBYL

The nuclear power plant disaster on 26 April 1986 in Chernobyl, U.S.S.R. caused a renewed concern about radioactive fallout. Of the isotopes released, one is of particular concern. Iodine 131 is an isotope of iodine that is radioactive. It accumulates in the thyroid where it can kill thyroid cells or cause cancer. In Europe, many people were advised not to eat fresh garden vegetables because iodine 131 could have fallen onto the plants. Similarly, people were advised not to drink milk from cows that had been grazing in open pastures, because the iodine 131 they consumed could show up in their milk.

■ Understand that organisms interact in a variety of ways within communities.

■ Recognize the differences between community, habitat, and niche.

■ Describe how atoms are cycled in communities.

■ Appreciate that humans alter and interfere with natural ecological processes.

■ Recognize that communities proceed through a series of stages to stable climax communities.

Community, Habitat, and Niche

There are two major ways that people approach the study of organism interactions. Many people look at interrelationships from the broad ecosystem point of view, while others focus on the individual organisms and the specific things that affect them in their daily lives. The first of these involves the study of all of the organisms that interact with one another—the community—and usually looks at general relationships among the organisms. In chapter 16 we talked about categories like producers, consumers, and decomposers as different kinds of activities performed by organisms.

Another way of looking at interrelationships is to study in detail the ecological relationships of particular species of organisms. Each organism has particular requirements for life and lives where the environment provides what is needed. The environmental requirements of a whale include large amounts of salt water. So you would not expect to find a whale in the desert nor an elephant in the ocean. The kind of place or part of an ecosystem occupied by an organism is known as its **habitat.** Habitats are usually described in terms of conspicuous or particularly significant features in the area where the organism lives. For example, the habitat of a prairie dog is usually described as a grassland, while the habitat of a tuna is described as the open ocean. The habitat of the fiddler crab is sandy ocean shores, and the habitat of various kinds of cacti is the desert. The key thing to keep in mind when you think of habitat is the *place* a particular kind of organism lives. When describing habitats of organisms, we sometimes use the terminology of the major biomes of the world. However, it is also possible to describe the habitat of the bacterium *Escherichia coli* as the human gut, or the habitat of a fungus as a rotting log. Organisms that have very specific places in which they live simply have more restricted habitats.

Each species of an organism has particular requirements for life and places specific demands on the habitat in which it lives. The specific functional role of an organism is its **niche.** Its niche is the way it goes about living its life. Just as the word *place* is the key to understanding the concept of habitat, the word *function* is the key to understanding the concept of niche. The niche of an organism involves a detailed understanding of the impacts an organism has on its biotic and abiotic surroundings, as well as everything that has an effect on the organism. For example, the niche of an earthworm includes abiotic items such as soil particle size, soil texture, and the moisture, pH, and temperature of the soil. The same niche also includes biotic impacts such as serving as food for birds, moles, and shrews, as bait for anglers, or as a consumer of dead plant organic matter (figure 17.1). In addition, it serves as a home for a variety of parasites, transports minerals and nutrients from deeper soil layers to the surface, and creates burrows that allow air and water to penetrate the soil more easily.

Some organisms have rather broad niches, while others are quite narrow with very specialized requirements and limited roles to play. The opossum is an animal with a very broad niche. It will eat a wide variety of plant and animal foods, can adjust to a wide variety of different climates, is used as food by many kinds of carnivores, including humans, and reproduces large numbers of offspring. By contrast, the koala of Australia has a very narrow niche. It can only live in areas of Australia with specific species of *Eucalyptus* trees, because the only things it eats are the leaves of a few kinds of these trees. Furthermore, it cannot tolerate low temperature and does not reproduce large numbers of offspring (figure 17.2). As you might guess, the opossum is expanding its range and the koala is endangered in much of its range.

Figure 17.1
The Niche of An Earthworm.

The niche of an earthworm involves a large number of factors. It includes the fact that the earthworm is a consumer of dead organic matter, is food for other animals, is a host to parasites, and is bait for an angler. Furthermore, it includes the fact that the earthworm loosens the soil by its burrowing and "plows" the soil when it deposits materials on the surface. Additionally, the pH, texture, and moisture content of the soil have an impact on the earthworm. (Please keep in mind that this list is only a small part of what the niche of the earthworm includes.)

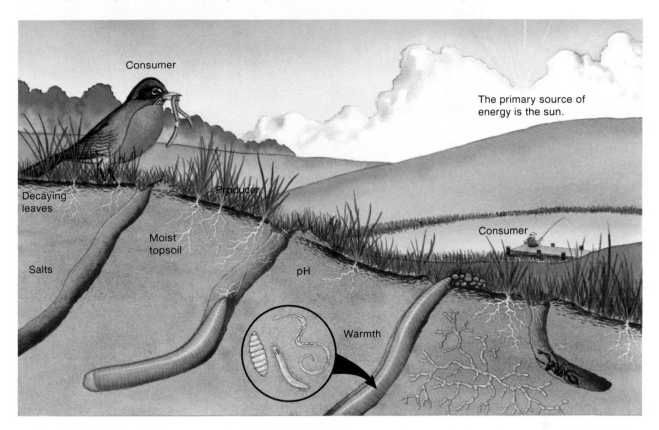

Figure 17.2
Broad and Narrow Niches.

The opossum has a very broad niche. It eats a variety of foods, is able to live in a variety of habitats, and has a large reproductive capacity. It is generally extending its range in the United States. The koala has a narrow niche. It only feeds on the leaves of specific *Eucalyptus* trees, is restricted to relatively warm, forested areas, and is generally endangered in much of its habitat.

Figure 17.3
Predator-Prey Relationship.
Many predators capture prey by making use of speed. Since strength is needed to kill the prey, in general the predator is larger than the prey. Predators obviously benefit from the food they obtain and the prey organism is harmed.

The complete description of an organism's niche is a very detailed inventory of influences, activities, and impacts. It involves what the organism does and what is done to the organism. Some of the impacts are abiotic, others are biotic. Since the niche of an organism is a complex set of items, it is often easy to overlook important roles played by organisms.

For example, in Australia when Europeans introduced cattle into a continent where there had previously been no large hoofed mammals, they did not think about the impact of cow manure or the significance of a group of beetles called dung beetles. These beetles rapidly colonize fresh dung and cause it to be broken down. No such beetles existed in Australia; therefore, a significant amount of land was covered with accumulated cow dung. The problem was eventually solved by the importation of several species of dung beetles from Africa, where large hoofed mammals are common. The cattle were consumers of plant material and the dung beetles made use of what the cattle did not digest, returning it to a form that the plant could more easily recycle into plant biomass. There are many other ways organisms can interact that can be divided into major categories.

Kinds of Organism Interactions

When organisms encounter one another in their habitats there are a number of ways they can influence one another. One kind of organism interaction is predation. **Predation** occurs when one animal captures, kills, and eats another animal. The organism that is killed is called the **prey** and the organism that does the killing is called the **predator.** The predator obviously benefits from the relationship, while the prey organism is harmed. Most predators are relatively large compared to their prey and have specific adaptations that aid them in catching prey. Many spiders build webs that serve as nets to catch flying insects. The prey are quickly paralyzed by the spider's bite and wrapped in a tangle of silk threads. Other rapidly moving spiders, like wolf spiders and jumping spiders, have large eyes that allow them to find prey without webs. Dragonflies patrol areas where they capture flying insects. Hawks and owls have excellent eyesight that allows them to find their prey. (Bald eagles can see the numbers on a pocket watch at 8 kilometers.) Many predators, like leopards, lions, and cheetahs, use speed to run down their prey (figure 17.3).

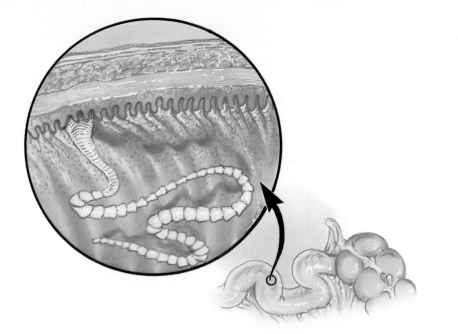

Figure 17.4
Parasite-Host Relationship.
Parasites benefit from the
relationship because they live in or
on the host and get nourishment
from it. The host may not be killed
directly by the relationship but is
often weakened, thus becoming
more vulnerable to predators.
There are more parasites in the
world than organisms that are not
parasites.

Many kinds of predators are useful to us, since they control the populations of organisms that do us harm. For example, snakes eat many kinds of rodents that eat stored grain and other agricultural products. Many birds eat insects that are agricultural pests. It is even possible to think of a predator as having a beneficial effect on the prey species. Certainly the *individual* organism that is killed is harmed, but the *population* can benefit. Predators can prevent starvation by preventing overpopulation in prey species or reduce the likelihood of epidemic disease by eating sick or diseased individuals. Furthermore, predators are acting as selecting agents, since the prey individuals who fall to them are likely to be those that are less well adapted than the individuals who escape predation. Usually the predators kill the slow, stupid, sick, or injured individuals. The genes that may have contributed to slowness, stupidity, illness, or likelihood of being injured are removed from the gene pool and a better adapted population remains. Because predators eliminate poorly adapted *individuals,* the *species* benefits. What is bad for the individual can be good for the species.

Another kind of interaction in which one organism is harmed and the other aided is the relationship of parasitism. This is a very common kind of relationship, since there are more species of parasites in the world than there are nonparasites. **Parasitism** involves one organism living in or on another living organism from which it derives nourishment. The **parasite** derives the benefit and harms the **host,** the organism it lives in or on. Many kinds of fungi live on trees and other kinds of plants doing harm to them. Dutch elm disease is caused by a fungus that infects the living sap-carrying parts of the tree. Many kinds of commercially valuable plants have fungus parasites that harm them. There are many kinds of worms, protozoa, bacteria, and viruses that are important parasites. Parasites that live on the outside of their host are called **external parasites.** For example, rats have fleas that live on the outside of them where they suck blood and do harm to the rat. At the same time, the rat could have a tapeworm in its intestine (figure 17.4). Since the tapeworm lives inside the host, it is called an **internal parasite.** Another kind of parasite found in the blood of the rat may be the bacterium *Yersinia pestis.* It does little harm to

Figure 17.5

Figure 17.5
Impact of Black Death on World Human Population.
Black death (plague) is caused by the bacterium *Yersinia pestis*, which is transmitted by the bite of a flea. Fleas were common on rats and people of the time, so the combination of poor living conditions, large rat populations, and large flea populations combined to allow epidemics of black death in which millions of people died. In some regions of Europe 50% of the population died.

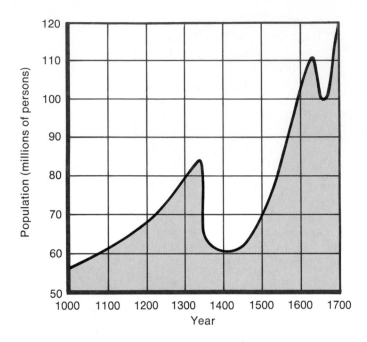

the rat but causes a disease known as "plague" or "black death" if it is transmitted to humans. The flea can serve as a carrier of the bacterium between rats and humans. An organism that can carry a disease from one individual to another is called a **vector.** The elm bark beetle is a vector for the Dutch elm disease fungus. If a flea sucks blood from an infected rat and then bites a human, it is possible for the bacterium to enter the human bloodstream causing "plague." During the mid-1300s when living conditions were poor and rats and fleas were common, epidemics of "plague" killed millions of people. In some countries in Western Europe, 50% of the population was killed by this disease (figure 17.5).

Predation and parasitism are both relationships in which one member of the pair is helped and the other is harmed. There are several related kinds of interactions that are similar but don't fit our definitions very well. When a cow eats grass, it is certainly harming the grass while deriving benefit from it. We could call cows grass predators but we usually refer to them as herbivores. Likewise, there are animals like mosquitoes, biting flies, vampire bats, and ticks that take blood meals but don't usually live permanently on the host nor do they kill it. Are they temporary parasites, or specialized predators? Finally, there are birds like cowbirds and cuckoos that lay their eggs in the nests of other species of birds who raise these foster young rather than their own. The adult cowbird and cuckoo or their offspring remove the eggs or the young of the host bird species so that it is usually only the cowbird or cuckoo that is raised by the parents. This kind of relationship has been called nest parasitism. There are many kinds of interactions between organisms that don't fit neatly into the classification scheme dreamed up by scientists.

Another specialized kind of interrelationship is one called amensalism. **Amensalism** exists when one member of the pair of interacting organisms is harmed, but the other is not affected either positively or negatively. The classic example of an amensal relationship is the inhibitory influence of the chemical penicillin on the growth of certain bacteria. Penicillin is a product of the mold *Penicillium*. It is obvious the bacteria are harmed by the relationship since they cannot grow and reproduce, but if there is a benefit for the mold, it must be a very subtle benefit.

There is also a kind of relationship in which one organism is benefited, while the other is not affected. This is known as **commensalism.** Sharks often have a different kind of fish attached to them. The remora has a sucker on the top side of its head that allows it to attach to the ventral (belly) side of the shark and, therefore, get a free ride (figure 17.6). While the remora benefits from the free ride and by eating the leftovers from the shark's meals, the shark does not appear to be troubled by this uninvited guest nor does it benefit from the presence of the remora. Another example of commensalism is the relationship between trees and epiphytic plants. **Epiphytes** are plants that live on the surface of other plants but do not derive nourishment from them. Many kinds of plants like orchids, ferns, mosses, and others use the surface of trees as a place to live. These kinds of organisms are particularly common in tropical rain forests. Many of these epiphytes derive benefit from the relationship because they are able to be located in the top of the tree where they receive better exposure to sunlight and more moisture. The trees receive no benefit from the relationship, nor are they harmed; they simply serve as a support surface for epiphytes.

So far in our examples, only one species has benefited from the association of two species. There are also many situations in which two species live in close association with one another and both benefit. This is called **mutualism.** One interesting example of mutualism involves digestion in rabbits. Rabbits eat plant material that is high in cellulose, even though they do not produce the enzymes capable of breaking down cellulose molecules into simple sugars. They manage to get energy out of these cellulose molecules with the help of special bacteria living in their digestive tract. The bacteria produce cellulose-digesting enzymes called cellulases that break down cellulose into smaller carbohydrate molecules that the rabbit's digestive enzymes can break down into smaller glucose molecules. The bacteria benefit because the gut of the rabbit provides them with a moist, warm, nourishing environment in which to live. The rabbit benefits because the bacteria provide them with a source of food. Similarly, many animals like cattle, termites, buffalo, and antelope have collections of bacteria and protozoa that live in their guts and help them to digest cellulose.

Another kind of mutualistic relationship exists between flowering plants and bees. Undoubtedly you have observed bees and other insects visiting flowers to obtain nectar from the blossoms. Usually the flowers are constructed in such a manner that the bees pick up pollen (sperm-containing packages) on their hairy bodies and transfer it to the female part of the next flower they visit (figure 17.7). Because bees

Figure 17.6
Commensalism.
Commensalism involves one organism benefiting while the other is not affected. The remora fish shown here hitchhike a ride on the shark. They eat scraps of food left over by the shark. The shark does not seem to be hindered in any way.

Figure 17.7
Mutualism.
Mutualism is an interaction between two organisms in which both benefit. The plant is benefited because cross-fertilization is more probable; the bee is benefited by the food it acquires.

Figure 17.8
Competition.
Whenever a needed resource is in limited supply, organisms compete for it. This competition may be between members of the same species (intraspecific) as shown in the photograph or may involve different species (interspecific).

normally visit the same species of flower for several minutes and ignore other species, they can serve as pollen carriers between two flowers of the same species. Plants pollinated in this manner produce less pollen than do plants that rely on the wind to transfer pollen. This saves the plant energy since it doesn't need to produce huge quantities of pollen. It does, however, need to transfer some of its energy savings into the production of showy flowers and nectar to attract the bees. The bees benefit from both the nectar and pollen because they use both for food.

One additional term that relates to parasitism, commensalism, and mutualism is symbiosis. **Symbiosis** literally means living together. Unfortunately this word is used in several ways, none of which are very clear. It is often used as a synonym for mutualism, but is also often used to refer to commensalistic relationships and parasitism. The emphasis, however, is on interactions that involve a close physical relationship between the two kinds of organisms.

So far in our discussion of organism interactions we have left out the most common one. **Competition** is a kind of interaction between organisms in which both organisms are harmed to some extent. Competition occurs whenever two organisms both need a vital resource that is in short supply (figure 17.8). The vital resource could be food, shelter, nesting sites, water, mates, or space. It can be a snarling tug-of-war between two dogs over a scrap of food, or it can be a silent struggle between plants for access to available light. If you have ever started tomato seeds (or other garden plants) in a garden and failed to eliminate the weeds, you have witnessed competition. If the weeds are not removed, they compete with the garden plants for available sunlight, water, and nutrients, resulting in poor growth of both the garden plants and the weeds.

The more similar the requirements of two species of organisms, the more intense the competition. The **competitive exclusion principle** says that no two species of organisms can occupy the same niche at the same time. If two species of organisms do occupy the same niche, the competition will be so intense that one will become extinct, one may be forced to migrate to a different area, or the two species may evolve into slightly different niches. Thus, even though both kinds of organisms are harmed during competition, there can still be winners and losers because one may be harmed more than the other. It is this competition that provides a major mechanism for natural selection. Because of this necessity to reduce interspecies competition, we often find organisms that have developed slight differences in their niches that prevent direct competition. For example, there are many birds that catch flying insects as food. However, they do not compete directly with each other because some feed at night, some feed high in the air, some feed only near the ground, while still others perch on branches and wait for insects to fly past.

Within any community of organisms, there are many similar examples of specialization to very specific niches. These niches overlap and interrelate into a community that is capable of using its resources very efficiently through a process of cycling materials.

Cycling of Materials in Ecosystems

The earth is a closed ecosystem, since no significant amount of new matter comes to the earth from space. Only sunlight energy comes to the earth in a continuous stream and even this is ultimately returned to outer space as heat energy. However, it is this flow of energy through the biosphere that drives all biological processes. Living systems have evolved ways of using this energy to continue life through growth and reproduction. Although some new atoms are being added to the earth from cosmic dust and meteorites, this amount is not significant to the entire biomass of the earth. Therefore, living things must reuse the existing atoms again and again. In this recycling process, inorganic molecules are combined to form the organic compounds of living things. If there were no way of recycling this organic matter back into its inorganic forms, organic material would build up as the bodies of dead organisms. Decomposers play a vital role in this recycling process if conditions allow them to operate. If they are kept from destroying organic matter, it will build up as deposits of organic matter. This is thought to have occurred millions of years ago when the present deposits of coal, oil, and natural gas were formed.

Living systems contain many kinds of atoms, but some are more common than others. Carbon, nitrogen, oxygen, hydrogen, and phosphorus are found in all living things and must be recycled when an organism dies. Let's look at some examples of this recycling process. Carbon and oxygen combine to form the molecule carbon dioxide (CO_2), which is a gas found in small quantities in the atmosphere. During photosynthesis, carbon dioxide (CO_2) combines with water (H_2O) to form complex organic molecules ($C_6H_{12}O_6$). At the same time, oxygen molecules (O_2) are released into the atmosphere. The organic matter in the bodies of plants may be used by herbivores as food. When an herbivore eats a plant, it breaks down the complex organic molecules into more simple molecules, like simple sugars, amino acids, glycerol, and fatty acids. These can be used as building blocks in the construction of its own body. Thus, the atoms in the body of the herbivore can be traced back to the herbivores that were eaten. Similarly, when herbivores are eaten by carnivores, these same atoms are transferred to them. The waste products of plants and animals and the remains of dead organisms are also used by decomposer organisms as a source of carbon and oxygen atoms. Finally, all of the organisms in this cycle including plants, herbivores, carnivores, and decomposers are involved in the process of respiration in which oxygen (O_2) is used to break down organic compounds into carbon dioxide (CO_2) and water (H_2O). Thus, the carbon atoms that started out as a component of carbon dioxide (CO_2) molecules have passed through the bodies of living organisms as parts of organic molecules in their bodies and returned to the atmo-

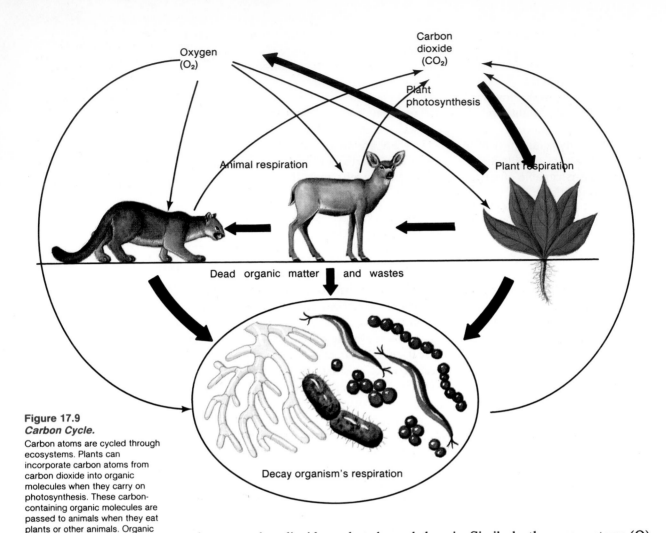

Oxygen
(O₂)

Carbon
dioxide
(CO₂)

Plant
photosynthesis

Animal respiration

Plant respiration

Dead organic matter and wastes

Decay organism's respiration

Figure 17.9
Carbon Cycle.
Carbon atoms are cycled through ecosystems. Plants can incorporate carbon atoms from carbon dioxide into organic molecules when they carry on photosynthesis. These carbon-containing organic molecules are passed to animals when they eat plants or other animals. Organic wastes or dead organisms are consumed by decay organisms. All organisms (plants, animals, and decomposers) return carbon atoms to the atmosphere as carbon dioxide when they carry on respiration. Oxygen atoms are being cycled at the same time that carbon is; oxygen is released to the atmosphere during photosynthesis and taken up during respiration.

sphere as carbon dioxide ready to be cycled again. Similarly, the oxygen atoms (O) released as oxygen molecules (O_2) during photosynthesis have been used during the process of respiration (figure 17.9).

Water molecules are essential for life. They are involved in the process of photosynthesis as raw material, since it is the hydrogen atoms (H) from water (H_2O) molecules that are added to the carbon atoms to make carbohydrates and other organic molecules. Furthermore, it is the oxygen in water molecules that is released during photosynthesis as oxygen molecules (O_2). All of the metabolic reactions that occur in organisms take place in a watery environment. The most common molecule in the body of any organism is water. This important molecule is circulated in a hydrologic cycle (figure 17.10).

Most of the forces that cause water to be cycled do not involve organisms, but are the result of normal physical processes. Because of the motion of molecules, liquid water is caused to evaporate into the atmosphere. This can occur wherever water is present. It will evaporate from lakes, rivers, soil, or the surface of organisms. Since the oceans contain most of the world's water, there is an extremely large amount of water entering the atmosphere from the oceans. Plants also transport water from the soil to the leaves where it evaporates in a process called **transpiration.** Once the water molecules are in the atmosphere, they are moved by prevailing wind

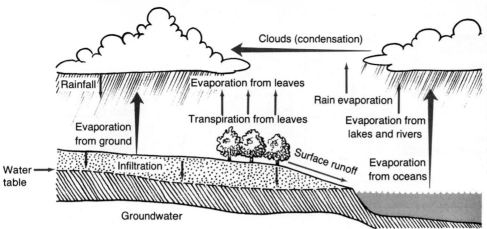

Figure 17.10
The Hydrologic Cycle.
The cycling of water through the environment follows a simple pattern. Moisture in the atmosphere condenses into droplets that fall to the earth as rain or snow where it supplies all living things with its life-sustaining properties. After flowing over the earth as surface water or through the soil as groundwater, water returns to the oceans where it evaporates back into the atmosphere to begin the cycle again.

patterns. If warm, moist air encounters cooler temperatures, which often happens over land masses, the water vapor will condense into droplets and fall as rain or snow. When the precipitation falls on land, some of it runs off the surface, some of it evaporates, and some penetrates into the soil. The water in the soil may be taken up by plants and transpired into the atmosphere or it may become groundwater. Much of the groundwater also eventually makes its way into lakes and streams and, therefore, ultimately arrives at the ocean from which it originated.

Another important element for living things is nitrogen (N). Nitrogen is essential in the formation of amino acids, which are needed to form proteins, and in the formation of nitrogenous bases, which are a part of the nucleic acids DNA and RNA. Nitrogen (N) is found as molecules of nitrogen gas (N_2) in the atmosphere. Although nitrogen gas (N_2) makes up approximately 80% of the earth's atmosphere, only a few kinds of bacteria are able to convert it into nitrogen compounds that other organisms can use. Therefore, in most ecosystems the amount of nitrogen available limits the amount of plant biomass that can be produced. Plants are able to obtain nitrogen atoms combined with other atoms into usable forms from several different sources (figure 17.11).

Symbiotic nitrogen-fixing bacteria live in the roots of certain kinds of plants where they convert nitrogen gas molecules into compounds that the plants can use to make amino acids and nucleic acids. The most common plants that enter into this mutualistic relationship with these bacteria are the legumes, such as beans, clover, peas, alfalfa, and locust trees. Some other organisms such as alder trees can also participate in this relationship. There are also **free-living nitrogen-fixing bacteria** in the soil that provide nitrogen compounds and can be taken up through the roots but do not live in a close physical union with plants.

Another way plants get usable nitrogen compounds involves a series of different bacteria. Decomposer bacteria convert organic nitrogen-containing compounds into ammonia (NH_3). **Nitrifying bacteria** can convert ammonia (NH_3) into nitrite- (NO_2^-) containing compounds, which can in turn be converted into nitrate- (NO_3^-) containing compounds. Many kinds of plants can use either ammonia (NH_3) or nitrate (NO_3^-) from the soil as building blocks for amino acids and nucleic acids.

All animals obtain their nitrogen in the food they eat. The ingested proteins are digested to amino acids, which can be assembled into new proteins. All dead organic matter and waste products of plants and animals are acted upon by decomposer organisms and the nitrogen is released as ammonia (NH_3), which is acted upon by nitrifying bacteria.

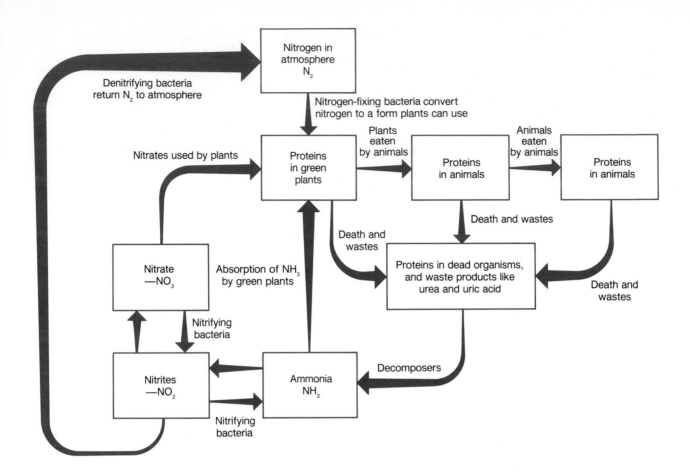

Figure 17.11
Nitrogen Cycle.
Nitrogen atoms are cycled in ecosystems. Atmospheric nitrogen is converted by nitrogen-fixing bacteria to nitrogen-containing compounds that plants can use to make proteins and other compounds. Proteins are passed to other organisms when one organism is eaten by another. Dead organisms and their waste products are acted upon by decay organisms to form ammonia, which may be reused by plants or converted to other nitrogen compounds by nitrifying bacteria. Denitrifying bacteria return nitrogen gas to the atmosphere.

Finally, other kinds of bacteria called **denitrifying bacteria** are capable of converting nitrite (NO_2^-) into nitrogen gas (N_2), which is released into the atmosphere. Thus, there is a nitrogen cycle in which nitrogen from the atmosphere is passed through a series of organisms, many of which are bacteria, and ultimately returned to the atmosphere to be cycled again. It is a much more complicated cycle than the carbon cycle and just as important.

Since nitrogen is in short supply in most ecosystems, farmers usually find it necessary to supplement the natural nitrogen sources in the soil to get maximum plant growth. This can be done in a number of ways. Alternating nitrogen-producing crops with nitrogen-demanding crops helps to maintain higher levels of usable nitrogen in the soil. One year, a crop can be planted that has symbiotic nitrogen-fixing bacteria associated with its roots. Beans and clover are good examples of these crops. The following year, the farmer can plant a nitrogen-demanding crop such as corn. The use of manure is another way of improving nitrogen levels. The waste products of animals are broken down by decomposer bacteria and nitrifying bacteria resulting in enhanced levels of ammonia and nitrate. Finally, the farmer can use industrially produced fertilizers containing ammonia or nitrate, since these compounds can be used directly by plants or be converted into other useful forms by nitrifying bacteria.

Fertilizers usually contain more than just nitrogen compounds. The numbers on a fertilizer bag tell you the percentage of nitrogen, phosphorus, and potassium in the fertilizer. For example a 6–24–24 fertilizer would have 6% nitrogen compounds, 24% phosphorus compounds, and 24% potassium-containing compounds. These other elements (phosphorus and potassium) are also cycled through ecosystems. In natural nonagricultural ecosystems these elements would be released by decomposers and enter the soil where they would be available for plant uptake through the roots. However, when crops are removed from fields, these elements are removed and, therefore, must be replaced by adding more fertilizer.

Community Consequences of Human Actions

As you can see from this discussion and from the discussion of food webs in chapter 16, all organisms are associated in a complex network of relationships. A community consists of all these sets of interrelations. Therefore, before one makes a decision to change a community, it is wise to analyze how the organisms are interrelated.

Predator Control

During the formative years of wildlife management it was thought that populations of game species could be increased if the populations of their predators were reduced. Consequently, many states passed laws that encouraged the killing of foxes, eagles, hawks, owls, coyotes, cougars, and other predators that use game animals as a source of food. Often bounties were paid to people who killed these predators. In South Dakota it was decided to increase the pheasant population by reducing the numbers of foxes and coyotes. However, when the supposed predator populations were significantly reduced, there was no increase in pheasant population, but rabbit and mouse populations increased rapidly and became serious pests. Evidently the foxes and coyotes were a major factor in keeping rabbit and mouse populations under control, but only had a minor impact on pheasants.

Habitat Destruction

Some communities are quite fragile, while others seem to be able to resist major human interference. Generally communities that have a wide variety of organisms and lots of different interactions are more resistant than those that have few organisms and interactions. In general, the more complex an ecosystem is, the more likely it will be able to recover after being disturbed. The tundra biome is an example of a community with relatively few organisms and interactions. It is not very resistant to change and because of the very slow rate of repair, damage caused by human activity may take hundreds of years to repair.

Some species are much more resistant to human activity than others. Rabbits, starlings, skunks, many kinds of insects, and many plants are able to maintain high populations despite human activity. Indeed, some may even be encouraged by human activity. Other organisms like whales, condors, eagles, and many plant and insect species are not able to resist human interference. For most of these endangered species it is not direct action of humans that causes the problem. There are very few organisms that have been driven to extinction by hunting or direct exploitation. Usually the human influence is an indirect one. Habitat destruction is the main

Table 17.1 *Endangered and Threatened Species*

Species	Reason for Endangerment
Hawaiian crow *Corvis hawaiinsis*	Predation by cat and mongoose, disease, habitat destruction
American alligator *Alligator mississippiensis*	Hunting
Sonora Chub *Gila ditaenia*	Competition with introduced species
Black-footed ferret *Mustela nigripes*	Poisoning of prairie dogs (their primary food)
Snail kite *Rostrhamus sociabilis*	Specialized eating habits (only eat apple snails), draining of marshes
Grizzly bear *Ursus arctos*	Loss of wilderness areas
California condor *Gymnogyps californianus*	Slow breeding, lead poisoning
Ringed sawback turtle *Graptemys oculifera*	Habitat modified by the construction of reservoir that reduced their primary food source
Scrub mint *Dicerandra frutescens*	Conversion of habitat to citrus groves and housing
Peter's mountain mallow *Iliamna corel*	Competition with introduced species and browsing by deer
Bald eagle *Haliaeetus leucocephalus*	Pesticides resulted in fragile eggs
Loach minnow *Tiaroga cobitis*	Impoundments, water diversion, introduction of exotic species
Red-cockaded woodpecker *Picoides borealis*	Needs diseased, live pines to excavate nesting cavities—logging activities reduced the number of diseased pines
Short's goldenrod *Solidago shortii*	Habitat modification
Slender rush-pea *Hoffmannseggia tenella*	Habitat modification
Brady pincushion cactus *Pediocactus bradyi*	Restricted habitat collectors

cause of extinction and the endangering of species. As humans convert land to farming, grazing, commercial forestry, and special wildlife management areas, the natural ecosystem is disrupted and those plants and animals with narrow niches tend to be eliminated because they lose critical resources in their environment. Table 17.1 lists several endangered species and the probable cause for their difficulties.

The DDT Story

Humans have developed a variety of chemicals to control specific pest organisms. One of these was the insecticide DDT.

DDT is an abbreviation for the chemical name dichloro-diphenyl-trichloro-ethane. The chemical structure of this compound is shown in figure 17.12. DDT is one of a group of organic compounds called chlorinated hydrocarbons. Because DDT

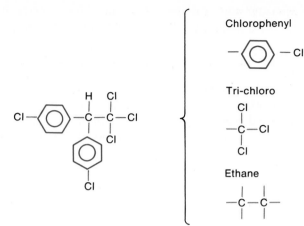

Figure 17.12
Structural Formula of DDT.
This diagram shows the
arrangement of atoms in a
molecule of DDT.

Chlorophenyl

Tri-chloro

Ethane

is a poison that was used to kill a variety of insects, it was called an **insecticide.**
Another term that is sometimes used is **pesticide,** which implies that the poison is
effective against pests. Although it is no longer used in the United States (its use
was banned in the early 1970s) DDT is still manufactured and used in many parts
of the world including Mexico.

DDT was a very valuable insecticide for the United States Armed Forces during
World War II. It was sprayed on clothing and dusted on the bodies of soldiers, ref-
ugees, and prisoners to kill body lice and other insects. Lice, besides being a nuisance, carry the bacteria that cause a disease known as typhus fever. When bitten
by a louse, the person can develop typhus fever. Because body lice could be transferred from one person to another by contact or by wearing infested clothing, DDT
was important in maintaining the health of millions of people. Since DDT was so
useful in controlling these insects, people could see the end of pesky mosquitoes and
flies, as well as the elimination of many disease-carrying insects.

Although DDT was originally a very effective insecticide, many species of insects developed resistance to DDT. All species have genetic variety. Some of this
genetic variety relates to how sensitive an organism is to many environmental factors, including manufactured environmental factors such as DDT. When DDT or
any pesticide is applied to a population of insects, those individuals that are susceptible die and those with some degree of resistance have a greater chance of living.
Now the reproducing population consists of many individuals that have DDT-
resistant genes, which are passed on to the offspring. When this happens repeatedly
over a long period of time, a DDT-resistant population develops and the insecticide
is no longer useful. DDT acts as a selecting agent, killing the normal insects but
allowing the resistant individuals to live. This has happened in the orange groves of
California. Many populations of pests became DDT resistant. Similarly, throughout
the world DDT was and in many areas still is used to control malaria-carrying mosquitoes. Many of these populations have become resistant to DDT and other kinds
of insecticides. The people who had foreseen the elimination of insect pests had not
reckoned with the genetic diversity of the gene pools of these insects.

Another problem that was not anticipated was the consequences of pesticide
use to the food chain. DDT was a very effective insecticide, since it was extremely
toxic to insects but not very toxic to animals like birds and mammals. It was also a
very stable compound, which meant that once it was applied it would remain present
and effective for a long period of time. It sounds like an ideal insecticide. What went
wrong? Why was its use banned?

When DDT was applied to an area to get rid of insect pests, it was usually dissolved in an oil or fatty compound. It was then sprayed over an area and fell on the insects and on the plants that the insects used for food. Eventually the DDT entered the insect either directly through the body wall or through the food it was eating. When ingested with food, DDT interferes with the normal metabolism of the insect. If small quantities were taken in, the insect could digest and break down the DDT just like any other large organic molecule. Since DDT is soluble in fat or oil, the DDT or its breakdown products were stored in the fat deposits of the insect. Some insects could break down and store all of the DDT they encountered and, therefore, they survived. If an area had been lightly sprayed with DDT, some insects died, some were able to tolerate the DDT, and others broke down and stored non-lethal quantities of DDT. As much as one part DDT per one billion parts of insect tissue could be stored in this manner. This is not much DDT! It is equivalent to one drop of DDT in one hundred railroad tank cars. However, when an aquatic area is sprayed with a small concentration of DDT, it is possible for many kinds of organisms in the area to accumulate such tiny quantities in their bodies. Even algae and protozoa found in aquatic ecosystems accumulate pesticides. They may accumulate concentrations in their cells that are 250 times more concentrated than the amount sprayed on the ecosystem. The algae and protozoa are eaten by insects, which are in turn eaten by frogs, fish, or other carnivores.

If you measure the concentration in frogs, it may be two thousand times what was sprayed. If you sample the birds that feed on the frogs and fish, the concentration in the birds may accumulate to as much as eighty thousand times the original amount. What was originally a dilute concentration accumulated in the food chain because DDT is relatively stable and is stored in the fat deposits of the organisms that take it in.

Many animals at higher trophic levels died as a result of this accumulation of pesticide in food chains. This process is called **biological amplification** (figure 17.13). Even if birds were not killed directly by DDT, many birds such as eagles, pelicans, and the osprey at higher trophic levels suffered reduced populations because the DDT interfered with the female bird's ability to produce eggshells. Thin eggshells are easily broken and there were no live young hatched.

What was originally used as an insecticide to control insect pests has been shown to have many harmful community consequences. Instead of controlling the pest, it selects for resistance and creates populations that can tolerate the poison. Instead of harming only the pest species, it accumulates in food chains to kill nontarget species as well. Furthermore, it is not specific to pest species only, but kills many beneficial species of insects. DDT has not been the gift to humankind that we originally thought it would be and its sale and use has been banned in the United States and many other countries.

Other Problems with Pesticides

A number of factors determine how successful you will be in controlling a pest with a pesticide. You must choose a pesticide that will cause the least amount of damage to the harmless or beneficial organisms in the community. The ideal pesticide or insecticide would only affect the target pest. Because many of the insects we consider pests are herbivores, you would expect there to be carnivores in the community that would use the pest species as prey, and parasites that would use the pest as a host. These predators and parasites would have an important role in controlling the

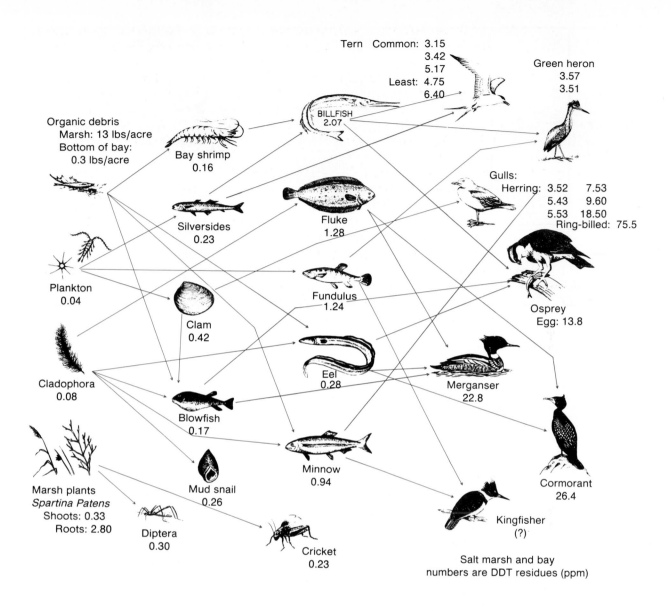

Tern   Common: 3.15
3.42
5.17
Least: 4.75
6.40

Green heron
3.57
3.51

BILLFISH
2.07

Organic debris
Marsh: 13 lbs/acre
Bottom of bay:
0.3 lbs/acre

Bay shrimp
0.16

Gulls:
Herring: 3.52   7.53
5.43   9.60
5.53   18.50
Ring-billed: 75.5

Fluke
1.28

Silversides
0.23

Plankton
0.04

Fundulus
1.24

Osprey
Egg: 13.8

Clam
0.42

Cladophora
0.08

Eel
0.28

Merganser
22.8

Blowfish
0.17

Cormorant
26.4

Minnow
0.94

Marsh plants
Spartina Patens
Shoots: 0.33
Roots: 2.80

Mud snail
0.26

Kingfisher
(?)

Diptera
0.30

Cricket
0.23

Salt marsh and bay
numbers are DDT residues (ppm)

numbers of a pest species. Generally the predators and parasites reproduce more slowly than their prey or host species. Because of this, the use of a nonspecific pesticide may actually make matters worse. If a nonspecific pesticide is applied to an area, the pest is killed but so are its predators and parasites. Since the herbivore pest reproduces faster than its predators and parasites, the pest population rebounds quickly, unchecked by natural predation and parasitism. This may necessitate more frequent and more concentrated applications of pesticides. This has actually happened in many cases of pesticide use. The use of pesticides made the problem worse and it cost money to apply the pesticides.

We have just looked at how human actions can cause changes in ecosystems. However, you should not think that all changes are abnormal. Within communities of organisms there are normal fluctuations in numbers and sometimes slow progressive changes that convert one kind of community into another.

Figure 17.13
Biological Amplification
of DDT.

All of the numbers present are in parts per million (ppm). A concentration of one part per million means that in a million equal parts of the organism, one of the parts would be DDT. Notice how the amount of DDT in the bodies of the organisms increases as we go from producers, to herbivores, to carnivores. Since DDT is persistent, it builds up in the top trophic levels of the food chain.

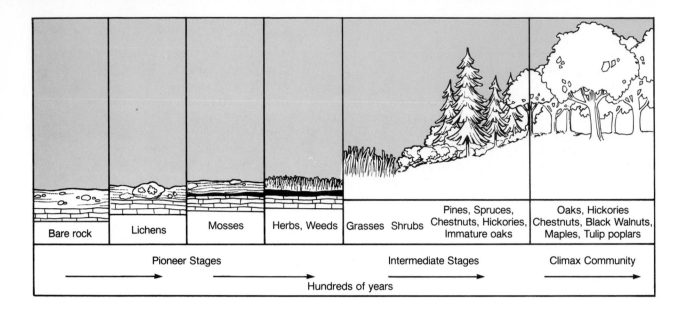

| Bare rock | Lichens | Mosses | Herbs, Weeds | Grasses | Shrubs | Pines, Spruces, Chestnuts, Hickories, Immature oaks | Oaks, Hickories Chestnuts, Black Walnuts, Maples, Tulip poplars |

Pioneer Stages · Intermediate Stages · Climax Community

Hundreds of years

Figure 17.14
Primary Succession.
The formation of soil is a major step in primary succession. Until soil is formed, the area is unable to support large amounts of vegetation. The vegetation modifies the harsh environment and increases the amount of organic matter that can build up in the area. The presence of plants eliminates the earlier pioneer stages of succession. If given enough time, a climax community may develop.

Succession

Many communities like the biomes we discussed in chapter 16 are relatively stable over long periods of time. Such a relatively stable, long-lasting community is called a **climax community.** The word climax implies the final step in a series of events. That is just what the word means in this context, because communities can go through a series of predictable, temporary stages that eventually result in a long-lasting stable community. The process of changing from one type of community to another is called **succession** and each intermediate stage leading to the climax community is known as a **successional stage, successional community,** or **sere.**

Two different kinds of succession are recognized. **Primary succession,** in which a community of terrestrial plants and animals develops where none existed previously and **secondary succession,** in which a community of organisms is disturbed by a natural or human-related event (hurricane, fire, or forest harvest) and returned to a previous stage in succession. Primary succession is much more difficult to observe than secondary succession because there are relatively few places on earth that lack communities of organisms. The tops of mountains, newly formed volcanic rock, and rock newly exposed by erosion or glaciers can be said to lack life. However, very quickly bacteria, algae, fungi, and lichens begin to grow on the bare rock surface and the process of succession has begun. The first organisms to colonize an area are often referred to as **pioneer organisms** and the community is called a **pioneer community.** Lichens are unusual organisms that consist of a combination of algae cells and fungi cells, but this combination is very hardy and is able to grow on the surface of bare rock (figure 17.14). Since algae cells are present, the lichen is capable of photosynthesis and new organic matter can be formed. Furthermore, many tiny consumer organisms can make use of the lichens as a source of food and a sheltered place to live. The action of the lichens also tends to cause the breakdown of the rock surface upon which they grow. This fragmentation of rock by lichens is aided by the physical weathering processes of freezing and thawing, dissolution by water, and movement by wind. It is the first step in the development of soil. Lichens trap dust particles, small rock particles, and the dead remains of lichens and other organisms that live in and on them, resulting in a thin layer of soil.

As the soil layer becomes thicker, some small plants such as mosses may become established, increasing the rate at which energy is trapped and adding more organic matter to the soil. Eventually the soil may be able to support larger plants that are even more efficient at trapping sunlight and the soil-building process continues at a more rapid pace. Associated with each of the producers in each successional stage is a variety of small animals, fungi, and bacteria. Each change in the community makes it more difficult for the previous group of organisms to maintain themselves. Tall plants shade smaller producers and consequently, the smaller organisms become less common. In some cases they may disappear entirely. Only shade-tolerant species are able to compete successfully. One stage has succeeded the other. Depending on the physical environment and the availability of new colonizing species, succession from this point can lead to different kinds of climax communities. If the area is dry, it might stop at a grassland stage. If it is cold and wet, a coniferous forest might be the climax community. If it is warm and wet, it may be a tropical rain forest. The rate at which this successional process takes place is variable. In some warm, moist, fertile areas the entire process might take place in less than one hundred years. In harsh environments, like mountain tops or very dry areas, it may take thousands of years.

Another situation that is often called primary succession is the progression from an aquatic community to a terrestrial community. Lakes, ponds, and slow-moving parts of rivers accumulate organic matter. Where the water is shallow, this organic matter will support the development of rooted plants. In deeper water, we find only floating plants like water lilies that send their roots down to the mucky bottom. In shallower water, upright rooted plants like cattails and rushes develop. The cattail community contributes more organic matter and the water level becomes more shallow. Eventually a mat of mosses, grasses, and even small trees may develop on the surface along the edge of the water. If this continues for perhaps one hundred to two hundred years, an entire pond or lake will become filled in. More organic matter accumulates because of the large number of producers and because the depression that was originally filled with water becomes dry land. This will usually result in a wet grassland but in many areas it will be replaced by the climax forest community typical of the area (figure 17.15).

Secondary succession occurs when a climax community or one of the successional stages leading to it is changed to an earlier stage. A good example of this is what happens when agricultural land is abandoned. One obvious difference between secondary succession and primary succession is that there is no need to develop a soil layer. If we begin with bare soil the first year, it is likely to be invaded by a pioneer community of weed species that are annual plants. Within a year or two, perennial plants like grasses become established. Since most of the weed species need bare soil for seed germination, they are replaced by the perennial grasses and other plants that live in association with grasses. The more permanent grassland community is able to support more insects, small mammals, and birds than the weed community could. If rainfall is adequate, several species of shrubs and fast-growing trees that require lots of sunlight (birch, aspen, juniper, hawthorn, sumac, pine, spruce, dogwood) will become common. As the trees become larger, the grasses fail

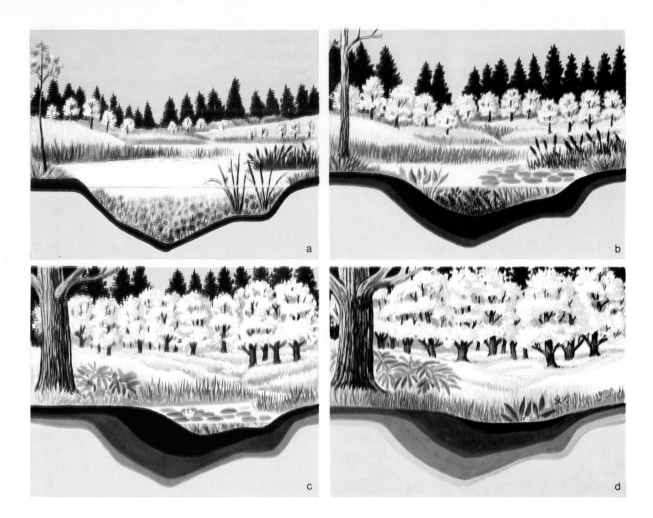

a

b

c

d

Figure 17.15
Succession from a Pond to a Wet Meadow.
A shallow pond will slowly fill with organic matter from the producers in the pond. Eventually a floating mat will form over the pond and grasses will become established. In many areas this will be succeeded by a climax forest community.

to get sufficient sunlight and die out. Eventually, shade-tolerant species of trees (beech, maple, hickory, oak, hemlock, cedar) will replace the shade-intolerant species and a climax community results (figure 17.16).

Most human use of ecosystems involves replacing the natural climax community with an artificial early successional stage. Agriculture involves replacing natural forest or prairie communities with specialized grasses such as wheat, corn, rice, and sorghum. This requires considerable effort on our part because the natural process of succession tends toward the original climax community. This is certainly true if refuges of the original natural community are still locally available to colonize agricultural land. Small woodlots in agricultural areas of the eastern United States serve this purpose. Much of the work and expense of farming is necessary to prevent succession to the natural climax community. It takes a lot of energy to fight nature.

In forestry practices, we often seek to simplify the forest and plant single species forests of the same age. This certainly makes management and harvest practices easier and more efficient, but these kinds of communities do not contain the variety of plants, animals, fungi, and other organisms typically found in natural ecosystems.

Human-constructed lakes or farm ponds often have weed problems because they are shallow and provide ideal conditions for the normal successional processes that will lead to the lake being filled in. Often we do not recognize what a powerful force succession is.

Figure 17.16
Secondary Succession.
When a plowed field is abandoned, a predictable series of communities will follow one another. First, annual weeds will dominate the community, followed by perennial herbs and grasses. In areas that develop forest ecosystems, this stage will be followed by shrubs and shade-intolerant trees. Eventually a forest of shade-tolerant species of trees will develop as the climax forest.

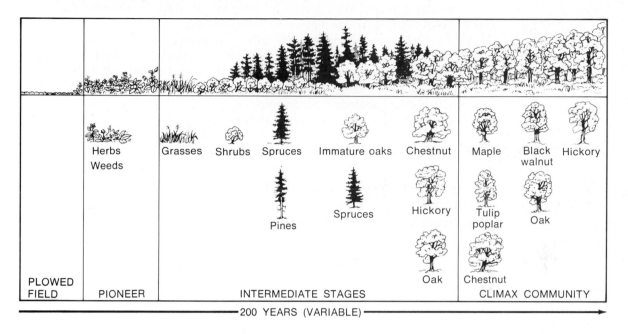

Each organism in a community occupies a specific space known as its habitat and has a specific functional role to play known as its niche. An organism's habitat is usually described in terms of some conspicuous element of its surroundings. The niche is very difficult to describe because it involves so many different interactions with the physical environment and other living things. Interactions between organisms fit into several categories. Predation involves one organism benefiting (predator) at the expense of the organism killed and eaten (prey). Parasitism involves one organism benefiting (parasite) by living in or on another organism (host) and deriving nourishment from it. Organisms that carry parasites from one host to another are called vectors. Amensal relationships exist when one organism is harmed but the other is not affected. Commensal relationships exist when one organism is helped but the other is not affected. Mutualistic relationships involve benefit to both organisms. Symbiosis is any interaction in which two organisms live together in a close physical relationship. Competition causes harm to both of the organisms involved, although one may be harmed more than the other and force it to migrate, become extinct, or evolve into a different niche.

Many atoms are cycled through ecosystems. The carbon atoms of living things are trapped by photosynthesis, passed from organism to organism as food and released to the atmosphere by respiration. Water is necessary as a raw material for photosynthesis and as the medium in which all metabolic reactions take place. Water is cycled by the physical processes of evaporation and condensation. Nitrogen originates in the atmosphere, is trapped by nitrogen-fixing bacteria, passes through a series of organisms, and is ultimately released to the atmosphere by denitrifying bacteria.

Summary

Organisms within a community are interrelated with one another in very sensitive ways. Changing one part of a community can lead to unexpected consequences. Predator control practices, habitat destruction, and pesticide use have all caused changes in parts of communities not directly associated with the part changed.

Succession occurs when a series of communities replace one another as each community changes the environment to make conditions favorable for a subsequent community and unfavorable for itself. Most successional processes result in a relatively stable stage called the climax community. The stages leading to the climax community are called successional stages. If the process begins with bare rock or water, it is called primary succession. If it begins as a disturbed portion of a community, it is called secondary succession.

Consider This

This is a thought puzzle—put it together! Here are the pieces:

People are starving.

Commercial fertilizer production requires temperatures of 900° C.

Geneticists have developed plants that grow very rapidly and require high amounts of nitrogen to germinate during the normal growing season.

Fossil fuels are stored organic matter.

The rate of the nitrogen cycle depends on the activity of bacteria.

The sun is expected to last for several million years.

Crop rotation is becoming a thing of the past.

The clearing of forests for agriculture changes weather in the area.

Experience This

Take a walk in a park, zoo, or natural area. Look for examples of predation, parasitism, mutualism, and competition. Write a short description of the niche of one of the animals you observe.

Questions

1. Describe your niche.
2. What is the difference between a habitat and a niche?
3. What do parasites, commensal organisms, and mutualistic organisms have in common? How are they different?
4. Describe two situations in which competition may involve combat and two that do not involve combat.
5. Trace the flow of carbon atoms through a community that contains plants, herbivores, decomposers, and parasites.
6. Describe four different roles played by bacteria in the nitrogen cycle.
7. Describe the flow of water through the hydrologic cycle.
8. How does primary succession differ from secondary succession?
9. Describe the impact of DDT on communities.
10. How does a climax community differ from a successional community?

Chapter Glossary

amensalism (a-men'sal-izm) A relationship between two organisms in which one organism is harmed and the other is not affected.

biological amplification (bi-o-loj'i-cal am''pli-fi-ka'shun) The accumulation of a compound in increasing concentrations in organisms at successively higher trophic levels.

climax community (kli-maks com-miu′nĭ-te) A relatively stable, long-lasting community.

commensalism (com-men′sal-izm) A relationship between two organisms in which one organism is helped and the other is not affected.

competition (com-pe-tĭ′shun) A relationship between two organisms in which both organisms are harmed.

competitive exclusion principle (com-pĕ′tĭ-tiv eks-klu′zhun prin′sĭ-pul) No two species can occupy the same niche at the same time.

denitrifying bacteria (de-ni′trĭ-fi-ing bak-te′re-ah) Several kinds of bacteria capable of converting nitrite to nitrogen gas.

epiphyte (ep′e-fīt) A plant that lives on the surface of another plant.

external parasite (eks-tur′nal pār′uh-sīt) A parasite that lives on the outside of its host.

free-living nitrogen-fixing bacteria (ni′tro-jen-fik-sing bak-te′re-ah) Soil bacteria that convert nitrogen gas molecules into nitrogen compounds that plants can use.

habitat (hab′ĭ-tat) The place or part of an ecosystem occupied by an organism.

host (hōst) An organism that a parasite lives in or on.

insecticide (in-sek′tĭ-sīd) A poison used to kill insects.

internal parasite (in-tur′nal pār′uh-sīt) A parasite that lives inside its host.

mutualism (miu′chu-al-izm) A relationship between two organisms in which both organisms benefit.

niche (nitch) The functional role of an organism.

nitrifying bacteria (ni′tri-fi-ing bak-te′re-ah) Several kinds of bacteria capable of converting ammonia to nitrite or nitrite to nitrate.

parasite (pār′uh-sīt) An organism that lives in or on another organism and derives nourishment from it.

parasitism (pār′uh-sit-izm) A relationship between two organisms that involves one organism living in or on another organism and deriving nourishment from it.

pesticide (pes′tĭ-sīd) A poison used to kill pests. This term is often used interchangeably with insecticide.

pioneer community (pi-o-nēr com-miu′nĭ-te) The first community of organisms in the successional process established in a previously uninhabited area.

pioneer organisms (pi-o-nēr ′or-gu-nizm) The first organisms in the successional process.

predation (pre-da′shun) A relationship between two organisms that involves the capturing, killing, and eating of another animal.

predator (pred′uh-tōr) An organism that captures, kills, and eats another animal.

prey (prā) An organism captured, killed, and eaten by a predator.

primary succession (pri′ma-re suk-sĕ′shun) The orderly series of changes leading to a climax community that begins in a previously uninhabited area.

secondary succession (sek′un-da-re suk-sĕ′shun) Orderly series of changes leading to a climax community that begins with the disturbance of an existing community.

sere (sēr) An intermediate stage in succession leading to a climax community.

succession (suk-sĕ′shun) The process of changing one type of community to another.

successional community (suk-sĕ′shun-al com-miu′nĭ-te) An intermediate stage in succession.

successional stage (suk-sĕ′shun-al stāj) An intermediate stage in succession.

symbiosis (sim-be-o′sis) A close physical relationship between two kinds of organisms. It usually includes parasitism, commensalism, and mutualism.

symbiotic nitrogen-fixing bacteria (sim-be-ah′tik ni-tro-jen-fik′sing bak-te′re-ah) Bacteria that live in the roots of certain kinds of plants where they convert nitrogen gas molecules into compounds that plants can use.

transpiration (trans″pi-ra′shun) The process of water evaporation from the leaves of a plant.

vector (vek′tōr) An organism that carries a disease or parasite from one host to the next.

18

Population Ecology

Chapter Outline

Purpose

Populations of organisms show many kinds of characteristics. An introduction to some of these characteristics will help you see how populations grow and how this growth is controlled. While not specifically about growth of the human population, much of the material in this chapter relates to the problems associated with the human population explosion.

For Your Information

China is the most populous country in the world with over one billion people. This is approximately one-fourth of the world's population. The government has tried to limit the size of families by encouraging later marriages, providing conception-control information, making small family size a patriotic duty, and giving financial incentives to couples who have no children or limit themselves to one child. Most families desire a male child. Consequently, many baby girls are left to die so the parents can have another opportunity to have a male child. The government must now deal with the problems of infanticide and an aging work force.

Learning Objectives

■ Recognize that populations vary in size, sex ratio, age distribution, and density.

■ Describe the characteristics of a typical population growth curve.

■ Understand why populations grow.

■ Recognize pressures that ultimately limit population size.

■ Understand that human population growth obeys the same rules as other kinds of organisms.

Population Characteristics

A **population** is a group of organisms of the same species located in the same place at the same time. Examples are the number of dandelions in your front yard, the rat population in the sewers of your city, or the number of people in your biology class. On a larger scale, all the people of the world constitute the human population. Since all the members of a population are of the same species, they are capable of reproducing among themselves. The terms *species* and *population* are interrelated, since a species is a population—the largest possible population of a particular kind of organism. Population, however, is often used to refer to portions of the species by specifying the space and time. For example, the size of the human population in a city changes from hour to hour during the day and is different depending on where you set the boundaries of the city.

Since members of a population are of the same species, sexual reproduction can occur and genes can flow from one generation to the next. Genes can also flow from one place to another as organisms migrate or are carried from one geographic location to another. **Gene flow** is used to refer to both the movement of genes within a population due to migration and the movement from one generation to the next as a result of gene replication and sexual reproduction (figure 18.1).

Populations of a species have specific characteristics. Since most of the populations we are familiar with are small portions of the species, we should expect these populations to show differences. One of the ways these populations can differ is in gene frequency. In chapter 13 on population genetics you were introduced to the concept of **gene frequency,** which is a measure of how often a specific gene shows up in the gametes of a population. Two populations of the same species often have quite different gene frequencies. For example, many populations of mosquitoes have

Figure 18.1
Distribution of Blood Type.
The gene for type B blood is not evenly distributed in the world. This map shows that the type B gene is most common in parts of Asia and has been dispersed to the Middle East and parts of Europe and Africa. There has been very little flow of the gene to the Americas.

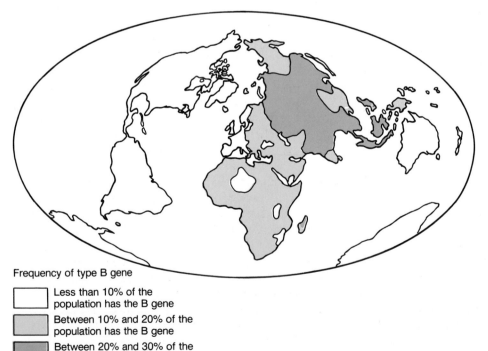

Frequency of type B gene

☐ Less than 10% of the population has the B gene

▨ Between 10% and 20% of the population has the B gene

▦ Between 20% and 30% of the population has the B gene

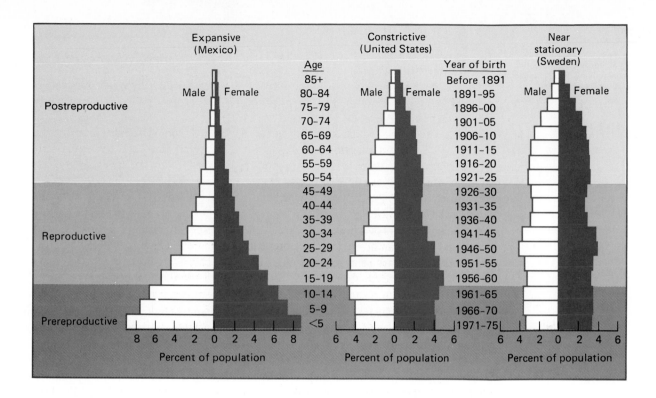

Figure 18.2
Age Distribution in Human Populations.

The relative number of individuals in each of the three categories (prereproductive, reproductive, and postreproductive) can give a good clue to the future of the population. Mexico has a large number of young individuals who will become reproducing adults. Therefore, this population is likely to grow rapidly. The United States has a declining proportion of prereproductive individuals, but a relatively large reproductive population. Therefore, it will continue to grow for a time, but will probably begin to decline in the future. Sweden's population has a large proportion of postreproductive individuals and a small proportion of prereproductive individuals. It will begin to decline before the United States does.

high frequencies of insecticide-resistant genes, while others do not. The frequency of the genes for tallness in humans is more common in certain African tribes than in any other human population. As you can recall from chapter 14 on variation and selection, natural selection will tend to increase the frequency of advantageous genes and decrease the frequency of genes that are disadvantageous. Furthermore, the same gene may be valuable in one environment and harmful in another. The genes that cause dark coat color in mice are good if the soil is dark, but bad if the soil is white sand. The frequency of a particular gene is unique to a particular population and is related to how the organisms adapt to their environment.

Another feature of a population is its **age distribution,** which is the number of organisms of each age in the population. Organisms are grouped into the following categories: (1) prereproductive juveniles—insect larvae, plant seedlings, or babies; (2) reproductive adults—mature insects, plants producing seeds, or humans in early adulthood; or (3) postreproductive adults no longer capable of reproduction—annual plants that have shed their seeds, salmon that have spawned, and many elderly humans. A population is not necessarily divided into equal thirds (figure 18.2). In some situations, a population may be made up of a majority of one age group. If the majority of the population is prereproductive, then a "baby boom" should be anticipated in the future. If a majority of the population is reproductive, the population should be growing rapidly. If the majority of the population is postreproductive a population decline should be anticipated, and fewer new gene combinations would be formed because fewer offspring are being produced.

Population density is the number of organisms of a species found per unit area. Some populations are extremely concentrated into a limited space, while others are well dispersed. As the population density increases, the competition among members of the population for the necessities of life increases. This increases the likelihood that some individuals will explore new habitats and migrate to new areas.

Figure 18.3
Sex Ratio in Elk.
Some male animals defend a
harem of females; therefore, the
sex ratio in these groups is several
females per male.

This increase in the intensity of competition that causes changes in the environment and leads to dispersal is often referred to as **population pressure.** The dispersal of individuals to new areas can relieve the pressure on the home area and lead to the establishment of new populations. Often among animals, it is the juvenile individuals who participate in this dispersal process. If dispersal cannot relieve population pressure, there is usually an increase in the rate at which individuals die due to predation, parasitism, starvation, and accidents. In plant populations, dispersal is not very useful for relieving population density, but death of weaker individuals usually results in reduced population density.

Populations can also differ in their sex ratio. The **sex ratio** is the number of males in a population compared to the number of females. In bird and mammal species where strong pair-bonding occurs, the sex ratio may be nearly one to one (1:1). Many mammals and birds that do not have strong pair-bonding may have sex ratios in which the number of females is larger than the number of males. This is particularly true among game species where the males are shot and the females are not. Since one male can fertilize several females, the population can remain large. However, if the population becomes so large that there is a problem, it becomes necessary to harvest some of the females as well, since the number of females determines how much reproduction can take place. In addition to these examples, many species of animals like bison, horses, and elk have mating systems in which one male maintains a harem of females. The sex ratio in these small groups is quite different from a 1:1 ratio (figure 18.3). There are very few situations in which the number of males exceeds the number of females. In some human and other populations, there may be sex ratios in which the males dominate if female mortality is unusually high or some special mechanism separates most of one sex from the other. Regardless of the specific sex ratio in a population, most species have the ability to produce large numbers of offspring.

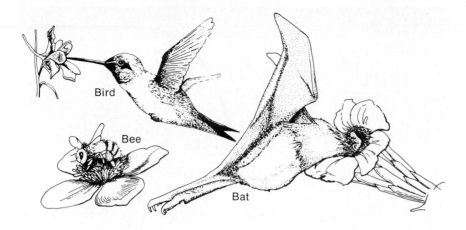

Figure 18.4
Adaptations for Pollination.
Many flowering plants make use of animals to assure that they will be pollinated, thus assuring a high reproductive potential.

Bird

Bee

Bat

Reproductive Capacity

Sex ratios and age distributions within a population have a direct bearing on the rate of reproduction. Each species has an inherent **reproductive capacity** or **biotic potential,** which is the theoretical maximum rate of reproduction. Generally this biotic potential is many times larger than the number of offspring needed to just maintain the population. There are many mechanisms that ensure the overreproduction of the species. Some of these are physical adaptations, and some are behavioral adaptations. Many flowering plants rely on bees to carry the sperm-containing pollen from one flower to the next. This process is called pollination. The flowers have special structures and color patterns that help direct the insect toward the reproductive parts of the plant. Clover and orchid petals, for example, have specific colors and color patterns as well as odors and nectar rewards that encourage insects to visit them and bring about pollination. Other kinds of flowers are adapted to bird or bat pollination (figure 18.4). Behavior patterns such as courtship and nest building in birds ensures that mating occurs and that the young are protected while they are helpless.

Overreproduction is valuable to a species because it provides many opportunities for survival. It also provides many slightly different individuals for the environment to select among. With most plants and animals, many of the potential gametes are never fertilized. An oyster may produce a million eggs a year, but not all of them are fertilized and most that are fertilized die. An apple tree may have thousands of flowers but only produce a few apples, because the pollen that contains the sperm cells may not be transferred to the female part of the flower in the process of pollination. Even after the new individuals are formed, there is usually high mortality of the young individuals. Most seeds that fall to the earth do not grow and most young animals die as well. But usually enough survive to assure continuance of the species. Organisms that reproduce in this way spend large amounts of energy on the production of potential young, with the probability that a small number of them will reach reproductive age.

Figure 18.5
A Typical Population Growth Curve.
In this mouse population, the period of time in which there is little growth is known as the lag phase. This is followed by a rapid increase in population as the offspring of the originating population begin to reproduce themselves; this is known as the exponential growth phase. Eventually the population reaches a stationary growth phase during which time the birth rate equals the death rate.

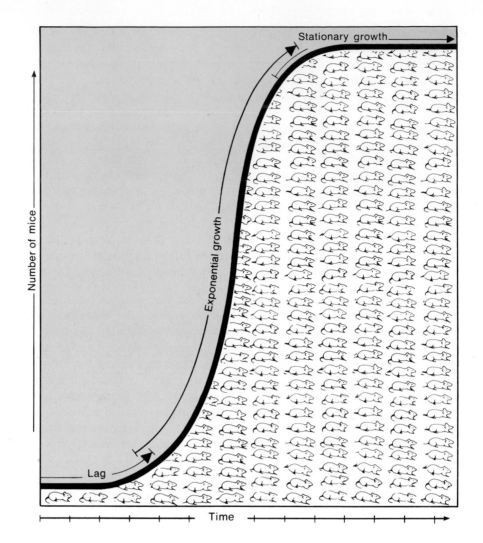

A second way of approaching reproduction is to produce relatively fewer individuals but to provide care and protection that assures a higher probability that the young will become reproductive adults. Humans generally produce a single offspring per pregnancy, but nearly all of them live. In effect, the energy has been channeled into the care and protection of the young produced rather than the production of incredibly large numbers of potential young. Even though fewer young are produced by animals like birds and mammals, they still have a reproductive capacity that exceeds the number required to just replace the parents when they die.

The Population Growth Curve

Because most species of organisms have a high reproductive capacity, there is a tendency for populations to grow. For example, if the usual litter size for a pair of mice is four, the four would produce eight, which in turn would produce sixteen and so forth. Figure 18.5 shows a graph of the change in population size over time known as a **population growth curve.** This kind of curve is typical for situations where a population has just become established in an area. The change in the size of a population depends on the rate at which new organisms enter the population compared to the rate at which individuals leave the population. The number of new individuals added to the population by reproduction per thousand individuals is called **natality.** The number of individuals leaving the population by death per thousand individuals is called **mortality.** When a small number of organisms (two mice) first invades an area, there is a period of time before reproduction takes place when the population remains small and relatively constant. This part of the population growth curve is known as the **lag phase.** Mortality and natality are similar during this period of time. In organisms that take a long time to mature and produce young, such as elephants, deer, or many kinds of plants, the lag phase may be measured in years. With the mice in our example, it will be measured in weeks. The first litter of young will be reproducing themselves in a matter of weeks. Furthermore, the original parents have probably produced an additional litter or two during this time period. Now we have several pairs of mice reproducing more than just once. With several pairs of mice reproducing, natality increases and mortality remains low; therefore, the size of the population begins to grow at an ever-increasing (accelerating) rate. This portion of the population growth curve is known as the **exponential growth phase.** The number of mice (or any other organism) cannot continue to increase at a faster and faster rate, because eventually something in the environment will cause an increase in the number of deaths. Eventually the number of individuals entering the population will equal the number of individuals leaving the population (deaths or migration), and the population size becomes stable. Often there is both a decrease in natality and an increase in mortality at this point. This portion of the population growth curve is known as the **stationary growth phase.** Reproduction will still be going on and the birthrate will still be high, but the death rate will increase and larger numbers of individuals will migrate away from the area.

Control of Population Size

Populations cannot continue to increase indefinitely. Eventually something happens that brings them under control. This observation has led to the development of a concept called the carrying capacity.

Carrying Capacity

It appears that for any species, an area has an optimum population size that can be supported over an extended period, which has been called the **carrying capacity.** The concept of carrying capacity is a general rule of thumb and not fixed for all time, since the environment is constantly altered by successional changes, climatic changes, and natural changes like disease epidemics, forest fires, or floods. The carrying capacity for a species may vary from time to time. Thus, at the same time populations of some species may be allowed to increase, others may be forced to get smaller. The size of the organisms in a population also has an impact on the carrying capacity. For example, an aquarium of a certain size can only support a limited number of fish, but the size of the fish makes a difference. If all the fish are tiny, a large number can be supported and the carrying capacity is high; however, the same aquarium may only be able to support one large fish. In other words, the biomass of the population makes a difference (figure 18.6). Similarly when an area is planted with small trees, the population size is high. But as the trees get larger, competition for nutrients and sunlight becomes more intense and the number of trees present declines, while the biomass increases.

Figure 18.6
Effect of Biomass on Population Size.
Each aquarium can support a biomass of 2 kilograms of fish. The size of the population is influenced by the body size of the fish in the population.

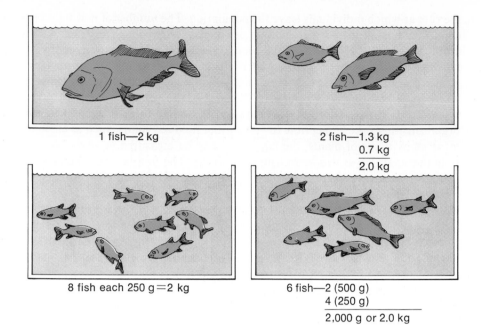

1 fish—2 kg

2 fish—1.3 kg
0.7 kg
_____
2.0 kg

8 fish each 250 g = 2 kg

6 fish—2 (500 g)
4 (250 g)
_____
2,000 g or 2.0 kg

Limiting Factors

The specific individual factors that limit population size are called **limiting factors.** All of the different limiting factors that act on a population are collectively known as **environmental resistance** (figure 18.7). Limiting factors can be placed in four broad categories: (1) availability of raw materials; (2) availability of energy; (3) production and disposal of waste products; and (4) interaction with other organisms.

The first category of limiting factors is the availability of raw materials. For plants, magnesium is necessary for the manufacture of chlorophyll, nitrogen is necessary for protein production, and water is necessary for the transport of materials and as a raw material for photosynthesis. If these are not present in the soil, the growth and reproduction of plants will be inhibited. However, if fertilizer supplies these nutrients or irrigation is used to supply the water, the effects of these limiting factors can be removed and some other factor becomes limiting. For animals, the amount of water, minerals, materials for nesting, suitable burrow sites, or food may be limiting factors. Food for animals really fits into this category and the next, since it supplies both raw materials and energy.

The second major type of limiting factor is availability of energy. In many cases, the amount of light available is a limiting factor, since plants require light as an energy source for photosynthesis. Since all animals use other living things as a source of energy, as well as a source of raw materials, a major limiting factor for any animal is its food source.

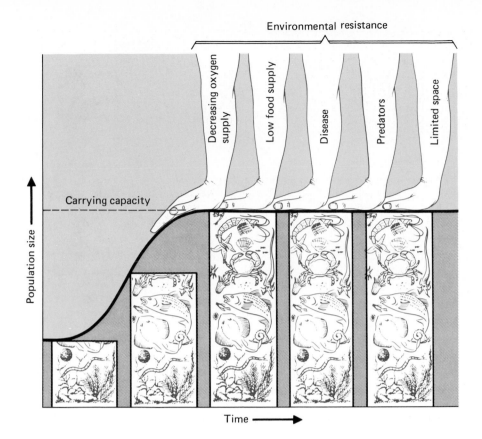

Figure 18.7
Carrying Capacity.
A number of factors in the environment, such as raw materials, energy, waste products, and other organisms, determine the number of a species that can live in an area over an extended time. The number of organisms that can be supported is called the carrying capacity of the area. Limiting factors are those environmental influences that determine population size. The collective effects of several limiting factors is called environmental resistance.

The accumulation of waste products is the third general category of limiting factors. It does not usually have an impact on limiting plant populations because they produce relatively few wastes. However, the buildup of high levels of self-generated waste products is a problem for some populations. Many bacterial populations or populations of tiny aquatic organisms produce waste products as a result of their normal metabolic processes. As these build up, they become more and more toxic and eventually reproduction stops or the population may even die out. When a few bacteria are introduced into a solution containing a source of food, they go through the kind of population growth curve typical of all organisms. As expected, the number of bacteria begins to increase during the lag phase, increases rapidly during the exponential growth phase, and eventually reaches stability in the stationary growth phase. But as the waste products increase, the bacteria literally drown in their own wastes. Space for disposal is limited and no other organisms are present that can convert the harmful wastes to less harmful products and a population decline known as the **death phase** follows (figure 18.8).

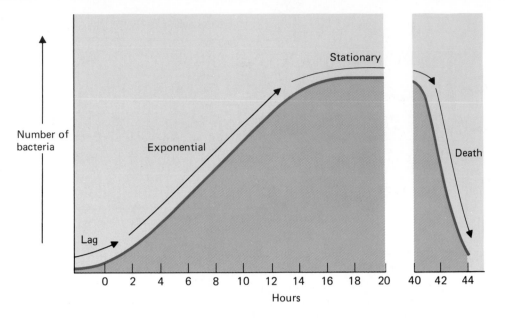

Figure 18.8
Bacterial Population Growth Curve.
The rate of increase in the population of these bacteria is typical of population growth in a favorable environment. As the environmental conditions change due to an increase in the amount of waste products, the population first levels off, then begins to decrease. This period of decreasing population size is known as the death phase.

Wine makers deal with this same situation. When yeasts ferment the sugar in grape juice, they produce ethyl alcohol. When the alcohol concentration reaches a certain level, the yeast population stops growing and eventually declines. Therefore, wine can naturally reach an alcohol concentration of only 12% to 15%. To make any drink stronger than that (of a higher alcohol content), water must be removed (to distill) or alcohol must be added (to fortify). In small aquatic pools like aquariums it is often difficult to keep populations of organisms healthy because of the buildup of ammonia in the water from the waste products of the animals. This is the primary reason that activated charcoal filters are commonly used with aquariums. The charcoal removes many kinds of toxic compounds and prevents the buildup of waste products.

The fourth set of limiting factors is organism interaction. As we learned in chapter 17 on community interaction, there are many different ways that organisms influence each other. Some organisms are harmed and others are benefited. The population size of any organism would be negatively affected by parasitism, predation, or competition. Parasitism and predation usually involve interactions between two different species. Cannibalism is rare, however competition among members of the same population is often very intense. Many kinds of organisms perform services for others that have beneficial effects on the population. For example, decomposer organisms destroy toxic waste products, thus benefiting populations of animals. They also recycle materials needed in the growth and development of all organisms. Mutualistic relationships benefit both of the populations involved.

Interactions Between Populations

Often the population size of two kinds of organisms will be locked together because each is a primary limiting factor of the other. This is most often seen in parasite-host relationships and predator-prey relationships. A good example of this interaction is the relationship of the *lynx* (a predator) and the *varying hare* (the prey) as it was studied in Canada. The varying hare has a high reproductive capacity that the lynx helps to control by using the varying hare as food. The lynx can capture

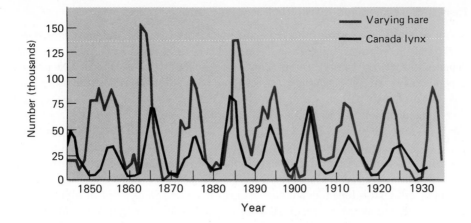

Figure 18.9
Organism Interaction.
The interaction between predator and prey species is complex and often difficult to interpret. These data were collected from the records of the number of pelts purchased by The Hudson Bay Company. It shows that the two populations fluctuate with the changes in the lynx population usually following the changes in varying hare population.

and kill the weak, the old, the diseased and the unwary varying hares. Thus, leaving stronger, healthier varying hares to reproduce. Because the lynx is removing unfit varying hares, it benefits the varying hare population by reducing the spread of disease and reducing the amount of competition between varying hares. At the same time, the varying hare gene pool benefits because those individuals that are less fit have their genes removed from the gene pool. While the lynx is helping to limit the varying hare population, the size of the varying hare population determines how many lynx can live in the area, since varying hares are their primary food source. If anything such as disease epidemics, or unusual weather conditions causes a decline in the varying hare population, the population of the lynx will also fall (figure 18.9).

Extrinsic and Intrinsic Limiting Factors

Some of the factors that help control populations come from outside the population and are known as **extrinsic factors.** Number of predators, loss of a food source, lack of sunlight, or accidents of nature are all extrinsic factors. However, many kinds of organisms show self-regulation of population size. The mechanisms that allow them to do this are called **intrinsic factors.** For example, a study of rats under crowded living conditions showed that as conditions became more crowded, abnormal social behavior became common. There was a decrease in litter size, fewer litters per year were produced, mothers were more likely to ignore their young, and many young were killed by adults. Thus, changes in behavior of the members of the rat population itself resulted in lower birthrates and higher death rates, leading to a reduction in the population growth rate. In another example, trees that are stressed by physical injury or disease often produce extremely large numbers of seeds (offspring) the following year. The trees themselves altered their reproductive rate. An opposite situation is found among populations of white-tailed deer. It is well known that reproductive success is reduced when the deer experience a series of severe winters. When times are bad, the female deer are more likely to have single births rather than twins.

Density-Dependent and Density-Independent Limiting Factors

Many populations are controlled by limiting factors that become more effective as the size of the population increases. These kinds of factors are referred to as **density-dependent factors.** Many of the things we have already discussed are density-dependent factors. For example, the larger a population becomes, the more likely it is that predators will have a chance to catch some of the individuals. Furthermore, a prolonged period of increasing population allows the size of the predator population to increase as well. Large populations with high population density are more likely to be affected by epidemics of parasites than are small populations of widely dispersed individuals, since dense populations allow for easy spread of parasites from one individual to another. The rat example discussed previously is another good example of a density-dependent factor operating, since the amount of abnormal behavior increased as the size of the population increased. In general, whenever there is competition among members of the same population, the intensity of the competition increases as the population increases. Large organisms that tend to live a long time and have relatively few young are most likely to be controlled by density-dependent factors.

A second category, made up of population-controlling influences that are not related to the size of the population, is known as **density-independent factors.** Density-independent factors are usually accidental or occasional extrinsic factors in nature that happen regardless of the size or density of a population. A sudden rainstorm may drown many small plant seedlings and soil organisms. Many plants and animals are killed by frosts that come late in spring or early in the fall. A small pond may dry up resulting in the death of many organisms. The organisms most likely to be controlled by density-independent factors are small, short-lived organisms that can reproduce very fast.

So far we have looked at populations primarily from a nonhuman point of view. Now it is time to focus on the human species and the current world human population problem.

Human Population Growth

It is important to realize that human populations follow the same patterns of growth and are acted upon by the same kinds of limiting factors as populations of other organisms. When we look at the curve of population growth over the past several thousand years, it has been estimated that human populations remained low and constant for long periods of time. For example, it has been estimated that when Columbus discovered America, the Native American population was about one million. Following colonization of North America by Europeans, this slow beginning lead to a very sharp and rapid exponential growth phase that continues today (figure 18.10). Does this mean that humans are different from other animal species? Can the human population continue to grow forever? The human species is no different from other animals. It has a carrying capacity, but has been able to continuously shift the carrying capacity upward by the use of technology and the displacement of other species. Much of the exponential growth phase of the human population can be attributed to the removal of diseases, improvement in agricultural methods, and destruction of natural ecosystems in favor of artificial agricultural ecosystems. But even this must have its limits. There must be some limiting factors that will eventually cause a leveling off of our population growth curve. We cannot increase beyond our ability to get raw materials and energy, nor can we ignore the waste products we produce, or the other organisms with which we interact.

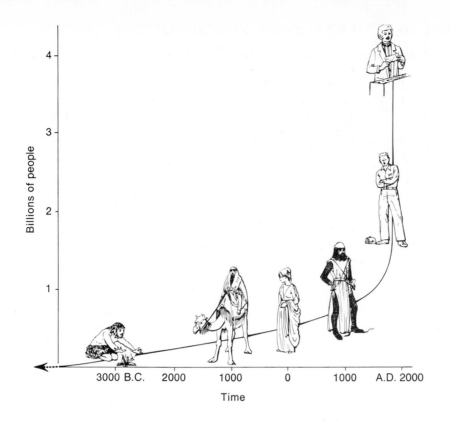

Figure 18.10
Human Population Growth.
The number of humans doubled from A.D. 1850 to 1930 (from one billion to two billion) and then doubled again by 1975 (four billion) and could double again (eight billion) by the year 2010. How long can the human population continue to double in ever shorter periods of time?

To many of us, raw materials simply consist of the amount of food available, but we should not forget that in a technological society, iron ore, lumber, irrigation water, and silicon chips are also raw materials. However, for many of the people of the world the needs are more basic. They do not have enough food. Since most people would die if they did not have food, shelter, safe drinking water, or transportation to bring them food, we are dependent on a continued supply of these materials. In the 1980s food production has become a limiting factor for some segments of the human population. Although it is biologically accurate to say that the world can currently produce enough food to feed all the people of the world, there are many reasons why people can't get food or won't eat it. Many cultures have food taboos or traditions that prevent available food sources from being utilized. For example, pork is forbidden in some cultures. Certain groups of people find it almost impossible to digest milk. Some African cultures use a mixture of cow's milk and cow's blood as food. Have you tried any lately? Would you if you were starving?

In addition, there are complex political, economic, and social problems related to the production and distribution of food. In some cultures, farming is a low-status job, which means that people would rather buy their food from someone else than grow it themselves. This can result in underutilization of agricultural resources. Food is sometimes used as a political weapon when governments want to control certain groups of people. But probably most important is the fact that transportation of food from centers of excess to centers of need is often very difficult and expensive.

A more fundamental question is, Can the world continue to produce enough food? In 1987 the world population was growing at a rate of 1.87% per year. This amounts to two new people being added to the world population every second and will result in a doubling of the world population in forty years. With a continuing

Box 18.1
Thomas Malthus and His Essay on Population

In 1798 Thomas Robert Malthus, an Englishman, published an essay on human population. It presented an idea that was contrary to popular opinion. His basic thesis was that human population increased in a geometric or exponential manner (2, 4, 8, 16, 32, 64, etc.), while the ability to produce food only increased in an arithmetic manner (1, 2, 3, 4, 5, 6, etc.). The ultimate outcome of these different rates would be that population would outgrow the ability of the land to produce food. He concluded that wars, famines, plagues, and natural disasters would be the means of controlling the size of the human population. His predictions were hotly debated by the intellectual community of his day. His assumptions and conclusions were attacked as being erroneous and against the best interest of society. At the time he wrote the essay, the popular opinion was that human knowledge and ''moral constraint'' would be able to create a world that would supply all human needs in abundance. One of Malthus's basic postulates was that ''commerce between the sexes'' (sexual intercourse) would continue unchanged, while other philosophers of the day believed that sexual behavior would take less procreative forms and human population would be limited. Only within the last forty years have really effective conception control mechanisms become widely accepted and used, and their use is primarily in developed countries of the world.

Malthus did not foresee the use of contraception, major changes in agricultural production techniques, or the exporting of excess people to colonies in the Americas. These factors, as well as high death rates, prevented the most devastating of his predictions from coming true. However, in many parts of the world today, people are experiencing the forms of population control (famine, epidemic disease, wars, and natural disasters) predicted by Malthus in 1798. Many people feel that his original predictions were valid, but his time scale was not correct and we are seeing his predictions come true today.

Another important impact of his essay was the effect it had on the young Charles Darwin. When Darwin read it, he saw what was true for the human population could be applied to the whole of the plant and animal kingdoms. As overreproduction took place, there would be increased competition for food, resulting in the death of the less fit organisms. This theory he called natural selection.

increase in the number of mouths to feed, it is unlikely that food production will be able to keep pace with the growth in human population. The severity of this problem was finally recognized on a worldwide basis in 1974 when the first World Food Conference was held to discuss possible ways of solving the problem. A primary indicator of the adequacy of the world food situation is the amount of grain eaten on the average by each person in the world (per capita grain consumption). During the 1960s until 1973, world per capita grain consumption increased from 264 kilograms to 310 kilograms, a 17.5% increase. Since 1973, world per capita grain consumption has increased very little. The less-developed nations of the world have had a disproportionately large increase in population and a decline in production because they are less able to afford costly fertilizer, machinery, and the energy necessary to run the machines and irrigate the land to produce their own cereals. Nearly all of the less-developed countries and many of the countries of the developed world have become net food importers. They rely on imports of grain from the North American breadbasket to supplement their own efforts to raise food.

With an increasing demand for cereals, high costs of production for fuel, fertilizer, machinery, and transportation, and the inability of less-developed countries to pay for the food they need, North America is faced with difficult political and economic decisions. If production is reduced, prices will increase for both American consumers and those countries that have extreme food needs. Do we use our food production capacity as an economic and political weapon as the Organization of Petroleum Exporting Countries (OPEC) did? What is the morality of international bargaining with limited food supplies? By refusing to sell surplus grain to a country, are we going to doom millions of malnourished children to mental retardation and death? Even now an estimated eleven million infant children die each year due to a combination of malnutrition and infectious diseases. If we do not solve the human population problem, natural forces will result in an increased death rate as the carrying capacity is reached.

Availability of energy is the second broad category of limiting factors that affects human populations as well as other kinds of organisms. All species on earth are ultimately dependent on sunlight for energy—so are we. Whether one produces electrical power from a hydroelectric dam, burns fossil fuels, or uses a solar cell, the energy is derived from the sun. Energy is needed for transportation, building and maintaining homes, and food production. It is very difficult to develop unbiased, reasonably accurate estimates of global energy "reserves" in the form of petroleum, natural gas, and coal. Therefore, it is difficult to predict how long these "reserves" might last. However, we do know that the quantities are limited and that the rate of use has been increasing, particularly in the developed and developing countries. If the less-developed countries were to attain a standard of living equal to the developed nations, the global energy "reserves" would disappear overnight. Since the United States constitutes 4.85% of the world's population and consumes approximately 25% of the world's energy resources, raising the standard of living of the entire world population to that of the United States would result in a 500% increase in the rate of consumption of energy and reduce theoretical reserves by an equivalent 500%. Humans should realize there is a limit to our energy resources, that we are living on solar energy that was stored over millions of years and we are using it at a rate that could deplete it in hundreds of years. Will energy availability be the limiting factor that determines the ultimate carrying capacity for humans or will it involve the problems of waste disposal?

One of the most talked about aspects of human activity is the problem associated with waste disposal. Not only do we have normal biological wastes that can be dealt with by decomposer organisms, but we generate a variety of technological wastes and by products that cannot be efficiently degraded by decomposers. Most of what we call pollution are the waste products of technology. The biological wastes can usually be dealt with fairly efficiently by the building of waste water treatment plants and other sewage facilities. Certainly these facilities take energy to run, but they rely on decomposers to degrade unwanted organic matter to carbon dioxide and water. Earlier in this chapter we disussed the problem bacteria and yeasts face when their metabolic waste products accumulate. In this situation, the organisms so "befouled their nest" that their wastes poisoned them. Are humans in a similar situation on a much larger scale? Are we dumping so much technological waste, much of which is toxic, into the environment that we are being poisoned? Some people believe that disregard for the quality of our environment will be a major factor in decreasing our population growth rate. In any case, it makes good sense to do everything possible to stop pollution and work toward cleaning our nest.

The fourth category of limiting factors that determines carrying capacity is interaction among organisms. Humans interact with other organisms in as many ways as other animals do. We have parasites, and occasionally predators. We are predators on a variety of animals both domesticated and wild. We have mutualistic relationships with many of our domesticated plants and animals, since they could not survive without our agricultural practices and we would not survive without the food they provide. Competition is also very important. Insects and rodents compete for the food we raise and we are in direct competition with many kinds of animals for the use of ecosystems. As humans convert more and more land to agricultural and other purposes, many other organisms are displaced. Many of these displaced organisms are not able to compete successfully and must leave the area, have their populations reduced, or become extinct. The American bison (buffalo), African and Asian elephants, the panda, and grizzly bear are a few species that are much reduced in number because they were not able to successfully compete with the human species. The passenger pigeon, Carolina parakeet, and great auk are a few that have become extinct. Our parks and natural areas have become tiny refuges for plants and animals that once occupied vast expanses of the world. If these refuges are lost, many organisms will become extinct. What today might seem to be an insignificant organism that we can easily do without, may tomorrow be seen as a link to our very survival. Humans have been extremely successful in their efforts to convert ecosystems to their own uses at the expense of other species.

Competition with one another (intraspecific competition), however, is a different matter. Since competition is negative to both organisms, humans must be harmed. We are not displacing another species, we are displacing some of our own kind. Certainly when resources are in short supply, there is competition. Unfortunately, it is usually the young that are least able to compete and high infant mortality is the result.

Humans are different from most other organisms in a fundamental way. Humans are able to predict the outcome of a specific course of action. Current technology and medical knowledge are available to control human population and improve the health and well-being of the people of the world. Why does the human population continue to grow, resulting in poverty among its members and stressing the environment in which we live? Since humans are social animals that have freedom of choice, they frequently do not do what is considered "best" from an unemotional, unselfish point of view. People make decisions based on historical, social, cultural, ethical, and personal considerations. What is best for the population as a whole may be bad for you as an individual. The biggest problems associated with control of the human population are not biological problems, but require the efforts of philosophers, theologians, politicians, sociologists, and others. As population increases, so will political, social, and biological problems. There will be less individual freedom.

Herd politics will prevail. The knowledge and technology necessary to control population is available, the will is not. What will eventually bring about the limiting of the human population? Will it be lack of resources, lack of energy, accumulated waste products, competition among ourselves, or rational planning of family size?

Summary

A population is a group of organisms of the same species in a particular place at a particular time. Populations differ from one another in gene frequency, age distribution, sex ratio, and population density. Organisms typically have a reproductive capacity that exceeds what is necessary to replace the parent organisms when they die. This inherent capacity to overreproduce causes a rapid increase in population size when a new area is colonized. A typical population growth curve consists of a lag phase in which population rises very slowly, followed by an exponential growth phase in which the population increases at an accelerating rate, and is followed by a leveling off of the population in a stationary growth phase. In some populations, a fourth phase may occur known as the death phase. This is typical of bacterial and yeast populations.

The carrying capacity is the number of organisms that can be sustained in an area over a long period of time. It is set by a variety of limiting factors. Availability of energy, availability of raw materials, accumulation of wastes, and interactions with other organisms are all categories of limiting factors. Because organisms are interrelated, sometimes population changes in one species will affect the size of other populations. This is particularly true when one organism uses the other as a source of food. Some limiting factors operate from outside the population and are known as extrinsic factors. Others are properties of the species itself and are called intrinsic factors. Some limiting factors become more intense as the size of the population increases and are known as density-dependent factors. Other limiting factors that are more accidental and not related to population size are called density-independent factors.

Humans as a species have the same limits and influences that other organisms do. Our current problems of food production, energy need, pollution, and habitat destruction are outcomes of uncontrolled population growth. However, humans can think and predict, thus providing the possibility of population control through conscious population limitation.

Consider This

If you return to figure 18.10, you will note that it has very little in common with the population growth curve shown in figure 18.5. What factors have allowed the human population to grow so rapidly? What natural limiting factors will eventually bring this population under control?

What is the ultimate carrying capacity of the world? What alternatives to the natural processes of population limitation could bring human population under control?

Consider the following in your answer: reproduction, death, diseases, food supply, energy, farming practices, food distribution, cultural biases, and anything else you consider to be appropriate.

Place a male and a female fruit fly in a bottle with half a banana. (You may use prepared fruit fly medium if you have it available.) The fruit flies can be wild or from cultures. Males have solid black tail ends, while females have striped rear ends. Count the fruit flies each day for two weeks and plot them on a graph.

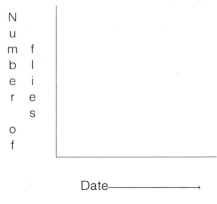

Number of flies

Date⟶

Questions

1. Draw the population growth curve of a yeast culture during the wine-making process. Label the lag, exponential growth, stationary growth, and death phases.
2. List four ways that two populations of the same species could be different.
3. Why do populations grow?
4. List four kinds of limiting factors that help to set the carrying capacity for a species.
5. How do the concepts of biomass and population size differ?
6. Differentiate between density-dependent and density-independent limiting factors. Give an example of each.
7. Differentiate between intrinsic and extrinsic limiting factors. Give an example of each.
8. As the human population continues to grow, what should we expect to happen to other species?
9. How does the population growth curve of humans compare with that of other kinds of animals?
10. All organisms overreproduce. What advantage does this give to the species? What disadvantages?

Chapter Glossary

age distribution (āj dis''tri-biu'shun) The number of organisms of each age in the population.

biotic potential (bi-ah'tik po-ten'shul) The theoretical maximum rate of reproduction.

carrying capacity (kār're-ing kuh-pas'ĭ-te) The optimum population size an area can support over an extended period of time.

death phase (deth fāz) The portion of some population growth curves in which the size of the population declines.

density-dependent factors (den'si-te de-pen'dent fak'tŏrz) Population-limiting factors that become more effective as the size of the population increases.

density-independent factors (den'si-te in''de-pen'dent fak'tŏrz) Population-controlling factors that are not related to the size of the population.

environmental resistance (en-vi-ron-men'tal re-zis'tants) The collective set of factors that limit population growth.

exponential growth phase (eks-po-nent'shul grōth fāz) A period of time during population growth when the population increases at an accelerating rate.

extrinsic factors (eks-trin'sik fak'tōrz) Population-controlling factors that arise outside the population.

gene flow (jēn flo) The movement of genes within a population due to migration and the movement of genes from one generation to the next by gene replication and sexual reproduction.

gene frequency (jēn fre'kwen-se) How often a specific gene shows up in the gametes of a population.

intrinsic factors (in-trin'sik fak'tōrz) Population-controlling factors that arise from within the population.

lag phase (lag fāz) A period of time during population growth when the population remains small, but increases slowly.

limiting factors (lim'ĭ-ting fak'tōrz) Environmental influences that limit biological activity.

mortality (mor-tal'ĭ-te) The number of individuals leaving the population by death per thousand individuals in the population.

natality (na-tal'ĭ-te) The number of individuals entering the population by reproduction per thousand individuals in the population.

population (pop''u-la'shun) Group of organisms of the same species located in the same place at the same time.

population density (pop''u-la'shun den'si-te) The number of organisms of a species found per unit area.

population growth curve (pop''u-la'shun grōth kurv) A graph of the change in population size over time.

population pressure (pop''u-la'shun presh-yur) Intense competition that leads to changes in the environment and dispersal of organisms.

reproductive capacity (re-pro-duk'tiv kuh-pas'ĭ-te) The theoretical maximum rate of reproduction.

sex ratio (seks ra-sho) The number of males in a population compared to the number of females.

stationary growth phase (sta-shun-a-re grōth fāz) A period of time during population growth when the number of individuals entering the population and the number leaving the population are equal, resulting in a stable population.

Chapter Outline

Purpose

This chapter focuses on the activities of individual organisms. We look at several examples of behavior and the significance of that behavior for the welfare of the species. The dangers of misinterpreting observed behavior is discussed and the ecological importance of behavior is a central theme.

The two most popular exhibits at the zoo are the primates and the reptiles. The fascination with reptiles is generally based on fear and hate, while the interest in primates is based on their similarity to humans. Many of the bizarre behaviors seen in captive animals are not normal. The pacing of many zoo animals does not occur when they are in the wild. Some other learned behaviors are the begging behavior of bears and elephants.

Learning Objectives

- Understand that behavior has evolutionary and ecological significance.

- Distinguish between instinctive and learned behaviors.

- Recognize there are several kinds of learning.

- Know that animals use sight, sound, and chemicals to communicate for reproductive purposes.

- Appreciate that territoriality and dominance hierarchies allocate resources.

- Know several methods used by animals in navigation.

- Describe why the evolution of social animals is different from nonsocial animals.

Understanding Behavior

Behavior is how an organism acts, what it does, and how it does it. When we think about the behavior of an animal, we need to understand that behavior is like any other characteristic displayed by the animal. It has a value or significance to the animal as it goes about exploiting resources and reproducing more of its own species. Many behaviors are inherited and consequently have evolved just like structures. In this respect, behavioral characteristics are no different from structural characteristics. The evolution of behavior is much more difficult to study, however, since behavior is transient and does not leave good fossils like structures do.

Behavior is a very important part of the ecological role of any animal, since behavior allows animals to escape predators, seek out mates, gain dominance over others of the same species, and respond to changes in the environment. Plants must, for the most part, use structures, physiological changes, or chance to accomplish the same ends. For example, a rabbit can run away from a predator—a plant cannot, but it may have developed thorns or toxic compounds within its leaves that discourage animals from eating it. Mate selection in animals often involves elaborate behaviors that assist animals in identifying the species and sex of the potential mate. Most plants rely on a much more random method for transferring male gametes to the female plant. Dominance in plants is often achieved by depriving competitors of essential nutrients or by inhibiting the development of seeds of other plants. Animals have a variety of behaviors that allow them to exert dominance over members of the same species.

Figure 19.1
Animal Behavior.

Baby herring gulls cause their parents to regurgitate food onto the ground when the baby bird pecks at the red spot on the parent's bill. The parent then picks up the food and feeds it to the baby.

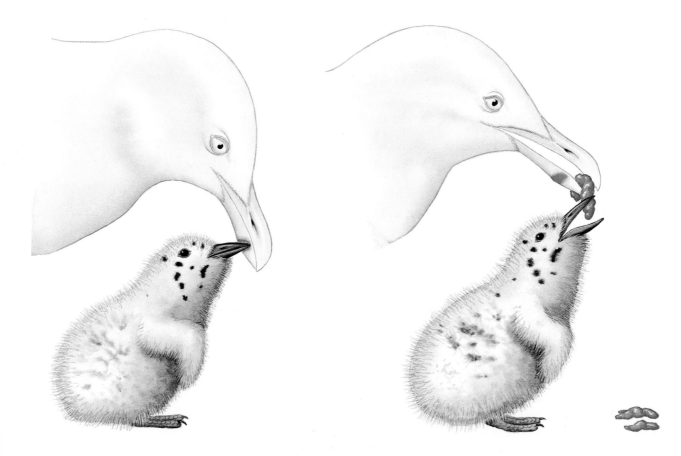

Behaviors are often subtle, so it is not always easy to identify the significance of a behavior without careful study of the behavior pattern and the impact it has on other organisms. For example, a hungry baby herring gull pecks at a red spot on its parent's bill. What possible value can this behavior have for either the chick or the parent? If we watch, we see that when the chick pecks at the spot, the parent regurgitates food onto the ground and the chick feeds (figure 19.1). This looks like a simple behavior, but there is more to it than meets the eye. Why did the chick peck to begin with? How did it know to peck at that particular spot? Why did the pecking cause the parent to regurgitate food? These questions are not easy to answer, and many people assume that the actions have the same motivation and direction as similar human behaviors. For example, when a human child points to a piece of candy and makes appropriate noises, it is indicating to its parent that it wants some candy. Is that what the herring gull chick is doing?

Some people believe that a bird singing on a warm, sunny spring day is making that beautiful sound because it is so happy. Students of animal behavior do not accept this idea and have demonstrated that a bird sings to tell other birds to keep out of its territory.

The barbed stinger of a honeybee remains in your skin after you are stung, and the bee tears the stinger out of its body when it flies away. The damage to its body is so great that it dies. Has the bee performed a noble deed of heroism and self-sacrifice? Was it defending its hive from you? We need to know a great deal more about the behavior of bees to understand the value of such behavior to the success of the bee species. The fact that bees are social animals like us makes it particularly tempting to think that bees are doing things for the same reasons we do.

The idea that we can ascribe human feelings, meanings, and emotions to the behavior of animals is called **anthropomorphism.** The fable of the grasshopper and the ant is another example of crediting animals with human qualities. The ant is pictured as an animal that despite temptations, works hard from morning 'til night, storing away food for the winter (figure 19.2). The grasshopper, on the other hand, is represented as a lazy good-for-nothing that fools away the summer when it really ought to be saving up for the tough times ahead. If one is looking for parallels to human behavior, these are pretty good illustrations. But they really are not accurate statements about the lives of the animals from an ecological point of view. Both the ant and the grasshopper are very successful organisms, but each has a different way of satisfying its needs and assuring that some of its offspring will be able to provide another generation of organisms. One method of survival is not necessarily better than another as long as both animals are successful. This is what the study of behavior is all about—looking at the activities of organisms during their life cycles and determining the value of the behavior in the ecological niche of the organism. The scientific study of the nature of behavior, and its ecological and evolutionary significance in its natural setting is known as **ethology.**

Before we go much further, we need to discuss how animals generate specific behaviors. Both instinct and learning are involved in behavior patterns of most organisms.

Instinct

Many behaviors of animals are automatic, preprogrammed, genetically inherited behaviors. Such behaviors are called **instinctive behavior.** These kinds of behaviors are found in a wide range of organisms from simple one-celled protozoans to complex vertebrates. These behaviors are performed correctly the first time without previous experience when the proper stimulus is given. A **stimulus** is some change in

Figure 19.2
The Fable of the Ant and the Grasshopper.
In many ways we give human meaning to the actions of animals. The ant is portrayed as an industrious individual who prepares for the future and the grasshopper is a lazy fellow who sits in the sun and sings all day. This kind of view of animal behavior is anthropomorphic.

Figure 19.3
Egg-Rolling Behavior in Geese.

Geese will use a specific set of head movements to roll any reasonably round object back into the nest. There are several components to the behavior including recognition of the object and head-tucking movements. If the egg is removed during the head-tucking movements, the behavior continues as if the egg were still there.

the internal or external environment of the organism. The reaction of the organism to the stimulus is called a **response.**

In our example of the herring gull chick, the red spot on the bill of the adult bird serves as a stimulus to the chick. The chick responds to this spot in a very specific, genetically programmed way. The behavior is innate—it is done correctly the first time without prior experience. The pecking behavior of the chick is in turn the stimulus for the adult bird to regurgitate food. It is obvious that these behaviors have adaptive value for the gull species, since it leaves little to chance. Instinctive behavior has great value because it allows correct, perfect, and necessary behavior to occur without prior experience.

Instinctive behavior is not capable of modification when a new situation presents itself, but it can be very effective for the survival of a species if it is involved in fundamental, essential activities that rarely require modification. Instinctive behavior is most common in animals that have short life cycles, small nervous systems, and little contact with parents. Over long periods of evolutionary time, these genetically determined behaviors have been selected for and have been useful to most of the individuals of the species. However, there may be some instances of inappropriate behavior generated by unusual stimuli or circumstances in which the stimulus is given. For example, many insects use the sun, moon, or stars as aids to navigation. Over the millions of years of insect evolution, this has been a valuable and useful tool for the animal. However, the human species has invented a variety of artificial lights that generate totally inappropriate behavior, such as when insects collect at a light and batter themselves to death. This mindless, mechanical behavior seems incredibly stupid to us, but it is still valuable for the species since the majority do not encounter artificial lights and complete their life cycles normally.

Certain species of geese will go through a behavior pattern that involves rolling eggs back into the nest. If the egg is taken from the goose when it is in the middle of egg-rolling behavior, it will continue its egg rolling until it gets back to the nest, even though there is no egg to roll (figure 19.3). It was also discovered that many other somewhat egg-shaped structures would generate the same behavior. For example, beer cans and baseballs were good triggers for the egg-rolling behavior. So not only was the bird unable to stop the egg-rolling behavior in midstride, but several non-egg objects generated inappropriate behavior because they had approximately the correct shape.

Some activities are so complex that it seems impossible for the organism to be born with such abilities. For example, you may have seen a caterpillar spin its cocoon. This is not just a careless jumble of silk threads. A cocoon is so precisely made that you can recognize what species of caterpillar made it. But cocoon spinning is not a learned ability. A caterpillar has no opportunity to learn how to spin a cocoon, since it never observes others doing it. Furthermore, they do not practice several

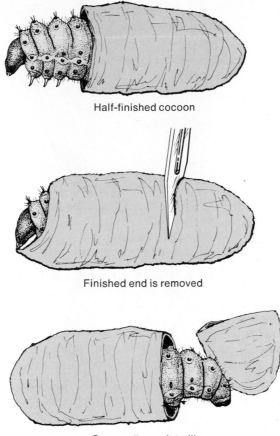

Half-finished cocoon

Finished end is removed

Cocoon "completed"

Figure 19.4
Inflexible Instinctive Behavior.
If part of the caterpillar's half-complete cocoon is removed while it is still spinning, the animal will finish the job but never repair the damaged end.

times before they get a proper, workable cocoon. It is as if a "program" for making a cocoon is in the caterpillar's "computer." If you interrupt the caterpillar's cocoon-making effort in the middle and remove most of the finished part, the caterpillar will go right on making the last half of the cocoon. The caterpillar's "program" does not allow for this kind of incident (figure 19.4). This inability to modify behavior as circumstances change is a prominent characteristic of instinctive behavior.

Could this behavior pattern be the result of natural selection? It is well established that many kinds of behaviors are controlled by genes. The "computer" in our example is really the DNA of the organism, and the "program" consists of a specific package of genes. Through the millions of years that insects have been in existence, natural selection has modified the cocoon-making "program" to refine the process. Certain genes of the "program" have undergone mutation, resulting in changes in behavior. Imagine various ancestral caterpillars, each with a slightly different "program." The inherited "program" that gave the caterpillar the best chance of living long enough to produce a new generation is the "program" selected for and most likely to be passed to the next generation.

Learned Behavior

The alternative to preprogrammed, instinctive behavior is learned behavior. **Learning** is a change in behavior as a result of experience. (Your behavior will be different in some way as a result of reading this chapter.) Many kinds of birds must learn parts of their songs. Experimenters have raised young song sparrows in the absence

Figure 19.5
Distribution of Learned and
Instinctive Behavior.

Different groups of animals show
different proportions of instinctive
and learned behavior in their
behavior patterns.

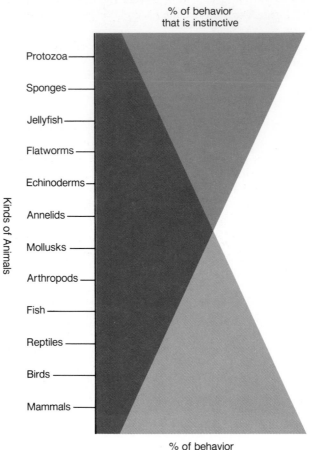

of any adult birds so there was no song for the young birds to imitate. These isolated
birds would sing a series of notes that was like the normal song of the species, but
not exactly correct. Birds from the same nest that were raised with their parents
developed a song nearly identical to their parents. If bird songs were totally instinctive, there would be no difference between these two groups. It appears that the
basic melody of the song was inherited by the birds and that the refinements of the
song were the result of experience. Therefore, the characteristic song of that species
was partly learned behavior (a change in behavior as a result of experience) and
partly unlearned (instinctive). This is probably true of the behavior of many organisms; they show complex behaviors that are a mixture of instinct and learning.
It is important to note that there are many kinds of birds that learn most of their
song with very few innate components. A mockingbird is very good at imitating the
songs of a wide variety of bird species.

This mixture of learned and instinctive behavior is not the same for all species.
Many invertebrate animals rely on instinct for the majority of their behavior patterns, while many of the vertebrates (particularly birds and mammals) make use of
a great deal of learning (figure 19.5).

Learning becomes more significant in long-lived animals that care for their
young, because it is possible for young to imitate their parents and develop behaviors
that are appropriate to local conditions. This takes time to develop but has the advantage of adaptability. In order for learning to become dominant in the life of an

animal, it must also have a large brain to store the new information it is learning. It is probably for this reason that learning is a major part of life for only a few kinds of animals like the vertebrates.

Learning is not just one kind of activity, but is subdivided into several categories: conditioning, imprinting, and insight learning.

Conditioning

A Russian physiologist, Ivan Pavlov (1849–1936), was investigating the physiology of digestion when he discovered that dogs can associate an unusual stimulus with a natural stimulus. He was studying the production of saliva by dogs and he knew that a natural stimulus, such as the presence or smell of food, would cause the dogs to start salivating. Then he rang a bell just prior to the presentation of the food. After a training period, the dogs would begin to salivate when the bell was rung even though no food was present. This kind of learning in which a "neutral" stimulus (the sound of the bell) is associated with a "natural" stimulus (the taste of food) is called **classical conditioning** or **associative learning.** The response produced by the neutral stimulus is called a **conditioned response.**

In this case the dogs were receiving positive reinforcement, but it is also possible to have negative reinforcement in which a painful or distressing stimulus is received. For example, if a dog was stimulated on the right foot with an electrical stimulus at the same time that a bell was rung, it would soon associate the bell with the painful stimulus and lift its foot even though no electrical stimulus was given.

In the real world of animals, this kind of learning takes place as well. If certain kinds of fruits or insects have unpleasant tastes, animals will learn to associate the bad taste with the color and shape of the offending object and avoid them in the future (figure 19.6).

Figure 19.6
Associative Learning.
Many animals learn to associate unpleasant experiences with the color or shape of offensive objects and avoid them in the future.

Figure 19.7
Imprinting.
Imprinting is a special kind of
relatively irreversible learning that
occurs during a very specific part
of the life of an animal. These
geese have been imprinted on
Konrad Lorenz and exhibit the
"following response" that is typical
of birds of this type.

Imprinting

Imprinting is a special kind of learning in which a very young animal is genetically primed to learn a specific behavior in a very short period. This type of learning was originally recognized by Konrad Lorenz (b. 1903) in his experiments with geese and ducks. He determined that shortly after hatching, a duckling would follow an object if the object was fairly large, moved, and made noise. In one of his books, Lorenz described himself squatting in the lawn one day, waddling and quacking, followed by newly hatched ducklings. He was busy being a "mother duck." He was surprised to see a group of tourists on the other side of the fence watching him in amazement. They couldn't see the ducklings hidden by the tall grass. All they could see was this strange performance by a big man with a beard!

Ducklings will follow only the object on which they were originally imprinted. Under normal conditions, the first large, noisy, moving object the newly hatched ducklings see is their mother. Imprinting assures that the immature birds will follow her and learn, by example, appropriate feeding, defensive tactics, and other behaviors. Since they are always near their mother, she can also protect them from enemies or bad weather. If animals imprint on the wrong objects, they are not likely to survive. Since these experiments by Lorenz in the early 1930s, we have discovered that many young animals can be imprinted by several types of stimuli and that there are responses other than following (figure 19.7).

In song sparrows, the learning of their song appears to be a kind of imprinting. It has been discovered that the young birds must hear the correct song during a specific part of their youth or they will never be able to perform the song correctly as adults. This is true even if they are surrounded by other adult song sparrows that are singing the correct song. Furthermore, the period of time when they learn the song is prior to the time that they begin singing. Recognizing and performing the correct song is important because it has particular meaning to other song sparrows. For males it conveys the information that a male song sparrow has a space reserved for himself. For females, the male song is an advertisement of the location of a male of the correct species that could be a possible mate.

Insight Learning

Insight learning is a special kind of learning in which past experiences are reorganized to solve new problems. When you are faced with a new problem, whether it is a crossword puzzle, a math problem, or any one of a hundred other everyday problems, you sort through your past experiences and locate those that apply. You may not even realize that you are doing it, but you put these past experiences together in a new way that may give the solution to your problem. Because this process is an internal one and can be demonstrated only through some response, it is very difficult to understand exactly what goes on during insight learning. Behavioral scientists have explored this area for many years, but the study of insight learning is still in its infancy.

Insight learning in animals is particularly difficult to study since it is impossible to know for sure whether a novel solution to a problem is the result of "thinking it through" or an accidental occurrence. For example, a small group of Japanese macaques (monkeys) was studied on an island. They were fed by simply dumping food such as sweet potatoes or wheat onto the beach. Eventually one of the macaques discovered that she could get the sand off the sweet potato by washing it in a nearby stream. She also discovered that she could sort the wheat from the sand by putting the mixture into water because the wheat would float. Is this an example of insight learning? We will probably never know, but it is tempting to think so.

In the examples we have used so far, some were laboratory studies, some were field studies, and some included aspects of both. Often they overlap into the field of psychology. This is particularly true for many of the laboratory studies. You can see that the science of animal behavior is a broad one that draws on information from several fields of study, and can be used to explore many different kinds of questions. In the topics that follow we will avoid the field of psychology and concentrate on the significance of the behavior from both an ecological and an evolutionary point of view.

Now that we have some understanding of how organisms generate behavior, we can look at a variety of kinds of behavior in several different kinds of animals and see how the behavior is useful to the animal in filling its ecological niche.

Reproductive Behavior

It is obvious that in order for any species to survive, it must reproduce. There are many stages in any successful reproductive strategy. First of all, animals must be able to recognize individuals of the same species that are of the opposite sex. Several different techniques are used for this purpose. For instance, frogs of different species produce sounds that are just as distinct as the calls of different species of birds. The call is a code system that delivers a very private message, since it is only meant for one species. It is, however, meant for any member of that species near enough to hear. The sending mechanism is the call produced by male frogs, which both male and female frogs can receive by hearing. This results in frogs of both sexes congregating in a small area. Once they gather in a small pond, it is much easier to have the further communication necessary for mating to take place.

Chemicals can also serve to attract animals. **Pheromones** are chemicals produced by animals and released into the environment that trigger behavioral or developmental changes in other animals of the same species. They have the same effect as sound but are a different code system. The classic example of a pheromone is the chemical that female moths release into the air. The large, fuzzy antennae of the

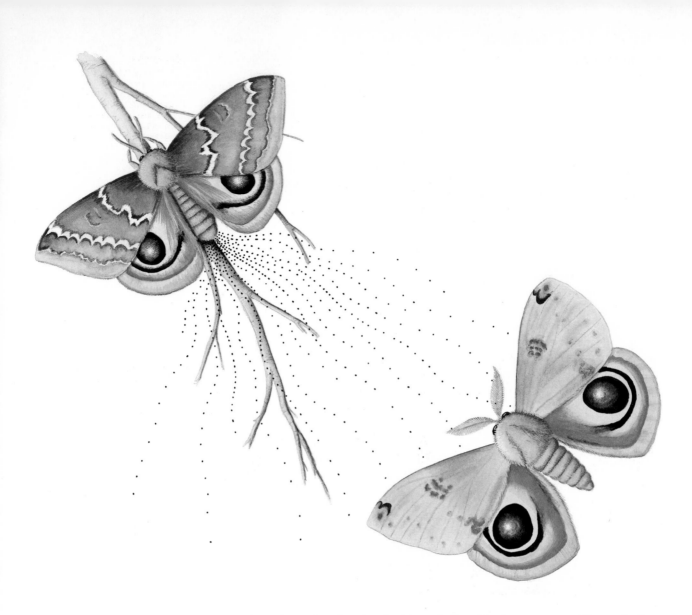

Figure 19.8
Communication.
The female moth signals her
readiness to mate and attracts
males by releasing a pheromone
that attracts males from long
distances downwind.

male moths can receive the chemical in unbelievably tiny amounts. The male then changes its direction of flight and flies upwind to the source of the pheromone, which is the female (figure 19.8). Some of these sex-attractant pheromones have been synthesized in the laboratory. One of these, called Disparlure[R], is widely used to attract and trap male gypsy moths (figure 19.9).

Since gypsy moths cause considerable damage to trees by feeding on the leaves, the sex attractant is used to estimate population size so that control measures can be taken to prevent large population outbreaks.

The firefly is probably the most familiar organism that uses light signals to bring together males and females. Several different species may live in the same area, but each species flashes its own code. The code is based on the length of the flashes, the frequency of the flashes, and the overall pattern of flashes (figure 19.10). There is also a difference between the signals given by males and females. For the most part, males are attracted to and mate with females of their own species. However, in one species of firefly, the female has the remarkable ability to signal the correct answering code to species other than her own. After she has mated, she will

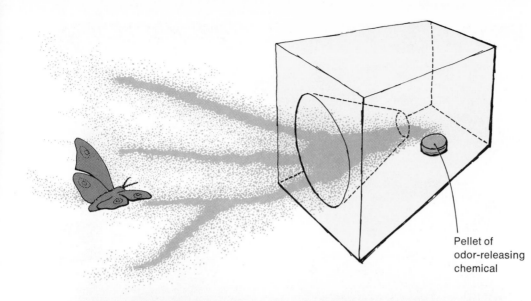

Figure 19.9
Artificial Attractants.
One method of controlling the population of certain insects is to use the normally produced sex pheromones to lure males into traps. In some instances, the traps are used to assess the size of the population rather than as a direct control measure.

Pellet of
odor-releasing
chemical

Figure 19.10
Firefly Communication.
The pattern of light flashes, the location of light flashes, and the duration of light flashes all help fireflies identify members of the opposite sex who are of the appropriate species.

Figure 19.11
Territoriality.
In colonial nesting seabirds it is typical to have very small nest territories. Each territory is just out of pecking range of the neighbors.

continue to signal to passing males of other species. She is not hungry for sex, she is just hungry. The luckless male who responds to her "come on" is going to be her dinner. Once male and female animals have attracted one another's attention, the second stage in successful reproduction takes place.

The second important activity in reproduction is fertilizing eggs. Many marine organisms simply release their gametes into the sea simultaneously and allow fertilization and further development to take place without any input from the parents. Sponges, jellyfishes, and many other marine animals fit into this category. Other aquatic animals congregate so that the chances of fertilization are enhanced by the male and female being near one another as the gametes are shed. This is typical of many fish and some amphibians such as frogs. Most terrestrial organisms have internal fertilization in which the sperm are introduced into the reproductive tract of the female. Terrestrial animals, like some spiders, produce packages of sperm that the female picks up with her reproductive structures. Many of these mating behaviors require elaborate species-specific communication prior to the mating act. Several examples were given in the previous paragraphs.

A third element in successful reproduction is providing the young with the resources they need to live to adulthood. Many invertebrate animals spend little energy on the care of the young. The young develop on their own. Usually the young become free-living larvae that eat and grow rapidly. In some species, females make preparations for the young by laying their eggs in particularly suitable sites. Many insects lay their eggs on the particular species of plant that the larva will use as food as it develops. Many parasitic species seek out the required host in which they lay their eggs. The eggs of others may be placed in spots that provide safety until the young hatch from the egg. Turtles, many fish, and some insects fit into this category. In most of these cases, however, the female lays large numbers of eggs and most of the young die before reaching adulthood. This is an enormously expensive process, since the female invested considerable energy in the production of the eggs with such a low success rate.

An alternative to this "wasteful" loss of potential young is to produce fewer young, but to invest large amounts of energy in their care. This is typical of birds and mammals. They build nests, parents share in the feeding and protection of the young, and they often assist the young in learning appropriate behavior. Many insects like bees, ants, and termites have elaborate social organizations in which one or a few females produce large numbers of young that are cared for by sterile offspring of the fertile females. Some of the female's offspring will be fertile, reproducing individuals. The activity of caring for young involves many complex behavior patterns. It appears that most animals that feed and raise young are able to recognize their own young from those of other nearby families and may even kill the young of another family unit. Usually elaborate greeting ceremonies are performed between animals when they return to the nest or the den. Perhaps this has something to do with being able to identify individual young. Often this behavior is shared among adults as well. This is true for many colonial nesting birds such as gulls and penguins, and among many carnivorous mammals such as wolves, dogs, and hyenas.

Allocating Resources

For an animal to be successful it must receive sufficient resources to live and reproduce. Therefore, we find many kinds of behaviors that divide the available resources so that the species as a whole is benefited, even though some individuals may be harmed. One kind of behavior that is often tied to successful reproduction is territoriality. **Territoriality** is the process of setting aside space for the exclusive use of an animal for food, mating, or other purposes. A **territory** is the space an animal defends against others of the same species. This territory has great importance, since it reserves exclusive rights to the use of a specific piece of space. When territories are first being established, there is much conflict between individuals. This eventually gives way to the use of a series of signals that defines the territory and communicates to others that the territory is occupied. The male redwing blackbird has red shoulder patches, but the female does not. The male will perch on a high spot, flash his red shoulder patches, and sing to other male redwing blackbirds that happen to venture into his territory. Most other males get the message and leave his territory; those that do not leave, he attacks. He will also attack a stuffed, dead, male redwing blackbird in his territory, or even a small piece of red cloth. Clearly, the spot of red is the characteristic that stimulates the male to defend his territory. Such key characteristics that trigger specific behavior patterns are called **sign stimuli.** (Refer back to figure 19.1 for another example of a sign stimulus.)

During the mating period, many animals are highly territorial. Within a gull colony, each nest is in a territory of about one square meter (figure 19.11). When one gull walks or lands on the territory of another, the defender walks toward the other in the upright threat posture. The head is pointed down with the neck stretched outward and upward. The folded wings are raised slightly as if to be used as clubs. The upright threat posture is one of a number of **intention movements** that signal what an animal is likely to do in the near future. The bird shows an intention to do something, to fight in this case, but it may not carry out the intention. If the invader shows no sign of retreating, then one or both gulls may start pulling up the grass very vigorously with their beaks. This seems to make no sense. The gulls were ready to fight one moment; the next moment they apparently have forgotten about the conflict and are pulling grass. But the struggle has not been forgotten. Pulling grass is an example of redirected aggression. In **redirected aggression,** the animal attacks something other than the natural opponent. If the intruding gull doesn't leave at

Figure 19.12
Dominance Hierarchy.
Many animals maintain order within their groups by establishing a dominance hierarchy. Whenever you see a group of animals like cows or sheep walking in single file, it is likely that the dominant animal is at the head of the line, while the lowest ranking individual is at the end.

this point, there will be an actual battle. (A person who starts pounding the desk during an argument is showing redirected aggression. Look for examples of this behavior in your neighborhood cats, dogs, classmates—maybe even in yourself!)

The possession of a territory is often a requirement for reproductive success. In a way, then, territorial behavior has the effect of allocating breeding space and limiting population size to that which the ecosystem can support. This kind of behavior is widespread in the animal kingdom and can be seen in such diverse groups as insects, fish, spiders, reptiles, birds, and mammals.

Another way of allocating resources is by the establishment of a **dominance hierarchy** in which a relatively stable, mutually understood order of priority within the group is maintained. A dominance hierarchy is often established in animals that form social groups. One individual in the group dominates all others. A second-ranking individual dominates all but the highest-ranking individual, and so forth until the lowest-ranking individual must give way to all others within the group. This kind of behavior is seen in barnyard chickens where it is known as a pecking order. Figure 19.12 shows a dominance hierarchy. The lead animal is the highest ranking and the last animal is the lowest ranking.

A dominance hierarchy allows certain individuals to get preferential treatment when resources are scarce. The dominant individual will have first choice of food, mates, shelter, water, and other resources because of the position occupied. Less-favored animals may fail to mate or be malnourished in times of scarcity. In many social animals like wolves, only the dominant males and females reproduce. This assures that the most favorable genes are passed to the next generation. Poorly adapted animals with low rank may never reproduce. Once a dominance hierarchy is established, it results in a more stable social unit with little conflict. There may be an occasional altercation that reinforces the knowledge of which position an animal occupies in the hierarchy. Such a hierarchy frequently results in low-ranking individuals emigrating to new areas. Migrating individuals are often subject to heavy predation. Thus, the dominance hierarchy serves as a population control mechanism and a way of allocating resources. Resource allocation becomes most critical during periods of scarcity. In some areas, the dry part of the year is most stressful. In temperate areas, winter reduces many sources of food and forces organisms to adjust. Animals have several ways of coping with seasonal stress.

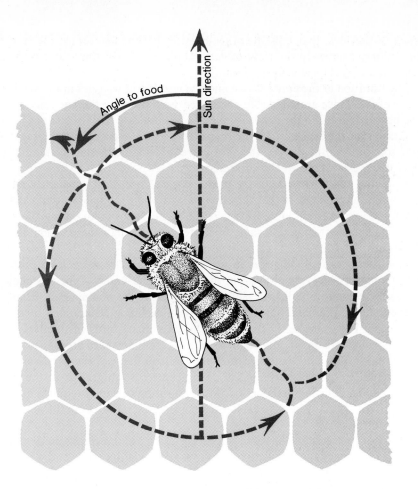

Angle to food

Sun direction

Figure 19.13
*Honeybee Communication
and Navigation.*
The direction of the straight, tail-
wagging part of the dance
indicates the direction to a source
of food. The angle this straight run
has to the vertical is the same
angle that the bee must fly to the
sun to find the food source. The
length of the straight run and the
duration of each dance cycle
indicates the flying time necessary
to reach the food source.

Some animals simply avoid the stress by hibernating. Hibernation is a physiological slowing of all body processes that allows the animal to survive on food it
has stored within its body. Hibernation is typical of many insects, bats, marmots,
and some squirrels. Other animals have built-in behavior patterns that cause them
to store food during seasons of plenty for periods of scarcity. These behaviors are
instinctive and are seen in a variety of animals. Squirrels bury nuts, acorns, and
other seeds. (They also plant trees because they never find all the seeds they bury.)
Chickadees stash seeds in cracks and crevices when seeds are plentiful and spend
many hours during the winter exploring similar places for food. Some of the food
they find is food they stored. Honeybees store honey, which allows them to live
through the winter when nectar is not available. This requires a rather complicated
set of behaviors that coordinates the activities of thousands of bees in the hive.

Navigation

The activities of honeybees involve communication among the various individuals
foraging for nectar. The bees are able to communicate information about the direction and distance of the nectar source from the hive. If the source of nectar is
some distance from the hive, the scout bee performs a "wagging dance" in the hive.
The bee walks in a straight line for a short distance, wagging its rear end from side
to side. It then circles around back to its starting position and walks the same path
as before (figure 19.13). This dance is repeated many times. The direction of the
straight path portion of the dance indicates the direction of the nectar in relationship
to the position of the sun. For instance, if the bee walks straight upward on a vertical

surface in the hive, that tells the other bees to fly directly toward the sun. If the path is thirty degrees to the right of vertical, the source of the nectar is thirty degrees to the right of the sun's position.

The duration of the entire dance and the number of waggles in the straight run portion of the dance are positively correlated with the time the bee must fly to get to the nectar source. So the dance is able to communicate the direction of flight and the duration of the flight. Since the recruited bees have picked up the scent of the nectar source from the dancer, they also have information about the kind of flower to visit when they arrive at the correct spot.

Since the sun is not stationary in the sky, the bee must constantly adjust its angle to the sun. It appears that they do this with some kind of internal clock. Bees that are prevented from going to the source of nectar or from seeing the sun will still fly in the proper direction sometime later, even though the position of the sun is different.

The ability to sense changes in time is often used by animals to prepare for seasonal changes. In areas away from the equator, the length of the day changes as the seasons change. The length of the day is called the **photoperiod.** Many birds prepare for migration and have their migration direction determined by changing photoperiod. For example, in the fall of the year many birds instinctively change their behavior, store up fat, and begin to migrate from northern areas to areas closer to the equator. This seasonal migration allows them to avoid the harsh winter conditions signalled by the shortening of days. The return migration in the spring is triggered by lengthening photoperiod. This migration certainly requires a lot of energy, but it allows many birds to exploit temporary food resources in the north during the summer months.

Like honeybees, some daytime migrating birds use the sun to guide them. We need two instruments to navigate by the sun—an accurate clock and a sextant for measuring the angle between the sun and the horizon. Can a bird perform such measurements without instruments when we, with our much bigger brains, need these instruments to help us? It is unquestionably true! For nighttime migration, some birds use the stars to help them find their way. In one interesting experiment, warblers, which migrate at night, were placed in a planetarium. The pattern of stars as they appear at any season could be projected onto a large domed ceiling. During autumn, when these birds would normally migrate southward, the stars of the autumn sky were shown on the ceiling. The birds responded with much fluttering activity at the south side of the cage, as if they were trying to migrate southward. Then the experimenters tried projecting the stars of the spring sky, even though it was autumn. Now the birds tended to try and fly northward. There was less unity in their efforts to head north; the birds seemed somewhat confused. Nevertheless, the experiment showed that the birds recognized star patterns and were influenced by them.

There is evidence that some birds navigate by compass direction, that is—they fly as if they had a compass in their heads. They seem to be able to sense magnetic north. Their ability to sense magnetic fields has been proven at the U.S. Navy's test facility in Wisconsin. The weak magnetism radiated from this test site has changed the flight pattern of migrating birds. But it is yet to be proven that birds use the magnetism of the earth to guide their migration.

Many animals besides birds and bees have a time sense built into their bodies. For instance, you have one. Travelers who fly part way around the world by nonstop jet plane need some time to recover from jet lag. Their digestion, sleep, or both may be upset. Their discomfort is not caused by altitude, water, or food, but by having rapidly crossed several time zones. There is a great difference in the time as measured by the sun or local clocks and that measured by the body. The body's clock adjusts more slowly.

In the animal world, mating is the most obviously timed event. In the Pacific Ocean, off some of the tropical islands, lives a marine worm known as the palolo worm. Its habit of making a well-timed brief appearance in enormous swarms is a striking example of a biological clock phenomenon. At mating time they swarm into the shallows of these islands and discharge sperm and eggs. There are so many worms that the sea looks like noodle soup. The people of the islands find this an excellent time to change their diet. They dip up the worms much as North Americans dip up smelt or other small fish making a spawning run. The worms appear around the third quarter of the moon in October or November, the time varying somewhat according to local environmental conditions.

Social Behavior

Many species of animals are found in interacting groups that show division of labor called **societies.** Societies differ from simple collections of organisms by the amount of specialization of the group's individuals. The individuals performing one function cooperate with others having different special abilities. As a result of specialization and cooperation, the society has characteristics not found in any one member of the group. The whole is more than the sum of its parts. But if cooperation and division of labor are to occur, there must be communication among individuals and coordination of effort.

Honeybees, for example, have an elaborate communication system and are specialized for specific functions. A few individuals known as queens and drones specialize in reproduction, while large numbers of worker honeybees are involved in collecting food, defending the hive, and caring for the larvae. These roles are quite rigidly determined by inherited behavior patterns. Each individual worker honeybee has a specific task, and all tasks must be fulfilled for the group to survive and prosper. As they age, the worker honeybees move through a series of tasks over a period of weeks. When they first emerge from their wax cells, they clean the cells. Several days later their job is to feed the larvae. Next they build cells. Later they become guards that challenge all insects that land near the entrance to the hive. Finally they become foragers who find and bring back nectar and pollen to feed the other bees in the hive. Foraging is usually the last job before the worker honeybee dies. Although this progression of tasks is the usual order, workers can shift from their main task to others if there is a need. Both the tasks performed and the progression of tasks are instinctively (genetically) determined (figure 19.14).

A hive of bees may contain thousands of individuals, but under normal conditions only the queen bee and the males called drones are capable of reproduction. None of the thousands of workers who are also females will reproduce. This does not seem to make sense because they appear to be giving up their chance to reproduce and pass their genes on to the next generation. Is this some kind of self-sacrifice on the part of the workers or is there another explanation? In general, the workers in the hive are the daughters or sisters of the queen and, therefore, share a large number of her genes. This means they are really helping a portion of their genes get to the next generation by assisting in the raising of their own sisters, some of whom will become new queens. This argument has been used to partially explain behaviors in societies that might be bad for the individual but advantageous for the society as a whole.

Within animal societies are levels of complexity. The type of social organization differs from species to species. Some societies show little specialization of individuals other than those determined by sexual differences or differences in physical size and endurance. The African wild dog illustrates such a flexible social organization. These animals are nomadic and hunt in packs. Although an individual wild dog can kill prey about its own size, groups are able to kill fairly large animals if they cooperate in the chase and the kill. The kill often involves a chase of several kilometers. When the dogs are young, they do not follow the pack. When adults return from a successful hunt they regurgitate food if the proper begging signal is presented to them. Therefore, the young and adults that remained behind to guard the young are fed by the hunters. The young are the responsibility of the entire pack, which cooperates in their feeding and protection. During the time that the young are at the den site, the pack must give up its nomadic way of life. Therefore, the young are born during the time of year when prey are most abundant. Only one or two of the females in the pack have young each year. If every female had young, the pack couldn't feed them all. At about two months of age, the young begin traveling with the pack and the pack can return to its nomadic way of life.

In many ways the honeybee and African wild dog societies are similar. Not all females reproduce, the raising of young is a shared responsibility, and there is some specialization of roles. The analysis and comparison of animal societies has led to the thought that there may be fundamental processes that shape all societies. The systematic study of all forms of social behavior, both animal and human, is a newly emerging study called **sociobiology.**

How did various types of societies develop? What selective advantage does a member of a social group have? In what ways are social groups better adapted to their environment than nonsocial organisms? How does social organization affect the way populations grow and change? These are difficult questions because, although evolution occurs at the population level, it is individual organisms who are selected. Thus, we need new ways of looking at evolutionary processes when describing the evolution of social structures.

The ultimate step in this new science is to analyze human societies according to sociobiological principles. But such an analysis is difficult and controversial since humans have a much greater ability to modify behavior than other animals. Human social structure and culture changes very rapidly compared to that of other animals. Sociobiology will continue to explore the basis of social organization and behavior, and will continue to be an interesting and controversial area of study.

Summary

Behavior is how an organism acts, what it does, and how it does it. The kinds of responses that organisms make to environmental changes (stimuli) may be simple reflexes, very complex instinctive behavior patterns, or learned responses.

From an evolutionary viewpoint, behaviors represent adaptations to the environment. They increase in complexity and variety the more highly specialized and developed the organism is. All organisms have inborn or instinctive behavior, while higher animals also have one or more ways of learning. These include conditioning, imprinting, and insight. Communication for purposes of courtship and mating is accomplished by sounds, visual displays, and chemicals called pheromones. Many animals have special behavior patterns that are useful in the care and raising of young.

Territoriality is communicated through a series of behaviors involving aggressive displays, redirected aggression, intention movements, and sign stimuli. Dominance hierarchies and territorial behavior are both involved in the allocation of scarce resources. To escape from seasonal stress, some animals hibernate, others store food, and others migrate. Migration to avoid seasonal extremes involves a timing sense and some way of determining direction. Animals navigate by means of sound, celestial light cues, and magnetic fields.

Societies consist of groups of animals that specialize and cooperate. Sociobiology attempts to analyze all social behavior in terms of evolutionary principles, ecological principles, and population dynamics.

Consider This

If you were going to teach an animal to communicate a message new to that animal, what message would you select? How would you teach the animal to communicate the message at the appropriate time?

Experience This

Spend ten minutes watching a specific individual animal. Birds, squirrels, and most insects are good subjects because many are active during the daytime. See if you can identify territorial behavior, sign stimuli, aggression, dominance, or any other behavior that was discussed in this chapter.

Questions

1. Why do students of animal behavior not accept the idea that a singing bird is a happy bird?
2. Briefly describe some animal's behavior that is learned. Name the animal.
3. Briefly describe some animal's behavior that is unlearned. Name the animal.
4. Describe an example of a conditioned response. Can you describe one that is not mentioned in this chapter?
5. Name three behaviors typically associated with reproduction.
6. How do territorial behavior and dominance hierarchies help to allocate scarce resources?

7. How do animals use chemicals, light, and sound to communicate?
8. What is sociobiology? Ethology? Anthropomorphism?
9. What is imprinting and what value does it have to the organism?
10. Describe how honeybees communicate the location of a nectar source.

Chapter Glossary

anthropomorphism (an-thro-po-mōr'fism) The ascribing of human feelings, emotions, or meanings to the behavior of animals.

associative learning (ă-so'shuh-tiv lur'ning) A kind of learning in which a neutral stimulus is associated with a natural stimulus to produce a particular response.

behavior (be-hav'yur) How an organism acts, what it does, and how it does it.

classical conditioning (klas'sĭ-kul kon-dĭ'shun-ing) A kind of learning in which a neutral stimulus is associated with a natural stimulus to produce a particular response.

conditioned response (kon-dĭ'shund re-sponts') The behavior displayed when the neutral stimulus is given after association has occurred.

dominance hierarchy (dom'in-ants hi'ur-ar-ke) A relatively stable, mutually understood order of priority within a group.

ethology (e-thol'uh-je) The scientific study of the nature of behavior and its ecological and evolutionary significance in its natural setting.

imprinting (im-prin'ting) Learning in which a very young animal is genetically primed to learn a specific behavior in a very short period.

insight learning (in-sīt lur'ning) Learning in which past experiences are reorganized to solve new problems.

instinctive behavior (in-stink'tiv be-hāv'yur) Automatic, preprogrammed, or genetically inherited behavior.

intention movements (in-ten'shun moov-ments) Behavior that signals what the animal is likely to do in the near future.

learning (lurn'ing) A change in behavior as a result of experience.

pheromone (fēr'o-mon) A chemical produced by an animal and released into the environment to trigger behavioral or developmental processes in some other animal of the same species.

photoperiod (fo-to-pe're-ud) The length of the light part of the day.

redirected aggression (re-di-rek'ted ă-grĕ'shun) A behavior in which the aggression of an animal is directed away from the opponent and to some other animal or object.

response (re-sponts') The reaction of an organism to a stimulus.

sign stimulus (sīn stim'yu-lus) A specific object or behavior that triggers a specific behavioral response.

society (so-si'uh-te) Interacting groups of animals of the same species that show division of labor.

sociobiology (so-sho-bi-ol'o-je) The systematic study of all forms of social behavior, both human and nonhuman.

stimulus (stim'yu-lus) Some change in the internal or external environment of an organism.

territoriality (tār''-rĭ-to-re-al'ĭ-te) A behavioral process in which an animal protects space for its exclusive use for food, mating, or other purposes.

territory (tār'rĭ-to-re) A space that an animal defends against others of the same species.

The Five Kingdoms

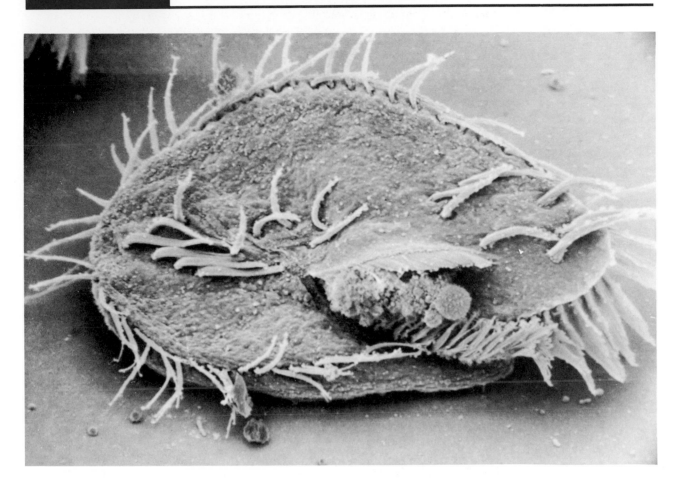

The origin of life on earth has long puzzled scientists. Many hypotheses have been proposed, but none have been validated. Evidence from such fields as chemistry, physics, astrophysics, biochemistry, and biology have provided support for some hypotheses; however, no one will ever know which hypothesis is true. It is known that since the first living things appeared, there has been an explosion of numbers and kinds of organisms on earth. Biologists have attempted to make their study of organisms easier by classifying them into five major groups. The five kingdoms of life include Prokaryotae, Protista, Mycetae, Plantae, and Animalia. While this system reflects our current understanding of life on earth, it continues to be modified to account for new information discovered every day. ∎

Purpose

This chapter deals with theories of the origin of cellular life, which begins with the formation of our solar system, and involves the changing of inorganic materials to organic materials. These organic materials may have been combined into living units called cells.

For Your Information

In 1975 an unmanned spacecraft, the Viking, landed on the planet Mars. One of its missions was to determine if life existed on the planet. Soil samples generated carbon dioxide, which suggests that some living organisms might have been present. It is also possible that the carbon dioxide came from inorganic sources. Recently scientists have postulated that life could have developed on Mars if liquid water was available. Iron-rich clay particles would have allowed organic molecules to be concentrated on their surfaces as a prelude to the formation of primitive cell types. However, water does not exist on Mars in the liquid form, but is found at the poles as ice caps.

Learning Objectives

- State the significance of Pasteur's experiment.

- Describe the formation of our solar system, including the earth.

- Describe the physical conditions on early earth and the changes thought to have happened before life could exist.

- Beginning with the gases in the early atmosphere, trace the events that led to the first living cells.

- Explain why spontaneous generation can only occur in a reducing atmosphere.

- Explain why the evolution of autotrophs assured a continuation of life on earth.

- Explain the evolution of eukaryotic cells.

Spontaneous Generation Versus Biogenesis

For centuries curiosity has spurred humans to study the basic nature of their environment. The vast amount of chemical and biological information presented in previous chapters indicates our ability to gather and analyze information. These efforts have resulted in solutions to many problems and have simultaneously revealed new and more challenging areas of concern. Despite these efforts, two questions have continued to be a subject of speculation: What is the nature of life and How did it originate?

In earlier times, no one ever doubted that life originated from nonliving things. The Greeks, Romans, Chinese, and many other ancient peoples believed that maggots arose from decaying meat, mice developed from wheat stored in dark, damp places, lice formed from sweat, and frogs originated from damp mud. The concept of **spontaneous generation,** the theory that living organisms arise from nonliving material, was widely believed until the seventeenth century (figure 20.1). However, there were some who did not believe this theory. These people subscribed to an opposing theory called **biogenesis.** Biogenesis is the concept that life originates only from preexisting life. One of the earliest challenges to the theory of spontaneous generation happened in 1668.

Francesco Redi, an Italian physician, set up a controlled experiment designed to disprove the theory of spontaneous generation (figure 20.2). He used two sets of jars that were identical except for one aspect. Both sets of jars contained decaying meat and both were exposed to the atmosphere. However, one set of jars was covered by a gauze and the other was uncovered. Redi observed that flies settled on the meat in the open jar, but the gauze blocked their access to the covered jars. When maggots appeared on the meat in the uncovered jars but not on the meat in the covered containers, Redi concluded that the maggots arose from the eggs of the flies and not from spontaneous generation in the meat.

Figure 20.1
Life from Nonlife.
Many works of art, such as shown here, explore the idea that living things could originate from nonliving matter or even from very different types of organisms. M. C. Escher's work entitled "The Reptiles 1943" shows the life cycle of a little alligator. Amid all kinds of objects, a drawing book lies open at a drawing of a mosaic of reptilian figures in three contrasting shades. Evidently, one of them has tired from lying flat and rigid amongst his fellows, so he puts one plastic-looking leg over the edge of the book, wrenches himself free, and launches out into real life. He climbs up the back of the book on zoology and works his laborious way up the slippery slope of a set-square to the highest point of his existence. Then after a quick snort, tired but fulfilled, he goes downhill again, via an ashtray, to the level surface, to that flat drawing paper, and meekly rejoins his erstwhile friends, taking up once more his function as an element of surface division.

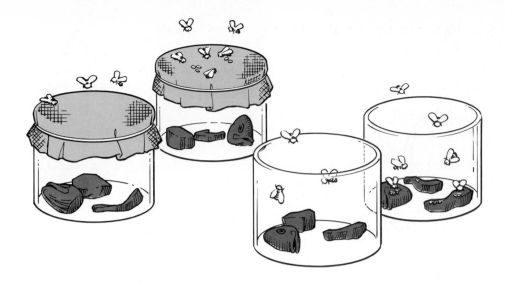

Figure 20.2
Redi's Experiment.
The two sets of jars here are identical in every way except one—the gauze covering. The set on the left is called the control group, the set on the right is the experimental group. Any differences seen between the control and the experimental group are the result of a single variable. In this manner, Redi concluded that the presence of maggots in meat was due to flies laying their eggs on the meat and not to spontaneous generation.

Even after Redi's experiment, there were still some who supported the theory of spontaneous generation. After all, a belief that has been prevalent for over two thousand years does not die a quick death. In 1748 John T. Needham, an English priest, placed a solution of boiled mutton broth in containers that he sealed with corks. Within several days, the broth became cloudy and contained a large population of microorganisms. Needham reasoned that boiling killed all the organisms and that the corks prevented any microorganisms from entering the broth. He concluded that life in the broth was the result of spontaneous generation.

In 1767 another Italian scientist, Abbe Lazzaro Spallanzani, challenged Needham's findings. Spallanzani boiled a meat and vegetable broth, placed this medium in clean glass containers, and sealed the openings by melting the glass over a flame. He placed the sealed containers in boiling water to make certain all microorganisms were destroyed. As a control, he set up the same conditions but did not seal the necks, allowing air to enter the flasks (figure 20.3). Two days later the open containers had a large population of microorganisms, but there were none in the sealed containers.

Spallanzani's experiment did not completely disprove the theory of spontaneous generation to everyone's satisfaction. The supporters of the theory attacked Spallanzani by stating that he excluded air, a factor believed necessary for spontaneous generation. Supporters also contested that boiling had destroyed a "vital element." When Joseph Priestly discovered oxygen in 1774, the proponents of spontaneous generation claimed that oxygen was the "vital element" that Spallanzani had excluded in his sealed containers.

In 1861 the French chemist Louis Pasteur convinced most scientists that spontaneous generation did not happen. He placed a fermentable sugar solution and yeast mixture in a flask that had a long swan neck. The mixture and the flask were boiled for a long time. The flask was left open to allow oxygen, the "vital element," to enter, but no organisms developed in the mixture. The organisms that did enter the flask settled on the bottom of the curved portion of the neck and could not reach the sugar water mixture. As a control, he cut off the swan neck (figure 20.4). This allowed microorganisms from the air to fall into the flask, and within two days the fermentable solution was supporting a population of microorganisms. In his address to the French Academy, Pasteur stated, "Never will the doctrine of spontaneous generation arise from this mortal blow."

Figure 20.3
Spallanzani's Experiment.
Spallanzani carried the experimental method of Redi one step further. He sealed one flask, the experimental group, after it had been boiled. He left the second flask, the control group, open after it had been boiled (a). Within two days, the open flask had a population of microorganisms (b). He demonstrated that spontaneous generation could not occur unless the broth was exposed to the "germs" in the air.

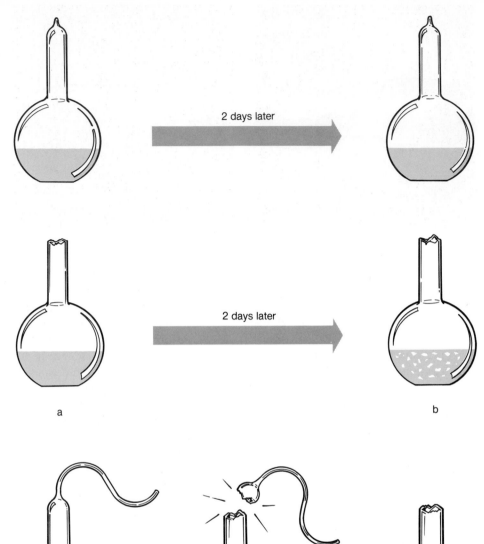

a

2 days later

2 days later

b

Figure 20.4
Pasteur's Experiment.
For an experimental group, Pasteur used the swan-neck flask (a). This allowed oxygen, but not the airborne organisms, to enter the flask. As a control, he broke the neck off of the flask (b). Within two days, there was growth in this flask (c). Pasteur demonstrated that germ-free air with its oxygen does not cause spontaneous generation.

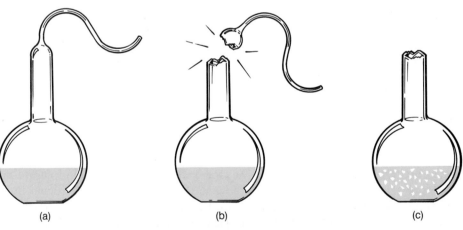

(a)   (b)   (c)

The Modern Theory of the Origin of Life

Pasteur's prediction regarding the theory of spontaneous generation was questioned some sixty years later. In the 1920s, a Russian biochemist, Alexander I. Oparin, and a British biologist, J. B. S. Haldane, working independently proposed the idea of spontaneous generation in a new form.

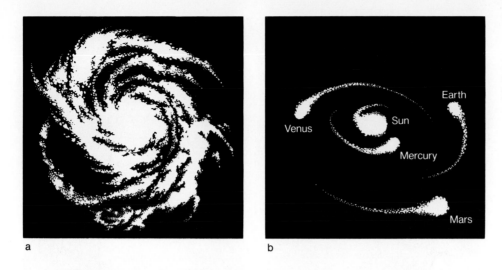

As the name implies, spontaneous generation proposes that nonliving material is naturally converted into living material. The first advocates of the theory of spontaneous generation believed that the creation of life was happening during their lifetime. They also believed that it happened in a matter of days. However, Oparin and Haldane proposed that it required about a billion years for the nonliving matter to be organized into units that could be called living. While early proponents thought of spontaneous generation as something that happened during their lifetime, modern supporters viewed it as something that could only have happened very early in earth's history when physical conditions were quite different from what exists today. In fact, Oparin and Haldane were endeavoring to answer the fundamental question posed at the beginning of this chapter: What is the nature of life and How did it originate?

Oparin first presented his ideas in a paper published in Moscow in 1924. However, his ideas did not come to the attention of the general scientific community because the work was never translated. A later work by Oparin, a book entitled *Origin of Life,* published in 1931 and translated into other languages, led to a renewed interest in developing theories that could account for the origin of life on earth.

The Early Earth

To understand Oparin's theory and variations of it, it is necessary to understand how our solar system and the earth are thought to have formed. The *solar nebula* theory proposes that the solar system was formed from a large cloud of gases that developed some ten to twenty billion years ago (figure 20.5). A gravitational force was created by the collection of particles within this cloud that caused other particles to be pulled from the outer edges to the center. As gravity caused these particles to collect, they formed a very large disk with our sun at its center. The condensation of atoms into the sun resulted in the release of large amounts of thermonuclear energy. Atoms were forced to fuse with one another because of their self-generated gravitational force.

As the particles within this solar nebula cloud were pulled into the center to form the sun, other gravitational centers developed. The material collected in these areas formed the sun's planets, including earth. Many scientists believe that earth was formed about four and a half billion years ago. A large amount of heat was

Figure 20.6
Primitive Earth.
The environment of the primitive earth was harsh and lifeless. But many scientists believe that it contained the necessary molecules to fashion the first living cell by the process of spontaneous generation.

generated as the particle became concentrated to form earth. While not as great as the heat within the sun, it formed a molten core that became encased by a thin outer crust as it cooled. In its early stages of formation, about four billion years ago, it is thought that there was a considerable amount of volcanic activity on earth (figure 20.6).

Physically, the earth was probably much different from what it is today. Because the surface was hot, there was no water on the earth's surface or in the atmosphere. In fact, the tremendous amount of heat probably prevented any atmosphere from forming around this early earth. The gases associated with our present atmosphere (nitrogen, oxygen, carbon dioxide, and water vapor) were contained in the molten core within the earth. These hostile conditions on early earth could not have supported any form of life.

Over hundreds of millions of years, the earth slowly changed. Volcanic activity caused the release of water vapor (H_2O), carbon dioxide (CO_2), methane (CH_4), ammonia (NH_3), and hydrogen (H_2) and the early atmosphere was formed. These gases formed a **reducing atmosphere,** an atmosphere that does not contain molecular oxygen (O_2). Further cooling enabled the water vapor in the atmosphere to condense into droplets of rain. The water ran over the land and collected to form the oceans we see today.

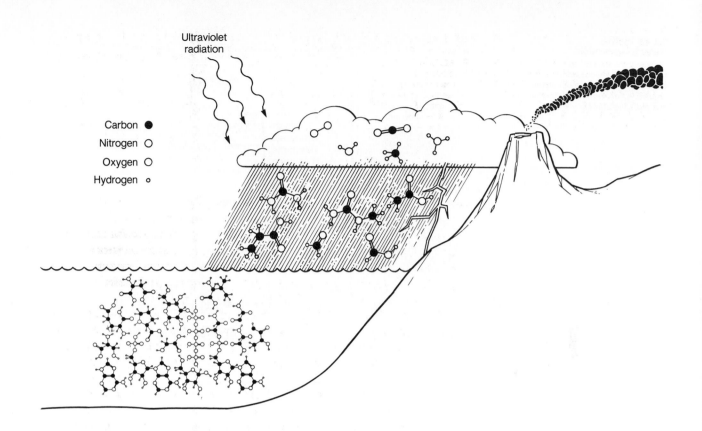

Ultraviolet
radiation

Carbon ●
Nitrogen ○
Oxygen ○
Hydrogen ∘

The First Organic Molecules

According to Oparin and Haldane, the first organic molecules were formed in this early reducing atmosphere. The molecules of water vapor, ammonia, methane, carbon dioxide, and hydrogen supplied the atoms of carbon, hydrogen, oxygen, and nitrogen necessary to form simple organic molecules. Lightning, heat from volcanoes, and ultraviolet radiation furnished the energy needed for these synthetic reactions (figure 20.7).

After these simple organic molecules were formed in the atmosphere, they were washed from the air and carried into the oceans by the rain. Here the molecules could have reacted with each other to form the more complex molecules of simple sugars, amino acids, and nucleic acids. This accumulation is thought to have occurred over half a billion years, resulting in oceans that were a dilute organic soup. These simple organic molecules in the ocean served as the building material for more complex organic macromolecules such as complex carbohydrates, proteins, lipids, and nucleic acids. But what scientific evidence is there to support these ideas?

In the early 1950s, support was growing for the idea that organic material was synthesized from inorganic material. One person who advanced this idea was Harold Urey of the University of Chicago. In 1953, one of Urey's students, Stanley L. Miller, conducted an experiment to test this idea. Miller constructed a model of early earth (figure 20.8). In this glass apparatus he placed distilled water to represent the early oceans. The reducing atmosphere was duplicated by adding hydrogen, methane, and ammonia to the water. Electrical sparks provided the energy needed to produce organic compounds. By heating parts of the apparatus and cooling others, he simulated the rains that are thought to have fallen into the early oceans. After a week

Figure 20.7
Formation of Organic Molecules in the Atmosphere.

The energy furnished by volcanoes, lightning, and ultraviolet light broke the bonds in the simple organic molecules in the atmosphere. These same sources of energy formed new bonds as the atoms from the smaller molecules were rearranged and bonded to form simple organic compounds in the atmosphere. The rain carried these chemicals into the oceans. Here they reacted with each other to form more complex organic molecules.

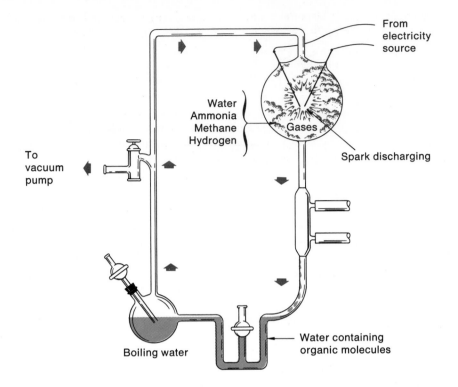

Figure 20.8
Miller's Apparatus.
Stanley Miller developed this apparatus to demonstrate that the spontaneous formation of complex organic molecules could take place in a reducing atmosphere.

From electricity source

Water
Ammonia
Methane
Hydrogen

Gases

Spark discharging

To vacuum pump

Boiling water

Water containing organic molecules

of operation, he removed some of the water that accumulated in the apparatus. When this water was analyzed, it was found to contain many simple organic compounds. Although Miller proved nonbiologic synthesis of simple organic molecules like amino acids and simple sugars, his results did not account for complex organic molecules like proteins and DNA.

Several ideas have been proposed for the concentration of simple organic molecules and their combining into macromolecules. A portion of the early ocean could have been separated from the main ocean by geological changes. The evaporation of water from this pool could have concentrated the molecules, which might have led to the manufacture of macromolecules by dehydration synthesis. It has also been proposed that freezing may have been the means of concentration. When a mixture of alcohol and water is placed in a freezer, the water freezes solid and the alcohol becomes concentrated into a small portion of liquid. A similar process could have occurred on earth's early surface, resulting in the concentration of simple organic molecules. In this concentrated solution, dehydration synthesis in a reducing atmosphere could have occurred, resulting in the formation of macromolecules.

A third theory proposes that clay particles may have been a factor in concentrating simple organic molecules. Small particles of clay have electrical charges that can attract and concentrate organic molecules such as protein from a watery solution. Once the molecules became concentrated, it would have been easier for them to interact to form larger macromolecules.

Coacervates and Microspheres

Typically, geologists and biologists measure the history of life by looking back from the present. Therefore, time scales are given in "years ago." It has been estimated that the formation of simple organic molecules in the atmosphere began about four billion years ago and lasted approximately one and one-half billion years. The oldest known fossils of living cells are thought to have formed three and one-half billion years ago. The question is How do you get from the spontaneous formation of macromolecules to primitive cells in a half billion years?

There are two theories proposed for the formation of **prebionts,** nonliving structures that led to the formation of the first living cells. Oparin speculated that a prebiont consisted of carbohydrates, proteins, lipids, and nucleic acids that accumulated to form a **coacervate.** Such a structure could have consisted of a collection of organic macromolecules surrounded by a film of water molecules. This arrangement of water molecules, while not a membrane, could have functioned as a physical barrier between the organic molecules and their surroundings.

Coacervates have been made in the laboratory. They can selectively absorb chemicals from the surrounding water and incorporate them into their structure. Also, the chemicals within coacervates have a specific arrangement—they are not a random collection of molecules. Some coacervates contain enzymes that direct a specific type of chemical reaction. No one claims coacervates are alive since they lack a definite membrane, but they do exhibit some lifelike traits. Coacervates are able to grow and divide if the environment is favorable.

More recently, it has been suggested that a possible prebiotic structure could have been a microsphere. A **microsphere** is a nonliving collection of organic macromolecules with a double-layered outer boundary. **Proteinoids** are proteinlike structures consisting of branched chains of amino acid. They are the basic ingredient of microspheres. Proteinoids are formed by the dehydration synthesis of amino acids at a temperature of 180° C. Sidney Fox, from the University of Miami, showed that it was feasible to combine single amino acids into polymers of proteinoids. He also demonstrated the ability to build microspheres from these proteinoids.

Microspheres can be formed when proteinoids are placed in boiling water and slowly allowed to cool. Some of the proteinoid material produces a double boundary structure that encloses the microsphere. Although these walls do not contain lipids, they do exhibit some membranelike characteristics and suggest the structure of a cellular membrane. Microspheres swell or shrink depending upon the osmotic potential in the surrounding solution. They also display a type of internal movement (streaming) similar to that exhibited by cells and contain some proteinoids that function as enzymes. Using ATP as a source of energy, microspheres can direct the formation of polypeptides and nucleic acids. Microspheres can absorb material from the surrounding medium and form buds. This results in a second generation of microspheres. Some investigators believe that these characteristics allow the microspheres to be considered **protocells,** the first living cells.

Other scientists throughout the world have duplicated, conducted variations of, and continued the works of Miller and Fox. Many different combinations of gases have been shown to furnish the raw materials necessary to produce organic compounds when supplied with a suitable source of energy. But these experiments had one common factor—they only worked in a reducing atmosphere. It is widely believed that spontaneous generation did happen, but that it could only occur under the past conditions of a reducing atmosphere and not in today's oxygen-rich atmosphere. The laboratory synthesis of coacervates and microspheres helps us understand how the first primitive living cells might have developed. However, there is still a large gap in our understanding because it does not explain how these first cells might have become the highly complex, living cells we see today.

Figure 20.9
"Feeding" Heterotrophs.

(a) When the heterotrophs were first formed, there probably was an ample supply of organic nutrients that they could take in by diffusion. With these conditions, the heterotrophs could survive and reproduce. (b) After the organic nutrients were exhausted, there would be no food and the heterotrophs would become extinct.

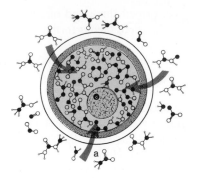
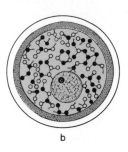

Heterotrophs to Autotrophs

Regardless of how it developed, the first living thing was probably the result of spontaneous generation. Fossil evidence indicates there were primitive forms of life on the earth about three and one-half billion years ago. These first primitive cells were thought to be **heterotrophs.** Heterotrophs require a source of organic material from their environment. Since the early heterotrophs are thought to have evolved in a reducing atmosphere that lacked oxygen, they were of necessity anaerobic organisms; therefore, they did not obtain the maximum amount of energy from the organic molecules in the environment. At first, this would not have been a problem. The organic molecules that had been accumulating in the ocean for millions of years served as an ample source of organic material for heterotrophs. However, as the population of heterotrophs increased through reproduction, the supply of organic material would have been consumed faster than it was being spontaneously produced in the atmosphere. If there was no other source of organic compounds, the heterotrophs would have eventually exhausted their nutrient supply, and they would have become extinct (figure 20.9).

Even though the early heterotrophs probably contained nucleic acids and were capable of producing enzymes that could regulate chemical reactions, they probably carried out a minimum of biochemical activity. The work of investigators such as Miller and Fox suggests that a wide variety of compounds were present in the early oceans. Some of these compounds could have been used unchanged by the heterotrophs. There was no need for the heterotrophs to modify the compounds to meet their needs.

Those compounds that could be easily used by heterotrophs would have been the first to become depleted from the early environment. However, some of the heterotrophs may have contained a mutated form of nucleic acid. Because of this mutation, these cells may have been capable of converting material that was not directly usable to the organism into a compound that the cell could use.

Heterotrophs with this mutation could have survived, while those without it would become extinct since the compounds they used as food became scarce. It has been suggested that through a series of mutations in the early heterotrophs, a more complex series of biochemical reactions originated within some of the cells. Such cells could use chemical reactions to convert ingestible chemicals into usable organic compounds. Possibly because of these kinds of mutations, new metabolic pathways evolved that led to the evolution of **autotrophs.**

Although it required over a billion years after the formation of the earth, once the heterotrophs had been formed by spontaneous generation and the autotrophs had evolved, the pattern of life on earth as it exists today was established (figure 20.10). A summary of events is shown in table 20.1.

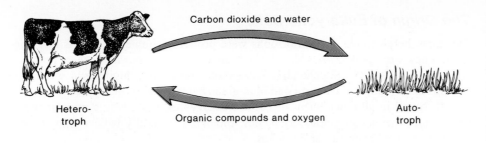

Carbon dioxide and water

Hetero-
troph

Organic compounds and oxygen

Auto-
troph

Figure 20.10
***Autotroph–Heterotroph
Interrelations.***
Autotrophs and heterotrophs have developed an interdependence. Today, the autotrophs depend upon the inorganic waste products of the heterotrophs—water and carbon dioxide—for the raw material to produce the end products of the organic compounds and oxygen essential for life. The heterotrophs also depend upon the organic material produced by the autotrophs as material for energy and structural growth.

Table 20.1   ***Spontaneous generation and evolution of autotrophs***

1. Conversion of inorganic compounds in the atmosphere into organic compounds.
2. The accumulation of these organic compounds in the oceans.
3. The concentration of some of these organic compounds into prebionts.
4. The formation of the heterotrophic protocells.
5. The evolution of the autotrophic cells.

An Oxidizing Atmosphere

Ever since its formation, the earth has undergone constant change. In the beginning, it was too hot to even support an atmosphere. Later, as the earth cooled and gases issued forth from volcanoes, a reducing atmosphere was formed. The emergence of heterotrophic cells led to the development of autotrophic cells. The metabolic activities of cells resulted in the release of waste products into the environment. This changed the earth's environment. One of the most significant changes to occur was the development of an **oxidizing atmosphere.** An oxidizing atmosphere contains molecular oxygen. The development of an oxidizing atmosphere created an environment unsuitable for spontaneous generation of organic molecules and life. Organic molecules tend to break down (oxidize) when oxygen is available to accept electrons, but the presence of molecular oxygen opened the door for the evolution of aerobic organisms.

As the result of autotrophic activity, an oxidizing atmosphere began to develop about two billion years ago. Although various chemical reactions released small amounts of molecular oxygen into the atmosphere, it was photosynthesis that generated most of the oxygen in the atmosphere. Since oxygen is a highly reactive atom, its appearance created an environment unfit for spontaneous generation of life. The oxygen molecules also reacted with one another to form *ozone,* O_3. Ozone collected in the upper reaches of the atmosphere and acted as a screen to prevent most of the ultraviolet light from reaching the surface of the earth. The lack of ultraviolet light diminished the spontaneous formation of complex organic molecules. It also reduced the number of mutations in primitive cells that produced a great variety of cellular life forms. In an oxidizing atmosphere, it was no longer possible for organic molecules to accumulate over millions of years to later be incorporated into living material.

The appearance of oxygen in the atmosphere also allowed for the evolution of aerobic respiration. Since the first heterotrophs were of necessity anaerobic organisms, they did not derive large amounts of energy from the organic materials available as food. With the evolution of aerobic heterotrophs, there could be a much more efficient conversion of food into usable energy. Aerobic organisms would have a significant advantage over anaerobic organisms. They could use the newly generated oxygen as a final hydrogen acceptor and, therefore, generate many more ATPs from the food molecules they consumed.

The Origin of Eukaryotic Cells

The early heterotrophs and autotrophs were probably simple organisms like bacteria. They were **prokaryotes** that lacked nuclear membranes and other membranous organelles such as mitochondria, endoplasmic reticulum, chloroplasts, and Golgi apparatus. Present-day bacteria and blue-green bacteria are prokaryotes. The types of cells found in all other forms of life are **eukaryotes,** possessing a nuclear membrane and other membranous organelles. Biologists generally believe that the eukaryotes evolved from the prokaryotes.

The **endosymbiotic theory** attempts to explain this evolution. This theory suggests that present-day eukaryotic cells evolved from the combining of several different types of primitive cells. It is thought that some organelles found in eukaryotic cells may have had their origin as free-living prokaryotes. Since mitochondria and chloroplasts contain bacterialike DNA and ribosomes, control their own reproduction, and synthesize their own enzymes, it has been suggested that they originated as free-living prokaryotic bacteria. These bacterial cells could have established a symbiotic relationship with another primitive nuclear membrane-containing cell type (figure 20.11).

If these cells adapted to one another, and were able to survive and reproduce better as a team, it is possible that a relationship may have evolved into present-day eukaryotic cells. If this relationship had only included a nuclear-containing cell and aerobic bacteria, the newly evolved cell would have been similar to present-day heterotrophic protozoa, fungi, and animal cells. If this relationship had included both aerobic bacteria and photosynthetic bacteria, the newly formed cell would have been similar to present-day autotrophic algae and plant cells.

Regardless of the type of cell (prokaryotic or eukaryotic), or if the organisms are heterotrophic or autotrophic, all organisms have a common basis. DNA is the universal genetic material; protein serves as structural material and enzymes, and ATP is the source of energy. Although there is a wide variety of organisms, they all are built from the same basic molecular building blocks. Therefore, it is probable that all life is derived from a single origin and that all the variety of living things seen today evolved from the first protocells. In this chapter we have studied how this protocell originated. It is thought that from these cells evolved the great diversity we see in living organisms today (table 20.2). The remaining chapters of this book are concerned with the study of this diversity.

Summary

The centuries of research outlined in this chapter illustrate the development of our attempts to understand the origin of life. The current theory of the origin of life speculates that the primitive earth's environment led to the spontaneous organization of chemicals that may have become organized into primitive cells. These basic units of life then changed through time as a result of mutation and in response to a changing environment. The likelihood of these occurrences is supported by experiments that have simulated primitive earth environments. Similarities between blue-green bacteria and chloroplasts and between bacteria and mitochondria suggest that eukaryotic cells may really be a combination of ancient cell ancestors that live together symbiotically. Despite volumes of information, the question of how life began remains unanswered. Although no one can prove how life began, a generally accepted scheme is presented in figure 20.12.

Table 20.2 *Time Table of Events*

20 billion years ago	*Solar nebula*
4–5 billion years ago	*Sun and planets*
3.5 billion years ago	*Prokaryotes*
2.3 billion years ago	*Oxidizing atmosphere*
1.5 billion years ago	*Eukaryotes*
700 million years ago	*Multicellular organisms*

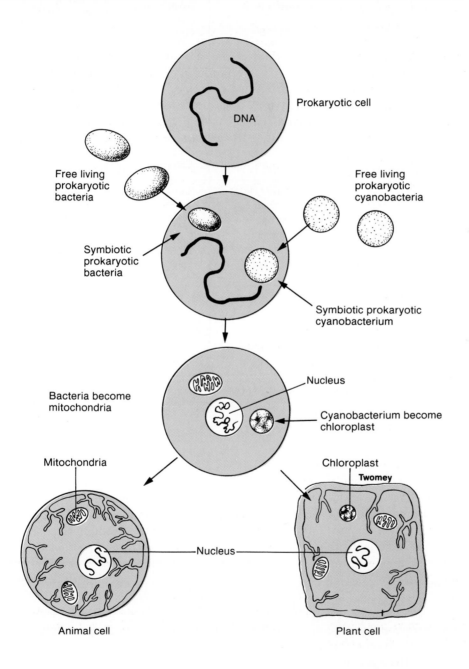

Figure 20.11
Endosymbiotic Theory.
This theory proposes that some free-living prokaryotic bacteria and cyanobacteria (blue-green algae) developed symbiotic relationships with a host cell. When the bacteria developed into a mitochondria and the cyanobacteria developed into chloroplasts, a eukaryotic cell evolved. These cells evolved into eukaryotic plant and animal cells.

Figure 20.12
Origin and Evolution of Cells.
This chart gives a summary of the events that are thought to have led to the origin and evolution of cells.

Billions of years ago	5 — 4.5	4 — 3.5	3 — 2.5	2 — 1.5	1.0 — 0.5 — 0
Energy sources	UV light high earth's heat high lightning abundant	UV light high earth's heat lower lightning less	UV light high earth's heat low lightning low	UV light low earth's heat low lightning low	UV light low earth's heat low lightning low
Types of gases in atmosphere	H_2 CH_4 NH_3 H_2O CO_2	H_2 CH_4 NH_3 H_2O CO_2	H_2 NH_3 H_2O	H_2 NH_3 H_2O ozone oxygen atmosphere forming CO_2	H_2O O_2 ozone N_2 CO_2
Molecules present in sea	simple organic molecules synthesized abiotically methane and other hydro-carbons, ammonia, acids and alcohols	complex organic molecules synthesized abiotically nucleotides, amino acids, sugars	complex organic molecules used by protobionts; some biotic synthesis protein, fat, carbohydrates in cells	Complex organic molecules resulting only from biotic synthesis	complex organic molecules resulting only from biotic synthesis
Types of life present	era of chemical evolution and protobiont cell forms	prokaryotes	prokaryotes	eukaryotes begin	multicellular organisms

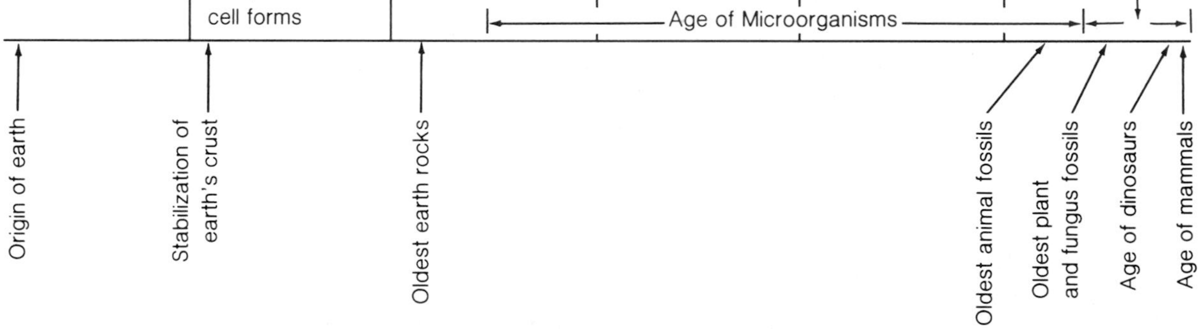

←————————— Age of Microorganisms —————————→

Origin of earth | Stabilization of earth's crust | Oldest earth rocks | Oldest animal fossils | Oldest plant and fungus fossils | Age of dinosaurs | Age of mammals

---

Consider This

It has been speculated that there is "life" on another planet in our galaxy. The following data concerning the nature of this life have been obtained from "reliable" sources. Using these data, what additional information is necessary, and how would you go about verifying these data in developing a theory on the origins of life on planet X? Data:

1. The age of the planet is ten billion years.
2. Water is present in the atmosphere.
3. The planet is farther from its sun than our earth is from our sun.
4. The molecules of various gases in the atmosphere are constantly being removed.
5. Chemical reactions on this planet occur at approximately one-half the rate they occur on earth.

Take a handful of straw, hay, grass clippings, or leaves and boil them to make an organic soup broth. After it has cooled, pour it through a strainer into a clear glass bottle. Add a teaspoon of soil and plug it with a cotton ball or piece of cloth. Place in a sunny spot. Observe this over the next two weeks and note the changes that occur. If you have a microscope, sample the microcosm with an eyedropper and try to identify the organisms. Was this spontaneous generation or biogenesis?

1. In what sequence did the following things happen: living cell, oxidizing atmosphere, autotrophy, heterotrophy, reducing atmosphere, first organic molecule?
2. What is meant by spontaneous generation? What is meant by biogenesis?
3. Which of the following scientists supplied evidence that supported the theory of spontaneous generation? Which supported biogenesis?: Spallanzani, Needham, Pasteur, Urey, Fox, Miller, Oparin
4. Can spontaneous generation occur today? Explain.
5. Why do scientists believe life originated in the seas?
6. What were the circumstances on primitive earth that favored the survival of an autotrophic type of organism?
7. The current theory of the spontaneous chemical generation of life on earth depends on our knowing something of earth's history. Why is this so?
8. List two important effects caused by the increase of oxygen in the atmosphere.
9. What evidence supports the theory that eukaryotic cells arose from the development of a symbiotic relationship between primitive prokaryotic cells and protocells?

autotroph (aw'to-trōf) An organism able to produce organic nutrients from inorganic materials.

biogenesis (bi-o-jen'uh-sis) The concept that life originates only from preexisting life.

coacervate (ko-as'ur-vāt) A collection of organic macromolecules surrounded by water molecules that are aligned to form a sphere.

endosymbiotic theory (en''do-sim-be-ot'ik the'o-re) The theory that suggests some organelles found in eukaryotic cells may have originated as free-living prokaryotes.

eukaryote (yu-kār'e-ōt) A cell possessing a nuclear membrane and other membranous organelles.

heterotroph (hĕ'tur-o-trōf) An organism unable to produce organic molecules from inorganic materials.

microsphere (mi'kro-sfēr) A collection of organic macromolecules in a structure with a double-layered outer boundary.

oxidizing atmosphere (ok'sĭ-di-zing at'mos-fēr) An atmosphere that contains molecular oxygen.

prebiont (pre''bi-ont) Nonliving structures that led to the formation of the first living cells.

prokaryote (pro-kār'e-ōt) A cell that lacks a nuclear membrane and other membranous organelles.

proteinoid (pro'te-in-oid) A proteinlike structure of branched amino acid chains that are the basic structure of a microsphere.

protocell (pro'to-sel) The first living cell.

reducing atmosphere (re-du'sing at'mos-fēr) An atmosphere that does not contain molecular oxygen (O_2).

spontaneous generation (spon-ta'ne-us jen-uh-ra'shun) The theory that living organisms arose from nonliving material.

The Classification of Life

Chapter Outline

Purpose

Evolution is a series of changes that occur in a population over a number of generations. These changes are the result of natural selection. For natural selection to occur, three things are needed: (1) a reproducing population, (2) genetically inheritable differences among individuals, and (3) some factor within the environment that selects certain individuals for survival because they possess favorable genes.

Although millions of forms of life exist on earth today, this was not always so. At one time there was that first living organism from which evolved millions of species. None of these early species exist today.

In this chapter we trace the evolutionary changes that are thought to have occurred from the first living thing through the following four billion years. The results of this evolutionary process are the various kinds of plants, animals, and microbes living today.

For Your Information

Humans have always tried to classify things so they could understand their environment better. By categorizing things in their environment, it is easier to focus on one particular feature while ignoring others that might cause confusion. All kinds of classifications are designed to be helpful in understanding how the world works. It is also important to communicate to others what is good and what is dangerous. Classification systems help you do this. The ancient Greeks divided their world into fire, water, earth, and sky. Eskimos have at least six different names for snow depending on where it is or how fluffy it is.

One of the earliest classifications was of living things into "good" or "harmful" categories. Subsequently people found such broad categories to be inadequate. Scientific classification began with two categories of living and nonliving (animate and inanimate). The category of animate objects was further subdivided into plants and animals. As more detailed information became available, further subdivisions followed. Living things were not simply classified as plant and animal, but were now split into three groups: plants, animals, and bacteria. This process resulted in the recognition of four groups: plants, animals, bacteria, and single-celled organisms. Currently, most biologists recognize five categories: plants, animals, bacteria, single-celled organisms, and fungi. Since this is an ongoing process, you should expect the way things are classified to continue to change. In all likelihood, biologists will subdivide the bacteria into two distinct groups in the near future, resulting in six major categories of living things.

Learning Objectives

■ Describe the criteria for constructing a classification system.

■ Name examples of the various classification groups and state their unique characteristics.

■ State the names of the various classification groups, but only as directed by the instructor.

■ State the characteristics needed for animals to exist on land.

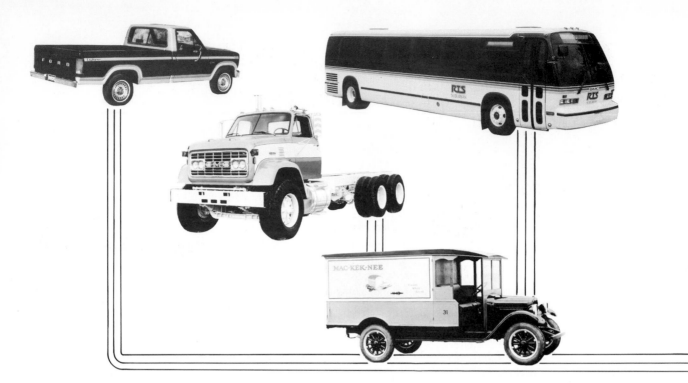

Figure 21.1
Automotive Evolution.

The evolution of various kinds of
automotive vehicles is similar to
the evolution of living things. Small
changes have been made on the
original car over many years,
resulting in a variety of types.

The Five Kingdoms

If we were to assemble before us one individual from every species on earth, there
would appear such diversity that one would think it impossible that all these individuals evolved from one common organism. But we should keep in mind that according to the theory of organic evolution, the organisms we see today were not
always present on the earth. Present-day organisms are the end products of about
four billion years of evolution. The kinds of changes that have occurred in the evolution of living things can be compared to the changes that have occurred in the
auto industry from the first vehicle to the present models. If we were to assemble
one representative of all the present-day cars, trucks, tractors, and buses made
throughout the world, there would be great variety. Yet, we know that at one time
there appeared on the earth that first contraption powered by an internal combustion engine, and from that grew all of the motor vehicles we see in the world today.
That first internal combustion engine allowed for the development of a large number
of vehicles that failed and are now **extinct,** such as the Kaiser, Edsel, and Henry J.
(figure 21.1). Some of these are preserved as "fossils." Likewise, there was a time
when nature produced a first model of living things (chapter 20). From this common
ancestor, through countless model changes, there came into being today's variety of
living things.

Figure 21.2
Prokaryotic Cells.

Some of the earliest organisms to evolve may have been similar to the present-day bacteria pictured here. These are examples of cells without a definite nucleus (prokaryotes).

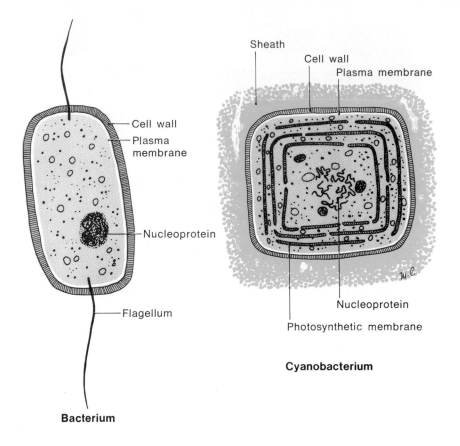

Cyanobacterium

Bacterium

You will recall that Oparin proposed the idea that life arose from a combination of water, ammonia, methane, and hydrogen united by energy. The resulting macromolecules then formed into protobionts, forerunners of the actual first species of life. This species is thought to have been a heterotroph. Since heterotrophs cannot make their own food, it would have used the organic materials in its ocean home as a source of food. These organic materials had been synthesized for millions of years prior to the appearance of the first form of life. Within this first species, as in all present species, there was genetic variation within the population. Some individuals were better able to compete in that environment. As the number of individuals increased, the supply of food decreased. The increasing competition favored those organisms that required little organic food. As a result of various mutations, some forms of early life synthesized enzymes that allowed them to produce their own food. These organisms were the world's first autotrophs. Now there were two models for life on earth, the original heterotrophs and the new autotrophs. This was an extremely important change, since there was now an efficient biological assembly line for the constant production of large amounts of organic molecules. Since the heterotrophs eat these organic molecules, they ultimately became dependent on the autotrophs.

Based on various lines of evidence, scientists think that these first autotrophs and heterotrophs were probably small, uncomplicated single cells. They may have been something like the present-day members of the microorganisms known as bacteria (figure 21.2). These **prokaryotic** cells lack a distinct nucleus and many of the complex organelles described in chapter 4. As competition continued and mutations caused new genes to appear in the population, some of the mutations may have

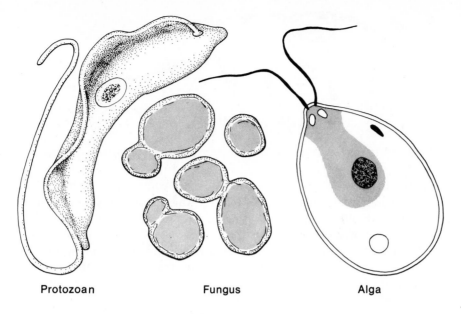

Figure 21.3
Eukaryotic Cells.
These single-celled organisms are examples of eukaryotic cells, cells with a definite nucleus.

Protozoan   Fungus   Alga

resulted in a more complex cell structure. This allowed these new cell types to expand into new niches and avoid competition with their ancestors. Such cell types may have resembled present-day, single-celled **eukaryotic** organisms. Examples of such present-day organisms are the protozoans, algae, other than cyanobacteria (blue-green), and fungi (figure 21.3). Eukaryotic cells contain that complex array of membranous cell organelles that are lacking in prokaryotic cells. The terms prokaryotic and eukaryotic divide all cellular life into two basic groups. It is the first step in building a system for classifying living things. In building such a system, the attempt is always to create a "natural" system of classification, groups based on a common ancestor. It is artificial if the groups do not have common ancestry. The ancient Greek classification of all plants into trees, shrubs, undershrubs, and herbs would be an example of an artificial system. It is based only on form and not on relationships.

Biologists have sorted organisms into five major groups called **kingdoms.** The first kingdom, **Prokaryotae,** contains a group of single-celled organisms commonly referred to as "bacteria." The word bacteria brings up the vision of tiny things that cause disease; things too small to be seen with the naked eye. It is certainly true that many diseases of both plants and animals are caused by bacteria. Crown Gall in plants, blackleg in cattle, and pneumonia in humans are all examples of infectious diseases caused by different bacteria. However, there are many bacteria that are beneficial. The Romans knew that bean plants somehow enriched the soil, but it was not until the late 1800's that bacteria were recognized as being the enriching agent. Also, some bacteria play a vital part in the disposal of organic wastes such as human wastes, dead animals, and leaves. Even with the help of bacteria, our communities have a growing problem of waste disposal.

Cyanobacteria are also part of this kingdom, as the word "bacteria" in the name suggests. The group includes a great variety of organisms from the colorful coatings in the boiling pools of Yellowstone National Park to the scums that cause bad odors and tastes in drinking water.

The second kingdom, **Protista,** contains the organisms commonly referred to as protozoans and algae. Unlike the previous kingdom, the Protista are eukaryotic. In this group the protozoans lack chlorophyll and are heterotrophic, while the algae

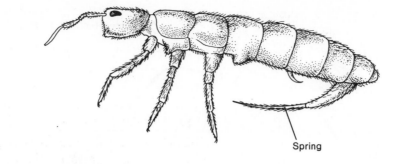

Spring

are autotrophic. Some of the protozoa are harmless, while others are deadly disease producers. For example, in one part of Central America 60% of the children do not live to their fifth birthday, partly because of a protozoan—a particular species of amoeba. This amoeba can thrive because human wastes are not removed in a sanitary way. The wastes enter the drinking water and amoebic dysentery is one of the terrible results. It causes an extreme loss of water from the body that is especially hard on small children.

The freshwater and marine algae are extremely important since they are major photosynthesizers. This, of course, means that they are major producers of oxygen.

The third kingdom is commonly called the fungi; officially it is called **Mycetae.** Members of this kingdom lack chlorophyll and therefore depend on an exterior organic source of food. Most of these grow on dead organisms but some are parasites. Give thanks to this kingdom if you enjoy any variety of cheese on the market. Medicines, beverages, and those delicious baked goods all owe a debt to the Mycetae. Nevertheless, it is a mixed bag! Many of these fungi are destructive to our valuables and dangerous to our health. Food will be destroyed by mold even in the refrigerator if stored too long. Athlete's foot, an all too common fungus infection, can be severe enough to cause hospitalization.

You might think that you need no introduction to the fourth kingdom, the **Plantae.** Perhaps that is true in part, but the plant kingdom is such a varied group and scattered to so many different habitats that there remains much to be learned, even for the experts. This group is found from the tropics to the polar regions and from a watery environment to desert conditions. The Plantae vary greatly in size from the tiny, aquatic duckweed to the giant sequoia tree. From the earliest times, members of this kingdom have provided food for humans.

The **Animalia,** our fifth kingdom, completes our classification of the organisms that are known, both from the fossil record as well as those living today. There is a magnificent variety just in the present-day living forms. They range in speed from the cheetah, clocked at up to 115 kilometers (70 miles) per hour, to the animal whose name means laziness, the sloth. The sizes extend from the great blue whale, up to 30 meters long and 150 metric tons, down to insects that are common but so tiny that they are largely unnoticed. One wingless insect, a springtail (figure 21.4), is so small that people usually do not notice it until there is a large gathering of them. For instance, one alarmed person called the college about some "dust that moved" on her recently cemented basement floor. It was a "dusting" of these harmless springtails attracted to the dampness of the floor. They are tiny individuals, but certainly not the smallest.

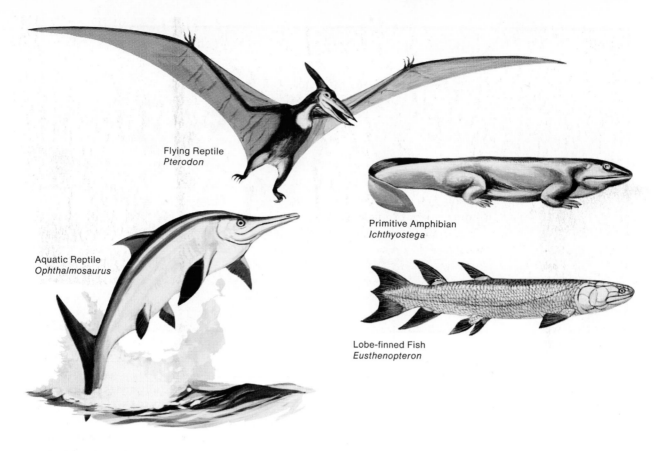

Figure 21.5
Extinct Animals.

These are examples of organisms that were unable to survive and are now extinct.

Flying Reptile
Pterodon

Primitive Amphibian
Ichthyostega

Aquatic Reptile
Ophthalmosaurus

Lobe-finned Fish
Eusthenopteron

Evolution to Land

Originally, organisms existed only in the oceans. Water provided a nearly constant environment with small variations in temperature and without drastic changes in light conditions. These conditions were more favorable for the primitive life forms than the harsher environment of the land. However, a continuing parade of mutations introduced new traits into the gene pools of these marine species. Characteristics developed that allowed some species to spend short periods of time on land. Eventually others could spend their entire lifetime on land. A variety of life forms appeared such as those pictured in figure 21.5. The land presented a very different type of environment. In the absence of being constantly surrounded by water, organisms needed the following characteristics: (1) a moist membrane that would allow adequate gas exchange between the atmosphere and the organism, (2) a means of support and locomotion suitable for land travel, (3) a covering that would conserve internal water, (4) a means of reproduction and early embryonic development that would not require a large amount of water, and (5) methods to survive the rapid and extreme climatic changes that characterize much of the land environment.

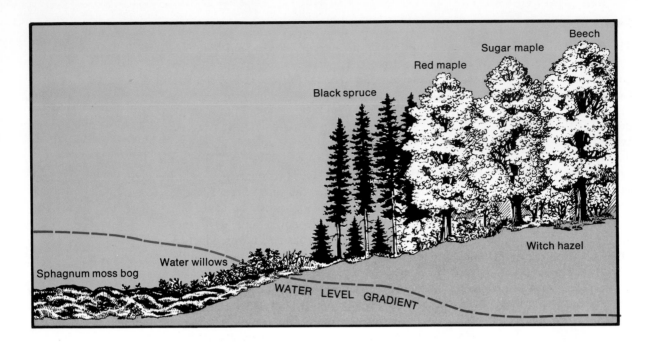

Labels in figure: Beech, Sugar maple, Red maple, Black spruce, Witch hazel, Water willows, Sphagnum moss bog, WATER LEVEL GRADIENT

Figure 21.6
Moisture Requirements.
Many organisms in the plant kingdom live on land, but terrestrial plants differ in the amount of moisture they require. The dashed line on the illustration indicates the relative amount of moisture required. At the left, where the line is highest, the organisms require more water; to the right, where the line is lowest, the water requirement is less.

Land Plants

Fossil evidence suggests that the first organisms to successfully adapt to a terrestrial environment were the plants. Today we recognize four major types of plants: mosses, ferns, cone-bearing trees, and the flowering plants (figure 21.6). The mosses were the first to live on land. However, these small plants lacked an efficient method for transporting water and other materials. They also produced sperm that must swim to the female part of the plant to fertilize the egg. These two conditions have limited the distribution of the mosses to continuously moist areas such as riverbanks, damp forests, and swamps. Over millions of years, there were a number of additional mutations and a trend toward a drier environment.

Plants that had developed tubes capable of transporting water through their bodies could increase in size and could better succeed in these drier places. Ferns are an example of this. Ferns were very successful on land because the tubes allowed them to transport water from the soil throughout the plant body. Since all cells contain a high percentage of water, this was a very important advancement. However, ferns still produce sperm that must swim to the egg. Therefore, at least part of their life must be spent in a wet area.

Further success on land was achieved when some plants began to produce a male gamete that did not require water to move from one plant to another. The structure capable of producing such a male gamete is known as **pollen.** Pollen can be carried for hundreds of miles through the air (ask anybody who suffers from hay fever!). It was the appearance of pollen that allowed plants to completely conquer the land.

Two kinds of pollen-producing plants developed: (1) the cone-bearing plants that rely primarily on the wind to transport the sperm-containing pollen (for example, pine trees, spruce trees, and the giant redwoods of California); and (2) the many flower-bearing plants that rely on insects or wind to carry the pollen (for example, maple trees, corn, roses, crabgrass, and orchids). To illustrate how effectively these two kinds of plants have adapted to drier conditions, you need only look at a desert. The cactus, tumbleweed, sagebrush, and the piñon pine are typical of our drier western states.

Land Animals

The animals had to solve these same problems if they were to achieve equal success on land. Maintaining body water, locomotion, and successful reproduction all required adaptive changes. Some of these adaptive changes can be seen in the fish. Certain fish possessed a combination of genes that produced strong fins, enabling them to crawl ashore for short periods of time. Some of these crawling fish also had the ability to swallow air by forcing it into outgrowths of their digestive tract. These fish did not lose as much water from their bodies as did fish that could breathe only through their gills. Those animals that could leave the water and spend some time on land could escape the intense competition in the aquatic environment. Further adaptations along these lines eventually led to the evolution of the amphibians. Present-day examples of amphibians are frogs, salamanders, toads, and newts.

With the appearance of the amphibians, vertebrate life moved onto "dry" land, but it still existed only near a constant source of water. The business of reproduction kept the amphibians from going too far inland. When these animals mate, the female releases eggs into the water and the male releases sperm amid the eggs. The sperm move and fertilize the eggs. These fertilized eggs must remain in water or they will dry and die. As a result, the process of reproduction closely tied the amphibians to water.

For some forty million years, the amphibians were the only vertebrate animals on land. During this time, mutations continued to take place. Some of these mutations resulted in the development of organs that would allow **copulation,** so the male could deposit sperm within the female. Because the sperm and egg could remain within the moist interior, it was not necessary to return to the water for reproduction.

Internal fertilization was not enough to completely free animals from water, since the developing young required a moist environment for their early growth. Some animals developed an egg that contained food and was enclosed by membranes and shells. These coverings on the egg sealed in the moisture and thus protected it from drying while allowing the exchange of gases. The first animals to do this were the reptiles. Snakes, dinosaurs, turtles, alligators, and lizards all produce shelled eggs. With the development of internal fertilization and the eggshell, the vertebrates had truly conquered land. The reptiles had evolved.

These developments allowed the reptiles to spread over much of the earth and occupy a large number of previously unfilled niches. For millions of years they were the only large animals on land. The evolution of reptiles increased competition for space and food. The amphibians generally lost in this competition, and consequently most became extinct. Some, however, were able to avoid competition because they had adaptations that allowed them to hide, live away from reptiles, and survive on a different diet. Those that were successful led to the present-day frogs, toads, and salamanders.

Even though the reptile had mastered the problem of coming ashore, mutations and natural selection continued and so did evolution. As good as the egg in a shell was, it did have drawbacks. It lacked protection from sudden environmental changes and from predators that used eggs as food.

Further mutations resulted in an animal that overcame the disadvantages of the egg by providing for internal development of the young. Such development allowed for a higher survival rate. The internal development of the young, along with milk gland development, a constant body temperature, and care of young by parents marked the emergence of mammals.

At about the same time the mammals were evolving from the reptiles, a second major group was evolving. In this group, the shelled egg remained the method of protecting the young. However, a series of mutations in some of the reptiles produced an animal with a more rapid metabolism, feathers, and adaptations for flight. These early birds also possessed behavioral instincts such as nest building, defense of their young, and feeding of the young. Because of these adaptations and their invasion of the air, a previously unoccupied niche, they became one of the very successful groups of animals.

Another extremely successful group of animals began to evolve at about the time the first amphibians began to move onto land. The insects, spiders, and their relatives solved the same problems of terrestrial existence with somewhat different adaptations. These animals have remained small and have tremendous reproductive capacity. They are very adaptable and have been able to compete successfully with the larger land animals. As a matter of fact, three quarters of all the animal species on the earth are insects, and they are our chief competitors for food.

Over a period of some three billion years, life that began in the water underwent mutations and natural selection to evolve into the complex array of plant and animal forms we see today. As these new forms of life emerged, they caused changes on the earth just by their presence. As a result of these changes, some species could not survive. They did not have the necessary genes to compete in the new environment and became extinct. There is nothing new about extinction; it is as old as life itself. Species have become extinct in the past, they are becoming extinct today, and they will become extinct in the future.

Summary

Early in this text the interrelationship of plants and animals was stressed. This interaction has occurred ever since the appearance of the first heterotrophs and autotrophs. The first organisms to evolve were single-celled organisms of the kingdom Prokaryotae. From these simple beginnings, more complex, many-celled organisms evolved, creating the kingdoms Protista, Mycetae, Plantae, and Animalia.

For millions of years, these plants and animals were confined to watery habitats. Through the processes of mutation, natural selection, and environmental change, these groups evolved into a variety of organisms capable of living in the more severe and changeable land environment. The features that allowed for success on land included a moist membrane for gas exchange, fertilization away from standing water, water-conserving methods, methods for protecting developing young, means of support and locomotion, and methods of surviving rapidly changing environmental conditions. The foldout (inside the back cover) represents a summary of the four billion years of evolution of these five major kingdoms of life.

Consider This

A minimum estimate of the number of species of insects is 750,000. Perhaps then it would not surprise you to see a fly with eyes on stalks as long as its wings, a dragonfly with a wingspread over two feet long, an insect that can revive after being frozen at 35° C below zero, and a wasp that can push its long, hairlike, egg-laying tool directly into a tree. Only the dragonfly is not presently living, but it once was!

What other curious features of this fascinating group can you discover? Have you tried to look at a common beetle under magnification? It will hold still if you chill it.

Besides the satisfaction and enjoyment of watching wildlife close up, there is still room for the amateur to make a contribution to the knowledge of animal behavior. This can be more easily accomplished if the animal is not aware of your presence. All of which leads up to the importance of having a blind (or hide, as our British cousins call it). A simple one can be made of four stout corner posts sharpened to be pushed into the soil. In addition, you will need four slats connecting the tops of these posts in a square to support the roof. This skeleton can be covered with some rough cloth such as canvas, sacking, or whatever is at hand. Cut one slit at the correct height for a camera, if desired, and one for your eye. Since watching animals takes patience, the eye hole should be at a comfortable height. Since the noise and motion of flapping fabric would be alarming to a wild animal, pegs may be necessary to avoid this.

It is necessary to set up the hide a day or so before you use it in order to let the animals get used to it. Then enter it before the animal's expected time of arrival. For instance, if you have selected the drumming log of a ruffed grouse that you know to be in use, it would be necessary to enter the blind in the morning while it is still dark.

Questions

1. What are the five kingdoms of living things?
2. What is the difference between the kingdom Prokaryotae and the kingdom Plantae?
3. List the series of changes that resulted in the evolution of living things as we know them today.
4. What major adjustments did land organisms need to make in order to live on land?
5. How did reptiles and flowering plants solve their problems of reproduction?
6. What is the difference between a heterotroph and an autotroph?

Chapter Glossary

Animalia (ahn''e-mahl 'e-uh) One of the five kingdoms. It consists of multicellular heterotrophic organisms that require a source of organic molecules as food.

copulation (kop''yu-la'shun) The act of transferring sperm from the male into the female reproductive tract.

eukaryotic (yu''kār-e-ot'ik) Having cells that contain a complex array of membranous cell organelles such as a nucleus, mitochondrion, and endoplasmic reticulum.

extinct (ek''stinkt') No longer existing.

kingdom (king'dom) The largest grouping used in the classification of organisms.

Mycetae (mi-se''te) One of the five kingdoms. It consists of heterotrophic fungi.

Plantae (plan'te) One of the five kingdoms. It consists of multicellular, autotrophic organisms.

pollen (pol'en) The male structures of the plant capable of forming sperm.

Prokaryotae (pro''kār-e-o̊'te) One of the five kingdoms. It consists of one-celled organisms commonly called bacteria.

prokaryotic (pro''kār-e-ot'ik) Having cells that lack a distinct nucleus and many of the complex organelles.

Protista (pro-tis'tah) One of the five kingdoms. It consists of one-celled organisms that possess a definite nucleus.

The Viruses and Bacteria

Chapter Outline

Purpose

In our everyday lives we usually categorize organisms as either plant or animal. But all organisms won't fit neatly into these two groups. Dogs, fish, snakes, and birds are animals, of course. Trees, grass, and raspberry bushes are plants. But there is a huge assortment of organisms that are not so easy to classify. This group has been investigated from their molecular structure to the nature of their metabolism and divided into additional groups. This chapter will give you a glimpse into the nature of two groups, the viruses and bacteria, and how they affect our lives.

For Your Information

Many of the unpleasant stories about America's early treatment of its Native American population have recently been repeated in Brazil, the islands of the Pacific, and Papua New Guinea. Although many forest dwellers from these areas appear to be immune to some serious infectious diseases such as yellow fever, others such as measles, tuberculosis, and influenza are deadly to them. These viral and bacterial diseases did not previously exist in these groups of people who were isolated from civilization. Efforts to "modernize" these groups have resulted in the accidental introduction of these previously unknown diseases causing widespread death. However, it is also possible for disease organisms to move in the opposite direction. The virus that causes AIDS is a disease organism that has been introduced into technologically advanced countries from less advanced areas of the world.

Learning Objectives

■ Know the names of the groups in the generalized classification system of organisms.

■ Know the characteristics of viruses.

■ State the importance, methods of transmission, and precautions regarding some viral diseases.

■ Discuss the characteristics and importance of members of the kingdom Prokaryotae.

Classification System

A basic difference among living things became apparent when biologists first became interested in the details of the structure of cells. Some cells were found to lack a clearly defined nucleus. Although a cell membrane was present, other membranous structures that might have been within the cell, such as mitochondria, were missing. All organisms that are composed of such cells (prokaryotes) have been placed into the kingdom Prokaryotae. The more complex cells, easily identified by the presence of a nucleus, are the eukaryotic type. Presently, organisms that are composed of eukaryotic cells have been divided into four additional kingdoms: Protista, Mycetae, Plantae, and Animalia.

It is convenient to divide each kingdom into smaller groups and give specific names to each of these. To see the relationship between these groups, refer to figure 22.1, which has examples from each of the five kingdoms. The subdivision under each kingdom is usually called a **phylum,** although microbiologists and botanists replace this term with the word *division.* Any one kingdom will have more than one phylum. For example, the kingdom Prokaryotae contains four of these subgroups. Each organism is investigated to determine the specific nature of its structure, metabolism, and biochemistry in order to determine its place in one of these phyla (divisions). You may have noticed the blanks in figure 22.1. Let us consider the reason for filling some blanks and not others in the group called **subphylum.** Within the phylum of the Animalia, the subphylum space is filled in with the name Vertebrata. It just happens that animals included in the phylum Chordata can be divided into several groups, one of which has vertebrae, or backbones. This is an important difference from other organisms within the phylum Chordata. Its placement here is a recognition that this difference was a major development in the evolution of chordates. There is no comparable difference in the phylum (division) Anthophyta of the Plantae kingdom, so no subphylum name is needed.

The names in the last two classifications, **genus** and **species,** together make up the scientific name of an organism. The name *Homo sapiens* is the **scientific name** of the animal more commonly called a human being. The scientific name has the advantage of being recognized all over the world.

Figure 22.1
Representatives of the Five Kingdoms of Life.
Prokaryotae is represented by the bacterium *Streptococcaceae Pyogenes;* Protista by one-celled *Amoeba proteus;* Mycetae by the yeast *Saccharomyces cerevisiae;* Plantae by the sugar maple, *Acer saccharum;* and the Kingdom Animalia by the human being, *Homo sapiens.*

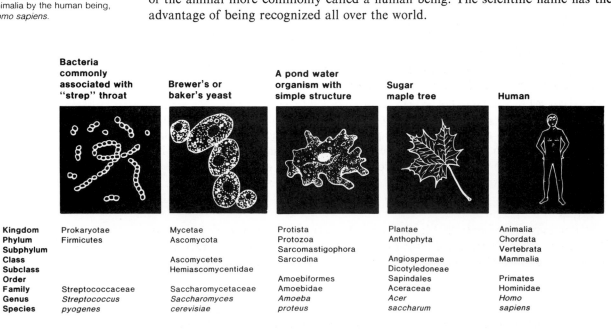

	Bacteria commonly associated with "strep" throat	Brewer's or baker's yeast	A pond water organism with simple structure	Sugar maple tree	Human
Kingdom	Prokaryotae	Mycetae	Protista	Plantae	Animalia
Phylum	Firmicutes	Ascomycota	Protozoa	Anthophyta	Chordata
Subphylum			Sarcomastigophora		Vertebrata
Class		Ascomycetes	Sarcodina	Angiospermae	Mammalia
Subclass		Hemiascomycentidae		Dicotyledoneae	
Order			Amoebiformes	Sapindales	Primates
Family	Streptococcaceae	Saccharomycetaceae	Amoebidae	Aceraceae	Hominidae
Genus	*Streptococcus*	*Saccharomyces*	*Amoeba*	*Acer*	*Homo*
Species	*pyogenes*	*cerevisiae*	*proteus*	*saccharum*	*sapiens*

Now that we have our classification system organized, let's face it: the organisms still don't quite fit *our* system. It is ours in the sense that it is artificial, or human-made. It doesn't completely show how the various organisms are related. Even worse, many systems of classification, including the one we are using, actually hide some relationships. This point will be made clearer as we become acquainted with the various groups of organisms, especially the viruses. In the case of bacteria such as *Streptococcus* in the Prokaryotae, the blanks also represent group names that are not commonly used.

Noncellular Microbes*

Viruses are peculiar—their structure is unlike all the other **microbes** mentioned so far. These parasites are not considered to be organisms at all since they are not cellular. Viruses consist of a nucleic acid enclosed by a covering of protein. Other components, typically found in prokaryotic or eukaryotic cells, are absent. One could say that they have no life until their nucleic acid enters another cell. Viruses are called **obligate intracellular parasites** since they are unable to carry out any of the typical life functions until they are inside a host cell. The virus sets up a new relationship with this **host** cell in which the virus (parasite) benefits and the host is harmed. Such a relationship is known as parasitism.

The **virus** parasite enters the host cell that contains all the materials needed by the virus. When the virus is still outside the host cell, it is called a **virion,** or virus particle. This word is comparable to the word *cell* in that both represent a single unit of that kind of microbe. Upon entering the host cell, the virion loses its protein coat. The nucleic acid portions may remain free in the cell or this viral nucleic acid may link with the host's genetic material. Some virions contain as few as three genes while other, larger virions contain as many as five hundred genes. Viral genes take command of the host's metabolic pathways and direct it to carry out the work of making new copies of the original virus. The virus makes use of the available enzymes and ATP to start producing copies of itself. When enough new viral nucleic acid and protein coat are produced, complete virus particles are assembled and released from the host (figure 22.2). The number of virions released ranges from tens to thousands. The virus that causes polio releases about ten thousand new virions after it has invaded its human host cell.

Figure 22.3 shows several types of virions that vary in size and shape. These two characteristics are of help when classifying viruses. Some are rod-shaped, others are round, while still others are in the shape of a coil or helix. Viruses are some of the smallest infecting agents known to humankind. Only a few can be seen with a standard laboratory microscope; they require an electron microscope to make them visible. The electron microscope has been of great value for this purpose, but it is not very practical in most college classrooms. A great deal of work is necessary to isolate virions from the environment and prepare them for observation with an electron microscope. For this reason, most viruses are more quickly identified by their activities in host cells.

When viruses infect host cells, they alter the normal metabolism of the cell (see box 22.1). If the alteration is extensive and the cell is harmed, the virus causes disease symptoms. A few of the diseases caused by viruses in animals and humans

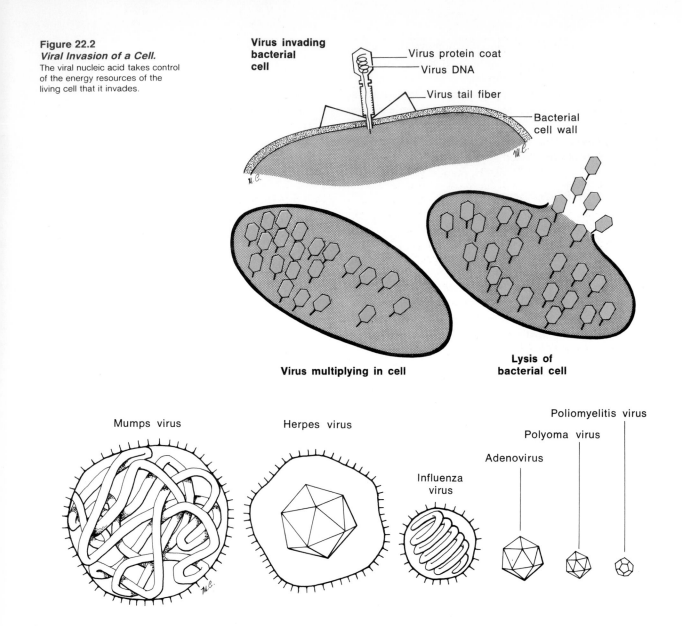

Figure 22.2
Viral Invasion of a Cell.
The viral nucleic acid takes control of the energy resources of the living cell that it invades.

Virus invading bacterial cell

Virus protein coat
Virus DNA
Virus tail fiber
Bacterial cell wall

Virus multiplying in cell

Lysis of bacterial cell

Mumps virus

Herpes virus

Influenza virus

Adenovirus

Polyoma virus

Poliomyelitis virus

Figure 22.3
Viral Types.

As large as these viral particles are, they are still smaller than the cells that they invade. Mumps is a common childhood disease, and influenza is that disagreeable wintertime visitor called flu. Cold sores are caused by the Herpes virus, while the Adenovirus is associated with respiratory disease. Poliomyelitis was a dreaded paralyzer, but with the development of vaccines can now be prevented. The threat of poliomyelitis has been so successfully met that some people have become careless about getting their children vaccinated. Tumors have been induced in mice by the Polyoma virus.

include AIDS, cold sores, warts, measles, and the "flu." In plants, viruses cause a mosaic pattern on tobacco leaves, color patterns in tulip petals, and celery leaf rot. Even microbes can be infected with microbes. Bacteria are infected by viruses called bacteriophage.

Viruses have been of great value in research related to cell biochemistry and molecular biology. The fundamental concept that deoxyribonucleic acid (DNA) serves as the genetic material was derived from work performed with certain viruses that infect bacteria. The names of bacteriophages are very different from the scientific names of plants and animals. T_2, ϕ X-174 (phi "x"-174), and phage λ (lambda) are viruses that infect bacterial cells. This naming system is different from the binomial system of nomenclature mentioned earlier because scientists are not sure of the evolutionary relationship of the viruses to cellular forms of life. However, scientists are attempting to group viruses in a uniform way by using fundamental chemical characteristics.

Box 22.1
Chicken Pox and Shingles

The virus varicella-zoster is responsible for two diseases, chicken pox (varicella) and shingles (zoster). Chicken pox is a highly contagious skin disease of children. About 75% of all children under fifteen years of age experience its symptoms, but it is rarely found in adults. The virus is acquired by contact with contaminated articles or infected children. Children with the disease can transmit it to others five days before the pox appear on the skin but not more than six days after the last pox appear. Once inside the body, it takes about two to three weeks before the first skin rash appears. The pox form in three to four days, erupt, and heal with no scars. Pox may form on the trunk of the body, in the mouth, arm-pit, vagina, and eyes. The pox usually appear in "crops" one after another. Sometimes infections are mild and only a few pox form. In other cases, no pox form, and the infection is not apparent.

A complication that occurs in some children following chicken pox and other virus diseases is known as *Reye's syndrome*. There is an acute swelling of the brain and a degeneration of internal organs such as the kidneys and liver. The symptoms occur two to three days after the child seems to have recovered from the viral infection. Children with Reye's develop a low-grade fever with repeated vomiting. This may change, as they become disoriented, show aggressive behavior, and lapse into a coma. These symptoms can be reversed by proper medical care, but the death rate is high. Reye's syndrome has also been associated with other virus infections, such as influenza, the common cold, and gastrointestinal upset.

In most cases a chicken pox infection passes without problems. However, this virus can establish a hidden infection that may reoccur in adulthood as shingles. This viral disease is more common in adults who have had chicken pox as children, but can also occur as a first varicella-zoster infection in adults or children. This disease begins with fever and a tired feeling, followed by skin pain and itching. The rash that follows is usually found on the sides and neck. The rash changes in a few days to a kind of pox that break and scab without scarring. Since most varicella-zoster infections are mild and have few side effects, no treatment is needed other than aspirin and calamine lotion.

Since 1974, Japanese researchers have been working with a pharmaceutical company on the development of a Oka-strain chicken pox vaccine. The vaccine is made of live herpes varicella virus that has been "attenuated" (made harmless). This vaccine is like numerous other live-virus vaccines commonly found in use the world over. The widespread introduction and use of this vaccine will be a valuable addition to the medical profession's arsenal of disease prevention weapons. Field testing results were first released in mid-1984 and indicated 100 percent effectiveness of the vaccine. While further field trials and research still need to be done, researchers agree that it will only be a short time before the Food and Drug Administration of the United States Department of Health gives its "OK" for general use.

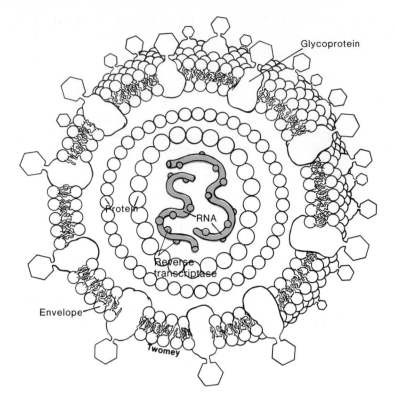

Some believe that the viruses were at one time parts of primitive cells. These parts displayed no functional characteristics until they again became located inside another whole cell. A change in their function allowed them to reproduce and then move on to another cell. Thus, the viruses may have been groups of genes that gained the ability to multiply independently of other genes in prokaryotic and eukaryotic cells. Their independent nature allowed them to move from one host to another producing more of their own kind. Others believe that the viruses have evolved from some of the first bacterial cells. As time passed, these cells became specialized and no longer manufactured their own surrounding cell structures; they relied on cells of others to perform the necessary life functions. Since no one was around at the time, it may never be known which theory is likely to be true.

The AIDS Pandemic

As noted in chapter 11, epidemiology is the study of the transmission of diseases through a population. Diseases that occur throughout the world population at extremely high rates are called **pandemics.** Influenza was the first great pandemic of the first part of the twentieth century and killed hundreds of thousands of people. AIDS has become the greatest pandemic of the second half. This viral disease has been identified in ninety-one countries around the world and is estimated to have infected over one and one half million people in African countries and the USA.

AIDS is an acronym for *a*cquire *i*mmuno*d*eficiency *s*yndrome and is caused by the human T-cell lymphotrophic virus III (HTLV-III), one of four members of the retrovirus family (figure 22.4). Evidence strongly supports that this RNA-containing virus originated through many mutations of a nonpathogenic African monkey virus sometime during the late 1950s or early 1960s. This has been determined by testing human blood serum for the presence of anti-AIDS antibodies taken from patients during that time. No blood has tested positive for HTLV-III prior to 1975. The virus probably moved from its original host, the African green monkey, to humans as a result of an accidental scratch or bite. It wasn't until the late 1970s that the virus took hold in native African populations. From there it spread to Haiti and then to the United States and Europe. The first reported case of full-blown AIDS was reported in the United States in January, 1981 at the UCLA Medical Center.

While the virus was moved to the United States through the homosexual population, it is not a disease unique to that group. In fact, in African populations where the virus originated and has had more time to spread, the ratio of males to females with AIDS is 1:1. This indicates that heterosexual transmission is a normal route for the virus to follow. No virus is known that shows a sexual preference. Many cases of heterosexual transmission have already been reported from many nations. Transmission from mother to child at birth has also been documented. Infection has also been reported to have occurred as a result of artificial insemination with infected semen.

Data from the Centers for Disease Control, Atlanta, Georgia, indicate that the mortality rate among those with the disease is over 50%! In early 1987 over thirty thousand cases had been diagnosed in the United States alone, and over seventeen thousand of those patients died. New full-blown AIDS cases are being diagnosed at the rate of over one thousand per month in the United States and at higher rates in Africa. If these figures are extended to the end of the century, an estimated one and one half million people around the world will have died needlessly. AIDS is the greatest health care concern we have ever had in the United States. The numbers of people infected are staggering and public health officials from around the globe expect them to get worse.

HTLV-III is a spherical virus containing an RNA genome, an enzyme called *reverse transcriptase,* a protein shell, and a lipid-protein envelope.

The virus gains entry into a suitable host cell through a very complex series of events involving the virus envelope and the host cell membrane. Human cells that can serve as hosts include brain cells and several types of cells belonging to the immune system, namely monocytes, macrophages, and T4-helper/inducer lymphocytes. Once inside the host, the RNA of the HTLV-III virus is used to make a DNA with the help of reverse transcriptase. This is the reverse of the normal transcription process in which a DNA template is used to manufacture an RNA molecule. When this reverse transcriptase has completed its job, the DNA genome is spliced into the host cell's DNA. In this form (viral DNA genome spliced into the host genome), the virus is called a *provirus.* As a provirus, it may remain inside the host cell for an extended period without causing any harm. In fact the host cell may reproduce without any interference from the HTLV-III provirus, resulting in a clone of infected cells. For some as yet unknown reason, the provirus in all but brain cells does not enter into its destructive phase until the body experiences another HTLV-III or other infection. This means that the virus has very long incubation and communicable periods—the times during which the viruses are transmitted to other cells in the body or to other persons. Once initiated into action, the provirus quickly replicates itself inside the host cell. HTLV-III is able to do this much more rapidly than other types of viruses. When the newly replicated virions leave, they cause huge holes to form in the host cell membrane that are unable to be repaired fast enough to prevent the host cell from dying. AIDS patients, therefore, have a decrease in the number of these cell types. One of the important diagnostic indicators of AIDS is a sharp decrease in the number of T4 lymphocytes. After replication, HTLV-III virions are free in the body and can be passed from one cell to another or from one person to another. There is also evidence that these virions can be passed directly from one host cell to another by direct contact and do not need to be released as free virions.

The other effect the virus has is on the brain cells that it invades. These cells probably become infected when the virus is transferred to them from infected macrophages (normally protective phagocytes) that move from other parts of the body into the spinal column and to the brain. Once inside the brain cells, the proviruses don't require a second infection to stimulate them into operation—they go to work immediately. The result of such an infection is progressive memory loss, mimicry of other neurological diseases such as multiple sclerosis, loss of coordination, dementia (senility), and ultimately death.

Since lymphocyte host cells are found in the blood and other body fluids, it is logical that these fluids serve as carriers for transmission of the virus. This has in fact been found to be true. HTLV-III virus is transmitted through contact with contaminated blood, semen, mucous secretions, serum, or on blood-contaminated hypodermic needles. If these body fluids contain the free virions or infected cells (monocytes, macrophages, T4-helper/inducer lymphocytes), they can be a source of infection. There is no evidence indicating the virus can be transmitted through the air, on toilet seats, by mosquitoes, or by casual contact such as shaking hands, hugging, touching, or closed mouth kissing. The virus is just too fragile to survive such trips.

When the virus infects cells of the immune system, it first invades the monocytes and macrophages. These cells then transfer them to T4-helper/inducer lymphocytes. T4-helper/inducer lymphocytes are only one of several types of lymphocytes found in the body and responsible for defending the body against a great variety of dangerous chemicals, tumor cells, and infectious organisms. These cells are responsible for six essential functions:

1. They help B lymphocytes change into plasma cells to produce protective antibody molecules against a specific dangerous agent.
2. They induce T8 lymphocytes to become "killer" cells that attack and kill invading microorganisms, body cells invaded by microorganisms, and tumor cells.
3. They induce T8 lymphocytes to stop maturing once the infection is brought under control.
4. They induce T4 lymphocytes to reproduce (clone).
5. They help T4 cells stop maturing once infection is brought under control.
6. They help plasma cells secrete antibodies.

When T4 lymphocytes are destroyed by HTLV-III, all these defensive efforts of the immune system are depressed. This leaves the body vulnerable to invasion by many types of infecting microbes or to be overtaken by body cells that have changed into tumor cells. This means that the HTLV-III virus does not directly cause the death of the infected individual. Illness is caused by otherwise harmless organisms that take advantage of an individual's decreased ability to defend itself against them. Some of the more common opportunistic infections include:

1. A rare lung infection *Pneumocysitis carinii* pneumonia (PCP) caused by a fungus.
2. Kaposi's sarcoma, a form of skin cancer seen as purple-red bruises.
3. Cytomegalovirus infections of the retina of the eye.

There appears to be a series of bodily changes that begin with changes in the brain cells and end in death as a result of secondary infections. The initial symptoms of the disease are known as ARC, or *A*IDS *R*elated *C*omplex.

At the present time there is hope of controlling the virus (but not curing an infection) by using an antiviral drug that can: 1) kill infected cells, 2) improve the body's immune system, or 3) selectively interfere with the life cycle of the virus. The life cycle may be disrupted when the virus enters the cell, when the reverse transcriptase allows the RNA to produce the DNA genome, or by interfering with a gene unique to HTLV-III, the trans-activator. The trans-activator gene controls the way the whole virus reproduces. If this gene does not operate, the virus is "dead in the water." The drug AZT, azidothymidine, is to date the most effective of the antiviral agents. It disrupts the operation of reverse transcriptase. However, there are two reasons why the use of this drug should be approached cautiously. AZT has not been tested enough to identify what long-term side effects may result from its use, and the drug is extremely expensive. One year's use will cost approximately $10,000.

What about the development of a vaccine to prevent the virus from infecting the body? An experimental vaccine has been developed based on the body's ability to produce antibodies against protein spikes on the surface of the virus. This vaccine, however, has only been shown to be effective in monkeys, and there is no evidence that it will also stimulate a very necessary protective response in killer T8-lymphocytes. In addition, it will be necessary to deal with the problem of genetic differences among the many kinds of HTLV-III viruses. The greater the variety of viruses, the greater the variety of vaccine types needed to prevent infection.

What must be done in order to control the spread of the virus is to recognize what is high-risk behavior and change it. This requires presenting the nature of the disease and how it is spread to the public as soon and as accurately as possible. What is high-risk behavior?

1. Promiscuous sexual behavior (i.e., sex with large numbers of partners). This increases the probability that one of the partners may be a carrier.
2. Intravenous drug administration with shared needles.
3. Contact with blood-contaminated articles.
4. Wet (French, open mouth) kissing.
5. Intercourse (vaginal, anal, oral) without the use of a condom.

Blood tests can be performed that indicate exposure to the virus. The test should be taken on a voluntary basis, absolutely anonymously, and with intensive counseling before and after. People who test positively should not expose anyone else or place themselves in a situation where they might be reinfected. They should do everything to maintain good health; exercise regularly, eat a balanced diet, get plenty of rest, and reduce stress. Don't place yourself in a situation where you can become infected. We cannot stop this pandemic in its tracks, but it can be slowed.

Kingdom Prokaryotae

The kingdom Prokaryotae contains organisms that are commonly referred to as **bacteria,** germs, or microorganisms (micro- = small). The word *microbe* is used in reference to the fact that these cells are so small that some type of magnification is required to see them. They are single-celled and lack an organized nucleus and other complex organelles. There is an enormous variety of bacteria in the world. *Bergey's*

Figure 22.5
A Trickling Filter in Action.
This equipment is a common sight in modern wastewater treatment plants where bacteria decompose wastes.

Manual of Determinative Bacteriology lists over seventeen hundred species of bacteria and describes the subtle differences between each. This manual is useful to microbiologists interested in the specific nature of bacteria, their classification, and identification. The latest classification system for the bacteria divides these microbes into four divisions (phyla): Gracilicutes, Firmicutes, Tenericutes, and Archaebacteria. The groups are recognized on the basis of cell structure, staining reactions, nature of their cell wall, and metabolic activities.

We are most familiar with bacteria that cause harm. Such diseases as tetanus, strep throat, acne, boils, scarlet fever, toxic shock syndrome, gonorrhea, dental cavities, and many others are all caused by bacteria. However, it would be a mistake to think of bacteria only in this negative sense. It is true that many serious diseases are caused by bacteria; however, one type found in the large intestine, *Escherichia coli,* is usually helpful. It provides certain vitamins (such as vitamin K) to humans. Other bacteria are used for making food, such as those used in making yogurt. Many thousands of gallons of vinegar are produced each year by bacterial fermentation.

The process of bacterial decay is both a nuisance and a great help to us. We lose much valuable food each year by bacterial spoilage, but we also have countless quantities of organic material recycled to a useful condition in ecosystems. Sewage treatment plants rely on the action of bacteria to break down wastes for recycling (figure 22.5). The bacteria mentioned thus far obtain their energy from food already made by some other organism. A few types of bacteria can make their own food by using light to carry on photosynthesis. Other bacteria can make use of inorganic materials by oxidizing them, such as some of the bacteria involved in the nitrogen cycle (figure 22.6). It should not be surprising to learn that bacteria occupy a wide variety of habitats, since they get food in many ways. Bacteria are classified according to the way they break down foods and according to other biochemical differences.

Bacteria can be identified by their shape (morphology). They can be placed into one of four morphological groups: **coccus** is spherical, **bacillus** is rod-shaped, **spirillum** is twisted or bent, and **pleomorphic** has many, or varied, shapes. The spiral bacteria can be further divided into three subgroups on the basis of their morphology. One subgroup is called the **vibrios**. These bacteria have only a slight twist

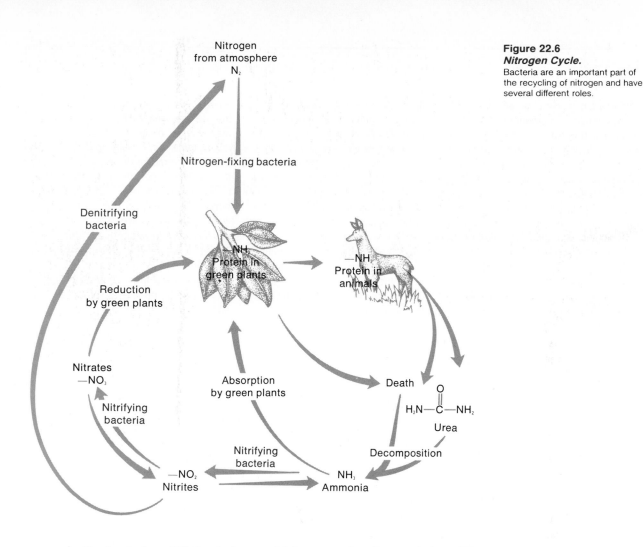

Nitrogen
from atmosphere
N₂

Nitrogen-fixing bacteria

Denitrifying
bacteria

Reduction
by green plants

—NH₂
Protein in
green plants

—NH₂
Protein in
animals

Nitrates
—NO₃

Nitrifying
bacteria

Absorption
by green plants

Death

$$H_2N-\overset{\displaystyle O}{\underset{\displaystyle C}{\|}}-NH_2$$

Urea

—NO₂
Nitrites

Nitrifying
bacteria

NH₃
Ammonia

Decomposition

Figure 22.6
Nitrogen Cycle.
Bacteria are an important part of
the recycling of nitrogen and have
several different roles.

as seen in the bacterium *Vibrio cholerae,* which causes the disease cholera. The second category of spiral bacteria is known as the spirilla, which are corkscrewlike in shape, and very rigid. Members of the genus *Campylobacter* have this morphology and are responsible for intestinal infections that resemble cholera. The third group of spiral bacteria, the **spirochetes,** are also corkscrew in shape, but are flexible and able to wiggle or twist. This is made possible by an axial filament that spirals from one end of the cell to the other. The venereal disease syphilis is caused by *Treponema pallidum,* which looks like a wiggling corkscrew under the microscope.

Bacteria sometimes adhere to one another in characteristic forms. Those that form chains are named using the prefix "strep-." For example, *Streptococcus pyogenes,* a bacterium that can cause scarlet fever, forms chains of spheres. Cells that can stick together in a random cluster are named using the prefix "staph-." The cells of *Staphylococcus aureus* are spheres that adhere to one another and resemble a bunch of grapes. *Streptobacillus* is another possible formation. A group's shape is determined by the way the bacilli remain connected to one another after they have gone through division. The bacterium *Bacillus subtilis* forms long chains after it divides. When it reproduces rapidly, cell separations are difficult to see. Another pattern used to classify the shape of bacteria is designated by the prefix "diplo-." Bacteria with this formation are usually linked together two at a time. A bacterium that typically takes this form is *Neisseria gonorrhoeae,* the cause of gonorrhea.

Box 22.2
Oxygen Requirements of Bacteria

Bacteria are grown in a semisolidified (agar-containing) nutrient medium. This medium is a mixture of carbohydrates, proteins, lipids, and growth factors. To demonstrate the differences in oxygen requirements, an oxygen gradient must be established for the bacteria being grown. This shows whether or not the bacteria were successful growing in an environment with a large amount of oxygen, or whether they were better able to grow in an environment that lacked this gas.

This procedure is done in a test tube. In addition to the agar and nutrients listed above, these special media also contain the chemicals sodium thioglycolate and resazurin.

Thioglycolate is a molecule that is able to capture oxygen from the medium and bind it tightly. This makes the oxygen unavailable to the microbes that are inoculated into the tubes.

The resazurin is a color indicator. When the oxygen is removed from the nutrient medium by the thioglycolate, the resazurin changes color, showing that the tube is ready to be inoculated.

Notice that the upper portion of the agar medium contains the greatest amount of free oxygen (O_2) because the gas may easily diffuse into the medium from the atmosphere. The larger center zone contains a decreasing amount of oxygen because the gas has more difficulty reaching this portion of the medium. The bottommost portion lacks oxygen. Any bacteria that would grow well in the upper portion of this tube culture would be aerobic. Those growing in the center portion of the tube would be facultative anaerobes, and those growing only at the bottom of the tube would be anaerobes.

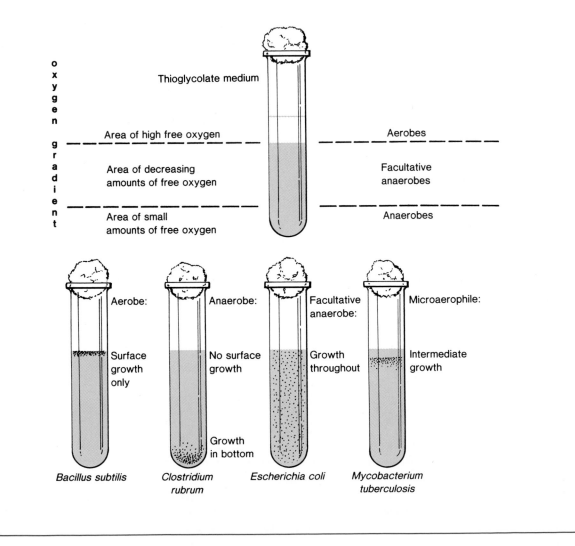

The microbial world is composed of a complex array of life forms, each maintaining the characteristics of life in its own unique way. To develop an understanding of the useful and harmful actions of microbes and apply this knowledge in beneficial ways, a clear and detailed understanding of their structure and function must be developed. The extremely small size of these organisms makes this task very difficult and demands special equipment and techniques. Since most microbes are too small to be seen with the naked eye, special microscopes have been developed that produce enlarged, clear images. Viewing specimens through a light microscope provides information on size and motility. However, this type of microscope has limits when investigating live microbes. The high percentage of water in cells (70–90%) allows visible light rays to pass through the cell very easily and little contrast is developed between the cell and its surroundings. To overcome this problem and highlight special cell components, dyes or stains are used to color the cell before viewing. In addition, there are types of microscopes that may be used to obtain other perspectives of microbes. Ultraviolet light microscopes and electron microscopes have advantages not found in standard light microscopes.

The small size of microbes also limits a researcher's ability to gather information about a single microbe. Laboratory methods used to study large multicellular organisms such as animals and plants are not readily adapted to the microbiology lab. An individual microbe is not easily dissected and studied apart from the whole organism. For this reason, microbiologists must approach the study of microbes using different, more effective, techniques. Microbes are usually studied in populations and not as individuals. Large numbers of a particular microbial species are grown in pure culture and studied collectively (box 22.2). Our understanding of a species is based on what the "group" does or how it behaves in a particular environment. If microbes of a different species become mixed with the study group, the results of the study may be inaccurate. Therefore, microbiologists spend much time and energy maintaining pure cultures for study. Many special procedures have been developed to prevent both contamination of pure cultures and the lab worker by invisible bacteria. Within the kingdom Prokaryotae are two groups that deserve our special attention. The first, the cyanobacteria, has ecological importance and the second, archaebacteria, is a unique cellular form.

Cyanobacteria

Cells of the **cyanobacteria** occur singly, as clusters, or as a series of cells organized into filaments. These bacteria have a unique form of the photosynthetic pigment chlorophyll different from that found in the algae and higher plants because it is blue-green. This pigment gives a unique color to colonies of cyanobacteria. Unfortunately for identification purposes, the blue-green color is not a reliable characteristic. Members of the kingdom Protista imitate this color, and some cyanobacteria show other colors: red, yellow, purple, brown, and black. This group of bacteria has received increased attention as interest in environmental quality has risen. Some cyanobacteria indicate organic pollution, while others are pollutants themselves. Some cause foul odors or tastes in drinking water or produce poisons that kill animals. Occasionally, the population of a species of cyanobacteria suddenly increases (a **bloom**) and causes trouble for water-living plants and animals. This bloom can shut off the light to submerged plants, causing their death and decay. The decay

Figure 22.7
An Algal Bloom.
The warm summer sun shining on
a pond rich in nutrients causes this
rapid multiplication of algae cells.

may then remove so much oxygen from the water that the fish and other animals suffocate (figure 22.7). Cyanobacteria are found in a variety of habitats, not just in fresh water. A small number of species are found in salt water. They are common in soil, on damp rocks, or tree bark—they even grow in the hot springs of Yellowstone National Park.

Archaebacteria

All members of this division are single-celled organisms that lack an organized nucleus and do not have the complex organelles typical of the more familiar types of organisms that are eukaryotic in nature. While these cells appear to be "true bacteria," they are significantly different in comparison to other prokaryotic cells. Archaebacteria have a different sequence of nucleotides in their ribosomal RNA, different kinds of lipids in their cell membranes, different types of transfer RNA, and a different type of sensitivity to antibiotics. The name **archaebacteria** has been given to indicate that they are considered to be an ancient form of cell and probably resemble the original cellular form of life on earth, the progenotes. These microorganisms have the ability to grow in environments that resemble the primitive earth atmosphere, (i.e., extremely hot, acid, and/or salty). Some archaebacteria can even produce the primitive atmospheric gas, methane. Bacteria that have these abilities are called thermoacidophiles (thermo = heat; phile = lover), extreme halophiles (halo- = salt), thermoplasmas, and methanogens. Present-day methanogens live only in anaerobic environments such as the ocean floor or stagnant ponds. These microbes are responsible for the production of "marsh gas" or "combustible air." The extreme halophiles grow in the Great Salt Lake, the Dead Sea, and along ocean borders where salt concentrations are in the range of 13–28% (the Pacific Ocean is about 4% salt). The thermoacidophiles live in hot sulfur springs where the temperature is about 80° C (174° F) and the pH is in the neighborhood of 2 (figure 22.8).

Figure 22.8
Hot Springs of Yellowstone National Park.
Some types of algae are able to grow in even this superheated environment.

Summary

Our classification system ranges from kingdoms through species. The kingdom is the broadest group and the species is the smallest. There are five kingdoms and an increasing number of categories as one moves toward the species. The names in the last two categories, the genus and species names, together form the scientific name of the organism being classified. The scientific name has several advantages, one of which is that it is used internationally.

Viruses are nucleic acid particles coated with protein that function only as intracellular parasites. They are responsible for a variety of diseases in animals, plants, and other microbes.

The Prokaryotae include all organisms that are usually thought of as bacteria. Bacteria are important as agents of disease, for carrying on fermentation, as decay organisms, and for lowering water quality, (i.e., the cyanobacteria).

The archaebacteria are a biochemically unique division and are believed to be a very ancient cell type. They are able to grow in extremely harsh environments of low pH and/or high temperatures.

Consider This

An ethical question resulting from the AIDS pandemic is Who should bear the cost of treating each case, the individual or society? Like those cancers blamed on smoking, most persons contract AIDS as a result of a personal choice. A conservative estimate of the cost of treatment of an AIDS case is $46,000; some would place it higher. It is expected that the number of cases will continue to climb and the total cost will be staggering. Select a side of this question and try to defend it.

Water Quality Testing

Of the many possible tests for water quality, one stands out for determining whether a given source of water is safe to drink or for swimming. You can easily perform this if you can get two items common in a bacteriological lab. You will need a sterilized petri dish containing a thin layer of EMB agar. The petri dish should have been sterilized in an autoclave (or pressure cooker) by raising the pressure to fifteen pounds per square inch and holding it there for fifteen minutes. The water sample to be tested should be collected in a similarly sterilized bottle and cap. A few drops of the water sample are swirled in the petri dish, trying to hold the cover off the petri dish as short a time as possible. Place the petri dish on a shelf or other safe place and look at it from day to day until you see some kind of growth. If the growth is smooth and has a greenish shine, it indicates the presence of "coliform" bacteria—bacteria present from fecal contamination. This would be unsafe to drink.

Questions

1. What kinds of organisms are found in the group known as the viruses?
2. Why are the archaebacteria different from other prokaryotes?
3. Why is a classification system necessary? What is meant by a scientific name?
4. What is meant by the phrase "obligate intracellular parasite"?
5. What is meant by the term "bloom"?
6. How does the term "virion" compare to the term "cell"?
7. Why do many people consider the virus to be nonliving?
8. Name two diseases caused by viruses.
9. Name two diseases caused by bacteria.
10. What are the four morphological forms of bacteria?

Chapter Glossary

archaebacteria (ar-ke-bak-te′re-ah) Microorganisms in the kingdom Prokaryotae comprised of species capable of living in environments of extremely low pH and high salt concentration. They differ significantly from other bacteria and eukaryotic cell types.

bacillus (bah-sil′us) A rod-shaped bacterium.

bacteria (bak″-te′re-ah) Forms of microorganisms that are characterized by their shape, biochemical characteristics, and genetic ability to function in various environments. They are members of the kingdom Prokaryotae.

bloom (blo͞om) A rapid increase in the number of microorganisms in a body of water.

coccus (kok′us) A spherical-shaped bacterium.

cyanobacteria (si″an-o-bak-te′re-ah) Blue-green bacteria belonging to the kingdom Prokaryotae.

genus (je͞′nus) One of the categories in the classification system, composed of groups of species that are very similar. The genus and species names together make up the scientific name of an organism.

host (ho͞st) An organism that provides the necessities for a parasite—generally food and place to live.

microbe (mi′kro͞b) Any single-celled organism. It commonly refers to members of the Prokaryotae, Protista, and Mycetae.

obligate intracellular parasite (ob′li-ga͞t in″trah-sel′yu-lar pa͞r′uh-sit) Any organism that must live inside the cell of another species in order to carry out its life functions.

pandemic (pan-dem′ik) Diseases that occur throughout the world at extremely high rates.

phylum (fi′lum) One of the categories in the classification system, composed of groups of organisms in the same kingdom.

pleomorphic (ple″o-mōr′fik) Refers to cells having many, varied shapes.

scientific name (si″en-tif′ik nām) The one name of an organism that is internationally recognized. It consists of the genus name written first and capitalized, followed by the species name usually written in lower case. It is printed in italics or underlined.

species (spe′shēz) One of the narrowest categories of the classification system, composed of members of a reproducing population.

spirillum (spi-ril′lum) A curve-shaped microbe of which three distinct forms are common; vibrios, spirilla, and spirochetes.

spirochete (spi″ro-kēt) One of the three forms of spirilla. Many curves make a spirochete resemble a corkscrew.

subphylum (sub′-fi′-lum) One of the categories in the classification system, composed of groups of organisms in the same phylum.

vibrio (vib′re-o) A type of spirillum characterized by a slightly twisted shape.

virion (vĭ′re-on) The unit of a virus particle composed of a nucleic acid core and a protein coat.

virus (vi′rus) A nucleic acid, protein-coated particle that shows some characteristics of life only when inside a living cell.

■ Chapter Outline

■ Purpose

Members of the kingdom *Mycetae* (the *fungi*) have a cellular structure that is intermediate between the kingdoms Plantae and Protista. These life forms have cell walls, methods of reproduction similar to plants, and tend to be multicellular, but differ completely in their mode of nutrition. The fungi are heterotrophic organisms since they do not contain chlorophyll. Many of them absorb food from dead plants and animals. This method of nutrition is called saprotrophism. It was once thought that the fungi were nonphotosynthetic plants; however, more recent investigations indicate that this group of organisms probably evolved from the protozoans into their own kingdom.

For Your Information

Throughout history, fungi and humans have had a close association. For humans it has been sometimes happy and sometimes not.

The Pharaohs of Egypt considered mushrooms too delicious to permit anyone but the ruling class to eat.

Roquefort cheese, mold ripened in the famous French caves of the same name, was considered a delicacy by the Romans before the first century A.D.

Fungi gave a boost to education for women in 1861 when Vassar Female College was formed from a fortune earned from beer.

Do you like to feed pigeons in the park? How do you feel about starlings? Did you know humans can become infected with a disease, Histoplasmosis, by breathing in spores from a fungus found in bird droppings?

A fungus disease of grain, ergot, has been used to aid childbirth, cause abortion, and unintentionally cause death.

This is only a very small sampling of the ways in which the fungi have been and still are affecting humans.

Learning Objectives

■ State the common names of the divisions of the Mycetae.

■ Name an example for each division of the kingdom Mycetae.

■ State an importance or characteristic for each of the examples in the divisions of the Mycetae.

■ Name and describe the effects of a harmful fungus.

■ Name and describe the effects of a beneficial fungus.

Classification and General Characteristics

Some **fungi** have cell walls composed of a very resistant material called **chitin,** while others contain the more common plant cell-wall material **cellulose.** Familiar examples of this nonphotosynthetic group are the molds, yeasts, bracket fungi, and mushrooms. None of these organisms have the plantlike features of roots, stems, or leaves. Indeed, the fungi have a much less specialized structure. The yeasts are essentially unicellular, while the molds and fleshy fungi are multicellular. In the group that includes the molds, the basic structural unit is the **hypha** (plural, hyphae). This is a slender filament composed of a series of cells (figure 23.1). Some fungi have so many filaments crowded into a firm mass that the individual filaments are difficult to recognize, such as in a mushroom.

In other fungi, the filaments form a vast network extending down into the material on which they are growing and appear as a cottonlike growth. An example of this type of growth can be seen in bread mold or sometimes as a fuzzy growth on dead insect bodies. This mass of filaments, in either type of growth habit, is called the **mycelium.**

The second major group of fungi, the yeasts, exist as single cells. An example of this type is *Saccharomyces cerevisiae,* brewer's yeast and baker's yeast. These fungi reproduce by a process known as budding. In this process, the bud is a gradually enlarging bubble on the cell. In a newly dividing cell, this bud is easy to recognize since it is smaller than the parent cell. Yeasts and molds are not official taxonomic groups, and many fungi may shift from one of these forms to the other. Under certain environmental conditions, yeast may change and grow into a cottonlike mass. The fungi that are capable of making this change are called dimorphic (two forms). The more acceptable classification for the fungi is handled by subdividing them into taxonomically smaller groups based on the kind of sexual spores that are produced. Six major divisions have been formed within the kingdom **Mycetae:**

Kingdom Mycetae *(fungi)*

Division: Slime molds (165+ species)—*Physarum, Dictydium, Stemonitis, Arcyria, Dictyostelium*

Division: Water molds (1800+ species)—*Saprolegnia, Rhizopus, Phytophthora*

Division: Sac fungi (25,000+ species)—*Neurospora,* yeasts, ergots, truffles, *Penicillium*

Division: Club fungi (25,000+ species)—*Fomes, Psilocybe, Agaricus, Amanita, Clavaria, Phallus, Marasmius,* corn smut, wheat rust

Division: Imperfect fungi (25,000+ species)—*Trichophyton, Dactylella, Fusarium*

Division: Lichens (25,000+ species)—crustose, foliose, and fruticose lichens

Slime Molds

There are two classes within this group: the acellular and the cellular **slime molds.** These organisms can be found growing on rotting damp logs, leaves, and soil. They appear as slimy masses or nets, or they may be seen as upright growths that have a variety of shapes (figure 23.2). Members of the acellular slime molds remind one of a giant amoeba whose nucleus and other organelles have divided repeatedly within a single large cell. No cell membranes partition this mass into separate segments. One species looks like whipped orange gelatin dessert and, like the others, creeps about feeding on organic materials. They vary in color from white to bright red or yellow, and may reach relatively large sizes (45 cm in length!) when the environment is optimum. Should the temperature, pH, or moisture change too drastically, this mass undergoes a change that results in the formation of a structure that either resists the severe environment or goes into a reproductive state. One of a variety of spore-bearing structures can be formed, depending on the species of slime mold (figure 23.2). Spores will be produced by meiosis and upon release will survive until more suitable conditions develop. The spores and other resistant structures are small,

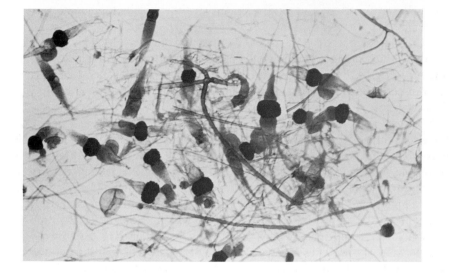

Figure 23.1
Fungal Hyphae.

This illustration shows filaments of fungi known as hyphae. When they grow together in a mass, they are collectively known as a mycelium. (Carolina Biological Supply Company)

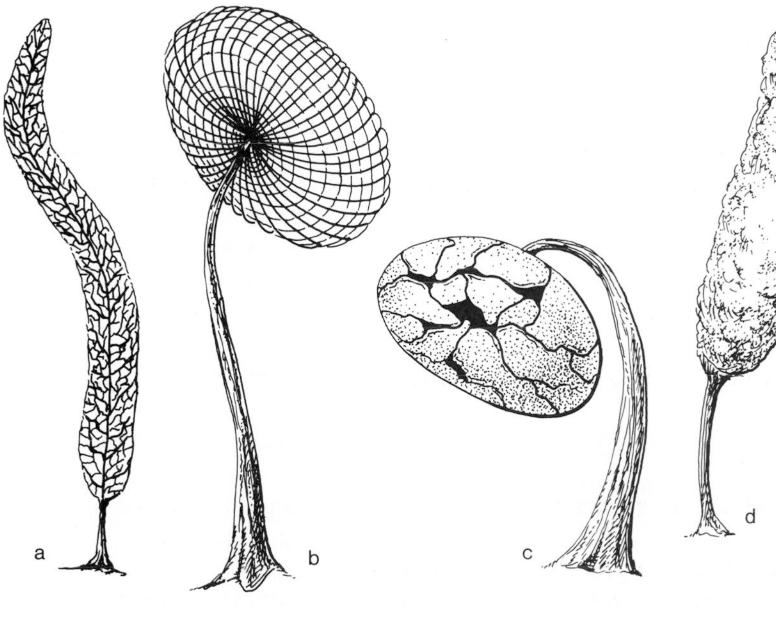

Figure 23.2
The Slime Molds.

(a) *Stemonitis*, (b) *Dictydium*, (c) *Physarum*, (d) *Arcyria*; the slime molds belong to the division Myxomycota. These particular ones may be found growing on decaying logs.

a   b   c   d

Figure 23.3
***The Active Stage
of Physarum.***
This is the creeping, feeding stage
of a species of slime mold.
(Carolina Biological Supply
Company)

dry, and light, and easily blown about in the wind. When they enter a suitable environment, they germinate into flagellated *swarm cells* that may either reproduce by mitosis or fuse together to form a new diploid, multinucleated mass (figure 23.3).

The cellular slime molds exist as large numbers of individual amoebalike cells. These haploid cells get food by engulfing microorganisms and reproduce by mitosis. When their environment becomes dry or otherwise unfavorable, the cells come together in an irregular mass. This mass glides along rather like an ordinary garden slug and is labeled the sluglike stage (figure 23.4). The cells in this mass do not fuse into a single, large mass of cytoplasm but remain as independent, individual cells. This sluglike form may flow about four hours before it forms the stalked, fruiting body. Within the fruiting body, haploid spores develop that will be released into the environment. These spores may be carried by the wind and, if they land in a favorable place, will start the cycle over again.

Water Molds

The wide variety of **water molds** are found growing in a range of environments from those that are merely moist to actual bodies of water. They differ in structure from other fungi in that some filaments have no cross walls, thus allowing cell contents to flow from cell to cell. Biologists call these nonseptate hyphae. Their cell walls may contain chitin, cellulose, or combinations of these two carbohydrates. Differences also exist in the details of their life cycles. It is because of these significant differences that many mycologists believe that water molds should be divided into several separate divisions. One group contains *Rhizopus* and *Mucor,* molds found in household dust (figure 23.5). *Rhizopus* is common black bread mold. It has cross walls and chitin is the chief component of its cell walls.

Another member of this fungal group, *Saprolegnia,* is often called the "fly fungus" since it is so easy to isolate and grow from pond water. This fungus grows easily on dead flies found floating in water from stagnant ditches or ponds. It only takes about two days for the animal's body to become covered with the soft, white, fuzzy mold that is the mycelium of *Saprolegnia.*

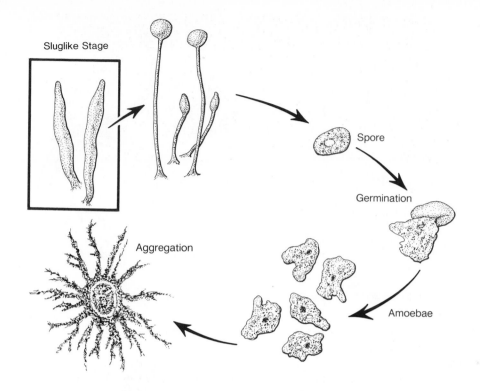

Sluglike Stage

Aggregation

Spore

Germination

Amoebae

Figure 23.4
The Life Cycle of a Cellular Slime Mold—Dictyostelium.
Bottom, from Carolina Biological Supply Company.

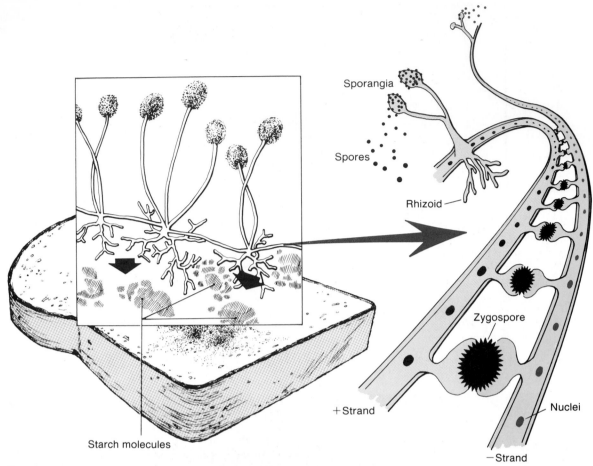

Sporangia

Spores

Rhizoid

Zygospore

+Strand

Nuclei

−Strand

Starch molecules

a

Figure 23.5
The Common Bread Mold.
Rhizopus stolonifer (a) and
another Phycomycete (b) *Mucor.*
(Carolina Biological Supply
Company)

b

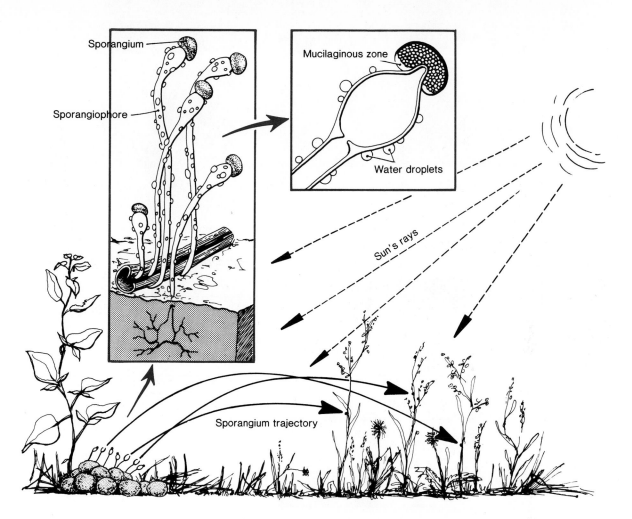

Figure 23.6
The Fungus Pilobolus.
This fungus is found growing on manure and has evolved a complex mechanism for shooting its spores toward the light and a more favorable growth environment.

Pilobolus is another interesting genus. This fungus may be found growing on the fecal material of grazing animals. The distribution of fungal spores from *Pilobolus* takes place from a thick-walled sporangium that has a slimy outer wall and a sticky bottom. This enables them to stick to blades of grass and horse dung. After the fecal material has been deposited, the spores germinate and form a mycelium. The sporangiophores developed by *Pilobolus* have a basal bulb filled with fluid that swells and builds up enough pressure to "fire off" the spores like a fungal cannon when it bursts. The spores are shot away from the dung several yards and at speeds that may reach 30 miles per hour! This dispersal mechanism allows the spores to reach fresh grasses and other plants that will serve as food for the grazing animals (figure 23.6).

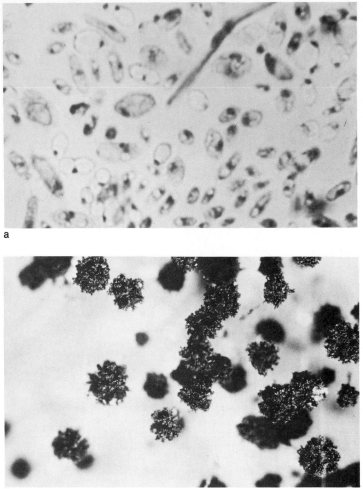

Figure 23.7
A Variety of Ascomycota.
(a) *Saccharomyces cerevisiae,*
common brewer's yeast (X 73);
(b) *Aspergillus niger,* the cause of
aspergillosis; (c) *Claviceps
purpurea;* the fungus responsible
for ergot poisoning in wheat. (**a**
and **b,** Carolina Biological Supply
Company; **c,** USDA.)

Sac Fungi

The **sac fungi** is the third division and has many different members (figure 23.7).
These organisms vary greatly from unicellular yeasts through powdery mildews to
cup fungi. They have cross walls in their hyphae and reproduce sexually by spores
formed in characteristic sacs. If a fungus produces just one sac in a parent cell, it
is called a yeast. Probably the most familiar of the sac fungi is the common brewer's
yeast and baker's yeast, *Saccharomyces cerevisiae,* used in the leavening of dough
before baking and for alcohol production in the brewing industry. However, not all
members of the sac fungi grow in such an uncomplicated way. *Penicillium,* which
is the source of the antibiotic penicillin, produces a very complex series of sacs.
Asexual reproduction can also occur through the formation of some spores distinc-
tively displayed as shown in figure 23.8. Another common member of this group is
the cuplike fungi, *Peziza.* These grow well on rich soils, rotten logs, and manure,
taking the form of small, brightly colored goblets or wine glasses. Members of the
sac fungi are responsible for a variety of beneficial and harmful effects. *Penicillium
roquefortii* is the mold used in the aging of Roquefort and blue cheeses, and
P. notatum produces penicillin. Some are responsible for such problems as Dutch
elm disease, ergot food poisoning in barley, meningitis, and mildew.

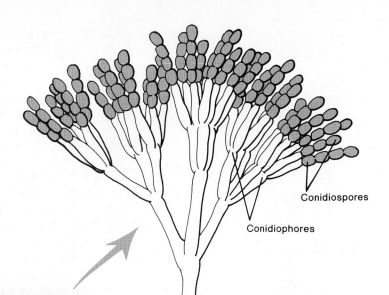

Conidiospores

Conidiophores

Figure 23.8
Penicillium.
This fungus is commonly found on many foods such as oranges. The green color is the result of spore coloration.

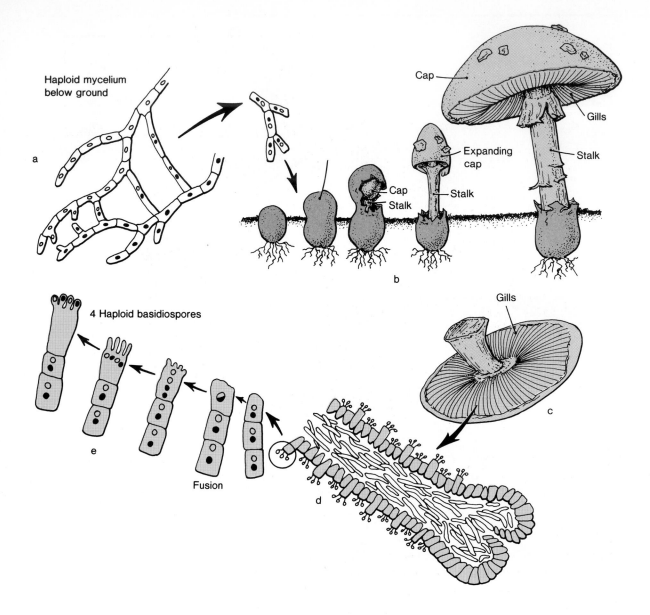

Haploid mycelium
below ground

a

Cap

Gills

Expanding
cap

Stalk

Cap

Stalk

Stalk

b

4 Haploid basidiospores

e

Fusion

Gills

c

d

Figure 23.9
Mushrooms, Basidiomycota.
These fungi are probably the most familiar. The portion of the organism that is edible is only one stage in the life cycle of the fungus. The other stages are found in the soil in which they grow.

Club Fungi

The fourth group of fungi, the **club fungi,** (figure 23.9) also have hyphae that possess cross walls. However, the spores are not the same as those produced by the other groups. Mushrooms, bracket fungi on trees, and puffballs produce their spores on a club-shaped structure. These spores can be seen as a cloud of dustlike particles leaving a mushroom when it has been dried and shaken.

One of the most interesting formations caused by mushroom growth can be seen in soil that is rich in mushroom mycelia such as lawns, fields, and the forest. These formations are known as "fairy rings" (figure 23.10) and result from the expanding growth of the mushrooms. The inner circle is normal grass and vegetation. The mushroom population originally began to grow here but left because of exhausting the soil nutrients necessary for fungal growth. As the microscopic hyphae grow outward from the center, they stunt the growth of the grass, forming a ring of short, inhibited grass. Just to the outside of this growth ring, the grass is luxuriant since the mycelia excrete enzymes that decompose soil material into rich nutrients for

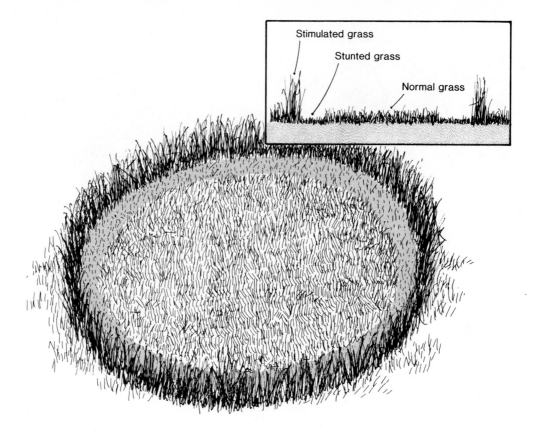

Stimulated grass

Stunted grass

Normal grass

growth. The name "fairy ring" comes from an old superstition that this ring has been formed by fairies, tramping down the grass while dancing in a circle. The mushroom best known for forming fairy rings is *Marasmius oreades*.

Imperfect Fungi

This group of fungi is known as the **imperfect fungi** since the sexual stage of reproduction either does not exist or has not yet been discovered. If and when this process is identified, the members of this group will be shifted into one of the other divisions. In this division, *Dactylella bembicoides* is one that actually entraps and kills soil roundworms (nematodes). Three cells of the hyphae form a loop or noose through which these tiny worms may pass. When they move into the loop, the cells swell to entrap the worm like a hangman's noose. The fungus then grows into the worm's body upon which it feeds.

Lichens

Fungi and algae come together in a peculiar group called **lichens.** Alga is believed to furnish food for the fungus, and the fungus supplies moisture and protection to the alga. Together, this association is important as a pioneer in ecological succession. Lichens are capable of growing on bare rock, where they begin the soil-making process. The so-called "reindeer moss" is actually a lichen that provides caribou with a source of food in northern Canada and Alaska.

The relationship formed between the algal and fungal components in a lichen is an excellent example of the symbiosis known as mutualism. When the two partners are separated from one another, they will actually reassemble into the original

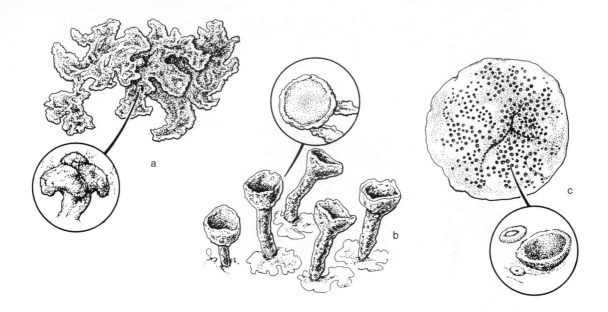

Figure 23.11
Lichen Forms.

Lichens are classified according to
their shapes (a) leaflike or foliose,
(b) branching or fruticose, and
(c) crustose. The inserts show the
reproductive structures.

lichen form. Since lichens are not true "organisms," as defined by most biologists, it is very difficult to classify them using the same criteria. They are placed in the kingdom Mycetae, based on the nature of their fungal symbiont. Classification within the lichens is also based on their physical form. Three types are commonly recognized. The foliose lichens are leaflike, the fruticose types are branching, and the crustose types resemble crusts growing on the surfaces of trees, shrubs, and rocks (figure 23.11).

The Molds as Harmful Fungi

Fungi are commercially important to us as destroyers. They cause losses of millions of dollars annually by destroying food and manufactured materials, and expensive chemical dusts and sprays are needed to control them. Because fungi produce millions of spores that can be easily and widely spread, their control is difficult. Many kinds of fungi have the ability to grow in very strange environments. Storing food in the refrigerator helps to keep it fresh and prevents spoilage, but many fungi are able to grow even in this cold environment. That fuzzy growth on the green beans on the back shelf of a refrigerator is from spores that contaminated the food. The fungi release millions of spores into the atmosphere where they are rapidly distributed by the wind currents. These spores may linger in the air or be inhaled by animals and humans. Many people show the symptoms of what is commonly referred to as "hay fever" when they are in contact with too many of these spores. This is an allergy to the fungal spores and not necessarily to hay. *Mucor* spores are a major cause of these upper respiratory symptoms.

Fungi are able to grow on the surface of the body and cause diseases such as athlete's foot and ringworm. Fungal infections can also occur inside the body. Liver, lung, and kidney infections cause severe damage and may be deadly if not quickly brought under control by proper medical therapy (see box 23.1).

A number of fungi produce deadly poisons called **mycotoxins.** There is no easy way to distinguish those which are poisonous from those that are safe to eat. The poisonous forms are sometimes called "toadstools" and the nonpoisonous ones

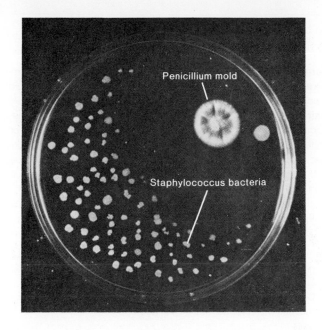

Figure 23.12
*The Beneficial Effects
of Penicillium.*
The zone of clearing around the
mold *Penicillium* seen in this photo
is the result of the antibiotic
penicillin diffusing from the mold.
The bacterium *Staphylococcus
aureus* growing at a distance from
the mold is the cause of boils and
other infectious diseases.

mushrooms. However, they are all properly called mushrooms. The origin of the name "toadstool" is unclear. One idea is that toadstools are mushrooms on which toads sit, while another states that the word is derived from the German "todstuhl"—seat of death. The most deadly of these, *Amanita verna,* is known as the "destroying angel" and can be found in woodlands during the summer. Mushroom hunters must learn to recognize this deadly, pure white species. This mushroom is believed to be so dangerous that food accidentally contaminated by its spores can cause illness and possibly death. Another mushroom, *Psilocybe mexicana,* has been used for centuries in religious ceremonies by certain Mexican tribes because of the hallucinogenic chemical that it produces. These mushrooms have been grown in culture, and the drug psilocybin has been isolated. In the past, it was used experimentally to study schizophrenia. Potatoes can be infected by a fungus known as *Phytophthora infestans.* When this fungus grows, it forms a scaly material on the surface of the potato that makes it inedible. In the past, this fungus was responsible for the loss of millions of dollars worth of potatoes and affected the lives of many people. The Irish potato famine of 1845 was the result of potato crop losses due to this fungus.

Some Advantageous Molds

Although mold on potatoes may spoil them, many cheeses, such as blue cheese and gorgonzola owe their distinct flavors to molds. Mold is also beneficial to such industries as brewing, baking, and drugmaking. Antibiotics like penicillin and other mold by-products are a great blessing.

In 1928, Dr. Alexander Fleming was working at St. Mary's Hospital in London. As he sorted through some old petri dishes on his bench, he noticed an unusual situation. The mold *Penicillium notatum* was growing on some of the petri dishes. Apparently, the mold had found its way through an open window and onto a culture of *Staphylococcus aureus.* The bacterial colonies that were growing at a distance from the fungus were typical but there was no growth close to the mold (figure 23.12). Fleming isolated the agent responsible for this destruction and named it

Box 23.1
Medical Mycology

The scientific study of pathogenic fungi began in 1839 when scrapings from a skin disease known as favus were grown in pure culture on potato slices. The fungal growth was then used to inoculate the scalp of a healthy child and resulted in a second case of the disease. This procedure demonstrated the fungus *Trichophyton schoenleinii* to be responsible for the disease. Over 100 years of research has uncovered more than 100,000 species of fungi.

Most are nonpathogenic, saprotrophic, yeastlike or filamentous microbes found in the soil. A few have evolved to be true pathogens and are able to actively infect, cause harm, and be transmitted from one suitable, susceptible host to another. Most fungi only demonstrate clinically recognizable symptoms of infection when, by chance, large numbers of spores enter and overwhelm a host; or when normal, opportunistic fungal flora are not kept

under control by defense mechanisms. The extensive use of antibiotics and immunosuppressive drugs has led to a sharp increase in infections caused by fungi normally found in or on the body.

Infections caused by fungi are known as **mycoses** (*singular,* mycosis). Three clinical types of mycoses have been identified: cutaneous, subcutaneous, and systemic (see table below). Fungi known as dermatophytes are able to cause cutaneous

Location of Infection	Disease	Causative Agent
Cutaneous (skin)	Candidiasis of skin, mucous membranes, and nails	*Candida albicans* and related species
	Ringworm of scalp, glabrous skin, nails	Dermatophytes (*Microsporum* sp., *Trichophyton* sp., *Epidermophyton* sp.)
Subcutaneous (under skin)	Chromomycosis	*Fonsecaea pedrosoi* and related forms
	Mycotic mycetoma	*Petriellidium boydii, Madeurella mycetomii,* etc.
	Sporotrichosis	*Sporothrix schenckii*
Systemic (internal organs)	Histoplasmosis	*Histoplasma capsulatum*
	Blastomycosis	*Blastomyces dermatitidis*
	Coccidioidomycosis	*Coccidioides immitis*
	Cryptococcosis	*Cryptococcus neoformans*
	Aspergillosis	*Aspergillus fumigatus*
	Candidiasis, systemic	*Candida albicans*

infections. Members of this group have the ability to use the protein keratin and become localized in the surface layers of the skin, hair, or nails. Seldom do they penetrate or extend into the deep tissues of the body. The dermatophytes are true pathogens capable of establishing an infection in a healthy host and may be transmitted from one host to another. The group of fungi responsible for subcutaneous infections is only acquired from the soil and enters the body through breaks in the skin caused by traumatic experiences. These grow very slowly once inside the skin and may show no visible signs of host damage for many years. Fungi responsible for systemic, or deep, infections may be acquired from the soil or are opportunistic, normal flora of the body. Minor infections caused by these fungi occur frequently but are kept in check by body defenses. Occasionally they are not controlled and cause the death of their host and themselves. These pathogens are not contagious under normal circumstances.

Thrush, or "black hairy tongue," is similar to other candidiasis infections of mucous membranes. Infections occur in infants during the establishment of normal flora (within the first three days after birth) or in older individuals after prolonged antibacterial antibiotic therapy. Fungal growth leads to the formation of white, curdy patches of pseudomembrane on the tongue and buccal mucosa. If the infection destroys papillae, the tongue takes on a blackened appearance. The disease also occurs in adults who experience mucosal irritation from poorly fitting dentures or in those who fail to maintain good oral hygiene. The infection may spread, causing chelitis (scattered infection on the lips) or angular stomatitis seen as pursed lips with grooves at the corners of the mouth. Vulvovaginitis is a candidal infection that runs a course similar to thrush but occurs on the mucosal membranes of the vagina. Symptoms include intense itching, irritation, and vaginal discharge. The condition is more likely to develop as the pH of vaginal secretions becomes more alkaline. Under normal circumstances, bacterial flora help maintain a low pH that limits the growth of *C. albicans*. *Candida* is part of the normal vaginal and internal flora; most individuals contain antibodies against this microbe, and it is not considered contagious.

Valley fever or coccidioidomycosis is the most feared and dangerous of all systemic mycoses. *Coccidioides immitis* occurs commonly in arid (desert) regions of the southwestern United States, and Central and South America. Spores are found in the soil surrounding the burrows of small desert rodents and are easily carried on the wind to susceptible hosts. They may be inhaled or enter the body through breaks in the skin. Most infections go unnoticed except for an increase in protective antibody titer. Others experience mild upper respiratory "flulike" symptoms known as primary coccidioidomycosis.

The symptoms of this disease form include fever, pain, coughing, and malaise. Primary infections are usually self-limiting and brought under control by natural defense mechanisms in a few days or weeks. Some of those experiencing primary infection can develop allergic reactions known as erythema nodosum, erythema multiforme, and "desert rheumatism." Erythema nodosum appears on the legs and feet as itching, painful, reddened nodules deep within the skin and lasts about a week. Erythema multiforme typically appears on the upper body as inflamed vesicles, pustules, or nodules that may hemorrhage. This allergic reaction usually lasts only four to seven days. The rheumatic reaction results in heat, pain, swelling, and stiffness of joints. Those who demonstrate an allergy rarely develop secondary coccidioidomycosis.

Secondary infections strongly resemble tuberculosis. Infection is not limited to the lungs and may spread to involve the skin, internal organs, bones, mucous membranes, and nervous system. This form of the disease is highly fatal. Tissue reactions result in the formation of tumors that later break down and spread. Spores can be isolated from the infection sites and, in some cases, their presence may be essential in distinguishing secondary coccidioidomycosis from tuberculosis. If diagnosed early, the infection might be controlled by amphotericin B, an antifungal antibiotic.

penicillin. Through his research efforts and those of several colleagues, the chemical was identified and used for about ten years in microbiological work in the laboratory. Many suspected that the penicillin might be used as a drug, but the fungus was not able to produce enough of the chemical to make it worthwhile. When World War II began and England was being fire bombed, there developed a great urgency for a drug that would control bacterial infections in burn wounds. Two scientists from England were sent to the United States to begin research into the mass production of penicillin. Their research in isolating new forms of *Penicillium* and purifying the drug were so successful that cultures of the mold now produce over one hundred times more of the drug than the original mold discovered by Fleming. In addition, the price of the drug dropped considerably, from a 1944 price of $10,000 per pound to a current price of less than $50.00. The species of *Penicillium* used to produce all the antibiotics today is *P. chrysogenum*. This was first isolated in Peoria, Illinois, from a mixture of molds found growing on a cantaloupe. The species name, *chrysogenum,* means golden and refers to the fact that the mold produces golden-yellow droplets of antibiotic on the surface of the mycelium. The spores of this mold were isolated and irradiated with high dosages of ultraviolet light, which caused mutations to occur in the genes. When some of these mutant spores were germinated, the new mycelia were found to produce much greater amounts of the antibiotic.

There are over one hundred species of *Penicillium* and each characteristically produces spores in a brushlike border; the word "penicillus" means *little brush*. Other members of this group do more than just produce antibiotics. Many people are familiar with the blue, cottony growth that sometimes occurs on citrus fruits. The *P. italicum* growing on the fruit appears to be blue because of the pigment produced in the spores. The blue cheeses, such as Danish, American, and the original Roquefort cheese, all have this color. Each has been aged with *Penicillium roquefortii* to produce the color, texture, and flavor. Differences in the cheese are determined by the kind of milk used and the conditions under which the aging occurs. Roquefort cheese is made from sheep's milk and aged in Roquefort, France, in special caves. American blue cheese is made from cow's milk and aged in many places around the United States. The blue color has become a very important feature of these cheeses. The same research laboratory that first isolated *P. chrysogenum* also found a mutant species of *P. roquefortii* that would produce spores having no blue color. The cheese made from this mold is "white" blue cheese. The flavor is exactly the same as "blue" blue cheese, but commercially it is worthless because people want the blue color.

A last example of a very beneficial fungus is *Aspergillus niger*. This mold produces very dark brown or black spores, which are in chains and may commonly infect and spoil refrigerated foods. But more importantly, this organism is the basis for the soft drink industry. The citric acid that gives a soft drink its sharp taste was originally produced by squeezing juice from lemons and purifying the acid. Today, however, *A. niger* is grown on nutrient media with table sugar (sucrose) to produce great quantities of citric acid.

Summary

The kingdom Mycetae contains organisms commonly known as the fungi. Fungi are saprotrophic, nonphotosynthetic organisms that occur in two basic morphological forms, yeast and molds. Classification of the Mycetae into divisions is based on the formation of the sexual production of spores. Six divisions have been formed including the Myxomycota (slime molds), Phycomycota (water molds), Ascomycota (sac fungi), Basidiomycota (club fungi), Deuteromycota (imperfect fungi), and the Lichenes (the lichens). Many of the members of this kingdom are harmful and destructive. They are responsible for the decay and decomposition of foods and many types of manufactured materials. They also cause diseases such as athlete's foot, ringworm, and food poisoning. Poisons released by some fungi are called mycotoxins. The study of medically important fungi is mycology. Some fungi are advantageous and widely used in such industries as brewing, baking, and drugmaking. One of the most universally used beneficial fungi is brewer's yeast or baker's yeast, *Saccharomyces cerevisiae*.

Consider This

Are fungi involved in diplomacy? This is a serious question since the U.S. State Department accused the U.S.S.R. of dropping the poison T-2, a toxin from the mold *Fusarium*, in Southeast Asia. Some scientists have said that this so-called yellow rain is only bee droppings. However, the U.S. State Department continues to be concerned with this alleged violation of the Biological Weapons Convention of 1972.

Who is correct? Look at the evidence for yourself. A short summary is in *Science*, July 4, 1986.

Experience This

Practical Application of Mycetae

Yeast Raised Sourdough Bread

This bread depends on a bacterial sourdough starter culture for its flavor and a commercial yeast to raise (leaven) the dough.

To prepare sourdough starter:

In a nonmetalic container using a nonmetalic spoon, mix the following:

3 cups milk
3 cups flour
¼ cup yogurt (unflavored, with active culture)

Cover and place mixture on top of the refrigerator for twenty-four hours to incubate. This will result in an active growing culture of acid-producing bacteria.

To make sourdough bread:

2 cups sourdough starter (see above)
1 cup scalded milk
3 tablespoons butter or oleo (softened)
3 tablespoons sugar
2 teaspoons salt
1 package yeast
¼ cup warm water
6½ cups flour
1 teaspoon baking soda

Mix scalded milk with butter to melt. Add sugar and salt. Stir yeast into warm water and let stand five minutes. Then add the yeast and sourdough starter to cooled milk mixture. Beat in 2 cups flour to make a smooth batter. Over this sprinkle baking soda, mix well, cover with a cloth, and set aside to rise for about thirty minutes.

Gradually mix in the rest of the flour until dough is stiff. Knead the dough on a floured surface until smooth and elastic, adding more flour to prevent excessive stickiness. Let stand about 10 minutes.

Oil two loaf pans, divide into two halves and place them into the prepared pans. Turn each loaf to coat lightly with the oil. Let rise about one hour. Bake at 375° for fifty minutes.

To maintain starter:

To the leftover starter add 1 cup milk and 1 cup flour, place on top of the refrigerator for twenty-four hours, after which it can be refrigerated until further use.

Before using the starter in another recipe, be sure to build it up (create additional volume) the night before by adding milk and flour in sufficient amounts to make enough for your recipe and have some leftover for future use.

Questions

1. What is the difference between hyphae and mycelia?
2. Name two positive results of fungal growth and activity.
3. What term is used to describe fungi that are able to grow as either yeast or molds?
4. Define the term saprotrophic.
5. Name the six divisions of the kingdom Mycetae and give an example of each.
6. What types of organic molecules are found as cell-wall materials in the fungi?
7. What is a mycotoxin? Give an example of a fungus that produces this material.
8. List three diseases caused by harmful fungi.
9. What is meant by the term "imperfect fungi"?
10. Why are Lichenes different from the other five divisions of the Mycetae?

Chapter Glossary

cellulose (sel'yu-lōs) An organic material produced by some fungi, Protists, and plants as a cell-wall material. It differs from the material known as chitin.

chitin (ki'tin) An organic material produced by some fungi as a wall material and by insects as an external skeleton.

club fungi (klub fun'ji) One of the six major divisions of the kingdom Mycetae (fungi) containing organisms producing sexual spores on clublike structures.

fungi (fun'ji) The common name for members of the kingdom Mycetae.

hypha (hi'fah) The basic unit of structure of a fungus that consists of a filament.

imperfect fungi (im-pur'fekt fun'ji) One of the six major divisions of the kingdom Mycetae (fungi) containing organisms in which sexual reproduction is absent or undiscovered.

lichens (li'kens) One of the six major divisions of the kingdom Mycetae (fungi) containing organisms that are an association of an alga with a fungus.

mycelium (mi-se'le-um) A mass of hyphae making up the body of a fungus.

Mycetae (mi-se'te) One of the major kingdoms of living things composed of nonphotosynthetic organisms commonly referred to as fungi and lichens. Members of this group include yeasts and molds.

mycoses (mi-ko'sēz) Diseases caused by fungi.

mycotoxin (mi''ko-tok'sin) One of a group of poisonous chemicals produced by certain fungi.

sac fungi (sak fun'ji) One of the six major divisions of the kingdom Mycetae (fungi) containing organisms that produce sexual spores in saclike structures.

slime molds (slīm mōlds) One of the six major divisions of the kingdom Mycetae (fungi) appearing as slimy masses.

water molds (watr mōlds) One of the six major divisions of the kingdom Mycetae (fungi) containing organisms that grow in water or very moist environments.

24 Protista

Chapter Outline

Purpose

We mentioned earlier how incomplete our classification system is. This applies especially to the kingdom Protista. Even from the brief description we give in this chapter, you will see that there are great differences among the organisms of this kingdom. The common name for the green (photosynthetic) Protista is *algae*. The nongreen (nonphotosynthetic) Protista are called *protozoa*.

For Your Information

A hot, sweaty hiker comes across a beautiful, sparkling, cold mountain stream. But a sign is posted: Don't Drink The Water!

Giardia lamblia, a protozoan found in these beautiful streams causes diarrhea. It does not seriously affect all adults, but who wants to find out if they are one of those who is affected seriously. It can be especially dangerous for children. Associated symptoms include abdominal cramps, bloating, fatigue, and loss of weight. While some of us might want to lose weight, this is indeed the hard way!

Giardia is carried in the intestine by many animals and they share a portion of these with us as they deposit their feces. In this way streams, lakes, and snow can be contaminated. The threat of Giardiasis is so great that national parks such as Grand Teton and others publish warnings to their visitors to boil any surface water or melted snow.

Learning Objectives

- Name six groups of algae.
- Name an important group of algae and discuss its importance.
- Name four groups of Protozoa.
- Describe the distinguishing feature by which the Protozoa are separated into groups.
- Name an important Protozoan and discuss its importance.

Algae

The **algae** are separated into phyla (divisions) on the basis of the different (1) kinds of pigments, (2) compounds used for storage, and (3) details of their reproductive cycles. Rather than consider these technical details, let's consider some facts relating more to our experience. The algae are "pond scum" or seaweed, and are generally considered a great nuisance. They can create a nuisance by fouling beaches and drinking water, as well as by producing poisonous substances. But they also have tremendous importance for all water-living animals. They are the first link in the food chain within all bodies of water. They are the producer organisms and, as such, deserve our attention, since we depend on them in many ways.

The green algae grow mostly in fresh water, but there are some in damp soil, salt water, and even in snowfields high in the mountains. They occur as single cells, as long filaments of single cells, as spherical clusters, or as lettucelike sheets (figure 24.1). Most of them are grass green and so live up to their common name, green algae.

The euglenoids are an interesting group of single-celled algae that have the plant characteristic of chlorophyll in addition to some animal characteristics. They pull themselves through the water by lashing **flagella** back and forth. The flagellum is a long, flexible whiplike filament. Euglenoids ingest food particles as well as carry on photosynthesis. They are equipped with a light-sensitive spot that enables them to move toward or away from light. They lack a cell wall. Most of this small group live in fresh water, but a few live in damp soil, and some in the food tubes of some animals (figure 24.2).

The diatoms are the most important producer in saltwater ecosystems. They are also common in fresh water. Animals of the sea depend on these organisms, which are so small that we need a microscope to see them. They occur mostly as individual cells, but some remain attached to each other after division and form a

Figure 24.1
Types of Green Algae.
Pictured are examples of the many kinds of green algae.

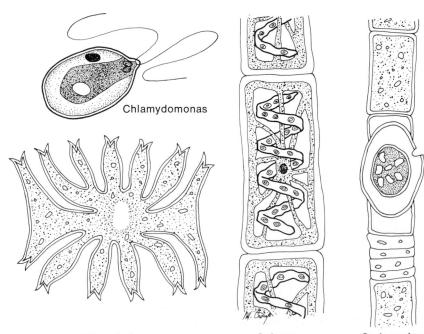

Chlamydomonas

Micrasterias

Spirogyra

Oedogonium

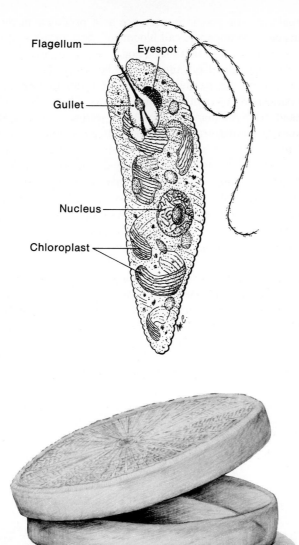

Figure 24.2
Euglena.
A cell with both plant and animal characteristics.

Flagellum

Eyespot

Gullet

Nucleus

Chloroplast

Figure 24.3
Diatoms.
Not all diatoms are cylindrical, but all have a "cover" and "bottom" made of silica deposited in their cell walls.

cluster called a **colony.** The diatoms are unique in that they deposit silicon dioxide in their cell walls. When the organisms die, everything breaks down and disappears except the silicon dioxide. These cell walls are arranged in pairs resembling a little dish and its cover (figure 24.3). The glassy walls of diatoms are decorated with grooves and holes so small that they are used to check the quality of microscopes (figure 24.4). Some of the markings are too small to be seen except under an electron microscope. The glassy remains of these algae collected on the floor of ancient seas in such quantity that it is commercially profitable to remove them with power shovels and trucks. The mass of the tiny diatoms has the texture of talcum powder and is used as a very fine abrasive in metal polishes. The cell walls of diatoms are able to pick up small particles of impurities in a liquid, and can be used as filters. The brewing industry has used diatoms to clarify beer. They also serve as an excellent insulator and are not destroyed by extremely high temperatures.

Figure 24.4
Types of Diatoms.
A few of the seemingly endless variety of diatoms. (Carolina Biological Supply Company)

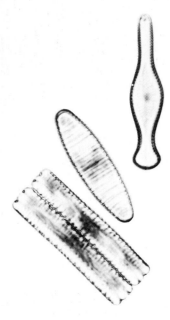

The dinoflagellates are next in importance as a producer in the oceans. They can also be a nuisance. Newspapers call blooms of these organisms *red tides,* since the cells occur in such quantities that they color the water. If the pinkish color added to ocean waves were the only effect, there would be little fuss—it might even be an attraction. Unfortunately, this particular species of dinoflagellate produces a poison deadly to fish, many smaller organisms, and humans. These blooms kill fish, which litter the beach and drive the tourists away. More seriously, people who eat the oysters or clams growing in such waters may become ill, since these animals concentrate the poison in their bodies as they filter food out of the water. There is an old saying that oysters and clams are in season only during the months containing the letter *R*—"Oysters *R* in season." This is true only if refrigeration is unreliable and to some extent because the red tide occurs more frequently during the warmer months (none of the names of the summer months contain the letter *R*). The red tides are confined to relatively small areas. Either the blooms are better reported in recent years or they have increased.

The brown algae are mostly saltwater organisms. The seaweed called kelp, which covers the rocks along the ocean coasts, are mostly in this group. Some grow to be quite large, up to one hundred meters in length (figure 24.5). Kelp is harvested mechanically and refined to commercially valuable products. Algin or alginates, which come from brown algae, are found in ice cream, cake frosting, pudding, cream-centered candies, and many cosmetics (figure 24.6).

The red algae are also saltwater organisms. They grow on rocks along the seacoast. Red algae have attracted scientific attention not only because they are unlike other algae but because they differ from the higher plants. They have been referred to as an evolutionary dead end. The red algae are used as an additive in foods for thickening and flavoring. The genus *Porphyra* is "farmed" along the coast of Japan on wooden racks (figure 24.7). You might look for the name *carrageenan* on packaged foods (figure 24.6). It is produced from red algae and used as a stabilizer to

Figure 24.5
Brown Algae.
This kelp is a common sight along our western seacoasts.

Figure 24.6
The Algae in Your Food.
The derivatives of algae, algin and carrageenan, are food additives, as seen on many food labels.

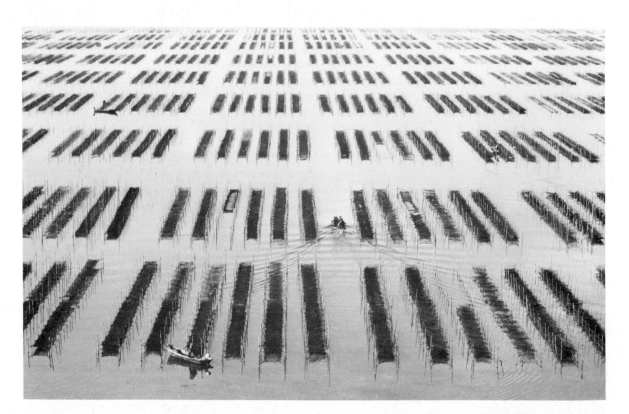

Figure 24.7
Algae Farming.
Algae are grown in Japan much as an agricultural crop is grown in this country.

Figure 24.8
Sarcodina.

These one-celled organisms all move by means of leglike extensions of protoplasm, the pseudopodia.

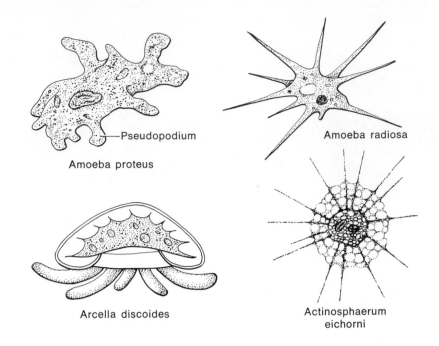

Pseudopodium

Amoeba proteus

Amoeba radiosa

Arcella discoides

Actinosphaerum eichorni

keep solids suspended in liquids, such as the particles of chocolate in chocolate milk. A high protein cattle food has been produced from the red algae grown along Scandinavian coasts. Agar, another product of red algae, has been used worldwide in hospital and bacteriological laboratories. It is commonly sold as a granular powder and added to hot water in much the same way gelatin dessert is made. When the mixture cools, it becomes semisolid, like gelatin, and is used principally as a medium on which to grow bacteria. Although some people eat agar, it is generally not used as a food. It acts as a laxative.

Protozoa

The second major group in the kingdom **Protista** is the phylum **Protozoa.** "Protozoa" is derived from Greek words meaning "first animal." It is still classified by many as an animal and is studied in zoology classes, just as algae are studied in botany classes. Most protozoans can move under their own power at some stage in their life cycle. The phylum is subdivided according to the particular method of moving around (locomotion).

Sarcodina

The *Amoeba* and its relatives flow without any permanent body form. This type of flowing movement is so characteristic that it is called amoeboid movement. Most amoebae live in water, and some are parasitic. One such parasite commonly causes amoebic dysentery in people whose drinking water has become contaminated. It results in death or serious illness (figure 24.8).

The name of this pathogen provides some basic information about the nature of the disease it causes: "ent" means *inside,* "amoeba" means *varied shape,* "histo" means *tissue-associated,* and "lytica" means *bursting. Entamoeba histolytica* is the cause of waterborne and foodborne amoebic dysentery. The amoeboid form of the parasite is highly susceptible to environmental influences and easily destroyed when

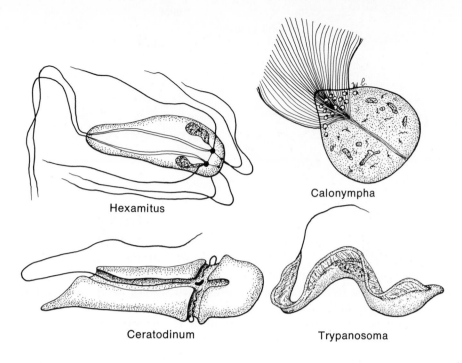

Figure 24.9
Flagellates.
These organisms have some characteristics of both plants and animals.

Hexamitus

Calonympha

Ceratodinum

Trypanosoma

released from the intestinal tract in fecal material. Transmission of this stage to other persons is highly unlikely; however, *E. histolytica* does form a more resistant cyst that is easily transmitted in water, food, or by houseflies. Normal levels of chlorine in drinking water do not destroy the cyst. Ingested cysts release four amoeboid trophozoites capable of establishing an intestinal tract infection. The amoebae multiply in the large intestine by binary fission and feed on epithelial tissue and intestinal bacteria. The severity of the disease depends on how many lesions there are and the extent of tissue damage. Many infected individuals are asymptomatic and become carriers. Those who develop symptoms may demonstrate only a mild, prolonged case of diarrhea, or they may have moderate diarrhea containing mucous and blood. Patients become weakened; and in extreme cases, amoebae may enter tissues, causing abscesses in the liver, lungs, or other organs. The usual course of infection involves a cycle of tissue destruction and repair at various points on the intestine. This prolongs the disease, which may last for months.

Immunity has not been demonstrated. The drug Flagyl is one of the most common treatments to control infections, but it can produce side effects in patients who have developed liver abscesses. Controlling the disease and preventing epidemics is based on eliminating water contaminated with cysts and the development of good personal hygiene.

Mastigophora

Another class moves by means of flagella. The flagellates have one or more flagella, no cell walls, and a relatively simple structure. The protozoan *Hexamitus* in figure 24.9 is an example of this kind of flagellate. Another of the flagellated protozoans causes the deadly African sleeping sickness.

The flagellated protozoan *Trichomonas vaginalis* is the cause of the venereal disease known as trichomoniasis or 'trich'. The parasite is transmitted by sexual intercourse but may also be acquired from contaminated objects such as towels,

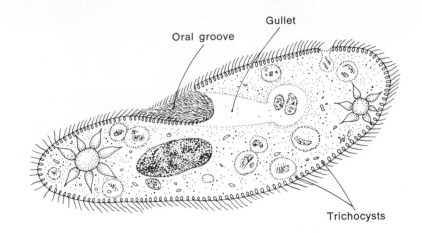

Figure 24.10
Paramecium.
This organism is one of the most highly evolved and specialized members of Protista.

Oral groove

Gullet

Trichocysts

catheters, douche equipment, and examination tools. Infection results in mild inflammation of the vagina, cervix, and vulva. A foul-smelling, yellow or cream-white discharge is produced from the sloughing of infected surface tissue. In males, the infection may extend to the prostate, seminal vesicle, and urethra, but only a thin, white discharge occurs. Infection can also result if *T. vaginalis,* which is found as opportunistic normal flora of the vagina, becomes the predominant microbe due to uncontrolled growth. An increase in the pH of vaginal fluids is primarily responsible for this form of infection. These pH changes may be due to a loss of normal acid-producing bacterial flora, hormonal imbalance, or the effects of drugs such as oral contraceptives. Infection is controlled by topical or oral treatment with antiprotozoal drugs and reestablishment of the normal vaginal pH between 3.8 and 4.4.

Ciliata

Still another class of Protozoa move by means of **cilia.** These are short, flexible filaments that frequently cover the cell completely. They beat in an organized, rhythmical manner and propel the cells very rapidly, so rapidly that it is sometimes difficult to keep them in focus under a microscope (figure 24.10). These ciliates are commonly seen in cultures of pond water and hay. In a freshwater ecosystem, the ciliates are the primary consumers of algae.

Sporozoa

The organism that causes malaria is representative of a fourth class of Protozoa. All members of this class, known as Sporozoa, are parasites. One of their major methods of reproducing is by producing resistant cells called spores. This group includes many important disease organisms. They are often transmitted by animals such as mosquitos, ticks, and flies (figure 24.11).

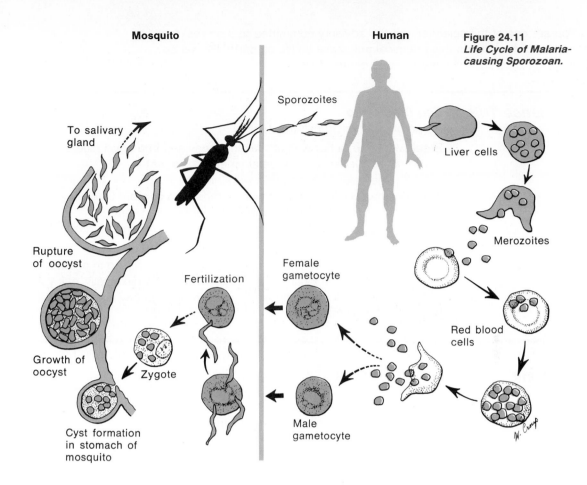

Mosquito

Human

Figure 24.11
Life Cycle of Malaria-causing Sporozoan.

Sporozoites

To salivary gland

Liver cells

Merozoites

Rupture of oocyst

Female gametocyte

Fertilization

Red blood cells

Growth of oocyst

Zygote

Cyst formation in stomach of mosquito

Male gametocyte

Summary

The Protista include those eukaryotes that lack specialized cells. Eukaryotes are organisms whose cells possess a nucleus with a nuclear membrane as well as other membranous structures, such as mitochondria and chloroplasts. The protists include the Protozoa and all the algae except the animal-like cyanobacteria. Algae are more plantlike, in general, while the protozoans are animal-like. Algae produce their own food by photosynthesis, while protozoa depend on organic materials produced by some other living thing. Algae are restricted in their living situation to a lighted area, but otherwise range widely from water to moist land habitats. Protozoa are largely aquatic, and some are parasites. Protists are very important since they cause disease, recycle organic materials, and are used in industrial processes and food production.

Consider This

Flash! "We interrupt this textbook to bring you an important announcement! A chemical plant in the Midwest has accidentally released an experimental compound that selectively kills Protista. It is feared that the chemicals have reached the jet stream and will be widely dispersed. A committee of scientists has been called to Washington to advise the President on what action to take. We will give you more information from the scene of the accident as soon as it becomes available."

You are one of the scientists on the advisory committee to the President. Advise him of the effects this chemical may have on life on earth. He will be interested in what we will have to do to continue life.

Experience This

Egg Drop Soup
Preparation time: 10 minutes

Break ½ ounce (a handful) of Wakame seaweed into small pieces, remove any tough riblike pieces, rinse several times in cold water.
Heat 2 cups clear chicken broth with 2 cups water.
Add the seaweed and season with ¼ teaspoon or more black pepper.
While soup is heating, beat 1 egg and 1 tablespoon water.
When the soup is boiling, slowly pour in the egg, stirring to make egg drops.
Serves four.

Wakame seaweed *(Undaria pinnatifida)* can be purchased in an oriental food store or perhaps in a local supermarket. If not available locally, ask your librarian for the telephone book of the nearest large city and consult the yellow pages. Order by mail.

Questions

1. What kinds of organisms are found in the group known as Protista?
2. What is the common name given to the photosynthetic Protista?
3. Give an example of a parasitic protozoan.
4. What is the major photosynthetic pigment in the algae?
5. What cell-wall chemical makes the diatoms unique?
6. How do the four classes of protozoans differ?
7. What is the cellular organelle of motility found in some algae?
8. Name three products that can be extracted from algae and used in the food industry.
9. Which members of the Protista are consumers? Which are producers?
10. Members of the kingdom Protista are of what cell type?

Chapter Glossary

algae (al′je) Protists that have cell walls and chlorophyll and, therefore, can carry on photosynthesis.

cilia (sil′e-ah) Short, flexible, hairlike filaments that beat rhythmically and propel the cell through the water or bring food to the cell.

colony (kol′o-ne) A cluster of independent cells produced by cell division.

flagella (flah-jel′lah) Long, flexible, hairlike filaments that lash back and forth and draw the cell through the water.

Protista (prō-tis′tah) The kingdom of eukaryotic organisms such as algae and protozoa.

protozoa (prō′′to-zō′ah) Single-celled protists that lack cell walls and chlorophyll.

Chapter Outline

Purpose

You know a plant when you see it. It's the green organism that doesn't run away from you. Their green pigment and nonmobility are just two of the interesting and important characteristics of plants. In this chapter, you will become familiar with some of these characteristics and begin to understand the basis for classifying and naming plants. Plant biologists, botanists, have a scheme that they think expresses the evolutionary advances that have occurred over millions of years. This chapter is organized around that scheme.

For Your Information

Myths of the Mandrake

Plants have been the subject of some fascinating myths, not the least of which are those concerning the mandrake. This is the genus *Mandragora,* a plant native to the Mediterranean area. We are not concerned here with the American mandrake or mayapple. It is from an entirely different family.

The mandrake has a taproot much like a carrot but it sometimes branches to form two leglike structures. With a little imagination, a human shape can be visualized in some of the roots, especially when someone has skillfully carved the root a little. In some of the ancient books on herbs, there are sketches of mandrake plants with the roots looking like male and female nude figures. All of this gave rise to a belief in the curative powers of the plant for all sorts of ailments, including help for the lovelorn. The belief that it counteracted sterility seems to be implied in the biblical account of Leah and Jacob in Genesis 30:14–16. Shakespeare reflected another myth in *Romeo and Juliet* with Juliet's soliloquy in Act IV. It was believed that when one dug up a mandrake plant, the plant gave a terrible shriek and released such a foul odor that anyone in the area would have difficulty withstanding it.

Learning Objectives

- Describe alternation of generations in the mosses.

- Discuss two distinctive ways in which mosses are dependent on water.

- State the advantage of vascular tissue to the ferns and diploid condition of the fern sporophyte.

- Describe the life cycle of gymnosperms.

- State the differences between the two major groups of angiosperms.

Mosses

The first classification of plants was based on their methods of reproduction. Carl von Linné later categorized plants according to the number and position of male parts in flowers. Although almost everyone has looked closely at a **flower,** few people realize that the flower is a structure associated with sexual reproduction in plants. To understand how and why the flower evolved, we need to go back in time and examine the primitive structures that were modified over and over again until the flower came into existence.

One group of plants that show many primitive characteristics are the **mosses.** Mosses grow as a carpet composed of many parts. Each individual moss plant is composed of a central stalk less than five centimeters tall with short, leaflike structures that are the sites of photosynthesis (figure 25.1). If you look at the individual cells in the leafy portion of a moss, you can distinguish the cytoplasm, cell wall, and chloroplasts (figure 25.2). You can also distinguish the nucleus of the cell. This nucleus is haploid, meaning that it has only one set of chromosomes. Every cell in the moss plant is haploid. Although all of the cells have the haploid number of chromosomes (the same as gametes), not all of them function as gametes. Because this plant produces cells that are capable of acting as gametes, it is called the **gametophyte generation,** or the gamete-producing plant. Special structures in the moss, called **antheridia,** produce mobile sperm cells capable of swimming to a female egg cell (figure 25.3). The sperm cells are enclosed within a jacket of cells that opens when the sperm are mature. The sperm swim by the undulating motion of flagella through a film of dew or rainwater toward the female structure, carrying their package of genetic information. Their destination is the egg cell of another moss plant with a different package of genetic information. The egg is produced within a jacket called the **archegonium** (figure 25.4). There is usually only one egg cell in each archegonium. The sperm and egg nuclei fuse, resulting in a diploid cell. The diploid zygote grows, divides, and differentiates into an embryo.

Figure 25.1
Moss Plants.
The insert shows one moss plant greatly enlarged, with both the male and female sex organs.

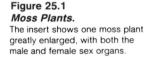

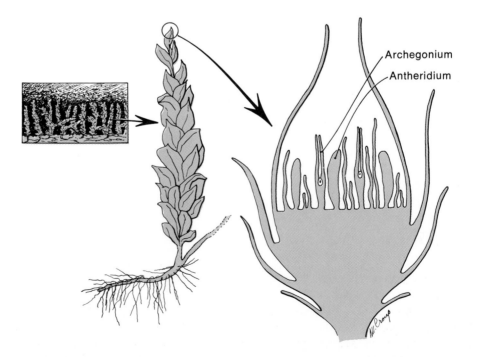

Archegonium

Antheridium

Chapter 25

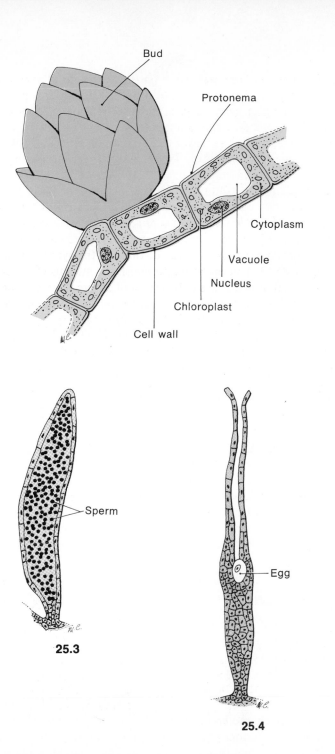

Bud

Protonema

Cytoplasm

Vacuole

Nucleus

Chloroplast

Cell wall

Figure 25.2
Cells of a Moss Plant.
These are typical plant cells: note the large central vacuole and cell wall. One characteristic that separates mosses from the photosynthetic protists is the presence of the individual chloroplasts.

Sperm

25.3

Egg

25.4

Figure 25.3
Antheridium of a Moss Plant.
This male portion of the moss gametophyte plant is composed of a series of cells surrounding the immature sperm cells.

Figure 25.4
Archegonium of a Moss Plant.
This female portion of the moss gametophyte plant is composed of an egg cell inside of a group of jacket cells.

Alternation of Generations

The embryo grows into a structure called the **sporophyte** plant. It is called the sporophyte because it is the **spore**-producing part of the **life cycle.** The sporophyte plant grows a stalk with a swollen tip, the **capsule** (figure 25.5). Inside the capsule, special cells undergo meiosis and form haploid spores. These spores are released and are carried by air currents until they reach the ground. Some land in areas that are too wet, too dry, or too sunny; these will not **germinate** or survive. Others land in a suitable environment and grow into gametophyte plants. Thus, the haploid plant

Figure 25.5
*Sporophyte Generation
of a Moss.*
The fertilized egg cell matures into
the sporophyte plant. The
sporophyte is composed of a stalk
and capsule. Certain cells inside of
the capsule can undergo meiosis
to become haploid spores.

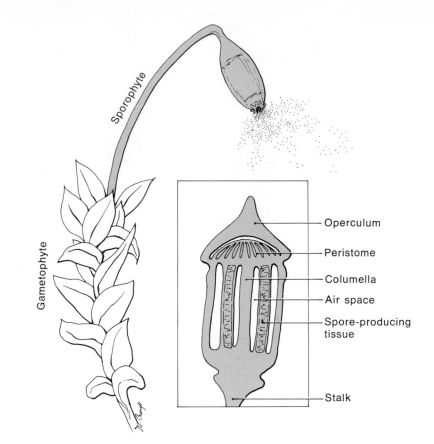

produces the diploid plant, which in turn produces the haploid plant. This aspect of the life cycle is called **alternation of generations** (figure 25.6). The gametophyte generation is dominant over the sporophyte generation in mosses. This means that the gametophyte generation is independent of the sporophyte generation and is more likely to be seen.

Adjustments to Land

What characteristics of mosses cause botanists to consider them the lowest step of the evolutionary ladder in the plant kingdom? First of all, they are considered primitive because they have not developed an efficient way of transporting water throughout their bodies. They must rely on the physical processes of diffusion and osmosis to move materials. The fact that mosses do not have a complex method of moving water limits their size to a few centimeters and their location to moist environments. Another characteristic of mosses points out how closely related they are to their aquatic ancestors. Water is required for fertilization. The sperm cells "swim" from the antheridia to the archegonia. The small size, moist habitat, and swimming sperm are considered characteristics of a primitive organism. In a primitive way, mosses have adapted to a terrestrial niche.

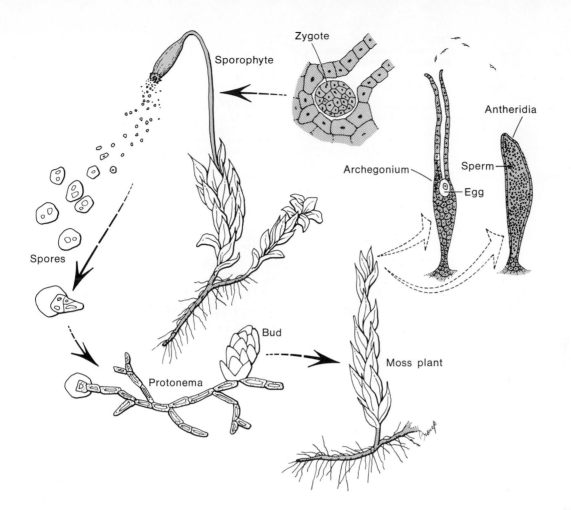

Labels on figure: Zygote, Sporophyte, Antheridia, Archegonium, Sperm, Egg, Spores, Bud, Protonema, Moss plant

Vascular Tissue

A small number of organisms in existence today show some of the unsuccessful directions of evolution. You might think of these evolutionary groups as experimental models that couldn't quite "cut it," but were important steps in the evolution of more successful plants. The advances all concern the greater specialization of some cells, which enables the plant to do a better job of acquiring, moving, and keeping water. Cells that are specialized to perform a particular function are called **tissues.** The tissue important in moving water is called **vascular tissue.** The vascular tissues in plants are of two types, **xylem** tissue and **phloem** tissue. Xylem is a series of hollow cells arranged end to end so that they form a tube. These cells carry water absorbed from the soil to the upper parts of the plant. Associated with these tubelike cells are cells with thickened cell walls that provide strength and support for the plant.

The second type of vascular tissue, phloem, carries the organic molecules produced in one part of the plant to storage areas in others. With the specialization of cells into vascular tissues, there has been a development of specialized parts of plants. **Roots, stems,** and **leaves** are examples of specialized parts that contain vascular

Figure 25.6
Moss Life Cycle.

In this life cycle of a moss, the portions in black represent cells that have haploid nuclei (gametophyte generation). Colored portions represent the diploid cells (sporophyte generation). Notice that the haploid and diploid portions of the life cycle alternate.

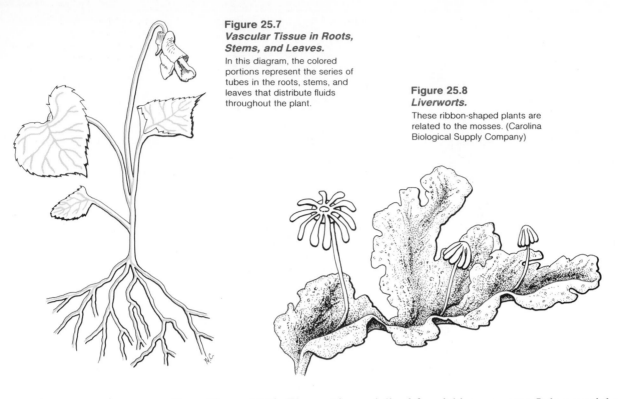

Figure 25.7
Vascular Tissue in Roots, Stems, and Leaves.
In this diagram, the colored portions represent the series of tubes in the roots, stems, and leaves that distribute fluids throughout the plant.

Figure 25.8
Liverworts.
These ribbon-shaped plants are related to the mosses. (Carolina Biological Supply Company)

tissue (figure 25.7). The root is specialized for picking up water. It has special outgrowths called **root hairs** that increase its efficiency in absorbing water from the soil. The stem has well-developed vascular tissue and transports water from roots to leaves as well as organic molecules from the leaves to storage areas in the roots. The leaf is the site of photosynthesis.

The experimental evolutionary groups have some vascular tissue and are links between the nonvascular mosses and the more successful land plants. They also show some specializations that are hints of true roots, stems, and leaves. But they tend to be small plants that require moist environments because they still have swimming sperm. Some of the experimental groups are the liverworts, club mosses, and horsetails.

Liverworts are usually overlooked by the casual observer because they are rather small, low-growing plants composed of a green ribbon of cells (figure 25.8). While they do not have well-developed roots or stems, the leaflike ribbon of tissue is well suited to absorb light for photosynthesis.

Club mosses are a group of low-growing plants that are somewhat more successful than liverworts in adapting to life on land. They have a stemlike structure that holds the leafy parts above the low-growing plants and allows them to better compete for the available sunlight. This allows them to be larger than mosses and not as closely tied to wet areas. While not as efficient as the stems in higher plants, the club moss stem with its vascular tissue is a hint of what's to come (figure 25.9).

Horsetails are very common plants that are considered primitive (figure 25.10). Horsetails have scalelike leaves, a well-developed stem, and a primitive root. Horsetails are interesting plants because they have silicon dioxide as an additive in their cell walls. This glassy addition is in rows of cells, which makes the stem of the plant sharp and rough. Sometimes called *scouring rushes,* horsetails can be used to scour a dirty pan when camping. (We have it on good authority, however, that a handful of sand does a better job of cleaning up the dirty pots and pans!)

Figure 25.9
Club Mosses.

These plants are sometimes called ground pines because of their slight resemblance to the evergreen trees. (Carolina Biological Supply Company)

Figure 25.10
Horsetails.

These unusual plants have silicon dioxide in their cell walls. They are sometimes called scouring rushes. (Carolina Biological Supply Company)

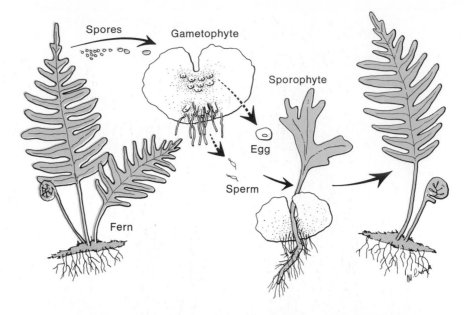

Figure 25.11
Life Cycle of a Fern.
In this life cycle of a fern, the portions in black represent cells that have haploid nuclei (gametophyte generation). Notice that the gametophyte and sporophyte generations alternate. Compare this life cycle with that of the moss (figure 25.6). In the moss, the gametophyte generation is considered dominant, whereas in the fern the sporophyte is dominant.

Spores

Gametophyte

Sporophyte

Egg

Sperm

Fern

Figure 25.12
A Typical Fern.
(Carolina Biological Supply Company)

Ferns

With vascular tissue, plants are no longer limited to wet areas. They can absorb water and distribute it to leaves many meters above the surface of the soil. The ferns are the most primitive vascular plants that are truly successful at terrestrial living. Not only do they have a wider range and greater size than mosses and club mosses, but they have one additional advantage. The sporophyte generation has assumed more importance, and the gametophyte generation has decreased in size and complexity. Figure 25.11 shows the life cycle of a fern. The diploid condition of the sporophyte is an advantage because a recessive gene can be masked until it is combined with another identical recessive gene. In other words, the plant does not suffer because it has one bad gene. The mutant may be a good change, but only time is lost by having it hidden in the heterozygous condition. In a haploid plant, any change, whether recessive or not, shows up. Not only is a diploid condition beneficial to an individual, but the population benefits when many alleles are available for selection.

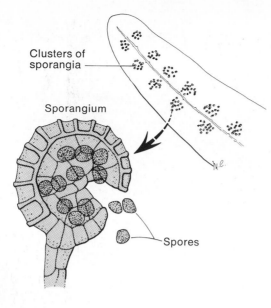

Clusters of
sporangia

Sporangium

Spores

Figure 25.13
Fern Sporangia.
On the back of some fern leaves,
there are some small brown dots.
These dots are the covering layer
of tissue that protects the spore-
producing structures called the
sporangia. The clusters of
sporangia and their covers are
called sori.

Ferns take many forms, including the delicate cloverlike fern, the maidenhair fern of northern wooded areas, the bushy growth of the bracken fern (figure 25.12), and the tree fern known primarily from the fossil record and seen today in tropical areas. In spite of all this variety, however, they still lack one tiny but very important structure—the **seed.** Without seeds, the ferns must rely on spores to spread the species from place to place (figure 25.13).

Gymnosperms

The next advance made in the plant kingdom was the evolution of the **seed.** A seed has an embryo enclosed in a protective covering called the **seed coat.** It also has some stored food for the embryo. The first attempt at seed production is exhibited by the conifers, which are cone-bearing plants such as pine trees. There are two types of **cones.** The male cone produces **pollen.** Pollen are miniaturized male gametophyte plants. These pollen grains are very small dustlike particles. Each pollen grain contains a sperm nucleus. The female cone is larger than the male cone and produces the female gametophyte plants. The archegonia in the female gametophyte contain eggs. The pollen is carried by wind to the female cone. The female cone holds the archegonium in a position to gather the airborne pollen. The process of getting the pollen from the male cone to the female cone is called **pollination.** The production of seeds and pollination by wind are features of conifers that place them higher on the evolutionary ladder than ferns.

Because seeds with their embryo are produced on the surface of the woody, leaflike structure (the female cone), they are said to be naked, or out in the open. The cone-producing plants are sometimes called **gymnosperms,** which means naked seed plants. Producing seeds out in the open makes this very important part of the life cycle vulnerable to adverse environmental influences, such as attack by insects, birds, and other organisms (figure 25.14).

Gymnosperms generally produce needle-shaped leaves. These leaves, or needles, do not all fall off at once. The tree is said to be **nondeciduous.** This term is misleading in that it suggests that the needles do not fall off at all. Actually, the needles are

Figure 25.14
Pine Cone with Seeds.
On the leaflike portions of the cone
are the seeds. Since these seeds
are produced out in the open, they
are aptly named the *naked seed
plants,* gymnosperms.

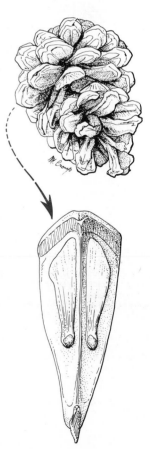

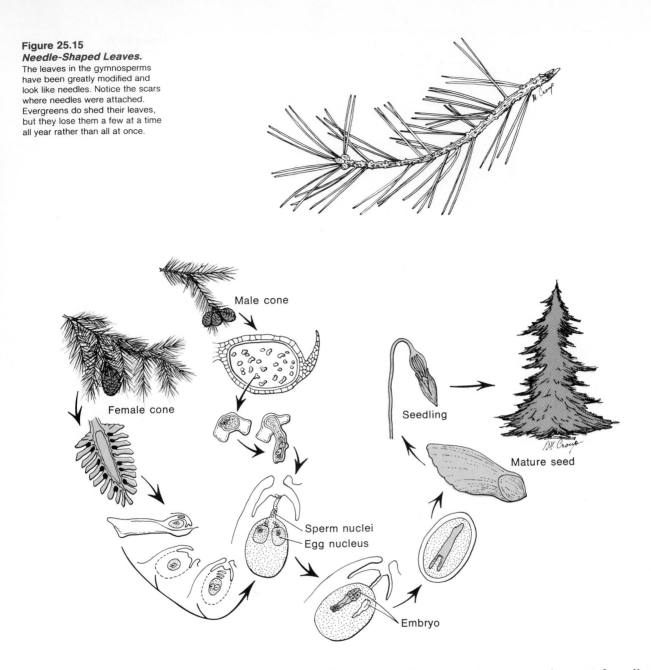

Figure 25.15
Needle-Shaped Leaves.
The leaves in the gymnosperms have been greatly modified and look like needles. Notice the scars where needles were attached. Evergreens do shed their leaves, but they lose them a few at a time all year rather than all at once.

Male cone

Female cone

Seedling

Mature seed

Sperm nuclei
Egg nucleus

Embryo

Figure 25.16
Life Cycle of a Pine.
In this life cycle of a pine tree, the portions in black represent cells that have haploid nuclei (gametophyte generation). Colored portions represent the diploid cells (sporophyte generation). Notice that the gametophyte and sporophyte generations alternate. Compare the life cycle of the pine with the life cycle of the moss (figure 25.6) and of the fern (figure 25.11). Notice the ever-increasing dominance in the sporophyte generation.

constantly being shed a few at a time. Perhaps you have seen the mat of needles under a conifer. The tree retains some leaves year-round, and therefore is called an evergreen (figure 25.15). That portion of the evergreen with which you are familiar is the sporophyte generation. The gametophyte or haploid stages have been reduced to only a few cells. Look closely at figure 25.16 which shows the life cycle of a pine with its alternation of haploid and diploid generations.

Gymnosperms are **perennials.** That is, they live year after year. Unlike **annuals,** which complete their life cycle in one year, gymnosperms take many years to grow from seeds to reproducing adults. The tree gets higher and bigger around each year, continually adding layers of strengthening cells and vascular tissue. As the tree becomes large, the strengthening tissue in the stem becomes more and more important.

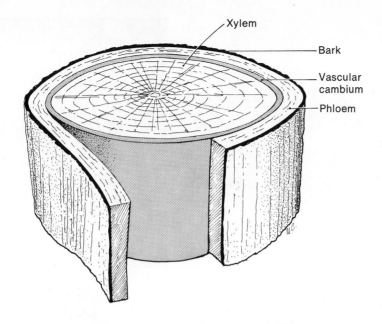

Xylem

Bark

Vascular cambium

Phloem

A layer of cells in the stem, called the **cambium,** is responsible for this increase in size. Xylem tissue is the innermost part of the tree trunk. Phloem is outside of the xylem. The cambium layer of cells is positioned between the xylem and the phloem. These cambium cells go through a mitotic cell division, and two cells form. One cell remains cambium tissue, and the other specializes to form vascular tissue. If the cell is inside the cambium ring, it becomes xylem; if it is outside the cambium ring, it becomes phloem. The one cambium cell then divides again and again. One cell always remains cambium; the other becomes vascular tissue. Thus, the tree constantly increases in diameter (figure 25.17).

The accumulation of the xylem in the trunk of gymnosperms is called **wood.** Wood is one of the most valuable biological resources of the world. We get lumber, paper products, turpentine, and many other valuable materials from gymnosperms. You are already familiar with many examples of gymnosperms, so rather than describe several of these to you, we refer you to figure 25.18.

Angiosperms

The group of plants considered most highly evolved are known as **angiosperms.** This name means that the seeds, rather than being produced naked, are contained within the surrounding tissues of the **ovary.** The ovary and other tissues mature into a protective structure known as the **fruit.** Many of the foods you eat are the seed-containing fruits of angiosperms. Green beans, melons, tomatoes, and apples are only a few of the many edible fruits (figure 25.19).

The flower is the structure that produces the sex cells and enables the sperm cells to get to the egg cells. The important parts of the flower are the **pistil** (female part) and the **stamen** (male part). Look at figure 25.20 and notice that the egg cell is located inside the ovary. Any flower that has both male and female parts is called **perfect.** A flower containing just female or male parts is **imperfect.** Any additional parts of the flower are called accessory structures, since fertilization can be accomplished without them. Many flowers have **accessory structures,** such as **petals, sepals,** and glands, that serve a protective function or increase the probability of

Figure 25.18
Several Gymnosperms.
How many of these do you
recognize? They are from left to
right (top row) *Sequoia, Pinon,
Taxus brevifolia* (Yew), (bottom
row) Torrey Pine, *Cupressus
goveniana* (Cypress), and
Juniperus virginiana (Cedar).

fertilization. Before the sperm cell (contained in the pollen) can join with the egg
cell, there must be some way of getting it to the egg. This is the process called
pollination. Some flowers with showy petals are adapted to attracting insects, who
accidentally carry the pollen to the pistil. Others have become adapted for wind
pollination. The important thing is to get the genetic information from the one parent
to the other parent.

All of the flowering plants have retained the evolutionary advances of previous
groups. That is, they have well-developed vascular tissue with true roots, stems, and
leaves. They have pollen and produce seeds within the protective structure of the
ovary.

There are thousands of kinds of plants that produce flowers, fruits, and seeds.
Almost any plant you can think of is an angiosperm. If you made a list of these
familiar plants, you would quickly see that they vary a great deal in structure and
habitat. The mighty oak, the delicate rose, the pesky dandelion, and the expensive
orchid are all flowering plants. How do we organize this diversity into some sensible
and workable arrangement? Botanists classify all of these plants into one of two
groups, **dicots** and **monocots.** The names dicot and monocot refer to a structure

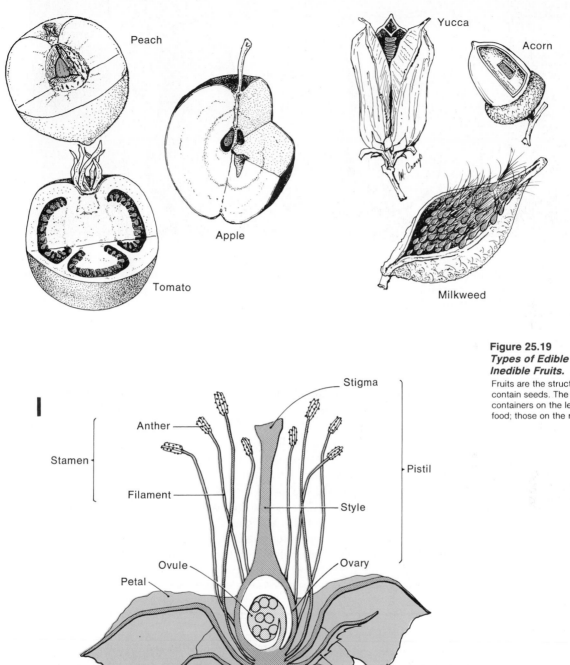

Peach

Apple

Tomato

Yucca

Acorn

Milkweed

Figure 25.19
*Types of Edible and
Inedible Fruits.*
Fruits are the structures that
contain seeds. The seed
containers on the left are used as
food; those on the right are not.

Stigma

Anther

Stamen

Filament

Pistil

Style

Ovule

Ovary

Petal

Sepal

Receptacle

Figure 25.20
Flower.
The flower is the structure in plants
that produces the sex cells. Notice
that the egg is produced within a
structure called the ovary. The
seeds, therefore, will not be naked
as in the gymnosperms, but will be
enclosed in fruit.

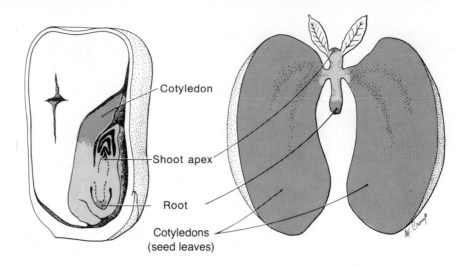

Figure 25.21
Embryos in Dicots and Monocots.
The number of seed leaves attached to an embryo is one of the characteristics used by botanists to classify flowering plants.

Cotyledon

Shoot apex

Root

Cotyledons
(seed leaves)

(called a cotyledon) in the seeds of these plants. If the embryo has two **seed leaves,** the plant is a dicot; those with only one seed leaf are the monocots (figure 25.21). A peanut is a dicot; a lima bean and an apple also are dicots. Grass, lilies, and orchids are all monocots. Even with this separation, the diversity is staggering. Characteristics used to classify and name plants are listed in figure 25.22. This figure also compares the extremes of these characteristics.

Summary

The plant kingdom is composed of organisms able to manufacture their own food by the process of photosynthesis. They have specialized structures for producing the male sex cell (sperm) and the female sex cell (the egg). The relative importance of the haploid gametophyte and the diploid sporophyte that alternate in plant life cycles is a major characteristic used to determine an evolutionary sequence. The extent and complexity of the vascular tissue and the degree to which plants rely on water for fertilization is also used to classify plants as primitive or complex. Within the gymnosperms and the angiosperms, the methods of production, protection, and dispersal of pollen are used to name and classify the organisms into an evolutionary sequence. Based on the information available, mosses are the most primitive. Liverworts, club mosses, and horsetails are experimental models. Ferns, seed-producing gymnosperms, and angiosperms are the most advanced and show development of roots, stems, and leaves.

Consider This

Some people say the ordinary "Irish" potato is poisonous when the skin is green. They are, at least, partly correct. A potato develops a green skin if the potato tuber grows so close to the surface of the soil that it is exposed to light. The alkaloid solanine also develops under this condition and may be present in toxic amounts. Eating such a potato raw may be dangerous. However, cooking breaks down the solanine molecule and makes that potato as edible and tasty as any other.

The so-called "Irish" potato has an interesting history in its relationship to humans. Its real country of origin is only one of these items. Check with your local library and find out about the potato and its relatives. Are all related organisms edible? Where did this group of plants develop? Why is it called the "Irish" potato?

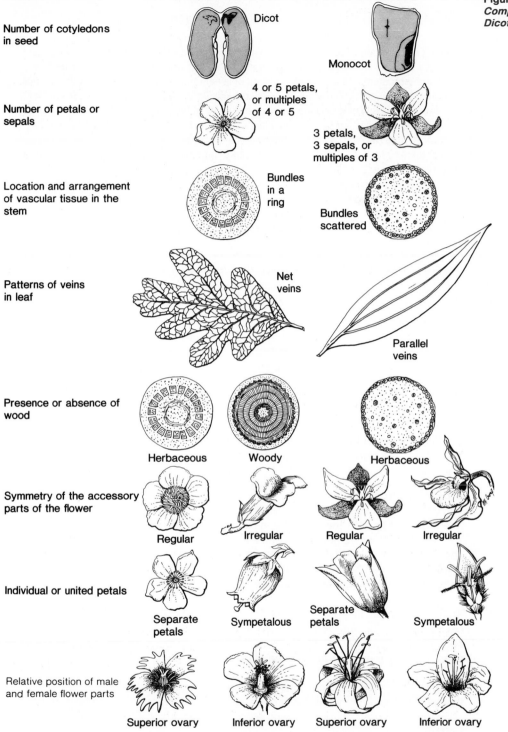

Figure 25.22
Comparison of Structures in Dicots and Monocots.

Number of cotyledons in seed

Dicot

Monocot

Number of petals or sepals

4 or 5 petals, or multiples of 4 or 5

3 petals, 3 sepals, or multiples of 3

Location and arrangement of vascular tissue in the stem

Bundles in a ring

Bundles scattered

Patterns of veins in leaf

Net veins

Parallel veins

Presence or absence of wood

Herbaceous

Woody

Herbaceous

Symmetry of the accessory parts of the flower

Regular

Irregular

Regular

Irregular

Individual or united petals

Separate petals

Sympetalous

Separate petals

Sympetalous

Relative position of male and female flower parts

Superior ovary

Inferior ovary

Superior ovary

Inferior ovary

Experience This   *New Plants from Old*

Anyone can have a variety of flowering plants with little expense by spending a small amount of time. One need only take a sharp knife and cut a piece of the stem of a desirable plant (with your neighbor's permission!). Woody stems are apt to be more difficult to root than plants with soft stems. Place some soil in a container, make a hole in the soil with a pencil, and insert the cut stem. The stem piece should include the tip and about five leaves. The container can be anything as long as there are a few holes in the bottom for drainage. Cover the container tightly with a transparent plastic bag to prevent water loss through the leaves before the roots develop. Also trim the lower leaves for the same purpose. Water the plant once and keep out of direct sun. Roots may take ten days or longer to develop. If the soil should feel dry to the touch, water it again.

If potting soil or vermiculite is available at a hardware or garden store, it might be better than the soil you could dig locally. Another aid would be to dip the cut end of the stem into some rooting hormone before planting. However, people have had success for years without these refinements.

Questions

1. What characteristics distinguish algae in the kingdom Protista from the organisms of the kingdom Plantae?
2. List three characteristics shared by mosses, ferns, and angiosperms.
3. What were the major advances that led to the development of angiosperms?
4. How is a seed different from pollen, and how do both of these differ from a spore?
5. How are cones and flowers different?
6. How are cones and flowers similar?
7. What is the dominant generation in mosses, ferns, gymnosperms, and angiosperms?
8. What is the difference between the xylem and the phloem?
9. Ferns have not been as successful as gymnosperms and angiosperms. Why?
10. What is the significance of the cambium tissue in perennials?

Chapter Glossary

accessory structures (ak-ses′o-re struk′churs) The parts of some flowers that are not directly involved in gamete production.

alternation of generations (awl″tur-na′shun uv jen″uh-ra′shunz) The cycling of a diploid sporophyte generation and a haploid gametophyte generation in plants.

angiosperms (an′je-o″spurmz) Plants that produce flowers and fruits.

annual (an′yu-uhl) A plant that completes its life cycle in one year.

antheridia (sing., antheridium) (an″thur-id′e-ah) The structures in lower plants that produce sperm.

archegonium (ar″ke-go′ne-um) The structure in lower plants that produces eggs.

cambium (kam′be-um) A tissue in higher plants that produces new xylem and phloem.

capsule (kap′sūl) Part of the sporophyte generation of mosses that contains spores.

cone (kōn) A reproductive structure of gymnosperms that produces pollen in males or eggs in females.

dicot (di′kot) An angiosperm whose embryo has two seed leaves.

flower (flow′er) A complex structure made from modified stems and leaves. It produces pollen in the males and eggs in the females.

fruit (froot) The structure in angiosperms that contains seeds.

gametophyte generation (gă-me′to-fit jen″uh-ra′shun) The haploid generation in plant life cycles. They produce gametes.

germinate (jur′min-āt) To begin to grow (as from a seed).

gymnosperms (jim′no-spurmz) Plants that produce their seeds in cones.

imperfect flowers (im′′pur′fekt flow′urs) Flowers that contain either male or female reproductive structures, but not both.

leaves (lēvs) Specialized portions of higher plants that are the sites of photosynthesis.

life cycle (līf sī′kl) The series of stages in the life of any organism.

monocot (mon′o-kot) An angiosperm whose embryo has one seed leaf.

mosses (mos′sez) Lower plants that have a dominant gametophyte generation, swimming sperm, spores and no vascular tissue.

nondeciduous (non′′de-sid′yu-us) Refers to trees that do not lose their leaves all at once.

ovary (o′vah-re) The female structure that produces eggs.

perennial (pur-en′e-uhl) A plant that requires many years to complete its life cycle.

perfect flowers (pur′fekt flow′urs) Flowers that contain both male and female reproductive structures.

petals (pĕ′tuls) Modified leaves of angiosperms; an accessory structure of a flower.

phloem (flo′em) One kind of vascular tissue found in higher plants. It transports food materials from the leaves to other parts of the plant.

pistil (pis′til) The female reproductive structure in flowers.

pollen (pol′en) The male gametophyte in gymnosperms and angiosperms.

pollination (pol′′ī-na′shun) The transferring of pollen in gymnosperms and angiosperms.

root (root) A specialized structure for the absorption of water and minerals in higher plants.

root hairs (root hārs) Tiny outgrowths of roots that improve the ability of plants to absorb water and minerals.

seed (seed) A specialized structure produced by gymnosperms and angiosperms that contains the embryo sporophyte.

seed coat (seed kōt) A protective layer around seeds.

seed leaves (seed lēvs) Embryonic leaves in seeds.

sepals (se′pals) Accessory structures of flowers.

spores (spōrz) Haploid structures produced by sporophytes.

sporophyte (spōr′o-fit) The diploid generation in the life cycle of plants that produces spores.

stamen (sta′men) The male reproductive structure of a flower.

stem (stem) The upright portion of a higher plant.

tissue (tish′yu) A group of specialized cells that work together to perform a particular function.

vascular tissue (vas′kyu-lar tish′yu) Specialized tissue that transports fluids in higher plants.

wood (wood) The xylem of gymnosperms and angiosperms.

xylem (zi′lem) A kind of vascular tissue that transports water from the roots to other parts of the plant.

Animalia

■ Chapter Outline

■ Purpose

Processes such as respiration, cell division, reproduction, and heredity are common to all forms of life. This was true for the first living organisms and is true for the many types of organisms that have evolved since. The evolution of the great variety of species was presented previously (chapter 21). At that point, only a passing reference was made to the various types of animals. This chapter will present the major groups of animals, giving some of the traits common to the members of each group.

For Your Information

Several years ago in Yosemite National Park, as a result of an administrative decision, a large number of black bears were systematically killed by park rangers. The park administration believed that the national park was set aside for people, not bears.

The Forest Service in Wyoming schedules clear cutting in one of the few remaining grizzly bear habitats. This forestry practice forces the grizzlies into areas where their activities are more likely to get them killed as dangerous nuisances.

Wolves introduced into the Upper Peninsula of Michigan by the Michigan Department of Natural Resources were killed by some outraged citizens. Those in favor of this introduction believe the stories of unprovoked attacks on humans are myths.

Learning Objectives

- Name eight phyla in the animal kingdom.

- State distinguishing characteristics of each phylum.

- Name seven classes of vertebrates and an example of each.

- State distinguishing characteristics of each of the seven classes.

- Describe the amniotic egg and discuss its importance.

Kinds of Animal Life Cycles

For this text, we will define animals as many-celled organisms that require food as an energy source. The life cycle of an animal has a number of stages. Within the cycle, reproduction may be either sexual or asexual. Asexual reproduction may occur at different stages of the life cycle and differs from sexual reproduction in that there is no mixing of genetic material. Organisms that reproduce asexually produce carbon copies of themselves in one of a number of ways. In **budding,** one method of asexual reproduction, an animal grows an offspring off its main body.

In sexual reproduction, the adult animal is capable of producing gametes. In a typical animal life cycle, the adult is capable of reproducing sexually to form a zygote (figure 26.1). In some species these zygotes undergo development and emerge fully formed. For example, a newly born human baby has the same basic features as the adult. In other forms of animals, the young have little resemblance to the adult. Such young are known as **larvae.** A larva is an incompletely developed animal; it must undergo many changes to develop into an adult. For instance, a caterpillar is the larval stage in the life cycle of a butterfly (figure 26.2).

Levels of Complexity

The first forms of animal life were not much more advanced than the one-celled organisms. In these first many-celled animals, each cell lived an almost independent life. As the animals evolved, certain cells of their bodies began to specialize and perform a particular function. Such a change resulted in the development of **tissues.** A tissue is a group of cells similar in function and structure. The cells that specialize in body movement make up muscle tissue. The cells that line the intestine are specialized for digestion and absorption.

As animals continued to evolve, **organs** developed. Organs are body units composed of many different tissues that operate together to perform a particular function. The stomach, brain, liver, and heart are examples of organs. Further evolution resulted in collections of organs known as **organ systems.** The circulatory system, the digestive system, the reproductive system, and the nervous system are examples. The present-day animal kingdom contains members displaying these various stages of evolution. One of the simplest kinds of animals present today are the **sponges.**

Figure 26.1
Animal Life Cycle.

The male and female produce the haploid number of chromosomes in sperm and eggs. These haploid cells combine to form a zygote, and another life begins.

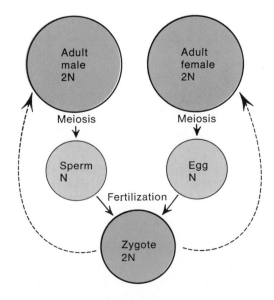

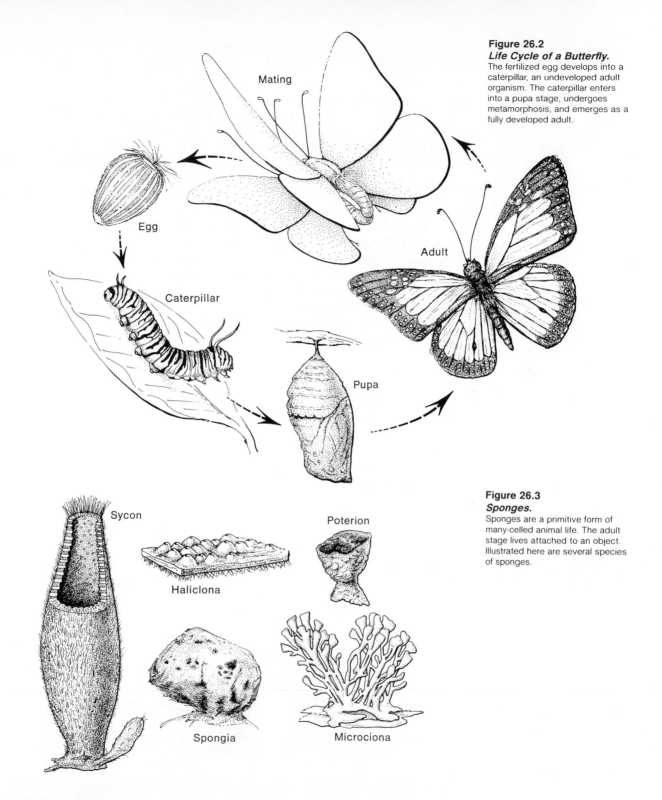

Mating

Egg

Caterpillar

Pupa

Adult

Sycon

Haliclona

Poterion

Spongia

Microciona

Porifera: Sponges

Most people think of a sponge as a pastel-colored, rectangular-shaped object that
comes wrapped in plastic. If you have seen natural sponges, you know they are
brownish-colored, irregular-shaped objects; you probably didn't think of them as
members of the animal kingdom. Sponges are one of the most primitive forms of
animal life (figure 26.3). They are mainly found in saltwater environments, but a

Figure 26.4
Circulation of Water through a Sponge.
Flagellated cells line the canals within a sponge and cause a constant flow of water through the canals.

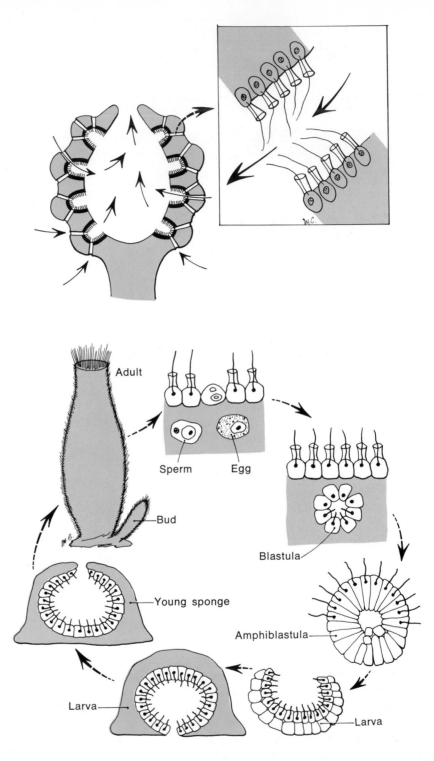

Figure 26.5
Life Cycle of a Sponge.
Sexual reproduction in sponges occurs when the flagellated sperm cell unites with an egg cell to form a zygote. The zygote undergoes several stages of development as it grows into an adult.

few species are found living in unpolluted fresh water. The adult sponge is attached to some object and does not have a great deal of cell specialization. The interior of a sponge is composed of numerous interconnecting tubes (figure 26.4). Lining these tubes are a number of **flagellated cells.** The flagella of these cells cause a current of water to flow through the tubes. Each sponge cell gets the nutrients it needs from

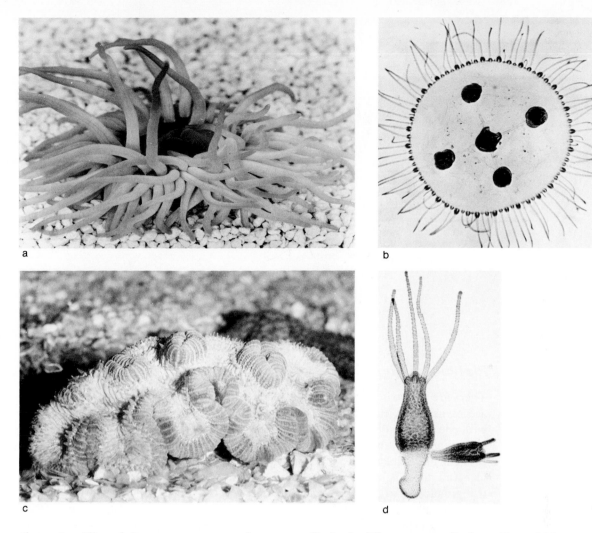

the water. The adult sponge can reproduce asexually by budding, or sexually by producing sperm and eggs. The fertilized egg gives rise to a free-swimming larva. This larva develops into a mature adult. Figure 26.5 illustrates the life cycle of a sponge.

Figure 26.6
Cnidaria.
(a) Sea anemone. (b) Jellyfish.
(c) Coral. (d) *Hydra.* (Carolina
Biological Supply Company)

Cnidaria: Jellyfish, Coral, and Sea Anemones

People who have been to the ocean have likely seen one or more species of **Cnidaria** (formerly the coelenterates). Common members of the Cnidaria include the *Hydra,* jellyfish, coral, and sea anemones (figure 26.6).

All Cnidaria have a single opening that leads into a saclike interior (figure 26.7). Surrounding the opening are a series of **tentacles.** These long, flexible, armlike tentacles possess specialized cells that can sting and paralyze small organisms—Cnidaria are carnivorous animals. The tentacles slowly move the paralyzed organisms into the animal. The Cnidaria, as evidenced by the tentacles, exhibit some degree of specialization.

A typical cnidarian has a free-swimming adult stage called the **medusa.** It gives rise by sexual reproduction to a **polyp** stage. This polyp is attached and reproduces by budding. Figure 26.7 illustrates the life cycle of a cnidarian.

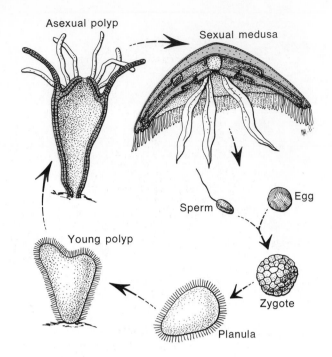

Figure 26.7
Life Cycle of a Jellyfish.
The polyp stage of a jellyfish reproduces asexually, and these organisms develop into the medusa stage. The medusa reproduces sexually, and the resulting zygote develops into a planula. The planula matures into the polyp.

Asexual polyp

Sexual medusa

Sperm

Egg

Young polyp

Zygote

Planula

Platyhelminthes: Flatworms

There are approximately ten thousand species of **flatworms,** and two-thirds of them are parasites. With such a variety, it is impossible to represent all the flatworms with one species. There are three basic kinds of flatworms: the free-living flatworms (like *Planaria*), tapeworms, and flukes (figure 26.8).

The blood fluke is a common human parasite in much of Asia, Africa, and tropical America. The adult lives in the blood vessels of the intestines and bladder. Here it reproduces, and the fluke's eggs are passed out with the human solid waste material or urine. If the eggs are deposited in water, they will hatch into a free-swimming larva. The larva enters the body of a snail, undergoes development, and asexually reproduces within the snail. A stage produced by asexual reproduction eventually leaves the snail and swims about in the water. If this stage comes into contact with the skin of a human, it drills through and enters a blood vessel. A sexually mature fluke develops, and the life cycle begins again (figure 26.9). The blood fluke may cause diarrhea, abdominal pain, and anemia in the infected human.

Each species of fluke requires a particular type of snail for the development of its larva, and a particular type of animal for the development of its adult stage. If any stage in the life cycle of a fluke enters the wrong animal, development will not proceed and that stage of the fluke dies. "Swimmer's itch" is caused when a larval stage of a duck blood fluke enters the human body. The larva cannot mature within the human body, but it does live for a few days before dying. This temporary infection causes an itchy, red rash.

Aschelminthes: Roundworms

Most people are not aware of the large number of **roundworms** present in the world; nor are they aware of the economic losses caused by roundworms. If you were to take a shovelful of rich soil, you would find millions of them. Each year in the United States, roundworms cause several billion dollars worth of damage to crops and livestock.

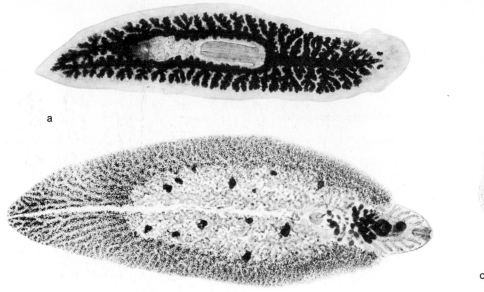

a

b

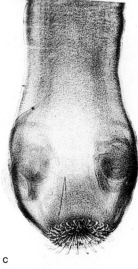

c

Figure 26.8
Flatworms.

(a) *Planaria.* (b) Fluke.
(c) Tapeworm. (Carolina Biological
Supply Company)

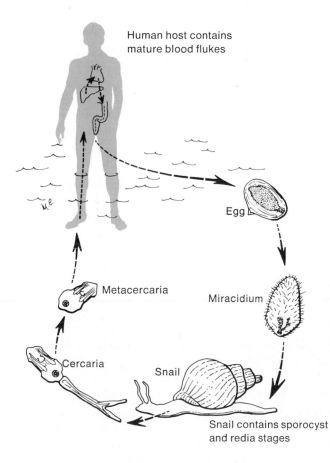

Human host contains
mature blood flukes

Egg

Miracidium

Snail

Snail contains sporocyst
and redia stages

Cercaria

Metacercaria

Figure 26.9
*Life Cycle of the Human
Blood Fluke.*

The fertilized blood fluke eggs
pass out of the human and
develop into the miracidium stage.
This stage enters a snail for
additional development (sporocyst
and redia), and the larvae leave the
snail as the cercaria stage. This
stage bores through the skin of a
human and then develops into a
mature organism.

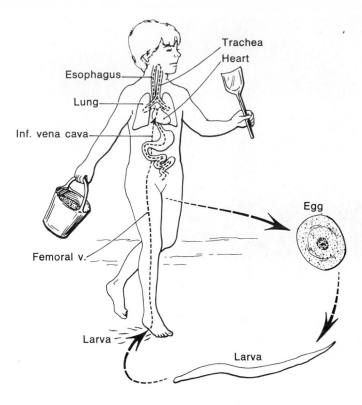

Figure 26.10
Life Cycle of a Hookworm.
Fertilized hookworm eggs pass out of the human and develop into larvae. The larva bores through the skin of a human and develops into an adult.

The roundworms are found throughout the world, inhabiting both soil and water environments. They can be found as free-living forms or as parasites. All species of vertebrates can be infected by some type of roundworm parasite. Figure 26.10 illustrates a roundworm parasite that infects humans. Other common roundworms are pinworms, *Ascaris,* and *Trichinella* (figure 26.11).

Annelids: Segmented Worms

The segmented worms, **annelids,** are known to almost everyone as earthworms, leeches, and clam worms; but they are especially well known to anglers because the common earthworm is a common fish bait. Most moist soils support at least some species of earthworms. In saltwater areas, the clam worm is commonly found in the sand and is also used for fishing (figure 26.12).

The members of this group of animals have evolved some fairly complex forms of specialization. There is a well-developed digestive system, circulatory system, nervous system, muscular system, reproductive system, and excretory system (figure 26.13). Annelid worms play a very important role in terrestrial ecosystems. Earthworms eat the soil as they burrow through it. The organic foods found in the soil are digested by the worm, and the undigestible materials are eliminated from the digestive system unchanged. Earthworms also come up to the surface to feed on fallen leaves. This loosens the soil and turns it over, and the organic material from dead plants and animals is mixed into it. All of these activities improve soil quality and bring about better plant growth.

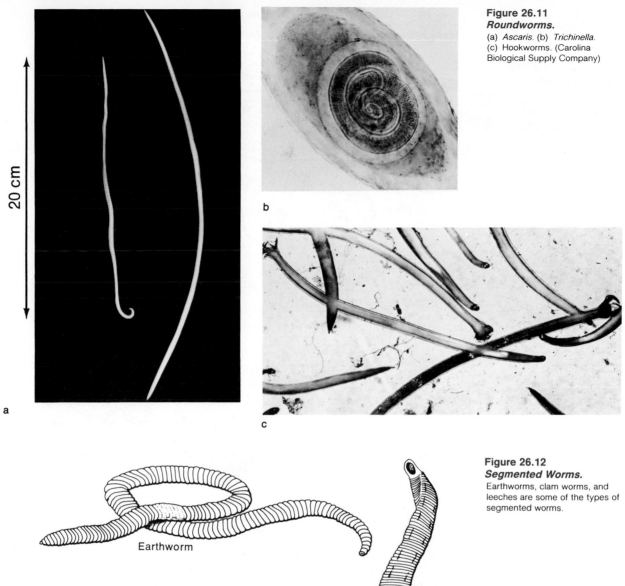

20 cm

a

b

c

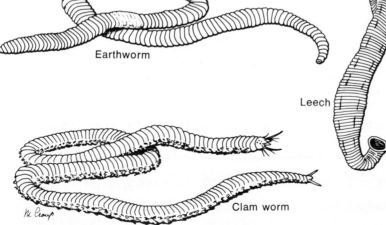

Earthworm

Leech

Clam worm

M. Croup

Figure 26.12
Segmented Worms.
Earthworms, clam worms, and
leeches are some of the types of
segmented worms.

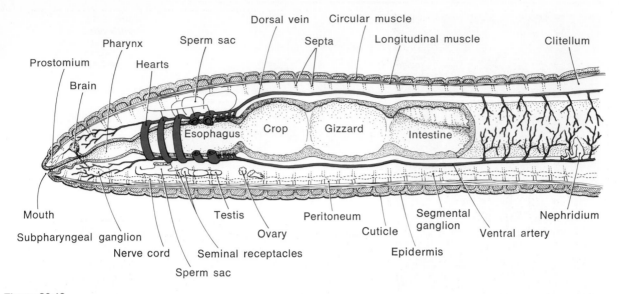

Figure 26.13
Internal Anatomy of the Earthworm.
This drawing shows the various organs within an earthworm.

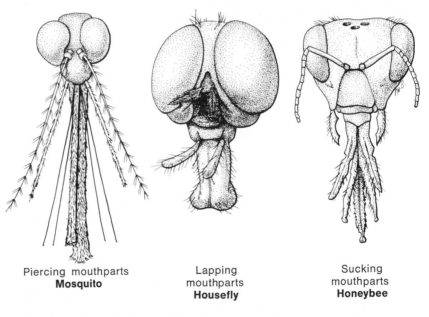

Figure 26.14
Arthropod Mouthparts.
Arthropods exhibit a number of adaptive traits, such as the piercing (mosquito), lapping (fly), and sucking (honeybee) mouthparts.

Piercing mouthparts
Mosquito

Lapping mouthparts
Housefly

Sucking mouthparts
Honeybee

Arthropods: Insects and Their Relatives

All **arthropods** have jointed legs and a hard **exoskeleton** on the outside of their body for protection. They have well-developed mouthparts and possess a highly developed nervous system (figure 26.14). This group has a greater diversity of animals than any other single phylum of animals. The arthropods are found in every type of habitat. They contain more species than are found in the rest of the animal kingdom. Figure 26.15 shows examples of the five major classes of arthropods.

Such common animals as crayfish, lobsters, barnacles, shrimp, and crabs are examples of **Crustacea** (figure 26.16). Several animals in this class are a source of food for humans. The crustaceans have gills that serve as respiratory organs. There

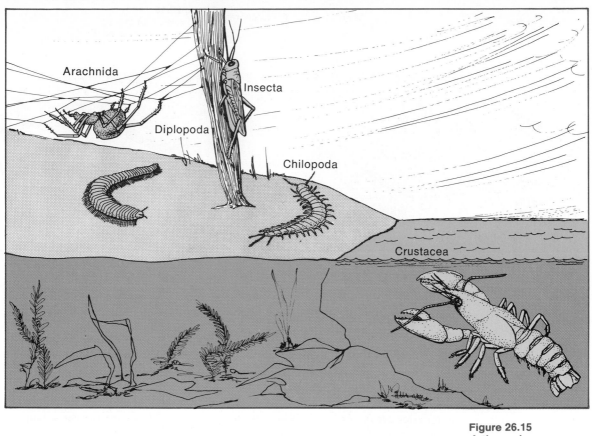

Figure 26.15
Arthropods.
Pictured are some of the types of
arthropods.

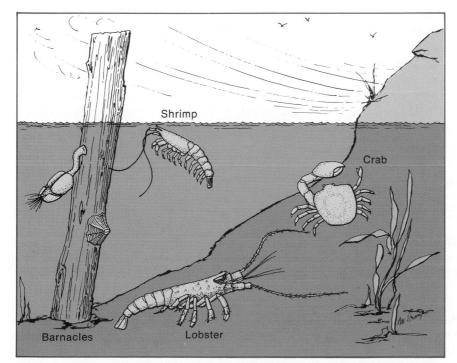

Figure 26.16
Crustaceans.
Crustaceans are a major grouping
of animals within the arthropods.

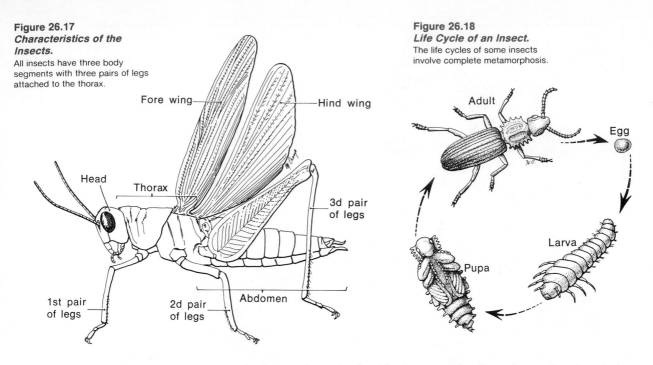

Figure 26.17
Characteristics of the Insects.
All insects have three body segments with three pairs of legs attached to the thorax.

Fore wing — — Hind wing

Head

Thorax

3d pair of legs

1st pair of legs

2d pair of legs

Abdomen

Figure 26.18
Life Cycle of an Insect.
The life cycles of some insects involve complete metamorphosis.

Adult

Egg

Larva

Pupa

are twenty-six thousand species in this class, ranging from free-swimming shrimp to fixed types such as barnacles. Barnacles attach themselves to the hulls of saltwater ships and slow them down.

The **insects** are the largest single class within the animal kingdom. Some characteristics are common to all insects (figure 26.17). They all have three body segments—head, thorax, and abdomen—and three pairs of legs attached to the thorax. A series of breathing tubes called **tracheae** serve to exchange gases. Insects are male or female. The young in some species develop as larva and after a period of time go into a **pupa** stage, undergoing additional development and emerging from the pupa as a sexually mature adult (figure 26.18). The change from egg to larva to pupa to adult is called **complete metamorphosis.**

Many people confuse the insects with the **arachnids.** The arachnids include spiders, ticks, and scorpions (figure 26.19). This group of arthropods has four or five pairs of legs, distinguishing them from the insects, which possess only three pairs of legs.

Mollusks

The **mollusks** include many common species of animals, such as the oyster, clam, snail, octopus, and squid (figure 26.20). All the members of this phylum have a soft body that houses the heart, digestive system, excretory system, nervous system, reproductive system, and other internal organs. A tissue called the **mantle** surrounds the body (figure 26.21). In many species, this mantle contains specialized glands that secrete material that forms a shell.

A common group of mollusks are the **bivalves**—animals that have a two-part shell. Many of these species are valued as food for humans. Most bivalves remain in one place and feed by circulating large amounts of water and filtering out the food. Oysters, clams, and scallops are examples of the bivalves.

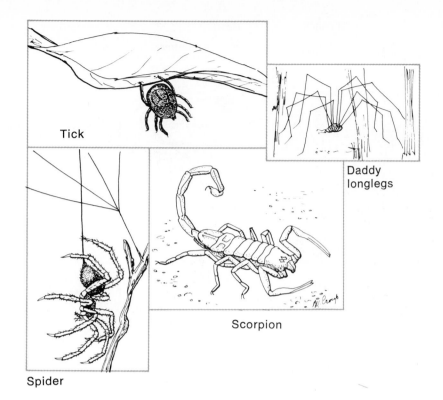

Figure 26.19
Arachnids.
Pictured are some of the types of arachnids. Arachnids are characterized by having four pairs of legs.

Tick

Daddy longlegs

Spider

Scorpion

Figure 26.20
Mollusks.
Pictured are some of the types of mollusks. Soft bodies are a characteristic of the mollusks.

Squid

Octopus

Oyster

Snail

Clam

Scallop

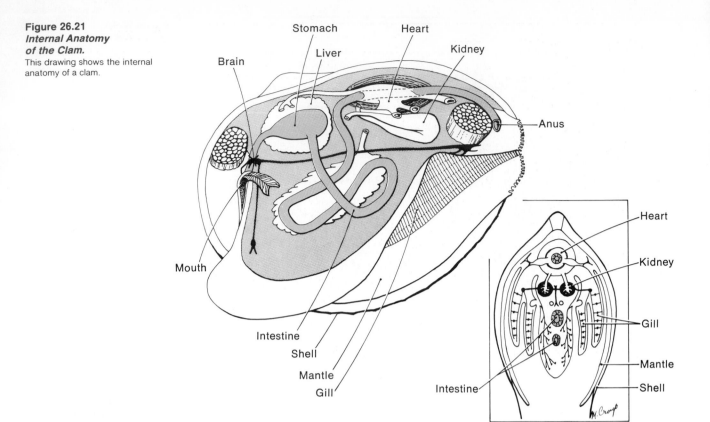

Figure 26.21
Internal Anatomy of the Clam.
This drawing shows the internal anatomy of a clam.

Cephalopods, mollusks with a distinct head, are the monsters of movies and television. The octopus and squid, in reality very timid animals, are cephalopods. These animals have a large head and a number of tentacles. The world's largest **invertebrate** animals (animals without a backbone) are members of this class. There are records of squid exceeding 20 meters in length and weighing nearly 9,000 kilograms.

Echinoderms: Starfish

Echinoderm means spiny skin. A common characteristic of this phylum of animals is an **endoskeleton.** The skeleton inside the body has a number of spiny projections. The echinoderms are found only in saltwater habitats.

Typical echinoderms are the starfish, serpent stars, sand dollars, sea urchins, sea cucumbers, and crinoids (figure 26.22). Characteristics of these animals include digestive, reproductive, and nervous systems. The **water-vascular system** is found only in the echinoderms (figure 26.23). This system is a series of tubes that radiates throughout the animal. The water enters through an opening and travels through canals. The **tube feet** located on the arms are used like little suction cups. By controlling the water pressure in the water-vascular system, the echinoderm can suck itself onto surfaces. By fixing and releasing these tube feet, it can move through its environment or capture food.

In the echinoderms, the sexes are separate and fertilization occurs outside the female's body. The eggs hatch into a larval form that must undergo metamorphosis to become an adult.

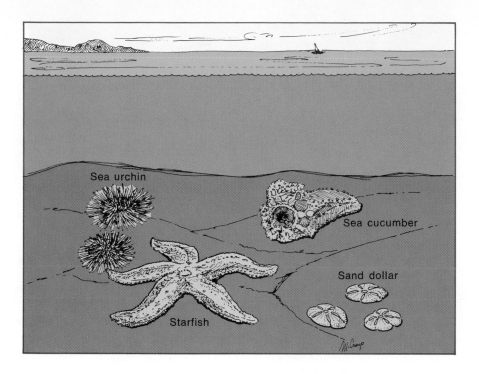

Figure 26.22
Echinoderms.
Pictured are some of the types of echinoderms.

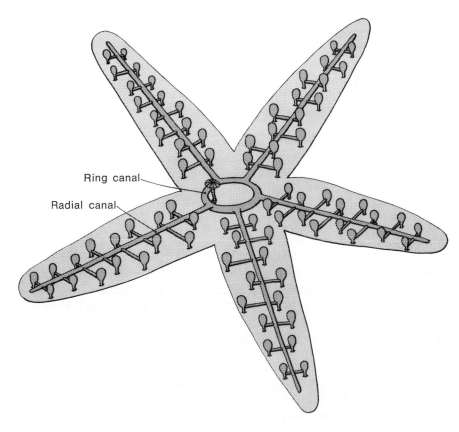

Figure 26.23
Water-Vascular System of a Starfish.
The water-vascular system is a characteristic unique to the echinoderms.

Figure 26.24
*Lamprey Feeding
on a Host Fish.*
Lampreys are parasitic organisms.

Chordates

The **chordates** are animals that at some time in their life have all three of the following characteristics: **gill slits, dorsal hollow nerve cord,** and **notochord.** Gill slits are openings that allow water to pass from the throat to the outside. In the aquatic chordates, these slits are a part of the respiratory system. They allow for the movement of water over the gills and the exchange of gases between the animal's blood and the water. The terrestrial chordates have gill slits early in their embryonic stage, but the slits do not carry over to the adult stage. In land animals, a pair of gill slits has become modified into **eustachian tubes,** connecting the ear to the throat. In higher forms of animals, the dorsal hollow nerve cord makes up the brain and spinal cord. The notochord is a stiff, rodlike structure along the back of the animal. It provides support for the animal. All of the animals presented in the remainder of this chapter are chordates.

The vertebrates are a group of chordates that includes most of the animals we see in our daily lives. In the **vertebrates,** the notochord is found in the developing embryo but is replaced by a series of vertebrae that collectively make up the backbone.

Agnatha: Jawless Fish

The **Agnatha** are the most primitive class of living vertebrates. An example of these animals is the sea lamprey (figure 26.24). These fishlike animals have no jawbones, but they do possess a **cartilaginous** skeleton. Cartilage is a dense type of supporting tissue that does not have the mineral deposits found in the harder supporting tissues known as bone. The fins of these animals are not in pairs. A large **caudal** (tail) **fin** is their main method of movement.

The adult lamprey spawns in rivers, and the eggs hatch into a larval form. This stage filters microscopic food particles out of the water. The larval stage remains in the river for as long as seven years before changing into the adult form. The opening of the locks in the canals around Niagara Falls to allow oceangoing vessels to enter the Great Lakes also allowed saltwater lamprey to enter. The parasitic lamprey adjusted to the freshwater environment. Since that time, millions of dollars have been spent to control this pest.

Ray

Shark

Figure 26.25
Sharks and Rays.
Pictured are some of the types of
cartilaginous fishes.

Chondrichthyes: Sharks and Rays

Movies and television have introduced most of us to sharks. But it would not be correct to think that all sharks are large man-eaters. Only seven of the forty species are known to have attacked people. The adults of some species seldom grow longer than three feet. Other forms of sharks can grow to over fifty feet long. Some large species of sharks feed on microscopic marine organisms. The sharks have a highly developed cartilaginous skeleton and paired fins. The sharks and rays are the principal members of the class **Chondrichthyes** (figure 26.25).

Osteichthyes: Bony Fish

Whenever the term fish is used, it usually applies to the group of animals called **Osteichthyes,** the **bony fish** (figure 26.26). Bony fish evolved in fresh water, and later some forms adapted to a saltwater environment. The gill slits are covered by a protective organ called the **operculum.** The body is covered by overlapping scales. The fish produce a slimy mucus that reduces friction and protects their bodies from infection. Many a fish has slipped through the fingers of an angler because of this mucus. Most of the bony fish have an **air bladder,** a sac that changes in size and regulates the density of the fish. This allows fish to remain at a given level without expending large amounts of energy to keep from sinking or floating. Sharks have no air bladder; they must swim constantly or they will sink.

Figure 26.26
Fishes.

Pictured are some of the types of fishes. They inhabit saltwater or freshwater environments. Some species can survive in either environment.

Amphibians: Frogs, Toads, and Salamanders

The **amphibians** were the first of the vertebrates to inhabit the terrestrial areas of the earth. The larval amphibians are almost fishlike in nature. They are freshwater organisms that possess external gills and a caudal fin for locomotion. Through a gradual series of changes, the gills disappear and lungs develop. Four appendages appear that develop into legs. Such changes enable the adult animal to partially adapt to life on land. However, the reproduction method of the amphibians is not suitable for land. As a result, they are bound to the water during the reproductive period of their lives. Figure 26.27 illustrates the life cycle of a frog. Frogs, toads, and salamanders are types of amphibians (and do not cause warts) (figure 26.28).

Reptiles

The **reptiles** were the first vertebrate animals to fully adapt to the terrestrial environment. This resulted from the evolution of the **amniotic egg** and a method of internal fertilization (figure 26.29). The **amnion** is a membrane that encloses a fluid-filled area surrounding the developing embryo. The embryo also develops an **allantois,** which is a storage area for waste material. Blood vessels in the allantois are used to exchange gases between the embryo's circulatory system and the atmosphere. A **yolk sac** is a membrane that develops from the young and surrounds the yolk of the egg. The **chorion** is a membrane that completely encloses the developing young and the other membranes. These membranes and a protective shell protect the embryo during its development. Internal fertilization freed reptiles from returning to water to reproduce.

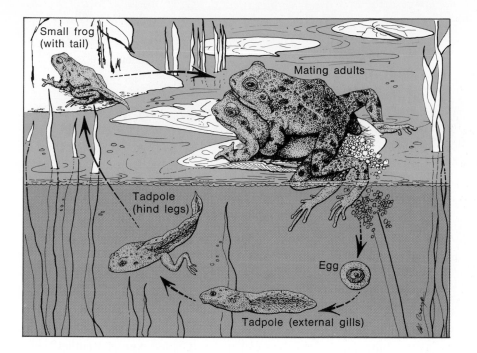

Figure 26.27
Life Cycle of a Frog.
Fertilized frog egg cells develop into the tadpole stage. This stage undergoes metamorphosis and develops into an adult frog.

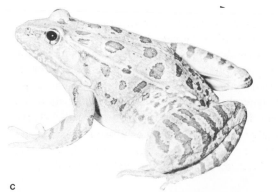

Figure 26.28
Amphibians.
Pictured are some of the types of amphibians. (a) Salamander. (b) Toad. (c) Grass Frog. (Carolina Biological Supply Company)

Figure 26.29
Amniotic Egg.

Amniotic eggs provide nutrients
and protection to the developing
embryo. This type of egg allows
embryonic development outside an
aquatic medium.

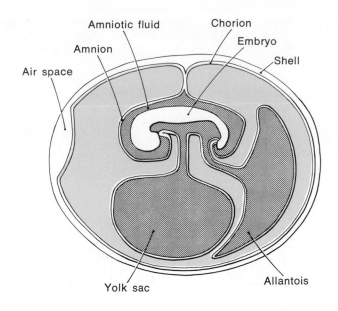

Since the reptiles were the first vertebrates to occupy the land, they filled many niches. As a result, there was a time called the "Golden Age of Reptiles" when reptiles were the dominant land species (figure 26.30). A series of geological and evolutionary events caused the decline of the reptiles. Today the reptiles are fewer in number and occupy only a small number of niches. Snakes, lizards, turtles, alligators, and crocodiles are all that remain as representatives of this class (figure 26.31).

Aves: Birds

The **birds** are a class (**Aves**) of vertebrates that are capable of flying. A constant body temperature provides an internal environment that produces the large amounts of energy required for flying. A high metabolic rate rapidly converts food into muscular energy. A highly developed circulatory system moves large amounts of oxygen throughout the body and supplies the muscles with the necessary amount of oxygenated blood. Feathers provide a light, efficient body insulation, and thin-walled bones reduce body weight. Many forms of beaks and feet have evolved as different species of birds specialized into specific niches (figure 26.32). The amniotic egg and internal means of fertilization found in the reptiles is also the method of reproduction and development in the birds.

Mammals

In the evolution of animals, the degree of complexity continued to increase and reached its most advanced form in the **mammals.** All the mammals except for the **marsupials** and the egg-laying mammals (**monotremes**) have a **placenta.** This organ allows the female and the developing embryo to constantly exchange materials throughout the period of development. The mammals also have fur that insulates the body and **mammary glands** that produce milk for their young.

The mammals have a constant body temperature and a highly developed circulatory system that moves blood to all parts of the body. The brain of the mammals is more highly developed than in other types of animals, and so mammals have a higher degree of intelligence. Typically, the mammals exhibit more care and protection of their young than do other types of animals.

There are three basic types of mammals. The egg-laying monotremes are the most primitive (figure 26.33). These animals have fur and primitive mammary glands that lack nipples. The young are hatched, not born. The marsupials have fur, developed mammary glands, and give birth to their young. But the young are born prematurely and crawl into a pouch to complete their development. The opossum

Figure 26.30
Prehistoric Reptiles.
These prehistoric reptiles could not successfully compete for survival and became extinct.

Figure 26.31
Reptiles.
(a) Water moccasin. (b) Turtle.
(Carolina Biological Supply
Company)

Figure 26.32
*Adaptive Features
of Some Birds.*
The evolution of various types of
beaks and feet have enabled the
birds to adapt to many types of
environments.

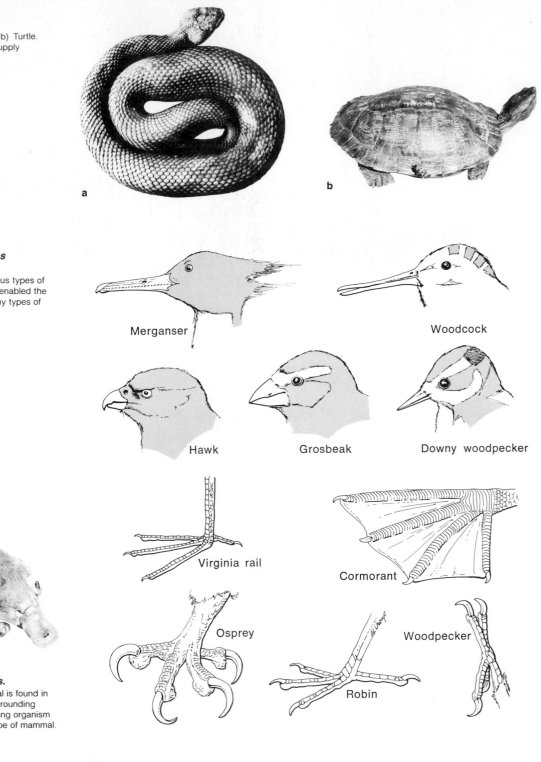

Merganser

Woodcock

Hawk

Grosbeak

Downy woodpecker

Virginia rail

Cormorant

Osprey

Robin

Woodpecker

Figure 26.33
Duckbill Platypus.
This unusual mammal is found in
Australia and the surrounding
islands. This egg-laying organism
is a very primitive type of mammal.

and wallaby are examples of marsupials (figure 26.34). The placental mammals give birth to fully formed young. Mammals occupy a wide variety of niches and habitats. They are found in terrestrial, freshwater, and saltwater environments and have many adaptive mechanisms that enable them to live in such diverse areas (figure 26.35).

Figure 26.34
Marsupials.

The young of these animals are not fully developed at birth. They undergo further development within the pouch of the female.

Summary

Animals are many-celled organisms that must eat food. They reproduce either sexually or asexually and show various degrees of structural complexity. Sponges represent a form of life only slightly above that of one-celled organisms. They filter sea water for nutrients. The Cnidaria are more complex and have an attached polyp form and free-swimming medusa form. They use stinging cells to help capture food. Flatworms are mostly parasitic and have complex life cycles that involve more than one host. Roundworms are extremely common organisms, and some are parasitic. The annelid worms are segmented and are much more complex than any of the previous organisms. The activities of the common earthworm improve the soil. Arthropods are extremely successful. They have an external skeleton, and have adapted to a great variety of niches. The mollusks are an important source of food for humans and usually have a shell for protection. Echinoderms also have a protective skeleton but within the body. They have a unique water-vascular system that serves many purposes. The chordates include those animals with gill slits, dorsal hollow nerve cords, and notochords. Among this group are the fish, amphibians, reptiles, birds, and mammals. The fish are adapted to an aquatic existence, while the reptiles, birds, and mammals are air breathers. The amphibians show features of both.

Figure 26.35
Mammals.

Pictured are some of the different
types of mammals.

Major Groups of Animals

Sponges (Phylum Porifera)

1. Attached to an object
2. Two cell layers
3. Body has many pores and canals
4. No movable appendages
5. Intercellular digestion
6. Aquatic organisms

Cnidaria (Phylum Cnidaria)

1. Digestive cavity
2. Two life stages: attached polyps and free-swimming medusae
3. Tentacles and stinging cells
4. Radial symmetry
5. Nerve net
6. Aquatic organisms

Flatworms (Phylum Platyhelminthes)

1. Flattened body
2. Bilateral symmetry
3. Three germ layers
4. Incomplete digestive system
5. Nervous system
6. Usually hermaphroditic

Roundworms (Phylum Aschelminthes)

1. Complete digestive tract
2. Unlined body cavity
3. Sexes usually separate
4. Unsegmented body
5. Cylindrical body with cuticle
6. Many forms are parasitic

Segmented Worms (Phylum Annelida)

1. Segmented body
2. Circular and longitudinal muscle layers
3. Closed circulatory system
4. Respiration by skin or gills
5. Organ systems present
6. Usually have a larval form

Arthropods (Phylum Arthropoda)

1. Jointed legs
2. Chitinous exoskeleton
3. Compound eyes
4. Three body regions: head, thorax (may be combined into a cephalothorax), and abdomen
5. Fertilization usually internal
6. Striated muscle fibers

Mollusks (Phylum Mollusca)

1. Body has anterior head region, dorsal visceral hump, and ventral foot
2. Usually has an open circulatory system
3. Chambered heart
4. Sexes usually separate
5. Many forms have a shell
6. Bilateral symmetry

Echinoderms (Phylum Echinodermata)

1. Spiny-skinned
2. Radial symmetry, usually five-parted
3. Only found in saltwater environment
4. No head
5. Larval stages
6. Water-vascular system

Jawless Fishes (Class Agnatha)

1. Sucking mouth
2. Unpaired fins
3. Two-chambered heart
4. Poikilotherms
5. Single gonad, no duct

Cartilaginous Fishes (Class Chondrichthyes)

1. Cartilaginous skeleton
2. No swim bladder
3. Lateral line
4. No operculum
5. Paired fins
6. Internal fertilization

Bony Fishes (Class Osteichthyes)

1. Bony skeleton, scales present
2. Operculum present
3. Swim bladder usually present
4. Well-developed jaws with teeth
5. Two-chambered heart
6. Usually external fertilization

Amphibians (Class Amphibia)

1. Usually moist skin, no scales
2. Aquatic larvae and terrestrial adults
3. Usually two pairs of legs
4. Lungs present in adults
5. Three-chambered heart
6. Usually external fertilization

Reptiles (Class Reptilia)

1. Scales present
2. Internal fertilization
3. Amniotic eggs
4. Imperfectly formed four-chambered heart
5. Respiration by lungs
6. Poikilotherms

Birds (Class Aves)

1. Body covered with feathers
2. Forelimbs usually adapted for flight
3. Complete four-chambered heart
4. Beak present, no teeth
5. Amniotic eggs
6. Homotherms

Mammals (Class Mammalia)

1. Body covered with hair
2. Mammary glands present
3. Male copulatory organ, internal fertilization
4. Placenta present
5. Limbs usually have five digits
6. Complete four-chambered heart

Consider This

Can Reindeer Monitor Radioactive Fallout?

Detection of the rare radioactive element Cesium 137 in fallout would indicate a very significant atomic event. However, its identification normally requires sophisticated laboratory equipment.

A news item on National Public Radio stated that the lichen, reindeer moss, selectively takes in Cesium 137. It went on to state that as a reindeer ate this lichen it would be concentrating a source of radioactivity in its tissues. Therefore, the detection of radioactivity from a reindeer would indicate the presence of this element. Further, such detection would indicate that a significant atomic event had occurred. Most importantly, this radioactivity could be detected with a simple handheld counter.

This sounds reasonable, yet, a little strange. What can you find out about this?

Tie a slip knot with a thread around the thorax of a living grasshopper, being careful not to interfere with its wings.

Lift the grasshopper into the air and blow gently at its head. What do you observe?

Suspend the grasshopper by the thread so that its feet touch the table top and again blow gently at the grasshopper's head. In what way does the grasshopper's behavior differ from the first trial?

Suspend the grasshopper above the table but allow its feet to grasp a piece of tissue paper. Blow gently at the grasshopper's head and observe its behavior once more.

What is your conclusion concerning grasshopper behavior as a result of these three experiments?

1. What are the major groupings of animals?
2. What is the importance of an amniotic egg?
3. List three characteristics of the chordates.
4. What is a larva?
5. How does sexual reproduction differ from asexual reproduction?
6. What is a tissue, an organ, and an organ system?
7. How does a polyp stage differ from a medusa stage?
8. Give two examples of mollusks.
9. List three characteristics of each of the following: Agnatha, sharks, bony fish, and amphibians.
10. List three similarities among reptiles, birds, and mammals.

Agnatha (ag-na'thah) The class of jawless vertebrates. Example: lamprey.

air bladder (ār blad'er) A sac that regulates the density of certain bony fish.

allantois (uh-lan'to-is) An embryonic membrane used in gas exchange and waste removal.

amnion (am'ne-on) An embryonic membrane that contains liquid and surrounds the embryo.

amniotic egg (am''ne-ot'ik eg) Any egg in reptiles, birds, and mammals that develops embryonic membranes.

amphibians (am-fib'e-uns) Vertebrates that must return to water to reproduce. Examples: frogs, toads, and salamanders.

annelids (an'uh-lids) Segmented worms. Examples: earthworms, leeches, and clam worms.

arachnids (uh-rak'nids) A class of arthropods. Examples: spiders, scorpions, and ticks.

arthropods (ar'thro-pods) A phylum of animals that have an external skeleton and jointed appendages. Examples: insects, spiders, and crustaceans.

Aves (a'vēz) The class of feathered chordates to which the birds belong.

birds (birds) Vertebrates that have wings, feathers, and a high metabolic rate. Examples: sparrows and eagles.

bivalve (bī'valv) Mollusks that have two parts to their shell. Examples: clams and oysters.

bony fish (bo′-ne fish) A class of fish with a skeleton hardened with mineral deposits.

budding (bud′ing) A kind of asexual reproduction in which the new organism is an outgrowth of the parent.

cartilaginous (car″ti-laj′in-us) Referring to a dense tissue that serves as a skeleton in many animals, particularly in sharks.

caudal fin (kaw′dul fin) A tail fin.

cephalopods (sef′uh-lo-pods) A group of mollusks that have a head and tentacles. Examples: squid and octopus.

Chondrichthyes (kon-drik′the-ēz) The class of chordates to which the sharks belong.

chordates (kōr′dāts) A phylum of animals characterized by a notochord, gill slits, and a dorsal hollow nerve cord. Examples: sharks, cats, frogs, and humans.

chorion (kōr′e-on) An embryonic membrane that encloses the embryo and all of the other embryonic membranes.

Cnidaria (ni-dah′re-ah) A phylum of animals that have stinging cells and only one opening to a digestive sac. Examples: jellyfish and coral.

complete metamorphosis (kum-plēt met-uh-mōr′fo-sis) The life cycle of an insect that includes an egg, larva, pupa, and adult.

crustaceans (krus-ta′shuns) A class of animals that belong to the arthropod phylum. Examples: lobster, crayfish, and barnacles.

dorsal hollow nerve cord (dōr′sul hol′lo nurv kōrd) The main part of the nervous system of the chordates.

echinoderm (e-ki′no-durm) A phylum of animals that have an internal skeleton with spines. Examples: starfish, sea urchins, and sea cucumbers.

endoskeleton (en″do-skel′uh-tun) A skeleton within the outer covering of the animal.

eustachian tube (yu-sta′shun tūb) A tube that connects the throat to the ear; a modified gill slit.

exoskeleton (ek″so-skel′uh-tun) A skeleton on the outside of an animal.

flagellated cells (flaj′uh-la″ted sels) Specialized cells found lining the canals that circulate water through the sponges.

flatworms (flat-werms) A phylum of animals (Platyhelminthes) with a flat appearance; many are parasites. Examples: tapeworms, blood flukes, and planarians.

gill slits (gil slits) Openings that allow water to flow from the throat of chordates to the outside (when functional).

insects (in′sekts) A six-legged class of animals belonging to the arthropod phylum. Examples: grasshoppers, butterflies, and mosquitos.

invertebrates (in-vur′tuh-brāts) Animals that lack a backbone.

larva (lar-vuh) An incompletely developed animal that must change its form before it becomes an adult.

mammals (mam′muls) A class of milk-producing animals, with hair, belonging to the phylum Chordata. Examples: kangaroos, humans, whales, and dogs.

mammary glands (mam′er-e glands) Milk-producing glands of the female mammals.

mantle (man′tul) A structure found in mollusks that produces the shell.

marsupial (mar-su′pe-ul) A mammal with a pouch in which the young develop after birth.

medusa (muh-du′sah) A free-swimming, umbrella-shaped stage in the life cycle of Cnidaria. Example: the jellyfish.

mollusks (mol′usks) A phylum of animals with a mantle and usually a shell. Examples: squids, clams, and snails.

monotreme (mon′o-trēm) One of a group of egg-laying mammals.

notochord (no′tuh-kōrd) A rodlike structure along the back of chordates, used for support.

operculum (o-pur′kyu-lum) A protective cover over the gills of bony fish.

organs (ōr′guns) Structures composed of groups of tissues that work together to perform a particular function. Examples: heart, stomach, and brain.

organ systems (ōr'gun sis'temz) Groups of organs that work together to perform a particular function. Examples: digestive system, reproductive system, and circulatory system.

Osteichthyes (os''te-ik'the-ēz) The class of chordates to which the bony fish belong.

placenta (plah-sen'tah) An organ for nourishing the developing mammal embryo.

polyp (pol'ip) An attached, vase-shaped stage in the life cycle of Cnidaria. Example: hydra.

pupa (pyu'pah) A developmental stage in the life cycle of some insects intermediate between the larva and adult.

reptiles (rep'tils) A class of animals of the phylum Chordata that has developed an amniotic egg and does not need to return to the water to reproduce. Examples: snakes, turtles, and dinosaurs.

roundworms (rownd-werms) A phylum of threadlike or cylindrical animals (Nematoda) that are very common in soils; some are parasitic. Example: pinworms.

sponges (spun'juz) A phylum of animals (Porifera) that are very primitive and contain many canals. They filter water for food.

tentacles (ten'tuh-kuls) Long flexible arms that can be used to obtain food and defend the animal.

tissue (tish'u) A group of specialized cells that work together to perform a particular function. Examples: muscle, nerve, and blood.

tracheae (tra'ke-e) Breathing tubes in insects.

tube feet (tūb feet) Suction-cuplike structures in echinoderms that are part of the water-vascular system.

vertebrates (vur'tuh-brāts) Animals with backbones.

water-vascular system (wah'tur vas'kyu-lar sis'tem) A series of water-filled tubes found in the phylum Echinodermata and used for movement.

yolk sac (yōk sac) An embryonic sac containing stored food for the embryo.

Glossary

abiotic factors (ā-bi-ah′tik fak′tōrz) The nonliving parts of an organism's environment.

abortion (ă-bor′shun) A medical procedure that causes the death and removal of the embryo from the uterus.

accessory structures (ak-ses′o-re struk′churs) The parts of some flowers that are not directly involved in gamete production.

acetyl (ă-sēt′l) The two-carbon remainder of the carbon skeleton of pyruvic acid that is able to enter the mitochondrion.

acid (ă′sid) Any compound that releases a hydrogen ion (or other ion that acts like a hydrogen ion) in a solution.

acquired characteristic (ă-kwīrd kār″ak-ter-iss′tik) A characteristic of an organism gained during its lifetime, not caused by its genes and, therefore, not transmitted to the offspring.

activation energy (ak″ti-va′shun en′ur-je) Energy required to start a reaction; it may be used to increase the number of effective collisions or required to form the transitional molecule in the progress of the reaction pathway.

active site (ak′tiv sīt) The point on the enzyme where it causes the substrate to change.

active transport (ak′tiv trans-port) The process of using a carrier molecule to move molecules across a cell membrane in a direction opposite of the concentration gradient. The carrier requires an input of energy other than the kinetic energy of the molecules.

adaptive radiation (uh-dap′tiv ra-de-a′shun) A specific evolutionary pattern in which there is a rapid increase in the number of kinds of closely related species.

adenine (ad′ĕ-nēn) A double-ring nitrogenous base molecule in DNA and RNA; it is the complementary base of thymine or uracil.

adenosine triphosphate (ATP) (uh-den′o-sēn tri-fos′fāt) A molecule formed from the building blocks adenine, ribose, and phosphates; it functions as the primary energy carrier in the cell.

aerobic cellular respiration (a″ro′bik sel′yu-lar res″pi-ra′shun) A series of reactions in the mitochondria that release usable energy from food molecules by combining them with oxygen molecules.

age distribution (āj dis″tri-biu′shun) The number of organisms of each age in the population.

Agnatha (ag-na′thah) The class of jawless vertebrates. Example: lamprey.

air bladder (ār blad′er) A sac that regulates the density of certain bony fish.

alcoholic fermentation (al-ko-hol′ik fur″men-ta′shun) The anaerobic respiration pathway in yeast cells when pyruvic acid from glycolysis is converted to ethanol and carbon dioxide.

algae (al′je) Protists that have cell walls and chlorophyll and, therefore, can carry on photosynthesis.

allantois (uh-lan′to-is) An embryonic membrane used in gas exchange and waste removal.

alleles (al′lēls) Alternative forms of a gene for a particular characteristic (e.g., attached earlobe genes and free earlobe genes are alternative alleles for ear shape).

alternation of generations (awl″tur-na′shun uv jen″uh-ra′shunz) The cycling of a diploid sporophyte generation and a haploid gametophyte generation in plants.

amensalism (a-men′sal-izm) A relationship between two organisms in which one organism is harmed and the other is not affected.

amino acid (ah-mēn′o ă′sid) A short carbon skeleton that contains an amino group, a carboxylic acid group, and one of various side groups.

amniocentesis (am″ne-o-sen-te′sis) A medical procedure that samples amniotic fluid. Fetal cells are examined to see if they are normal.

amnion (am′ne-on) An embryonic membrane that contains liquid and surrounds the embryo.

amniotic egg (am″ne-ot′ik eg) Any egg in reptiles, birds, and mammals that develops embryonic membranes.

amphibians (am-fib′e-uns) Vertebrates that must return to water to reproduce. Examples: frogs, toads, and salamanders.

amylase (am′i-lās) The enzyme in saliva that begins the hydrolysis of starch.

anaerobic respiration (an′uh-ro″bik res″pi-ra′shun) A biochemical pathway that does not require oxygen for the production of ATP.

anaphase (an′ă-fāz) The third stage of mitosis when the centromeres split and the chromosomes move to the poles.

angiosperms (an′je-o″spurmz) Plants that produce flowers and fruits.

Animalia (ahn″e-mahl′e-uh) One of the five kingdoms that consists of multicellular heterotrophic organisms that require a source of organic material as food.

annelids (an′uh-lids) Segmented worms. Examples: earthworms, leeches, and clam worms.

annual (an′yu-uhl) A plant that completes its life cycle in one year.

anorexia nervosa (an″o-rek′se-ah ner-vo′sah) A nutritional deficiency disease characterized by severe, prolonged weight loss due to fear of becoming fat. This eating disorder is thought to stem from sociocultural factors.

anther (an-ther) Sex organ in plants that produces the sperm.

antheridia (sing. antheridium) (an″thur-id′e-ah) The structures in lower plants that produce sperm.

anthropomorphism (an-thro-po-mor′fism) The ascribing of human feelings, emotions, or meanings to the behavior of animals.

antibiotics (an-te-bi-ot′iks) Drugs that selectively interfere with the function of prokaryotic ribosomes and prevent them from producing the proteins essential for the cell's survival.

anticodon (an″te-ko′don) A sequence of three nitrogenous bases on a tRNA molecule, capable of forming hydrogen bonds with three complementary bases on an mRNA codon during translation of protein synthesis.

anus (ān-us) The opening of the digestive system through which undigested food is eliminated.

applied science (ap-plīd si-ens) The science that makes practical use of the theories provided by scientists to solve everyday problems.

arachnids (uh-rak′nids) A class of arthropods. Examples: spiders, scorpions, and ticks.

archaebacteria (ar-ke-bak-te′re-ah) Microorganisms in the kingdom Prokaryotae comprised of species capable of living in environments of extremely low pH and high salt concentration. They differ significantly from other bacteria and eukaryotic cell types.

archegonium (ar″ke-go′ne-um) The structure in lower plants that produces eggs.

arthropods (ar′thro-pods) A phylum of animals that have an external skeleton and jointed appendages. Examples: insects, spiders, crustaceans.

asexual reproduction (a-sek′shu-al re″pro-duk′shun) Any method of reproduction that does not involve the mixing of genetic material by fertilization.

associative learning (ă-so′shuh-tiv lur′ning) A kind of learning in which a neutral stimulus is associated with a natural stimulus to produce a particular response.

atom (ă′tom) The smallest part of an element that still acts like that element.

atomic mass unit (ă-tom′ik mas yu-nit) An expression of the mass of one proton (equal to 1.67×10^{-24} grams).

atomic nucleus (ă-tom′ik nu′kle-us) The central region of the atom.

atomic number (ă-tom′ik num′bur) The number of protons in an atom.

attachment site (uh-tatch′munt sīt) A specific point on the surface of the enzyme where it can physically attach itself to the substrate.

autosomes (aw′to-sōmz) Chromosomes not involved in determining the sex of individuals.

autotroph (aw′to-trōf) An organism able to produce organic nutrients from inorganic materials.

Aves (a′vez) The class of feathered chordates to which birds belong.

bacillus (bah-sil′us) A rod-shaped bacterium.

bacteria (bak″te′re-ah) Forms of microorganisms that are characterized by their shapes, biochemical characteristics, and genetic ability to function in various environments. They are members of the kingdom Prokaryotae.

basal metabolism (ba′sal mĕ-tab′o-lizm) The amount of energy required to maintain normal body activity while at rest.

base (bās) Any compound that releases a hydroxyl group in a solution (or other ion that acts like a hydroxyl group).

behavior (be-hāv′yur) How an organism acts, what it does, and how it does it.

behavioral isolation (be-hāv′yu-ral i-so-la′shun) A genetic isolating mechanism that prevents interbreeding between species because of differences in behavior.

binding site (bīn′ding sīt) A specific point on the surface of the enzyme where it can physically attach itself to the substrate.

biochemical pathway (bi′o-kem″ĭ-kal path′wā) A major series of enzyme-controlled reactions linked together.

biochemistry (bi-o-kem′iss-tre) The chemistry of living things often called biological chemistry.

biogenesis (bi-o-jen′uh-sis) The concept that life originates only from preexisting life.

biological amplification (bi-o-loj′ĭ-cal am″pli-fĭ-ka′shun) The accumulation of a compound in increasing concentrations in organisms at successively higher trophic levels.

biology (bi-ol′o-je) The science that deals with life.

biomass (bi-o-mas) The dry weight of a collection of designated organisms.

biomes (bi-ōmz) Large, regional communities.

biosphere (bi-o-sfēr) A worldwide ecosystem.

biotechnology (bi-o-tek-nol′uh-je) The science of gene manipulation.

biotic factors (bi-ah′tik fak′tōrz) The living parts of an organism's environment.

biotic potential (bi-ah′tik po-ten′shul) The theoretical maximum rate of reproduction.

birds (birds) Vertebrates that have wings, feathers, and a high metabolic rate. Examples: sparrows and eagles.

birth-control pill (burth kon-trol′ pil) Pills containing hormones that prevent the release of eggs from the ovary.

bivalve (bī′valv) Mollusks that have two parts to their shell. Examples: clams and oysters.

blastula stage (blas′chu-lah stāj) An early embryological stage that follows the morula. It consists of a hollow ball of cells.

bloom (blōom) A rapid increase in the number of microorganisms in a body of water.

bony fish (bo′ne fish) A class of fish with a skeleton hardened with mineral deposits.

breech birth (brēch birth) A birth in which the head is not born first.

budding (bud′ing) A kind of asexual reproduction in which the new organism is an outgrowth of the parent.

bulbo-urethral gland (bul″bo-yu-re′thral gland) A part of the male reproductive system that produces a portion of the semen.

bulimia (bu-lim′e-ah) A nutritional deficiency disease brought on by repeated food binging and purging. It is thought to stem from psychological disorders.

caesarean (se-zār′e-an) The surgical removal of a baby through the mother's abdominal wall.

calorie (kal′o-re) The amount of heat necessary to raise the temperature of one gram of water one degree Celsius.

cambium (kam′be-um) A tissue in higher plants that produces new xylem and phloem.

cancer (kan′sur) A tumor that is malignant.

capsule (kap′sūl) Part of the sporophyte generation of mosses that contains spores.

carbohydrate (kar-bo-hi′drāt) One class of organic molecules composed of carbon, hydrogen, and oxygen in a ratio of 1:2:1. The basic building block of a carbohydrate is a simple sugar (= monosaccharide).

carbon dioxide conversion stage (kar′bon di-ok′sīd kon-vur′zhun stāj) A series of reactions that make up the second stage of photosynthesis during which inorganic carbon from carbon dioxide becomes incorporated into a sugar molecule.

carbon skeleton (kar′bon skel′uh-ton) The central portion of an organic molecule composed of rings or chains of carbon atoms.

carnivores (kar-ni-vōrz) Those animals that eat other animals.

carrier (kar′re-er) Any individual having a hidden recessive gene.

carrying capacity (kār′re-ing kuh-pas′i-te) The optimum population size an area can support over an extended period of time.

cartilaginous (car″ti-laj′in-us) Referring to a dense tissue that serves as a skeleton in many animals, particularly in sharks.

catalyst (cat′uh-list) A chemical that speeds up a reaction but is not used up in the reaction.

caudal fin (kaw′dul fin) A tail fin.

cell (sel) The basic structural unit that makes up all living things.

cell membrane (sel mem′brān) The outer boundary membrane of the cell also known as the plasma membrane.

cell plate (sel plāt) A plant cell structure that begins to form in the center of the cell and proceeds to the cell membrane, resulting in cytokinesis.

cellular membranes (sel′u-lar mem′brāns) Thin sheets of material composed of phospholipid and protein. Some of the proteins have attached carbohydrates or fats.

cellular respiration (sel′yu-lar res″pi-ra′shun) A major biochemical pathway during which cells release the chemical-bond energy from food and convert it into a usable form (ATP).

cellulose (sel′yu-lōs) An organic material produced by some fungi, some Protists, and plants as a cell wall material. It differs from the material known as chitin.

centriole (sen′tre-ōl) Two sets of nine short microtubules. Each set is arranged in a cylinder.

centromere (sen′tro-mēr) The region where two chromatids are joined.

cephalopods (sef′uh-lo-pods) A group of mollusks that have a head and tentacles. Examples: squid and octopus.

chemical bonds (kem′i-kal bonds) Forces that combine atoms or ions and hold them together.

chemical reaction (kem′i-kal re-ak′shun) The formation or rearrangement of chemical bonds, usually indicated in an equation by an arrow from the reactants to the products.

chemical symbol (kem′i-kal sim′bol) Represents one atom of an element, such as Al for aluminum or C for carbon.

chitin (ki′tin) An organic material produced by some fungi as a wall material and by insects as an external skeleton.

chlorophyll (klo′ro-fil) A molecule directly involved in the light-trapping and energy-conversion process of photosynthesis.

chloroplast (klo′ro-plast) An energy-converting, membranous, saclike organelle in plant cells containing the green pigment chlorophyll.

cholesterol (ko-les′ter-ol) A steroid molecule. May be found in the bile that can crystallize to form gallstones.

Chondrichthyes (kon-drik′the-ēz) The class of chordates to which the sharks belong.

chordates (kōr′dāts) A phylum of animals characterized by a notochord, gill slits, and a dorsal hollow nerve cord. Examples: sharks, cats, frogs, and humans.

chorion (kōr′e-on) An embryonic membrane that encloses the embryo and all of the other embryonic membranes.

chromatid (kro′mah-tid) A replicated chromosome physically attached to an identical chromatid at the centromere.

chromatin (kro′mah-tin) Areas or structures within the nucleus of a cell composed of long molecules of deoxyribonucleic acid (DNA) in association with proteins.

chromosomal mutation (kro-mo-sōm′al miu-ta′shun) A change in the gene arrangement in a cell as a result of breaks in the DNA molecule.

chromosomes (kro-mo-sōms) Structures composed of various kinds of histone proteins and DNA that contain the cell's genetic information.

cilia (sil′e-ah) Short, flexible, hairlike filaments that beat rhythmically and propel the cell through the water or bring food to the cell.

classical conditioning (klas′si-kul kon-dĭ′shun-ing) A kind of learning in which a neutral stimulus is associated with a natural stimulus to produce a particular response.

cleavage (kle′vuj) An early embryological stage during which the zygote divides by mitosis into smaller cells.

cleavage furrow (kle′vaj fur-ro) The indentation of the cell membrane of an animal cell that pinches the cytoplasm into two parts.

climax community (kli-maks com-miu′ni-te) A relatively stable, long-lasting community.

clones (klōnz) All of the individuals reproduced asexually that have exactly the same genes.

club fungi (klub fun′ji) One of the six major divisions of the kingdom Mycetae (fungi) containing organisms producing sexual spores on clublike structures.

Cnidaria (ni-dah′re-ah) A phylum of animals that have stinging cells and only one opening to a digestive sac. Examples: jellyfish and coral.

coacervate (ko-as′ur-vāt) A collection of organic macromolecules surrounded by water molecules that are aligned to form a sphere.

coccus (kok′us) A spherical-shaped bacterium.

codon (ko′don) Sequence of three nucleotides of an mRNA molecule that directs the placement of a particular amino acid during translation.

coenzyme (ko-en′zīm) A molecule that works with an enzyme to enable it to function as a catalyst.

colloid (kol-loid) A mixture containing dispersed particles that are larger than molecules but small enough that they do not settle out.

colon (ko′lon) The large intestine. Water and salts are absorbed from the contents of the colon.

colony (kol′o-ne) A cluster of independent cells produced by cell division.

commensalism (com-men′sal-izm) A relationship between two organisms in which one organism is helped and the other is not affected.

competitive exclusion principle (com-pē′tī-tiv eks-klu′zhun prin′sī-pul) No two species can occupy the same niche at the same time.

competitive inhibition (kum-pet′i-tiv in″hī-bi′shun) The formation of a temporary enzyme-inhibitor complex that interferes with the normal formation of enzyme-substrate complexes, resulting in a decreased turnover.

complementary base (kom″ple-men′tah-re bās) A base that can form hydrogen bonds with another base of a specific nucleotide.

complete metamorphosis (kum-plet met-uh-mōr′fo-sis) The life cycle of an insect that includes egg, larva, pupa, and adult.

complete protein (kom-plet pro′te-in) Protein molecules that provide all the essential amino acids.

complex carbohydrates (kom′plèks kar-bo-hi′drāts) Macromolecules composed of simple sugars combined by dehydration synthesis to form a polymer.

compound (kom-pound) A kind of matter that consists of a specific number of atoms (or ions) joined to each other in a particular way and held together by chemical bonds.

concentration gradient (kon″sen-tra′shun gra″de-ent) A gradual change in the number of molecules over distance.

conditioned response (kon-dī′shund re-sponts′) The behavior displayed when the neutral stimulus is given after association has occurred.

condom (kon′dum) A thin, rubber sheath placed over an erect penis before sexual intercourse.

cone (kōn) A reproductive structure of gymnosperms that produces pollen in males or eggs in females.

consumers (kon-soom′urs) Organisms that must obtain energy in the form of organic matter.

control group (con-trōl grŭp) The situation used as the basis for comparison in a controlled experiment.

controlled experiment (con-trōld′ ek-sper′i-ment) An experiment that allows for a comparison of two events that are identical in all but one respect.

control processes (con-trōl pro′ses-es) Mechanisms that ensure an organism will carry out all metabolic activities in the proper sequence (coordination) and the proper amount.

convergent evolution (kon-vur′jent ev-o-lu′shun) An evolutionary pattern in which widely different organisms show similar characteristics.

copulation (kop″yu-la′shun) The act of transferring sperm from the male into the female reproductive tract.

corpus luteum (kor′pus lu′te-um) The follicle after the primary oocyte has erupted. It becomes a glandular structure.

coupled reactions (kup′ld re-ak′shuns) The linkage of a set of energy-requiring reactions with energy-releasing reactions.

covalent bond (ko-va′lent bond) The attractive force formed between two atoms that share a pair of electrons.

cristae (kris′te) Folded surfaces of the inner membranes of mitochondria.

crossing-over (kros-sing o-ver) The exchange of a part of a chromatid from one chromosome with an equivalent part of a chromatid from a homologous chromosome.

crustaceans (krus-ta′shuns) A class of animals that belong to the arthropod phylum. Examples: lobster, crayfish, and barnacles.

cyanobacteria (si″an-o-bak-te′re-ah) Blue-green bacteria belonging to the kingdom Prokaryotae.

cytokinesis (si-to-ki-ne′sis) Division of the cytoplasm of one cell into two new cells.

cytoplasm (si″to-plazm) The more fluid portion of the protoplasm that surrounds the nucleus.

cytosine (si′to-sēn) A single-ring nitrogenous base molecule in DNA and RNA. It is complementary to guanine.

daughter cells (daw′tur sels) Two cells formed by cell division.

daughter chromosomes (daw′tur kro′mo-sōms) Chromosomes produced by DNA replication and containing identical genetic information; formed after chromosome division in anaphase.

daughter nuclei (daw′tur nu′kle-i) Two nuclei formed by mitosis.

death phase (deth fāz) The portion of some population growth curves in which the size of the population declines.

decomposers (de-kom-po′zurs) Organisms that use dead organic matter as a source of energy.

dehydration synthesis (de″hi-dra′shun sin′thē-sis) A reaction that results in the joining of smaller molecules to make larger molecules by the removal of water.

deme (dēm) A local, recognizable population that differs in gene frequencies from other local populations of the same species (also see subspecies).

denature (de-na′chur) To permanently change the protein structure of an enzyme so that it loses its ability to function.

denitrifying bacteria (de-ni′tri-fi-ing bak-te′re-ah) Several kinds of bacteria capable of converting nitrite to nitrogen gas.

density (den′si-te) The weight of a certain volume of a material.

density-dependent factors (den′si-te de-pen′dent fak′tōrz) Population-limiting factors that become more effective as the size of the population increases.

density-independent factors (den′si-te in″de-pen′dent fak′tōrz) Population-controlling factors that are not related to the size of the population.

deoxyribonucleic acid (DNA) (de-ok″se-ri-bo-nu-kle′ik ā′sid) A polymer of nucleotides that serves as genetic information. In prokaryotic cells, it is a duplex DNA (double-stranded) loop and contains no permanently attached proteins. In eukaryotic cells, it is found in strands with attached histone proteins. When tightly coiled, it is known as a chromosome.

deoxyribose (de-ok″se-ri′bōs) A five-carbon sugar molecule; a component of DNA.

diaphragm (di′uh-fram) A mechanical conception-control device that covers the entrance to the uterus.

dicot (di′kot) An angiosperm whose embryo has two seed leaves.

differential reproduction (dif″fur-ent′shul re-pro-duk′shun) A process by which those organisms that have better genetic information for a particular environment reproduce more than individuals with less desirable genetic information.

differentially permeable (dif″fur-ent′shul-le per′me-uh-bul) A membrane that allows certain molecules to pass through it but prevents the passage of others.

differentiation (dif″fur-ent-she-a′shun) The process of forming specialized cells within a multicellular organism.

diffusion (dī″fiu′zhun) Net movement of a kind of molecule from an area of higher concentration to an area of lesser concentration.

diffusion gradient (dī″fiu′zhun gra″de-ent) A difference in the concentration of diffusing molecules across an area.

digestive tract (di-jes′tiv trakt) The series of tubes and structures that break down complex food molecules and absorb nutrients into the body.

dihybrid cross (di-hi′brid kros) A genetic study in which two pairs of alleles are followed from the parental generation to the offspring.

diploid (dip′loid) A cell that has two sets of chromosomes; one set from the maternal parent and one set from the paternal parent.

divergent evolution (di-vur′jent ev-o-lu′shun) A basic evolutionary pattern in which individual speciation events cause many branches in the evolution of a group of organisms.

DNA code (D-N-A cōd) A sequence of three nucleotides of a DNA molecule.

DNA polymerase (po-lim′er-ās) An enzyme that brings new DNA triphosphate nucleotides into position for bonding on another DNA molecule.

DNA replication (rep″li-ka′shun) The process by which the genetic material (DNA) of the cell reproduces itself prior to its distribution to the next generation of cells.

dominance hierarchy (dom′in-ants hi′ur-ar-ke) A relatively stable, mutually understood order of priority within a group.

dominant allele (dom′in-ant al′ēl) A gene that expresses itself and masks the effect of other alleles for the trait.

dorsal hollow nerve cord (dōr′sul hol′lo nurv kōrd) The main part of the nervous system of the chordates.

double bond (dŭ′b′l bond) A pair of covalent bonds between two atoms formed when they share two pairs of electrons.

Down's syndrome (downs sin′drōm) A genetic disorder resulting from the presence of an extra chromosome number 21. Symptoms include thickened eyelids, a low level of intelligence, and faulty speech. Sometimes called mongolism.

duodenum (du″o-dēn′um) The upper portion of the small intestine into which flows various digestive enzymes and materials from the stomach, liver, and pancreas.

duplex DNA (du-pleks) DNA in a double-helical shape.

dynamic equilibrium (di-nam′ik e-kwi-lib′re-um) The condition where molecular motion continues but the molecules are equally dispersed, so movement is equal in all directions.

echinoderm (e-ki′no-durm) A phylum of animals that have an internal skeleton with spines. Examples: starfish, sea urchins, and sea cucumbers.

ecological isolation (e-ko-loj′i-kal i-so-la′shun) A genetic-isolating mechanism that prevents interbreeding between species because they live in different areas.

ecology (e-kol′o-je) A branch of biology that studies the relationships between organisms and their environment.

ecosystem (e″ko-sis-tum″) An interacting collection of organisms and the abiotic factors that affect them.

egg cells (eg sels) The haploid sex cell produced by sexually mature females.

electron transfer system (e-lek′tron trans′fur sis′tem) The series of oxidation-reduction reactions in aerobic cellular respiration in which the energy is removed from hydrogens and transferred to ATP.

electrons (e-lek'trons) The negatively charged particles moving at a distance from the nucleus of an atom that balance the positive charge of protons.

elements (el'ĕ-ments) A kind of matter that consists of only one kind of atom.

embryo (em'bre-o) The early stage in the development of a sexually reproduced organism.

empirical evidence (im-pir-i-kal ev-i-dens) The information gained by observing an event.

empirical formula (im-pir-i-kal fōr'miu-lah) Chemical shorthand that indicates the number of each kind of atom within a molecule.

endoplasmic reticulum (ER) (en''do-plaz'mik re-tik'yu-lum) Folded membranes and tubes throughout the eukaryotic cell that provide a large surface upon which chemical activities take place.

endoskeleton (en'do-skel''uh-tun) A skeleton within the outer covering of the animal.

endosymbiotic theory (en'do-sim-be-ot'ik the'o-re) The theory that suggests some organelles found in eukaryotic cells may have had their origin as free-living prokaryotes.

environment (en-vi'ron-ment) Anything that impacts on an organism during its lifetime.

environmental resistance (en-vi-ron-men'tal re-zis'tants) The collective set of factors that limit population growth.

enzymatic competition (en-zi-mă'tik com-pĕ-tĭ'shun) Several different enzymes available to combine with a given substrate material.

enzyme (en'zīm) A specific protein that acts as a catalyst to change the rate of a reaction.

enzyme-substrate complex (en'zīm sub'strāt kom'pleks) A temporary molecule formed when an enzyme attaches itself to a substrate molecule.

epiphyte (ep'e-fīt) A plant that lives on the surface of another plant.

esophagus (e-sof'ă-gus) A tube that conducts food from the mouth to the stomach.

essential amino acids (e-sen'shal ah-mēn'o ass-ids) Those amino acids that must be part of the diet, such as lysine, tryptophan, and valine.

ethology (e-thol'uh-je) The scientific study of the nature of behavior and its ecological and evolutionary significance in its natural setting.

eukaryote (yu-kār'e-ōt) A cell possessing a nuclear membrane and other membranous organelles.

eukaryotic cells (yu'kār-e-ah''tik sels) A type of cell with a true nucleus separated from the rest of the cytoplasm by a nuclear membrane.

eugenics laws (yu-jen'iks laws) Laws designed to eliminate "bad" genes from the human gene pool and encourage "good" gene combinations.

eustachian tube (yu-sta'shun tūb) A tube that connects the throat to the ear; a modified gill slit.

evolution (ĕ-vol-lu'shun) The adaptation of a population of organisms to its environment by way of change in the gene frequencies over generations.

exoskeleton (ek''so-skel'uh-tun) A skeleton on the outside of an animal.

experiment (ek-sper'ĭ-ment) A re-creation of an event in a way that enables a scientist to gain valid and reliable empirical evidence.

experimental group (ek-sper-i-men'tal grūp) The situation in a controlled experiment that is identical to the control group in all respects but one.

exponential growth phase (eks-po-nent'shul grōth fāz) A period of time during population growth when the population increases at an accelerating rate.

external parasite (eks-tur'nal pār'uh-sīt) A parasite that lives on the outside of its host.

extinct (ek''stinkt') No longer existing.

extrinsic factors (eks-trin'sik fak'tōrz) Population-controlling factors that arise outside the population.

facilitated diffusion (fah-sil'ĭ-ta''ted di-fiu'zhun) Diffusion assisted by carrier molecules.

FAD (**f**lavin **a**denine **d**inucleotide) A hydrogen carrier used in respiration.

fat (fat) A class of water-insoluble macromolecules composed of a glycerol and three fatty acids.

fatty acid (fat-te ă'sid) One of the building blocks of a fat, composed of a long chain carbon skeleton with a carboxylic acid functional group.

feces (fe-sēz) The undigested food material eliminated from the digestive tract through the anus.

fertilization (fer''ti-li-za'shun) The joining of haploid nuclei, usually from an egg and a sperm cell, resulting in a diploid cell called the zygote.

first law of thermodynamics (furst law uv thur-mo-di-nam'iks) Energy in the universe remains constant, it can neither be created nor destroyed.

flagella (flah-jel'lah) Long, flexible, hairlike filaments that lash back and forth and draw the cell through the water.

flagellated cells (flaj'uh-la''ted sels) Specialized cells found lining the canals that circulate water through the sponges.

flatworms (flat-werms) A phylum of animals (Platyhelminthes) with a flat appearance; many are parasites. Examples: tapeworms, blood flukes, and planarians.

flower (flow'ur) A complex structure made from modified stems and leaves. It produces pollen in the males and eggs in the females.

follicle (fol'lĭ-kul) The saclike structure in the ovary that contains the developing egg and develops into the corpus luteum.

food chain (food chān) Sequence of organisms that feed on one another, resulting in the flow of energy from a producer through a series of consumers.

food web (food web) A system of interlocking food chains.

formula (form'yu-lah) The group of chemical symbols that indicate what elements are in a compound and the number of each kind of atom present.

fraternal twins (frā-tur'nal twins) Two offspring, formed at the same time, that result from the fertilization of two separate eggs.

free-living nitrogen-fixing bacteria (ni'tro-jen-fik-sing bak-te're-ah) Soil bacteria that convert nitrogen gas molecules into nitrogen compounds that plants can use.

fruit (froot) The structure in angiosperms that contains seeds.

functional groups (fung'shun-al grūps) Specific combinations of atoms attached to the carbon skeleton that determine specific chemical properties.

fungi (fun'ji) The common name for members of the kingdom Mycetae.

gallbladder (gawl blad-der) A saclike structure that holds bile before it is released into the small intestine.

gallstones (gawl-stōns) Crystals of cholesterol formed in the gallbladder.

gamete (gam'ēt) A haploid sex cell.

gametogenesis (gă-me''to-jen'ĕ-sis) The generating of gametes. The meiotic cell division process that produces sex cells; oogenesis and spermatogenesis.

gametophyte generation (gă-me'to-fīt jen''uh-ra'shun) The haploid generation in plant life cycles. They produce gametes.

gas (gas) The state of matter in which the molecules are more energetic than the molecules of a liquid, resulting in only slight attraction for each other.

gastrula stage (gas'tru-lah stāj) The embryological stage following the blastula stage in which the primitive gut is formed.

gene (jēn) Any molecule, usually a segment of DNA, that is able to: (1) *replicate* by directing the manufacture of copies of itself; (2) *mutate* or chemically change and transmit these changes to future generations; (3) *store* information that determines the characteristics of cells and organisms; and (4) use this information to *direct* the synthesis of structural and regulatory proteins.

gene flow (jēn flo) The movement of genes within a population due to migration and the movement of genes from one generation to the next by gene replication and sexual reproduction.

gene frequency (jēn fre'kwen-se) The measure of the number of times a gene occurs in a population; the percentage of sex cells that contain a particular gene.

gene pool (jēn pool) All the genes of all the individuals of the same species.

generative processes (jen'uh-ra''tiv pros'es-es) Actions that result in an increase in the size of an individual organism (growth), or an increase in the number of individuals in a population (reproduction).

genetic counselor (jĕ-net'ik kown'sel-or) A professional biologist with specific training in human genetics.

genetic isolating mechanism (jĕ-net'ik i-so-la'ting mek'an-izm) A mechanism that prevents interbreeding between species.

genetic recombination (jĕ-net'ik re-kom-bĭ-na'shun) The sum total of all the gene mixing that occurs during sexual reproduction.

genetics (jĕ-net-iks) The study of genes, how genes produce characteristics, and how the characteristics are inherited.

genome (je'nōm) A set of all the genes necessary to specify an organism's complete list of characteristics.

genotype (je'no-tīp) The catalog of genes of an organism, whether or not these genes are expressed.

genus (jē'nus) One of the categories in the classification system composed of groups of species that are very similar. The genus and species names together make up the scientific name of an organism.

geographic barrier (je-o-graf'ik bār'yur) Geographic features that keep different portions of a species from exchanging genes.

geographic isolation (je-o-graf'ik i-so-la'shun) A condition in which part of the gene pool is separated by geographic barriers from the rest of the population.

germinate (jur'min-āt) To begin to grow (as from a seed).

gill slits (gil slits) Openings that allow water to flow from the throat of chordates to the outside (when functional).

glycerol (glis'er-ol) One of the building blocks of a fat; composed of a carbon skeleton that has three alcohol groups (— OH) attached to it.

glycolysis (gli-kol'ĭ-sis) The anaerobic first stage of cellular respiration that consists of the enzymatic breakdown of a sugar into two molecules of pyruvic acid.

Golgi apparatus (gol'je ap''pah-rat'us) A stack of flattened, smooth membrane sacks; the site of synthesis and packaging of certain molecules in eukaryotic cells.

gonad (go-nad) Animal organ that produces gametes.

grana (gra'nuh) Areas of concentrated chlorophyll within the chloroplast.

granules (gran'yūls) Materials whose structure is not defined as well as other organelles.

growth factor (grōth fak-tor) A nutrient needed by the body for proper functioning, but only in very small amounts. Examples: vitamins and minerals.

guanine (gwah'nēn) A double-ring nitrogenous base molecule in DNA and RNA. It is the complementary base of cytosine.

gymnosperms (jim'no-spurmz) Plants that produce their seeds in cones.

habitat (hab'ĭ-tat) The place or part of an ecosystem occupied by an organism.

habitat preference (hab-ĭ-tat pref'ur-ents) A genetic isolating mechanism that prevents interbreeding between species because they live in different areas.

haploid (hap'loid) A single set of chromosomes resulting from the reduction division of meiosis.

Hardy-Weinberg law (har-de wīn′burg) The law that states that populations of organisms will maintain constant gene frequencies from generation to generation as long as mating is random, the population is large, mutation does not occur, and no migration occurs.

herbivores (her′-bĭ-vŏrz) Those animals that feed directly on plants.

heterotroph (hĕ′tur-o-trŏf) Organism unable to produce organic molecules from inorganic materials.

heterozygous (hĕ″ter-o-zi′gus) A diploid organism that has two different allelic forms of a particular gene.

high-energy phosphate bond (hi en′ur-je fos-fāt bond) The bond between two phosphates in an ADF or ATP molecule that readily releases its energy for cellular processes.

homologous chromosomes (ho-mol′o-gus kro′mo-sōms) A pair of chromosomes in a diploid cell that contain similar genes on corresponding loci throughout their length.

homozygous (ho″mo-zi′gus) A diploid organism that has two identical genes for a particular characteristic.

hormone (hŏr′mōn) A chemical substance that is released from glands in the body to regulate other parts of the body.

host (hōst) An organism that provides the necessities for a parasite—generally food and a place to live.

hydrogen bond (hi′dro-jen bond) Weak attractive forces between molecules that are important in determining how groups of molecules are arranged.

hydrolysis (hi-drol′ĭ-sis) A process that occurs when a large molecule is broken down into smaller parts by the addition of water.

hydroxyl ion (hi-drok′sil i-on) A group of charged atoms that are released when a base is dissolved in water, (OH⁻).

hypha (hi′fah) The basic unit of structure of a fungus, consisting of a filament.

hypothesis (hi-poth′ē-sis) A possible answer to, or explanation of, a question that accounts for all the observed facts and is testable.

identical twins (i-den′ti-cal twins) Two offspring born at the same time as a result of the fertilization of one egg that separates into two individuals.

imperfect flowers (im″pur′fekt flow′ers) Flowers that contain either male or female reproductive structures, but not both.

imperfect fungi (im-pur′fekt fun′ji) One of the six major divisions of the kingdom Mycetae (fungi) containing organisms in which sexual reproduction is absent or undiscovered.

implantation (im-plan-ta′shun) The embedding of the early embryo in the lining of the uterus.

imprinting (im-prin′ting) Learning in which a very young animal is genetically primed to learn a specific behavior in a very short period.

inclusions (in-klu′zhuns) A general term referring to materials inside a cell that are usually not readily identifiable; stored material.

independent assortment (in″de-pen′dent ā-sort′ment) The segregation, or assortment, of one pair of homologous chromosomes is not dependent upon the segregation, or assortment, of any other pair of chromosomes.

inhibitor (in-hib′ĭ-tor) A molecule that temporarily attaches itself to an enzyme, thereby interfering with the enzyme's ability to form an enzyme-substrate complex.

initiation code (i-ni′she-a″shun cōd) The code on DNA that begins the process of translation.

inorganic molecules (in-or-gan′ik mol-uh-kiuls) Molecules that do not contain carbon atoms in rings or chains.

insecticide (in-sek′ti-sīd) A poison used to kill insects.

insects (in′sekts) A six-legged class of animals belonging to the arthropod phylum. Examples: grasshoppers, butterflies, and mosquitoes.

insight learning (in-sīt lur′ning) Learning in which past experiences are reorganized to solve new problems.

instinctive behavior (in-stink′tiv be-hāv′yur) Automatic, preprogrammed, or genetically inherited behavior.

intention movements (in-ten′shun moov-ments) Behavior that signals what the animal is likely to do in the near future.

internal parasite (in-tur′nal pār′uh-sīt) A parasite that lives inside its host.

interphase (in′tur-fāz) The stage between cell divisions when the cell is engaged in metabolic activities.

intrauterine device (IUD) (in-trah-yu′tur-īn de-vīs) A mechanical contraception-control device placed in the uterus to prevent embryo implantation.

intrinsic factors (in-trin′sik fak′tōrz) Population-controlling factors that arise from within the population.

invertebrates (in-vur′tuh-brāts) Animals that lack a backbone.

ionic bond (i-on′ik bond) The attractive force between ions of opposite charge.

ions (i′ons) Electrically unbalanced or charged atoms.

isotopes (i′so-tōps) Atoms of the same element that differ only in the number of neutrons.

kilocalorie (kil″o-kal′o-re) A measure of heat energy, one thousand times larger than a calorie.

kinetic energy (ki-net′ik en′er-je) Energy of motion.

kingdom (king′dom) The largest grouping used in the classification of organisms.

Klinefelter's syndrome (klīn′fel-turs sin′drōm) A genetic disorder caused by having two X chromosomes as well as a Y chromosome. Symptoms include tallness, sterility, and mental impairment.

Krebs cycle (krebs si′kl) The series of reactions in aerobic cellular respiration that results in the production of two carbon dioxides, the release of four pairs of hydrogens, and the formation of an ATP molecule.

kwashiorkor (kwash-e-or′kor) A protein deficiency disease.

labor (la′bor) The contractions of the uterus that result in the birth of the baby.

lack of dominance (lak uv dom′in-ans) The condition of two unlike alleles both expressing themselves, neither being dominant.

lactase (lak′tās) An enzyme produced by the cells lining the small intestine. It breaks the complex sugar, lactose, into simple sugars.

lactose intolerance (lak-tōs in-tol′er-ans) A condition resulting from the inability to digest lactose.

lag phase (lag fāz) A period of time during population growth when the population remains small, but increases slowly.

larva (lar′vuh) An incompletely developed animal that must change its form before it becomes an adult.

law of dominance (law uv dom′in-ans) When an organism has two different alleles for a trait, the allele that is expressed and overshadows the expression of the other allele is said to be dominant. The gene whose expression is overshadowed is said to be recessive.

law of independent assortment (law uv in″de-pen′dent ā-sort′ment) Members of one gene pair will separate from each other independently of the members of the other gene pairs.

law of segregation (law uv seg″re-ga′shun) When gametes are formed by a diploid organism, the alleles that control a trait separate from one another into different gametes, retaining their individuality.

learning (lurn′ing) A change in behavior as a result of experience.

leaves (lēvs) Specialized portions of higher plants that are the sites of photosynthesis.

lichens (li′kens) One of the six major divisions of the kingdom Mycetae (fungi) containing organisms that are an association of an alga with a fungus.

life cycle (līf si′kl) The series of stages in the life of any organism.

light energy conversion stage (līt en′ur-je kon-vur′zhun stāj) The first of the two stages of photosynthesis during which light energy is converted to chemical-bond energy.

limiting factors (lim′ī-ting fak′tōrz) Environmental influences that limit biological activity.

linkage group (lingk′ij grūp) Genes that are located on the same chromosome and tend to be inherited together.

lipids (li′pids) Large organic molecules that do not easily dissolve in water. Classes include fats, phospholipids, and steroids.

liquid (lik′wid) The state of matter in which the molecules are strongly attracted to each other, but they are farther apart than in a solid so they move about each other more freely.

locus (loci) (lo′kus) (lo′si) The spot on a chromosome where an allele is located.

lysosome (li′so-sōmz) A specialized organelle that holds a mixture of hydrolytic enzymes.

macroevolution (mā″kro-ev-o-lu′shun) The concept that large numbers of characteristics can change in a very short period of time to produce rapid evolutionary change.

mammals (mam′muls) A class of milk-producing animals, with hair, belonging to the phylum Chordata. Examples: kangaroos, humans, whales, and dogs.

mammary glands (mam′er-e glands) Milk-producing glands of female mammals.

mantle (man′tul) A structure found in mollusks that produces the shell.

marsupial (mar-su′pe-ul) A mammal with a pouch in which the young develop after birth.

mass number (mas num′ber) The weight of an atomic nucleus expressed in atomic mass units.

matter (mat′er) Anything that has weight (mass) and also takes up space (volume).

medusa (muh-du′sah) A free-swimming, umbrella-shaped stage in the life cycle of Cnidaria. Example: the jellyfish.

meiosis (mi-o′sis) The specialized pair of cell divisions that reduce the chromosome number from diploid (2N) to haploid (N).

Mendelian genetics (Men-de′le-an jĕ-net′iks) The pattern of inheriting characteristics that follows the laws formulated by Johann Gregor Mendel.

menses (period, menstrual flow) (men′sēz) The shedding of the lining of the uterus.

menstrual cycle (men′stru-al si′kl) The repeated buildup and shedding of the lining of the uterus.

messenger RNA (mRNA) (mes′en-jer) A molecule composed of ribonucleotides that functions as a copy of the gene and is used in the cytoplasm of the cell during protein synthesis.

metabolic processes (mĕ-tă-bol′ik pros′es-es) The total of all chemical reactions within an organism; for example, nutrient uptake and processing and waste elimination.

metaphase (mĕ′tah-fāz) The second stage in mitosis during which the chromosomes align at the equatorial plane.

microbe (mi′krōb) Any single-celled organism. It commonly refers to members of the Prokaryotae, Protista, and Mycetae.

microevolution (mi″kro-ev-o-lu′shun) The concept that evolution is a slow, gradual, progressive process.

microscope (mi′kro-skōp) An instrument that will produce an enlarged image of a small object.

microsphere (mi′kro-sfer) A collection of organic macromolecules in a structure with a double-layered outer boundary.

microtubules (mi′kro-tūb″yūls) Small, hollow tubes of protein that function throughout the cytoplasm to provide structural support and enable movement.

minerals (min′er-als) Growth factors, usually inorganic salts. Examples: calcium and magnesium.

mitochondrion (mi-to-kon′dre-on) A membranous organelle resembling a small bag with a larger bag inside that is folded back on itself; serves as the site of aerobic cellular respiration.

mitosis (mi-to′sis) A process that results in equal and identical distribution of replicated chromosomes into two newly formed nuclei.

mixture (miks′tūr) Several molecules physically near one another but not chemically bound.

molecule (mol′ĕ-kūl) The smallest particle of a chemical compound, also the smallest naturally occurring part of an element or compound.

mollusks (mol′usks) A phylum of animals with a mantle and usually a shell. Examples: squids, clams, and snails.

monocot (mon′o-kot) An angiosperm whose embryo has one seed leaf.

monoculture (mon″o-kul′chur) The agricultural practice of planting the same species over large expanses of land.

monohybrid cross (mon″o-hi′brid kros) A genetic study in which a single characteristic is followed from the parental generation to the offspring.

monotreme (mon′o-trēm) One of a group of egg-laying mammals.

morning sickness (mor′ning sik′nes) One of the symptoms of pregnancy characterized by nausea, vomiting, and dizziness.

mortality (mōr-tal′ĭ-te) The number of individuals leaving the population by death per thousand individuals in the population.

morula (mōr′yu-lah) An early embryological stage consisting of a solid ball of cells.

mosses (mos′sez) Lower plants that have a dominant gametophyte generation, no vascular tissue, swimming sperm, and spores.

mouth (mowth) The opening to the digestive tract.

mucus (miu-kus) A slimy material produced in various parts of the digestive tract. It aids the movement of food through the system and protects the lining of the digestive tract from being harmed by acids and enzymes.

multiple alleles (mul′tĭ-p′l al′lēls) The concept that there are several different forms of genes for a particular characteristic.

mutagenic agent (miu-tah-jen′ik a-jent) Anything that causes permanent change in DNA.

mutation (miu-ta′shun) Any change in the genetic information of a cell.

mycelium (mi-se′le-um) A mass of hyphae making up the body of a fungus.

Mycetae (mi-se′te) One of the major kingdoms of living things composed of nonphotosynthetic organisms commonly referred to as fungi and lichens. Members of this group include yeasts and molds.

mycoses (mi-ko′sēz) Diseases caused by fungi.

mycotoxin (mi″ko-tok′sin) One of a group of poisonous chemicals produced by certain fungi.

NAD (**n**icotinamide-**a**denine **d**inucleotide) An electron acceptor and hydrogen carrier used in respiration.

NADP (**n**icotinamide-**a**denine **d**inucleotide **p**hosphate) An electron acceptor and hydrogen carrier used in photosynthesis.

natality (na-tal′ĭ-te) The number of individuals entering the population by reproduction per thousand individuals in the population.

net movement (net muv′ment) The movement in one direction minus the movement in the opposite direction.

neutralization (nu′tral-i-za″shun) A chemical reaction involved in mixing an acid with a base, resulting in the formation of a salt and water.

neutrons (nu′trons) Particles in the nucleus of an atom that have no electrical charge. They were named neutrons to reflect their lack of electrical charge.

niche (nitch) The functional role of an organism.

nitrifying bacteria (ni′tri-fi-ing bak-te′re-ah) Several kinds of bacteria capable of converting ammonia to nitrite, or nitrite to nitrate.

nitrogenous base (ni-trah′jen-us bās) A category of organic molecules found as components of nucleic acids. There are five common types; thymine, guanine, cytosine, adenine, and uracil.

nondeciduous (non″de-sid′yu-us) Refers to trees that do not lose their leaves all at once.

nondisjunction (non″dis-junk′shun) An abnormal meiotic division that results in sex cells having too many or too few chromosomes.

notochord (no′tuh-kōrd) A rodlike structure along the back of chordates, used for support.

nuclear membrane (nu′kle-ar mem′brān) The structure surrounding the nucleus that separates the nucleoplasm from the cytoplasm.

nucleic acids (nu′kle-ik ă′sids) Complex molecules that store and transfer information within a cell. They are constructed of fundamental monomers known as nucleotides.

nucleoli (sing., nucleolus) (nu-kle′o-li) Nuclear structures composed of completed or partially completed ribosomes and the specific parts of chromosomes that contain the information for their construction.

nucleoplasm (nu′kle-o-plazm) The liquid matrix of the nucleus, composed of water and the molecules used in the construction of the rest of the nuclear structures.

nucleoproteins (**chromatin fibers**) (nu-kle-o-pro′te-inz) The duplex DNA strands with attached proteins.

nucleosomes (nu′kle-o-sōmz) Histone clusters with their encircling DNA.

nucleotide (nu′kle-o-tīd) The building block of nucleic acids. Each is composed of a five-carbon sugar, a phosphate, and a nitrogenous base.

nucleus (nu′kle-us) The central body of an atom or a cell.

obese (o-bēs) Extremely overweight.

obligate intracellular parasite (ob′li-gāt in″trah-sel′yu-lar pār′uh-sīt) Any organism that must live inside the cell of another species in order to carry out its life functions.

omnivores (om′ni-vōrz) Those animals that are carnivores at some times and herbivores at others.

oogenesis (o″o-jen′ĕ-sis) The specific name given to the gametogenesis process that leads to the formation of eggs.

operculum (o-pur'kyu-lum) A protective cover over the gills of bony fish.

oral contraceptive (ōr-al kon-trah-sep'tiv) A pill containing hormones that prevent the release of primary oocytes from the ovary.

orbital (ōr'bi-tal) The area of an atom able to hold a maximum of two electrons.

organ systems (ōr'gun sis'temz) Groups of organs that work together to perform a particular function. Examples: digestive system, reproductive system, and circulatory system.

organelles (ōr-gan-els') Cellular structures that have particular functions they perform in the cell. The function of an organelle is directly related to its structure.

organic molecules (ōr-gan'ik mol'uh-kiuls) Complex molecules whose basic building blocks are carbon atoms in chains or rings.

organs (ōr'guns) Structures composed of groups of tissues that work together to perform a particular function. Examples: heart, stomach, and brain.

osmosis (os-mo'sis) Net movement of water molecules through a differentially permeable membrane.

Osteichthyes (os''te-ik'the-ēz) The class of chordates to which the bony fish belong.

ovaries (o'var-ēz) The female sex organ that produces haploid sex cells; the eggs or ova.

oviduct (o'vi-dukt) The tube (Fallopian tube) that carries the primary oocyte to the uterus.

ovulation (ov-yu-la'shun) The release of a secondary oocyte from the ovary.

oxidation reactions (ok''si-da'shun re-ak'shuns) The loss of electrons from the reactant.

oxidation-reduction reactions (ok''si-da'shun re-duk'shun re-ak'shuns) Reactions that deal with the movement of electrons from one atom to another.

oxidizing atmosphere (ok'si-di-zing at'mos-fer) An atmosphere that contains molecular oxygen.

pandemic (pan dem' ik) Diseases that occur throughout the world at extremely high rates.

parasite (pār'uh-sīt) An organism that lives in or on another organism and derives nourishment from it.

parasitism (pār'uh-sit-izm) A relationship between two organisms that involves one organism living in or on another organism and deriving nourishment from it.

partial protein (par-shal pro'te-in) Protein molecules that do not provide all the essential amino acids.

penis (pe-nis) The portion of the male reproductive system that deposits sperm in the female reproductive tract.

pepsin (pep'sin) An enzyme produced by the cells lining the stomach. It begins the breakdown of proteins.

peptic ulcer (pep'tik ul-ser) A cavity formed in the wall of the digestive tract as a result of the action of the enzyme pepsin.

peptide bond (pep'tīd bond) A covalent bond between amino acids in a protein.

perennial (pur-en'e-uhl) A plant that requires many years to complete its life cycle.

perfect flowers (pur'fekt flow'urs) Flowers that contain both male and female reproductive structures.

periodic table of the elements (pēr-e-od'ik ta-bul uv the el'ĕ-ments) A list of all the elements in order of increasing atomic number (number of protons).

peristalsis (per''i-stal'sis) Wavelike contractions of the muscles of the digestive tract that move food through the tube.

pesticide (pes'ti-sīd) A poison used to kill pests. This term is often used interchangeably with insecticide.

petals (pĕ'tuls) Modified leaves of an angiosperm; an accessory structure of a flower.

PGAL (phosphoglyceraldehyde) The end product of the carbon dioxide conversion stage of photosynthesis. It is produced when a molecule of carbon dioxide is incorporated into a larger organic molecule.

pH (pe-āch) Scale used to indicate the strength of an acid or base.

phagocytosis (fā''jo-si-to'sis) A process by which the cell wraps around a particle and engulfs it.

phenotype (fen'o-tīp) The physical, chemical, and psychological expression of the genes possessed by an organism.

pheromone (fēr'o-mōn) A chemical produced by an animal and released into the environment to trigger behavioral or developmental processes in some other animal of the same species.

phloem (flo'em) One kind of vascular tissue found in higher plants. It transports food materials from the leaves to other parts of the plant.

phosphate (fos-fāt) A part of a nucleotide composed of phosphorus, oxygen, and hydrogen atoms.

phospholipid (fos''fo-lip'id) A class of water-insoluble molecules that resemble fats but contain a phosphate group (PO_4) in their structure.

photoperiod (fo-to-pe're-ud) The length of the light part of the day.

photosynthesis (fo-to-sin'thuh-sis) The major biochemical pathway in green plants that manufactures food molecules.

phylum (fi'lum) One of the categories in the classification system, composed of groups of organisms in the same kingdom.

pinocytosis (pi''no-si-to'sis) The process by which a cell engulfs some molecules dissolved in water.

pioneer community (pi-o-nēr com-miu'ni-te) The first community of organisms in the successional process established in a previously uninhabited area.

pioneer organisms (pi-o-nēr ōr'-gan-isms) The first organisms in the successional process.

pistil (pis'til) The female reproductive structure in flowers.

placenta (plah-sen'tah) An organ made up of tissues from the embryo and the uterus of the mother that allows the exchange of materials between the mother's bloodstream and the embryo's bloodstream. It also produces some hormones.

Plantae (plan'te) One of the five kingdoms, it consists of multicellular, autotrophic organisms.

plasma membrane (plaz'muh mem'brān) The outer boundary membrane of the cell, also known as the cell membrane.

pleiotropy (pli-ot'ro-pe) The multiple effects that a gene may have in the phenotype of an organism.

pleomorphic (ple''o-mor'fik) Refers to cells having many, varied shapes.

point mutation (point miu-ta'shun) A change in the DNA of a cell as a result of a loss or change in a nitrogenous base sequence.

polar body (po'lar bod-e) The smaller cell formed by an unequal meiotic division during oogenesis.

pollen (pol'en) The male gametophyte in gymnosperms and angiosperms.

pollination (pol''ī-na'shun) The transferring of pollen in gymnosperms and angiosperms.

polygenic inheritance (pol''e-jen'ik in-her'ī-tans) The concept that a number of different pairs of alleles may combine their efforts to determine a characteristic.

polyp (pol'ip) An attached, vase-shaped stage in the life cycle of Cnidaria. Example: hydra.

polypeptide chain (pol''e-pep'tīd chān) A macromolecule composed of a specific sequence of amino acids.

polyploidy (pah''lī-ploy'de) A condition in which cells contain multiple sets of chromosomes.

polysome (pah'le-sōm) A sequence of several translating ribosomes attached to the same mRNA.

population (pop''-u-la'shun) A group of organisms of the same species located in the same place at the same time.

population density (pop''u-la'shun den'si-te) The number of organisms of a species found per unit area.

population growth curve (pop''u-la'shun grōth kurv) A graph of the change in population size over time.

population pressure (pop''u-la'shun presh-yur) Intense competition that leads to changes in the environment and dispersal of organisms.

potential energy (po-ten'shul en'er-je) The energy an object has due to its position.

prebionts (pre''bi-onts) Nonliving structures that led to the formation of the first living cells.

predation (pre-da'shun) A relationship between two organisms that involves the capturing, killing, and eating of another animal.

predator (pred'uh-tōr) An organism that captures, kills, and eats another animal.

pregnancy (preg'nan-se) In mammals, the period of time during which the embryo is developing in the uterus of the female.

prey (pra) An organism captured, killed, and eaten by a predator.

primary carnivores (pri'ma-re kar'nī-vōrz) Those carnivores that eat herbivores and, therefore, are on the third trophic level.

primary consumers (pri'ma-re kon-su'merz) Those organisms that feed directly on plants—herbivores.

primary oocyte (pri'ma-re o'o-sīt) A diploid cell found in the ovary that undergoes the first meiotic division in the process of oogenesis.

primary spermatocyte (pri'ma-re spur-mat'o-sīt) A diploid cell in the testes that undergoes the first meiotic division in the process of spermatogenesis.

primary succession (pri'ma-re suk-sē'shun) The orderly series of changes leading to a climax community that begins in a previously uninhabited area.

probability (prob''a-bil'ī-te) The chance that an event will happen expressed as a percent or fraction.

producers (pro-du'surz) Organisms that produce new organic material from inorganic material with the aid of sunlight.

productivity (pro-duk-tiv'ī-te) The rate at which an ecosystem can accumulate new organic matter.

products (prŏ'dukts) New molecules resulting from a reaction.

Prokaryotae (pro''kār-e-o'te) One of the five kingdoms, it consists of one-celled organisms commonly called bacteria.

prokaryotic cells (pro'ka-re-ot''ik sels) One of the two major types of cells. They do not have a typical nucleus bound by a nuclear membrane and lack many of the other membranous cellular organelles: example, bacteria.

promoter (pro-mo'ter) A region of DNA at the beginning of each gene just ahead of an initiator code.

prophase (pro'fāz) The first stage of mitosis during which individual chromosomes become visible.

protein (pro'te-in) Macromolecules made up of amino acid subunits attached to each other by peptide bonds or groups of polypeptides.

protein synthesis (pro'te-in sin'thē-sis) The process whereby the tRNA utilizes the mRNA as a guide to arrange the amino acids in their proper sequence according to the genetic information in the chemical code of DNA.

proteinoid (pro'te-in-oid) A proteinlike structure of branched amino acid chains that are the basic structure of a microsphere.

Protista (prō-tis'tah) The kingdom of eukaryotic organisms such as algae and protozoa.

protocell (pro'to-sel) The first living cell.

protons (pro'tons) Particles in the nucleus of an atom that have a positive electrical charge.

protoplasm (pro'to-plazm) The living portion of a cell as distinguished from the nonliving cell wall.

protozoa (prō''to-zō'ah) Single-celled protists that lack cell walls and chlorophyll.

punctuated equilibrium (pung'chu-a-ted e-kwi-lib're-um) The theory that evolution has consisted of intermittent periods of rapid change interspersed with long periods of relative stability.

Punnett Square (pun'net sqwār) A method to determine probabilities of gene combinations in a zygote.

pupa (pyu'pah) A developmental stage in the life cycle of some insects intermediate between the larva and adult.

pyruvic acid (pi-ru′vik ă′sid) A three-carbon carbohydrate that is the end product of glycolysis.

radioactive (ra-de-o-ak′tiv) Property of releasing energy or particles from an unstable atom.

range (rānj) The geographical distribution of a species.

reactants (re-ak′tants) Materials that will be changed in a chemical reaction.

recessive allele (re-sĕ′siv al′lēl) An allele that when present with its homologue does not express itself and is masked by the effect of the other allele.

recombinant DNA (re-kom′bi-nant) DNA that has been constructed by inserting new pieces of DNA into the DNA of another organism such as a bacterium.

rectum (rek′tum) The final portion of the digestive tract in which undigested food is stored before being eliminated through the anus.

redirected aggression (re-di-rek′ted ă-grē′shun) A behavior in which the aggression of an animal is directed away from the opponent and to some other animal or object.

reducing atmosphere (re-du′sing at′mos-fēr) An atmosphere that does not contain molecular oxygen (O_2).

reduction division (re-duk′shun di-vi′zhun) A type of cell division in which daughter cells get only half the chromosomes from the parent cell.

reduction reaction (re-duk′shun re-ak′shun) The gain of electrons by a reactant.

regulator proteins (reg′yu-la-tor pro′te-ins) Proteins that influence the activities that occur in an organism; for example, enzymes and some hormones.

reliable (re-li′ă-bul) Giving the same result on successive trials.

reproductive capacity (re-pro-duk′tiv kuh-pas′i-te) The theoretical maximum rate of reproduction.

reproductive isolating mechanism (re-pro-duk′tiv i-so-la-ting mĕ′kan-ism) A mechanism that prevents interbreeding between species.

reptiles (rep′tīls) A class of animals of the phylum Chordata that has developed an amniotic egg and does not need to return to the water to reproduce. Examples: snakes, turtles, and dinosaurs.

response (re-sponts′) The reaction of an organism to a stimulus.

responsive processes (re-spon′siv prŏs′es-es) Those abilities to react to external and internal changes in the environment; for example, irritability, individual adaptation, and evolution.

rhythm method (rith′m meth′od) A method of conception-control in which couples avoid sexual intercourse when the egg is most likely to be present.

ribonucleic acid (ri-bo-nu-kle′ik ā′sid) A polymer of nucleotides formed on the template surface of DNA by transcription. Three forms have been identified; mRNA, rRNA, and tRNA.

ribose (ri′bōs) A five-carbon sugar molecule component of RNA.

ribosomal RNA (rRNA) (ri-bo-sōm′al) A globular form of RNA; a part of ribosomes.

ribosomes (ri-bo-sōmz) Small structures composed of two protein and ribonucleic acid subunits involved in the assembly of proteins from amino acids.

RNA polymerase (po-lim′er-ās) An enzyme that attaches to the DNA at the promoter region of a gene when the genetic information is transcribed into RNA.

root (root) A specialized structure for the absorption of water and minerals in higher plants.

root hairs (root hārs) Tiny outgrowths of roots that improve the ability of plants to absorb water and minerals.

roundworms (rownd-werms) A phylum of threadlike or cylindrical animals (Nematoda) that are very common in soils; some are parasitic. Example: pinworms.

sac fungi (sak fun′ji) One of the six major divisions of the kingdom Mycetae (fungi) containing organisms that produce sexual spores in saclike structures.

saliva (sah-li′vah) A digestive juice produced by the salivary glands that aids in the digestion of starches and also moistens the food.

salts (salts) Ionic compounds formed from a reaction between an acid and a base.

saturated (sat′yu-ra-ted) Carbon skeleton of a fatty acid that has as much hydrogen bonded to it as possible.

science (si′ens) A way of arriving at a solution to a problem or understanding an event in nature.

scientific law (si-en-tif′ik law) A uniform or constant feature of nature supported by several theories.

scientific method (si-en-tif′ik meth′ud) A way of gaining information (facts) about the world around you involving observation, hypothesis formation, experimentation, theory formation, and law formation.

scientific name (si-en-tif′ik nām) The one name of an organism that is internationally recognized. It consists of the genus name, written first and capitalized, followed by the species name, usually written in lower case. It is printed in italics or underlined.

scrotum (skro′tum) The sac that contains the testes.

seasonal isolation (se′zun-al i-so-la′shun) A genetic isolating mechanism that prevents interbreeding between species because reproductive periods differ.

seasonal reproductive pattern (se′zun-al re″pro-duk-tiv pat′ern) A behavior typical of most plants and animals in which breeding activities take place only during a particular time of the year.

second law of thermodynamics (sek′ond law uv ther″mo-di-nam′iks) Whenever energy is converted from one form to another, some useful energy is lost.

secondary carnivores (sek′un-da-re kar′ni-vōrz) Those carnivores that feed on primary carnivores and, therefore, are in the fourth trophic level.

secondary consumers (sek'un-da-re kon-su'murz) Those animals that eat other animals—carnivores.

secondary oocyte (sĕ-kun-da-re o'o-sīt) A haploid cell that goes through the second meiotic division to produce an egg.

secondary spermatocyte (sĕ-kun-da-re spur-mat'o-sīt) A haploid cell found in the testes that goes through the second meiotic division to provide spermatids.

secondary succession (sek'un-da-re suk-sĕ'shun) Orderly series of changes leading to a climax community that begins with the disturbance of an existing community.

seed (seed) A specialized structure produced by gymnosperms and angiosperms that contains the embryo sporophyte.

seed coat (seed kōt) A protective layer around seeds.

seed leaves (seed lēvs) Embryonic leaves in seeds.

segregation (seg''re-ga'shun) The separation and movement of homologous chromosomes to the poles of the cell.

selecting agent (se-lek'ting a-jent) Any factor that affects the probability that a gene will be passed to the next generation.

semen (se-men) The fluid produced by the seminal vesicle, prostate gland, and bulbo-urethral gland of males that carries sperm.

seminal vesicle (sem-in-al ves'ĭ-kul) A part of the male reproductive system that produces a portion of the semen.

seminiferous tubules (sem''in-if'ur-us tūb-yūls) Sperm-producing tubes in the testes.

sepals (se'pals) Accessory structures of flowers.

sere (sēr) An intermediate stage in succession leading to a climax community.

sex chromosomes (seks kro'mo-sōmz) Chromosomes that contain many of the genes that determine the sex characteristics of the individual.

sex ratio (seks ra-sho) The number of males in a population compared to the number of females.

sexual intercourse (copulation, coitus) (kop-yu-la'shun, ko'ĭ-tus) The mating male and female. The action of depositing sperm in the reproductive tract of the female.

sexual reproduction (sek'shu-al re''pro-duk'shun) The propagation of organisms involving the union of gametes from two parents.

sickle-cell anemia (sī-kul sel ah-ne'me-ah) A disease caused by a point mutation. This malfunction produces sickle-shaped red blood corpuscles.

sign stimulus (sīn stim'yu-lus) A specific object or behavior that triggers a specific behavioral response.

slime molds (slīm mōlds) One of the six major divisions of the kingdom Mycetae (fungi) appearing as slimy masses.

small intestine (smal in-test'in) The portion of the digestive tract in which most of the digestion and absorption of food occurs.

society (so-si'uh-te) Interacting groups of animals of the same species that show division of labor.

sociobiology (so-sho-bi-ol'o-je) The systematic study of all forms of social behavior, both human and nonhuman.

solid (sol'id) The state of matter in which the molecules are packed tightly together; they vibrate in place.

solution (so-lu'shun) A mixture that contains molecules dispersed in the system that will not settle out.

speciation (spe-she-a'shun) The process of generating new species.

species (spe-shēz) A group of organisms that can interbreed naturally to produce fertile offspring.

sperm (spurm) A male gamete.

sperm cell (spurm sel) The haploid sex cell produced by sexually mature males.

spermatids (spurm'ah-tids) Haploid cells produced by spermatogenesis that change into sperm.

spermatogenesis (spur-mat-o-jen'uh-sis) The specific name given to the gametogenesis process that leads to the formation of sperm.

sphincter muscle (sfingk'ter mus-el) A circular muscle that closes the digestive tube when contracted.

spindle (spin'dul) An array of microtubules extending from pole to pole and used in the movement of chromosomes.

spirillum (spi-ril'lum) A curve-shaped microbe; three distinct forms are common: vibrios, spirilla, and spirochetes.

spirochete (spi''ro-ket) One of the three forms of spirilla; many curves make a spirochete resemble a corkscrew.

sponge (spunj) A vaginal contraceptive device; operates by preventing sperm from reaching the egg.

sponges (spun'juz) A phylum of animals (Porifera) that are very primitive and contain many canals. They filter water for food.

spontaneous generation (spon-ta'ne-us jen-uh-ra'shun) The theory that living organisms arise from nonliving material.

spontaneous mutation (spon-ta'ne-us miu-ta'shun) Natural changes in the DNA caused by unidentified environmental factors.

spores (spōrz) Haploid structures produced by sporophytes.

sporophyte (spōr'o-fit) The diploid generation in the life cycle of plants that produces spores.

stamen (sta'men) The male reproductive structure of a flower.

states of matter (stātes uv mater) Physical conditions of matter (solid, liquid, and gas) determined by the relative amounts of energy in the molecules.

stationary growth phase (sta-shun-a-re grōth fāz) A period of time during population growth when the number of individuals entering the population and the number leaving the population are equal, resulting in a stable population.

stem (stem) The upright portion of a higher plant.

steroid (stēr'oid) One of the three kinds of lipid molecules characterized by its arrangement of interlocking rings of carbon.

stimulus (stim'yu-lus) Some change in the internal or external environment of an organism.

stomach (stum-ak) The portion of the digestive tract that holds the food while digestive enzymes are mixed with it.

stroma (stro-muh) The region within a chloroplast that has no chlorophyll.

structural formula (struk'chu-ral for'miu-lah) An illustration that shows the arrangement of the atoms and their bonding within a molecule.

structural proteins (struk'chu-ral pro'te-ins) Proteins that are important for holding cells and organisms together, such as the proteins that make up the cell membrane, muscles, tendons, and blood.

subphylum (sub'fi-lum) One of the categories in the classification system, composed of groups of organisms in the same phylum.

subspecies (**races, breeds, strains,** or **varieties**) (sub'spe-shēs) A number of more or less separate groups (demes) within the same gene pool. These groups differ from one another in gene frequency.

substrate (sub'strāt) A reactant molecule with which the enzyme combines.

succession (suk-sĕ'shun) The process of changing one type of community to another.

successional community (suk-se'shun-al com-miu'ni-te) An intermediate stage in succession.

successional stage (suk-sĕ'shun-al staj) An intermediate stage in succession.

suspension (sus-pen'shun) A type of mixture in which the dispersed particles are larger than molecular size, so they eventually settle out.

symbiosis (sim-be-o'sis) A close physical relationship between two kinds of organisms. It usually includes parasitism, commensalism, and mutualism.

symbiotic nitrogen-fixing bacteria (sim-be-ah'tik ni-tro-jen-fik'sing bak-te're-ah) Bacteria that live in the roots of certain kinds of plants where they convert nitrogen gas molecules into compounds that plants can use.

synapsis (sin-ap'sis) The condition in which the two members of a pair of homologous chromosomes come to lie close to one another.

telophase (tel'uh-fāz) The last stage in mitosis during which daughter nuclei are formed.

temperature (tem'per-ă-chiur) The measure of molecular energy of motion.

template (tem'plāt) A model from which a new structure can be made. This term has special reference to DNA as a model for both DNA replication and transcription.

tentacles (ten'tuh-kuls) Long flexible arms that can be used to obtain food and defend the animal.

termination codes (ter-mi-na'shun cōdz) The DNA nucleotide sequence just in back of a gene with the code ATT, ATC, or ACT, which signals "stop here."

territoriality (tār''-ri-to-re-al'i-te) A behavioral process in which an animal protects space for its exclusive use for food, mating, or other purposes.

territory (tār'ri-to-re) A space that an animal defends against others of the same species.

testes (tes'tēz) The male sex organ that produces haploid cells—the sperm.

testosterone (tes-tos'tur-ōn) The male sex hormone produced in the testes that controls the secondary sex characteristics.

theoretical science (the-o-ret'i-kul si-ens) The science interested in obtaining new information for its own sake.

theory (the'o-re) A plausible, scientifically acceptable generalization supported by several hypotheses and experimental trials.

theory of natural selection (the'o-re uv nat'chu-ral se-lek'shun) In a species of genetically differing organisms, the organisms with the genes that enable them to better survive in the environment and thus reproduce more offspring than others will transmit more of their genes to the next generation.

thymine (thi'mēn) A single-ring, nitrogenous base molecule of DNA but not RNA. It is complementary to adenine.

tissue (tish'u) A group of specialized cells that work together to perform a particular function. Examples: muscle, nerve, and blood.

tracheae (tra'ke-e) Breathing tubes in insects.

transcription (tran-skrip'shun) The process of manufacturing RNA from the template surface of DNA. Three forms of RNA may be produced; mRNA, rRNA, and tRNA.

transfer RNA (tRNA) (trans-fur) A molecule composed of ribonucleic acid. It is responsible for transporting a specific amino acid into a ribosome for assembly into a protein.

translation (trans-la'shun) The assembly of individual amino acids into a polypeptide.

transpiration (trans''pi-ra'shun) The process of water evaporation from the leaves of a plant.

trophic level (tro'fik le-vel) A step in the flow of energy through an ecosystem.

tube feet (tūb feet) Suction cup-like structures in echinoderms that are part of the water-vascular system.

Turner's syndrome (tur'nurz sin'drōm) A genetic disorder caused by the lack of one sex chromosome. Symptoms include difficulty with spacial relations, webbed neck, sterility, and usually shortness.

turnover number (turn'o-ver num'ber) The number of molecules of substrate that a single molecule of enzyme can react within a given time under ideal conditions.

ulcer (ul-ser) A wound that does not heal.

umbilical cord (um-bil'i-cal cord) The cord containing the blood vessels that transports materials between the placenta and the embryo.

unsaturated (un-sat'yu-ra-ted) A carbon skeleton of a fatty acid that contains carbons that are double bonded to each other at one or more points.

uracil (yu'rah-sil) A single-ring nitrogenous base molecule in RNA and not in DNA. It is complementary to adenine.

uterus (yu'tur-us) An organ found in mammals in which the embryo develops.

vacuole (vak'yu-ōl) A membranous storage container within the cytoplasm of a cell.

vagina (vuh-ji'nah) The passageway between the uterus and outside of the body; the birth canal.

valid (vă'-lid) Meaningful data that fits into the framework of scientific knowledge.

variable (va-r'e-ă-bul) The single factor that is allowed to be different in the experimental group in comparison to the control group.

vas deferens (vas def'ur-ens) The portion of the sperm duct that is cut and tied during a vasectomy.

vascular tissue (vas'kyu-lar tish'yu) Specialized tissue that transports fluids in higher plants.

vector (vek'tōr) An organism that carries a disease or parasite from one host to the next.

vertebrates (vur'tuh-brāts) Animals with backbones.

vesicle (vĕ'sĭ-kul) A membranous storage container within the cytoplasm of a cell.

vibrio (vib're-o) A type of spirillum characterized by a slightly twisted shape.

villus (vil-lus) A microscopic fingerlike projection from the lining of the small intestine that increases the surface area of the digestive tract.

virion (vi're-on) The unit of a virus particle composed of a nucleic acid core and a protein coat.

virus (vi'rus) A nucleic acid, protein-coated particle that shows some characteristics of life only when inside a living cell.

vitamin deficiency disease (vi-tah-min de-fish'en-se di-zēz) Poor health caused by the lack of a certain vitamin in the diet. Example: scurvy.

vitamins (vi-tah-mins) Growth factors needed in the diet in very small amounts.

vomiting (vom-i-ting) The forceful ejection of food from the stomach through the mouth.

water molds (wah'tur molds) One of the six major divisions of the kingdom Mycetae (fungi) containing organisms that grow in water or very moist environments.

water-vascular system (wah'tur vas'kyu-lar sis'tem) A series of water-filled tubes found in the phylum Echinodermata and used for movement.

wood (wood) The xylem of gymnosperms and angiosperms.

x-linked gene (eks lingt jēn) A gene located on one of the sex-determining chromosomes.

xylem (zi'lem) A kind of vascular tissue that transports water from the roots to other parts of the plant.

yolk sac (yōk sak) A small sac that is present as a rudiment in mammalian embryos.

zygote (zi'gōt) A diploid cell that results from the union of an egg and a sperm.

Credits

Photographs

Part Openers

Part One: © Bob Coyle; **Part Two:** © John D. Cunningham/Visuals Unlimited; **Part Three:** © David M. Phillips/Visuals Unlimited; **Part Four:** © John Dominis/Wheeler Pictures, Inc.; **Part Five:** K. G. Murti © 1987/Visuals Unlimited.

Chapter 1

1.1: Bob Coyle/Wm. C. Brown Company Publishers; **1.4:** © Jerry Schad 1987/Photo Researchers, Inc.; **1.6a:** from J. D. Watson, "The Double Helix" page 215 Atheneum, New York, 1968, by J. D. Watson; **b:** © Hank Morgan/Photo Researchers, Inc.; **1.7 TOP:** Bettmann News Photos, **BOTTOM:** © Bob Coyle; **1.9:** USDA Soil Conservation Service/Elmer Turnage.

Chapter 3

3.2 BOTH: courtesy of Goodyear Tire and Rubber Company.

Chapter 4

4.1 TOP RIGHT: Historical Pictures Service, Inc. Chicago, **BOTTOM:** National Library of Medicine; **4.10 TOP RIGHT:** © Richard Rodewald, University of Virginia/BPS, **BOTTOM LEFT:** © John D. Cunningham 1987/Visuals Unlimited, **BOTTOM MIDDLE:** © Warren Rosenberg, Iona College/BPS, **BOTTOM RIGHT:** K. G. Murti © 1987/Visuals Unlimited; **4.12 TOP LEFT:** Photo by Charles Havel/CPH Photography, **BOTTOM RIGHT:** © Herbert Israel, Cornell College; **4.13:** Courtesy of Dr. Keith Porter; **4.15:** © Warren Rosenberg, Iona College/BPS; **4.17:** Jensen, William and Park, Roderick B. CELL ULTRASTRUCTURE, 1967 by Wadsworth Publishing Co. Inc. page 57.

Chapter 6

6.13 LEFT: © Bob Coyle, **RIGHT:** © Paul Buddle.

Chapter 7

7.4: UNICEF; **7.5:** Courtesy of Hi-Bred International, Inc.; **page 154:** Photo from the *Reveille Till Taps:* Soldier Life at Ft. Mackinac 1780–1895 by Keither R. Wedder. Mackinac Island State Park Commission, Mackinac Island, Michigan.

Chapter 8

8.6a: © E. J. Dupaw, **c:** © Dr. Marta Walters.

Chapter 9

9.10 all: Carolina Biological Supply Company.

Chapter 11

11.8: Courtesy of March of Dimes; **11.10 BOTH:** Bartalos, M. and Baranski, T. A. MEDICAL CYTOGENETICS © 1967 Williams and Wilkins Company; **11.18a:** Pennsylvania State University Still Photography Services, **B-C:** © Bob Coyle, **D:** Pennsylvania State University Still Photography Services, **E:** © Russ Kinne/Photo Researchers, Inc.

Chapter 12

12.1 BOTH: Curtis, Francis D. and Urban, John: BIOLOGY, THE LIVING WORLD. 1958 Ginn and Co. (Xerox Corporation). Used with permission.

Chapter 13

13.3B (BOTH): © Isabelle Hunt Conant; **13.5:** National Park Service; **13.7 LEFT:** courtesy of Ball Seed Company, **RIGHT:** courtesy of Paul Ecke Poinsettia Ranch, Encinitas, CA; **13.9:** courtesy of Holland Dreves Reilly, Inc. Omaha, NE.

Chapter 14

14.1: Energy Research and Development Administration; **14.3 TOP:** AP/Wide World Photos; **14.4:** © Bob Coyle; **14.9 BOTH:** © Robert Fleming; **14.11:** Neg. #326866, Courtesy Department of Natural History.

Chapter 15

15.2 TOP LEFT: Courtesy of Georgia Forestry Commission, **BOTH RIGHT:** Courtesy of J. Enger; **15.6:** © Mary M. Termaine, National Audobon Society/Photo Researchers, Inc.; **15.7 TOP:** © A. D. Copley/Visuals Unlimited, **BOTTOM RIGHT:** Russ Kinne © 1987/COMSTOCK, Inc.; **15.8 BOTH:** USDA.

Chapter 16

16.14: C. P. Hickman © 1967/Visuals Unlimited; **16.15:** © G. R. Roberts/Nelson/New Zealand; **16.17:** © Michael Giannechini/Photo Researchers, Inc.; **16.18:** S. L. Pimm © 1987/Visuals Unlimited.

Chapter 17

17.2 LEFT: John D. Cunningham © 1987/Visuals Unlimited, **RIGHT:** © G. R. Roberts/Nelson/New Zealand; **17.3:** Envision/© David Waters; **17.6:** © Douglas Faulkner/Sally Faulkner Collection; **17.7:** © John D. Cunningham/Visuals Unlimited; **17.8:** Envision/© Susanna Pashko.

Chapter 18

18.3: Michigan Department of Natural Resources.

Chapter 19

19.6 BOTH: Courtesy of Lincoln P. Brower; **19.7:** © Sybille Kalas; **19.11:** © G. R. Roberts/Nelson/New Zealand; **19.12:** © Bob Coyle.

Chapter 20

20.1: Courtesy of Cordon Art, Holland.

Chapter 21

21.1 page 444 TOP LEFT: courtesy of Ford Motor Company, **TOP RIGHT and CENTER:** Courtesy of General Motors Truck Division, **BOTTOM:** Courtesy of Frederick Lewis Photography, New York; **page 445 BOTH TOP AND CENTER LEFT:** courtesy of Chrysler/Plymouth, **CENTER RIGHT:** courtesy Pontiac/GM, **BOTTOM THREE:** courtesy General Motors.

Chapter 22

22.5: courtesy of General Filter Company, Ames, Iowa; **22.7:** A. H. Gibson; **22.8:** National Park Service.

Chapter 23

23.1–23.7a and b: Carolina Biological Supply Company; **23.7 c:** USDA; **23.8:** © Bob Coyle; **23.12:** Hare, Ronald: THE BIRTH OF PENICILLIN © 1970 Allen and Unwin, LTD, London.

Chapter 24

24.4: Carolina Biological Supply Company; **24.5 BOTH:** Fred Ross; **24.6:** © Bob Coyle **24.7:** Japan National Tourist Organization.

Chapter 25

25.9–25.12: Carolina Biological Supply Company; **25.18 ALL:** Field Museum of Natural History, Chicago.

Chapter 26

26.6 all–26.11 all: Carolina Biological Supply Company; **26.24:** Michigan Department of Natural Resources; **26.28 all:** Carolina Biological Supply Company; **26.30 BOTH:** Field Museum of Natural History/Charles R. Knight, artist; **26.31 BOTH:** Carolina Biological Supply Company; **26.33:** Australian News and Information Service/W. Pederson; **26.34 LEFT:** © W. H. Hodge/Peter Arnold, Inc., **RIGHT and 26.35 ALL:** Michigan Department of Natural Resources.

Line Art and Text

Chapter 4

Figure 4.8: From Johnson, Leland G., *Biology*, 2d ed. © 1983, 1987 Wm. C. Brown Publishers, Dubuque, Iowa. All Rights Reserved. Reprinted by permission.

Chapter 8

Figure 8.7: From Kornberg, Arthur, *Trends in Biochemical Science*, 9, (4), Figure 1. © 1984 Elsevier Biomedical Press, NETHERLANDS. Reprinted by permission of the publisher and author.

Chapter 9

Figure 9.12: From Johnson, Leland G., *Biology*, 2d ed. © 1983, 1987 Wm. C. Brown Publishers, Dubuque, Iowa. All Rights Reserved. Reprinted by permission.

Chapter 11

Figure 11.4a: From Hole, John W., Jr., *Human Anatomy and Physiology*, 4th ed. © 1978, 1981, 1983, 1987 Wm. C. Brown Publishers, Dubuque, Iowa. All Rights Reserved. Reprinted by permission. Figure 11.6: From Wisniewski, L. P., and Hirschhorn, K: "A Guide to Human Chromosome Defects," 2d ed. White Plains: March of Dimes Birth Defects Foundation, BD:OAS XVI(6), 1980.

Chapter 12

Figure 12.6: From Wisniewski, L. P., and Hirschhorn, K: "A Guide to Human Chromosomes Defects," 2d ed. White Plains: March of Dimes Birth Defects Foundation, BD:OAS XVI(6), 1980.

Chapter 14

Figure 14.5: From Gunn, D. C., and J. G. H. Stevens, editors, "Pesticides and Human Welfare," *Science*, 226:1293. © 1976 Oxford University Press, Oxford, ENGLAND. Reprinted by permission.

Chapter 15

Figure 15.1: Reprinted with permission of the *Los Angeles Herald Examiner*. Figure 15.3: © Copyright 1982 TORI. Figures 15.10 and 15.11: Adapted from *Evolution*, 2d edition by Jay M. Savage. © 1963, 1969 by Holt, Rinehart, & Winston, Inc. Reprinted by permission of Holt, Rinehart, & Winston, Inc.

Chapter 16

Figure 16.3: From "Biotic Communities of the Aspen Parkland of Central Canada," by Ralph D. Bird, *Ecology* II, 1930, April, 410. © 1930 by the Ecological Society of America. Reprinted by permission.

Figures 16.9 and 16.10: Adapted from Ralph D. Bird, "Biotic Communities of the Aspen Parkland of Central Canada," *Ecology* II, April 1930:410. Reprinted by permission.

Chapter 17

Figure 17.5: From "The Black Death" by William L. Langer. Copyright © 1964 by Scientific American, Inc. All rights reserved.

Figures 17.9, 17.11, and 17.14: From Enger, Eldon D., et al., *Environmental Science: The Study of Interrelationships,* 2d ed. © 1983, 1986 Wm. C. Brown Publishers, Dubuque, Iowa. All Rights Reserved. Reprinted by permission.

Figure 17.13: Courtesy of Brookhaven National Laboratory.

Chapter 18

Figure 18.2: From Arthur Haupt and Thomas T. Kane, *Population Handbook.* Washington, D.C.: Population Reference Bureau, 1978, p. 14.

Figures 18.7 and 18.8: From Enger, Eldon, D., et al., *Environmental Science: The Study of Interrelationships,* 2d ed. © 1983, 1986 Wm. C. Brown Publishers, Dubuque, Iowa. All Rights Reserved. Reprinted by permission.

Figure 18.9: Data from D. A. Maclulich, *Fluctuations in the Numbers of the Varying Hare* (Lepus americanus). Toronto: University of Toronto Press, 1937, reprinted 1974.

Chapter 19

Figure 19.10: Courtesy of James E. Lloyd.

Chapter 23

Figure 23.4 top: From Pelczar, Michael J., and Roger D. Reid, *Microbiology.* © 1977 McGraw-Hill Book Company, New York, NY. Reprinted with permission.

Figure 23.10: From *Botany,* Second Edition by William A. Jensen and Frank B. Salisbury © 1984 by Wadsworth, Inc. Reprinted by permission of the publisher.

Chapter 25

Figure 25.8: Courtesy Carolina Biological Supply Company.

Chapter 26

Figure 26.4: Reprinted with permission of Macmillan Publishing Company from *Foundations of Biology,* 6/e, by Lorande L. Woodruff. Copyright © 1941 by Macmillan Publishing Company, renewed 1969 by Lorande M. Woodruff and Margaret Woodruff Wilson.

Index

*An italicized page number indicates that the term or phrase is defined in the glossary.

Hexose, 53
High blood pressure, 17
High-energy phosphate bond, 111, 112, *131,* 164
Histone, 190
Hitchhiker's thumb, 308, 309
HIV. *See* HTLV-III, 460–63
HMS *Beagle,* Darwin, 305
Hogs, 16
Homologous chromosome, 207, *223,* 254, 279
    nondisjunction, 230
Homo sapiens, 456
Homozygous, 256, *275,* 290
Homozygous genotype, 286
Hooke, Robert, 68, 69
Hormone, 12, 55, 166, 178, 179, 234, 236, 241, *250*
    control of fertility, 235, 236, 237, 238, 239, 240
    function, 55
    insulin, 61, 63
    oxytocin, 61
    progesterone, 56
    reproductive, 234, 235, 236
    structure, 55
    testosterone, 56
    thyroxin, 26
Horsetails (scouring rush), 510, 511
Host, 365, 366, *383,* 457, *470*
HTLV-III, 244, 460–63
Human lungs cilia, 84
Human milk production, 235
Human population growth, 396, 397, 398, 399, 400, 401
Human reproductive hormones, 235
    estrogen, 235
    follicle stimulating hormone, 235
    luteinizing, 235
    oxytocin, 235
    progesterone, 235
    prolactin, 235
    testosterone, 235
Human sexual behavior, 235
    hormonal control, 235
Hunger geography, 355
Hybrid, 284, 286, 288, *292*
    corn, 142
Hybrid orbitals, 49
Hydra, 527
Hydrogen, 27
    bonds of, 50
    isotope, 26
    reducing agent, 40
Hydrogen bond, 36, *43,* 165, 172
Hydrologic cycle, 370, 371
Hydrolysis, *43,* 52, 53
    digestion, 52, 53
Hydrolytic enzymes, 79
Hydroxyl ion, *43*
Hypha (hyphae), 474, 475, *490*
Hypothesis, 9, *20*
Hydrogen carrier, 120

Ice, 38
Identical twin, 246, *250*
Immunity, 17
Imperfect flower, 515, *521*
Imperfect fungi, 483, *491*
Implantation, 236, 237, *250*
Imprinting, 412, *424*
Inbreeding dog, 283
Inclusion, 69, 84, *92*

Independent assortment, 211, 219, 223
Inert gas, 31
Infectious disease, 11
Information storage, 162
Information systems, 160
Inhibition enzymatic, 103, *106*
Inhibitor arsenic, 104
    carbon monoxide, 104
    enzyme, 103
    organophosphate, 104
    pesticide, 103, 104
    sulfa drug, 104
Initiation code, 170, *182*
Initiator, 173
Inorganic molecules, 48, *65*
Insect, 16, 244
Insecticide, 48, 284, 286, 288, 302, 303, 346, 347, 374, 375, 376, *383*
    resistance, 302, 303, 386, 387
    tolerance, 302, 303, 386, 387
Insects, 522, 532, 534, *550*
Insight learning, 413, *424*
Instinctive behavior, 407, 408, 409, *424*
    cocoon spinning, 408, 409
    egg rolling, 408
    gull feeding, 406, 407, 408
Insulin, 48, 60, 63, 79, 175, 177, 178, 179
    structure, 60
Integrated pest control, 356
Intention movements, 417, *424*
Interbreed, 278
Interconversion membranous organells, 80
Interconversion of foods, 128
Interdependency photosynthesis, respiration, 127, 129
Interferon, 177, 178, 179
Internal parasite, 365, *383*
Interphase, 188, 190, *202,* 210
Intrauterine device (I.U.D.), 243, *250*
Intrinsic factor, 395, *403*
Introduced species, 316, 317
    African honeybee, 316
    dandelion, 316
    European rabbit, 317
    gypsy moth, 317
    house sparrow, 317
    kudzu, 316
    Melaleuca, 316
    starling, 317
    wild carrot, 316
Intron, 170, 171
Invertebrates, 536, *550*
In vitro fertilization, 235
In vivo fertilization, 235
Iodine isotope, 26
Ion, 31, 35, *44*
    chloride, 32
    fluoride, 32
    stable, 31
Ionic attraction, 32
Ionic bond, 32, *44*
    acid, 33
    base, 33
    dissociation of, 33
    salt, 33
Ionic dissociation, 32
Iron, 24, 27
Irritability, 15

Isotope, 25, *44*
    common, 26
    comparison of, 25, 26
    deuterium, 26
    energy, 26
    hydrogen, 26
    oxygen, 26
    radioactive, 26
    tritium, 26

Jawless fish, 538
Jellyfish, 522
Jenner, E. 2, 17
    smallpox, 17
Jumping gene, 176
Jungle rot, 88

Kidney, 55
Kilocalorie, 134, 139, *158*
    basal metabolism, 137, 138
Kinetic energy, *44*
Kingdom, 444, *453*
    Animalia, 448, 522–47
    Mycetae, 472–91
    Plantae, 448, 501–18
    Prokaryotae, 447, *453,* 456, 463–69
    Protista, 447, 492–501
Klinefelter's syndrome, 233, *250*
Krakatoa, 307, 308
Krebs cycle, 119, 123, *131*
Kudzu, 316
Kwashiorkor, 140, 141, *158*

Labor, *250*
Lack of dominance, 263, *275*
Lactase, 152, 158
Lactic acid, 122
    fermentation, 122
Lactose intolerance, 151, *158*
Lag phase, 391, *403*
Lamarck, Jean Baptiste de, 324
Lard, 54
Large intestine, 155
    anus, 155
    bacterium, 155
    colon, 155
    rectum, 155
Larva hookworm, 530
Larva (larvae), 524, *550*
Larynx, 8
Late prophase, 192
Law of dominance, 257, *275*
Law of independent assortment, 257, 260, *275*
Law of segregation, 257, *275*
Learned behavior, 409, 410, 411, 412, 413, *424*
    associative, 411
    conditioning, 411
    imprinting, 412
    insight, 413
Learning, 409, 410, 411, 412, 413, *424*
Leaves, 509, 516, *521*
Leech, 530, 531
Lens, 68
Leucocytes, 77
Leukemia, 199
Leuwenhoek's Anton Van, animalcules, 68

LH, 235, 236
Lichen, 472, 474, 483, 484, *491*
Life, legal definition, 13
Life cycle, 206, *521,* 524
    butterfly, 525
    fern, 512
    fluke, 528, 529
    frog, 541
    hookworm, 530
    jellyfish, 528
    moss, 507–9
    pine, 514
    sponge, 526, 527
Light energy conversion stage, *131*
    biochemical pathway, 113, 114
    chlorophyll, 113, 114
    cytochrome, 113, 114
    excited electrons, 113, 114
    light wave, 113, 114
    oxidation, 113, 114
Light microscope, 68
Limitations, 2
Limiting factor, 392, 393, 394, 395, 396, *403*
    density-dependent, 396, *402*
    density-independent, 396, *402*
    extrinsic, 395, *403*
    intrinsic, 395, *403*
Linkage, 268
Linkage group, 268, *275*
Lipase, 152, 153
Lipid, 63, *65,* 79, 134
    fatty acid, 54, 63
    glycerol, 54, 63
Liquid, *44*
    property of, 38
Lithium, 27
Liver, 56, 150, 151
    bile, 151
Locus, 216, 255, *275,* 279
Lorenz, Konrad, 412
LSD (lysergic acid diethylamide), 176
Lung, 72
Luteinizing hormone, 235, 236
Lymphatic system, 153
Lymphogranuloma venerum (LGV), 245
Lysosome, 89, *92*
    function, 79, 89
    structure, 79

Macroevolution, 51, 332, *334*
    formation of, 52
Magnesium, 27
Magnesium ion, 31
Magnification, 68
Magnifying glass, 68
Major cell types, 86
    eukaryotic, 86, 89
    prokaryotic, 86, 89
Major energy levels, 28
Malaria, 501
Malonic acid, 103
Maltase, 152
Malthus, Thomas, 398
Mammals, 522, 542–45, *550*
Mammary glands, 542, 543, *550*
Mannose, 51
Mantle, 534, *550*
Marsupial, 542–44, *550*
Mass, 24
    isotope, 25
    neutron, 25
    proton, 25, 28